**McGRAW-HILL
BOOK COMPANY**

New York
St. Louis
San Francisco
Auckland
Düsseldorf
Johannesburg
Kuala Lumpur
London
Mexico
Montreal
New Delhi
Panama
Paris
São Paulo
Singapore
Sydney
Tokyo
Toronto

MILTON B. DOBRIN

Professor of Geology
University of Houston

Introduction to Geophysical Prospecting

THIRD EDITION

To the memory of my teacher

W. MAURICE EWING
1906–1974

Library of Congress Cataloging in Publication Data

Dobrin, Milton Burnett.
    Introduction to geophysical prospecting.

    Includes index.
    1. Prospecting—Geophysical methods.   I.   Title.
TN269.D6   1976   622'.15   75-37603
ISBN 0-07-017195-5

INTRODUCTION TO
GEOPHYSICAL
PROSPECTING

567890DODO78321098

This book was set in Modern 8A by Maryland Composition Incorporated.
The editors were Robert H. Summersgill and Carol First;
the cover was designed by Victor Bastante;
the production supervisor was Sam Ratkewitch.
The drawings were done by J & R Services, Inc.
R. R. Donnelley & Sons Company was printer and binder.

CONTENTS

PREFACE

Revolutionary changes since 1960 in all aspects of geophysical technology have necessitated a much more thorough revision of this text than is usually required between successive editions of a book. The present edition is different in so many ways from the previous one that it can almost be looked upon as a new book rather than as a revision.

Three chapters (on acquisition of seismic data at sea, seismic data enhancement in digital processing centers, and direct detection of hydrocarbons using seismic data) are entirely new. The need for such chapters reflects the great advances made during the past decade or so in digital recording and processing of seismic data as well as in techniques for seismic prospecting in offshore areas.

The development during the 1960s of a new generation of high-speed digital computers has had an enormous impact on all phases of applied geophysics—from the acquisition of data in the field to its ultimate interpretation. The computer has not only made it possible to obtain better-quality seismic data but also to derive from them new kinds of geological information that until recently were not considered obtainable by geophysics at all. To give such developments proper coverage has required the introduction of material on many aspects of geophysics that were not even in existence when the second edition of this book was published in 1960.

A significant innovation in the present edition is the use of elementary calculus in presenting the basic principles of the various geophysical methods. In earlier editions the calculus was not employed because of concern that many geologists using the book might not have studied this subject. Now nearly all geological curricula leading to a bachelor's degree in geology require at least a year of calculus, and the restriction against its use observed in previous editions no longer appears necessary. It is unlikely that those readers who are not familiar with calculus will encounter any real difficulty if they skip over the equations which use it, assume them to be correct, and determine their significance from pertinent discussions in the text.

Applied geophysics has become so specialized in recent decades that few readers are likely to be equally interested in all its phases. Even in elementary courses in the subject, the emphasis on different topics will vary with the background and interests of the instructor. Among those working in the field of exploration some will be primarily concerned with techniques used in oil exploration and others with those most widely applied in the search for metallic minerals. Those involved with geophysics in either the oil or mining industry will probably have different areas of concentration, e.g., field operations, instruments, data processing, or geological interpretation. It is hoped that the needs of all such users will be met in this book.

To meet the needs of instructors offering a one-semester course in applied geophysics who would like to use this time for more intensive study of specific aspects of the subject rather than for broad coverage of the entire book, the individual chapters have been designed to be self-contained so far as that is possible.

The present edition, like previous editions, was written with the needs of geologists in mind and I have emphasized the geological applications of geophysics. It is not possible for the geologist to use geophysical tools most effectively unless he has a thorough understanding of the physical principles behind the various methods, particularly those involved in the recording of field data and its processing. Geological considerations should guide all aspects of geophysical prospecting from choice of field recording parameters and processing programs to final mapping of results in geological terms.

The book is also intended for students of geophysics who want a broad view of all phases of geophysics, particularly those outside their own areas of specialization. It is hoped that the book will be helpful to professional geophysicists who would like to review basic principles and at the same time keep up to date on new developments. Continuing education courses in geophysics have been well attended in recent years, indicating a widespread need for such updating.

It is hoped that this book will serve a broader purpose than the mere presentation of technical information. Geophysical prospecting is a field of activity that is of particular current importance because of its bearing on the maintenance of the world's industrial economy as well as living standards in many parts of the world. Both of these are highly dependent on the continued extraction of needed energy and mineral resources from the earth. As the challenge of finding such resources has grown, there has been an increasing need for dedicated individuals of exceptional ability to choose applied geophysics as a career. It would be most gratifying if this book could lead students with the necessary qualifications to consider careers in a field that can offer the dual satisfaction of helping meet society's material needs as well as of meeting the challenge, more and more exacting as time goes on, of unravelling the fragmentary clues left by nature to the location of the treasures still hidden in the earth.

Many people have helped me with this edition in numerous ways during the time when I was writing it. I should like to acknowledge such assistance from Terry Spencer, of Texas A and M University; Carl Savit, of Western Geophysical Co.; Harry Mayne, of Petty-Ray Geophysical; Leroy Brow, of Exxon, Inc.; Robert E. Sheriff, of Seiscom-Delta, Inc.; Harold Mooney, of the University of Minnesota; and Fred Hilterman, of the University of Houston; all of them have given me valuable information or material.

Several geophysicists have reviewed portions of the manuscript relating to their areas of specialization and have given me the benefit of their comments. Among these are Mr. Savit, Mr. Mayne, and Dr. Sheriff as well as Thomas R. LaFehr of Edcon, Inc., and Ralph C. Holmer and George V. Keller of Colorado School of Mines. Ralph B. Ross, a consultant and John C. Hollister, retired chairman of the Geophysics Department of Colorado School of Mines, have read the entire manuscript. All of these reviewers have made many valuable suggestions for improving the book and their assistance is deeply appreciated.

I am particularly indebted to Bernard F. Bash and John Hough, students at the University of Houston, for invaluable help in assembling the material for the book and to Mrs. Doris Segelhorst for her patient typing of what must have appeared to be an endless number of drafts of the text.

MILTON B. DOBRIN

THE PLACE OF GEOPHYSICS IN OIL AND MINERAL EXPLORATION

The extraction at a continually increasing rate of fossil fuels and useful minerals from the earth has raised the specter of impending shortages that could threaten the economy and way of life of the civilized world. Events of the middle 1970s have demonstrated how well founded this concern can be. The amounts of oil, gas, and metallic minerals that actually exist in the earth, both known and undiscovered, are of course limited, but the immediate problem as established reserves become scarce is to find new supplies in the earth that will replace those which have been consumed. The exploration for energy supplies and mineral resources has become increasingly difficult as sources easier to find at any time are discovered and exploited.

To meet the challenge, earth scientists have developed more and more sophisticated techniques of exploration. Early in this century the search for oil and solid minerals was confined to deposits directly observable on the surface in the form of seeps and outcrops or other exposures. When all accumulations in an area that could be discovered by such simple means had been found, it was necessary to deduce the presence of buried deposits indirectly by downward projection of geologic information observable on the surface. As this approach reached the point of diminishing returns, new methods of studying the subsurface were needed. They did not require any geological observations, but they did involve physical measurements at the earth's surface that would give information on the structure or composition of concealed rocks that might be useful for locating desired deposits.

1-1 GEOPHYSICS AND GEOLOGY

We designate the study of the earth using physical measurements at the surface as *geophysics*. While it is not always easy to establish a meaningful border line between geology and geophysics, the difference lies primarily in the type of data with which one begins. Geology involves the study of the earth by direct observations on rocks, either from surface exposure or boreholes, and the deduction of its structure, composition, or history by analysis of such observations. Geophysics, on the other hand, involves the study of those parts of the earth hidden from direct view by measuring their physical properties with appropriate instruments, usually on the surface. It also includes interpretation of the measurements to obtain useful information on the structure and composition of the concealed zones. The distinction between the two branches of earth science is not clear-cut. Well logs, for example, are widely used in geological studies, even though they present the results of purely instrumental observations. The term *borehole geophysics* is often used to designate such measurements.

In a broader sense, geophysics provides the tools for studying the structure and composition of the earth's interior. Most of what we know about the earth below the limited depths to which boreholes or mine shafts have penetrated has come from geophysical observations. The existence and properties of the earth's crust, mantle, and core have been determined primarily by observations upon seismic waves from earthquakes, as well as by measurements of the earth's gravitational, magnetic, and thermal properties. Many of the tools and techniques developed for such studies have been used in exploration for hydrocarbons and minerals. At the same time, geophysical methods devised for prospecting applications have been put to use in more academic research on the nature of the earth's interior. While this book will emphasize the economic applications of geophysics, it should be stressed that the areas of "pure" and "applied" geophysics have so much interdependence that the separation is artificial at best.

1-2 THE TECHNOLOGICAL CHALLENGE OF GEOPHYSICS

Geophysical exploration is a relatively new area of technology. Ferrous minerals were sought with magnetic compasses as early as the 1600s, but only during the past century have special instruments been put to use in mining exploration. Geophysical prospecting for oil and gas is only about half a century old, the first oil discovery attributable to geophysics having been made in 1924. Throughout its history, the tools and techniques of exploration geophysics have been continually improved, both in performance and economy. This progress has been in response to an unrelenting pressure to develop new capabilities after existing ones have become inadequate to find enough new deposits. Except in areas newly opened to exploration, most geophysical surveys are undertaken where previous ones have failed because the instruments, field techniques, or interpretational methods were not good enough. In other words, those accumulations which are capable of being located with existing technology are the only ones that will be discovered at a given time. Those remaining will not be found until the technology improves sufficiently to bring them to light.

Thus the exploration geophysicist finds himself in the same situation as a man on an accelerating treadmill who must run faster and faster just to stay where he is. This problem is also faced by others involved in the exploration process such as geologists and drilling engineers.

The technological improvements in geophysical exploration have been of several types. In some cases new developments have been necessitated by problems associated with the environment where exploration is to be carried out. In offshore areas, or in deserts, Arctic tundra, or lava-covered terrain, special logistics are needed. Moreover, unique types of noise in such areas often cause interference with desired geophysical information, and special techniques must be developed to suppress such interference. The introduction of computer technology in the 1950s and 1960s brought about new capabilities in the recording and processing of all kinds of geophysical data, making it possible to extract useful information otherwise concealed by undesired noise.

The technological revolution following World War II brought about many scientific developments which have contributed greatly to the effectiveness of geophysical exploration. Electronic computers, information-processing techniques, and navigation satellites, to cite some examples of pertinent space age developments, have all been put to extensive use by geophysicists searching for oil and other natural resources.

1-3   REVIEW OF GEOPHYSICAL PROSPECTING METHODS

The geophysical techniques most widely employed for exploration work are the seismic, gravity, magnetic, and electrical methods. Less common methods involve the measurement of radioactivity and temperature at or near the earth's surface and in the air.

Some of these methods are used almost entirely in the search for oil and gas. Others are used primarily in exploring for solid minerals. Most of them may be employed for either objective. Seismic and gravity prospecting are mainly tools for oil exploration; electrical methods are predominantly used for mineral exploration, although in the U.S.S.R. and in former French territories they have been applied routinely to the search for oil as well. Magnetic methods are employed for both types of prospecting.

Seismic reflection method With this method—by far the most widely used geophysical technique—the structure of subsurface formations is mapped by measuring the times required for a seismic wave (or pulse), generated in the earth by a near-surface explosion of dynamite, mechanical impact, or vibration, to return to the surface after reflection from interfaces between formations having different physical properties. The reflections are recorded by detecting instruments responsive to ground motion. They are laid along the ground at distances from the shot point which are generally small compared with the depth of the reflector. Variations in the reflection times from place to place on the surface usually indicate structural features in the strata below. Depths to reflecting interfaces can be determined from the times using velocity information that can be obtained either from the reflected signals themselves or from surveys in wells. Reflections from depths as great as 20,000 ft can normally be observed from a single explosion, so that in most areas geologic structure can be determined throughout the sedimentary section. In recent years reflection data have also been used for identifying lithology, generally from velocity and attenuation characteristics, and for detecting hydrocarbons, primarily gas, directly on the basis of reflection amplitudes and other seismic indicators.

The reflection method comes closer than any other prospecting technique to providing a structural picture of the subsurface comparable to what could be obtained from a great number of boreholes in close proximity. Modern reflection record sections are very similar in appearance to geologic cross sections, and geologists must sometimes be cautioned not to use them as such without taking into consideration some potential hazards that might lead to erroneous interpretation, even with good-quality reflection data. Under ideal conditions, structural relief can be determined with a precision of 10 to 20 ft.

The method makes it possible to produce structural maps of any geologic horizons which yield reflections, but the horizons themselves cannot be identified without independent geological information such as might be obtained from wells. Reflection data can be used to determine average velocities and, more important from a geological viewpoint, interval velocities over thicknesses of section no less than many hundreds of feet. This information provides at least a statistical indication of lithology, the usefulness depending on the layering as well as the problem at hand.

With reflection methods one can locate and map such features as anticlines, faults, salt domes, and reefs. Many of these are associated with the accumulation of oil and gas. Major convergences caused by depositional thinning can be detected from reflection sections, but the resolution of the method is not as favorable as we would usually like for finding stratigraphic traps such as pinchouts or facies changes. In any case, successful exploration for stratigraphic oil accumulations using reflection requires skillful coordination of geologic and seismic information.

While current technological improvements have made it possible to obtain usable reflection data in many areas where reflections were formerly too poor to map, there are still places where reflection does not yield reliable information even though highly sophisticated data acquisition and processing techniques are used. Where this occurs, other geophysical methods must be employed.

Seismic refraction method   In refraction shooting, the detecting instruments record seismic signals at a distance from the shot point that is large compared with the depth of the horizon to be mapped. The explosion waves must thus travel large horizontal distances through the earth, and the times required for the travel at various source-receiver distances give information on the velocities and depths of the subsurface formations along which they propagate. Although the refraction method does not give as much information or as precise a structural picture as reflection, it provides data on the velocity of the refracting beds which often allow the geophysicist to identify them or to specify their lithology. The method usually makes it possible to cover a given area in a shorter time and more economically than with the reflection method.

Refraction is particularly suitable where the structure of a high-speed surface, such as the basement or the top of a limestone layer, is a target of geological interest. If the problem is to determine the depth and shape of a sedimentary basin by mapping the basement surface, and if the sedimentary rocks have a consistently lower speed than the basement formations, refraction can be a highly effective and economical approach for achieving this objective. Because velocities in salt and evaporites are often greater than in surrounding formations, refraction has been useful in mapping diapiric features such as salt domes. Under favorable circumstances this technique can be used to detect and determine the throw of faults in high-speed formations such as dense limestone and basement materials.

In spite of its numerous advantages, refraction is employed much less extensively than reflection in oil exploration. This is probably attributable to the greater amounts of dynamite required, the larger scale of the field operations, and the lower precision in the structural information obtainable from the method.

Gravity method   In gravity prospecting one measures minute variations in the pull of gravity from rocks within the first few miles of the earth's surface. Different types of rocks have different densities, and the denser rocks have the greater gravitational attraction. If the higher-density formations are arched upward in a structural high, such as an anticline, the earth's gravitational field will be greater over the axis of the structure than along its flanks. A salt dome, on the other hand, which is generally less dense than the rocks into which it is intruded, can be detected from the low value of gravity recorded above it compared with that measured on either side. Anomalies in gravity which are sought in oil exploration may represent only one-millionth or even one-ten-millionth of the earth's total field. For this reason, gravity instruments are designed to measure *variations* in the force of gravity from one place to another rather than the absolute force itself. Modern gravimeters are so sensitive that they can detect variations in gravity to within one-hundred-millionth of the earth's total acceleration.

The gravity method is useful wherever the formations of interest have densities which are appreciably different from those of surrounding formations. It is an effective means of mapping sedimentary basins where the basement rocks have a consistently higher density than the sediments. It is also suitable for locating and mapping salt bodies because of the generally low density of salt compared with that of surrounding formations. Occasionally it can be used for direct detection of heavy minerals such as chromites.

Data from gravity surveys are more subject to ambiguity in interpretation than with seismic surveys, because any gravity field can be accounted for equally well by widely different mass distributions. Additional geophysical or geological information over a gravity anomaly will reduce the ambiguity and increase the usefulness of the gravity data.

Magnetic method   Magnetic prospecting maps variations in the magnetic field of the earth which are attributable to changes of structure or magnetic susceptibility in certain near-surface rocks. Sedimentary rocks generally have a very small susceptibility compared with igneous or metamorphic rocks, which tend to have a much higher magnetite content, and most magnetic surveys are designed to map structure on or inside the basement or to detect magnetic minerals directly. The magnetic method was initially used for petroleum exploration in areas where the structure in oil-bearing sedimentary layers appeared to be controlled by topographic features, such as ridges or faults, on the basement surface.

Since the development of aeromagnetic methods, most magnetic surveys undertaken for oil exploration are carried out to ascertain the thickness of the sedimentary section in areas where such information is not otherwise available. Interpretation of such data is complicated by the fact that intrabasement susceptibility changes usually have a much more significant effect on the observed magnetic field than structural relief on the basement surface itself.

In mining exploration, magnetic methods are employed for direct location of ores containing magnetic minerals such as magnetite. Intrusive bodies such as dikes can often be distinguished on the basis of magnetic observations alone.

Interpretation of magnetic data is subject to the same uncertainty as gravity work because of the lack of uniqueness inherent in all potential methods. Here again, the more geologic information available the less the uncertainty in the final interpretation.

Electrical methods   Electrical prospecting makes use of a large variety of techniques, each based on some different electrical property or characteristic of materials in the earth. The resistivity method is designed to yield information on formations or bodies having anomalous electric conductivity. The induced-polarization method, employed in the exploration for disseminated ore bodies such as sulfides, will give diagnostic readings where ionic exchanges take place on the surfaces of metallic grains. Such effects cause delays in the falloff of voltage across the ore mass when current passed through the mass from surface electrodes is suddenly cut off. The resistivity method has been used for a long time to map boundaries between layers having different conductivities. It is employed in engineering geophysics to map bedrock and in groundwater studies to determine salinity. Most recently it has been applied in the search for geothermal power as subterranean steam affects the resistivity of formations in a way that can often be diagnostic. Telluric current and magnetotelluric methods use natural earth currents (the latter involving natural alternating magnetic fields as well), and anomalies are sought in the passage of such currents through earth materials. In this respect these methods are different from resistivity and induced polarization, which require artificial introduction of electricity into the earth.

The self-potential method is used to detect the presence of certain minerals which react with electrolytes in the earth in such a way as to generate electrochemical potentials. A sulfide body oxidized to a greater extent on its top than along its bottom will give rise to such potentials, which are detectable with electrodes at the surface.

Electromagnetic methods detect anomalies in the inductive properties of the earth's subsurface rocks. An alternating voltage is introduced into the earth by induction from transmitting coils either on the surface or in the air, and the amplitude and phase shift of the induced potential generated in the subsurface are measured by detecting coils and recorded. Ore of base metals can often be detected by this technique.

The resistivity and magnetotelluric methods are used quite extensively in the U.S.S.R. for mapping sedimentary basins at the early stages of exploration for petroleum in new areas. Other electrical methods, such as the telluric, have been employed by French geophysicists in Europe and Africa. Elsewhere in the world, electrical techniques have been employed almost entirely in the search for solid minerals and for geothermal energy.

Radioactive methods Radioactive prospecting for minerals containing uranium has involved the use of geophysical tools (geiger counters and scintillometers) and must therefore be looked upon as a geophysical method. Much of the surface exploration for uranium is carried out by amateurs equipped with detecting instruments. Industrial prospecting involves radioactive logging of exploratory drill holes and airborne surveys with scintillometers. Earlier editions of this book contained chapters on exploration for radioactive minerals, but the subject will not be covered in the present edition because of the expanded space now needed to cover the state of the art in the more widely used areas of geophysics.

Well logging This involves probing the earth with instruments which give continuous readings recorded at the surface as they are lowered into boreholes. Among rock properties currently being logged with such instruments are electrical resistivity, self-potential, gamma-ray generation (both natural and in response to neutron bombardment), density, magnetic susceptibility, and acoustic velocity.

Although well logging is one of the most widely used of all geophysical techniques, it would require a book as long as the present one even to introduce this subject properly. For this reason, well logging will not be covered here except for special applications such as velocity or density measurement, and the reader interested in logging is directed to other publications devoted specifically to the subject.

1-4 GEOPHYSICS IN OIL EXPLORATION

Geophysics and Our Future Oil Supply

During the latter part of 1973, the matter of maintaining needed supplies of petroleum products became one of the most critical issues faced by most Western countries since World War II. As long ago as 1956, Hubbert predicted that the United States would reach its peak petroleum production in 1969 or 1970 and that from then on new reserves would not keep pace with increased consumption. His prediction for the United States was reaffirmed[1] in 1969, and statistics now available indicate that his projections were correct. Since 1973 the dependence of the United States upon imported oil has increased, but it was not until the embargos of 1973 that limitations on the availability of oil were really felt, not only in the United States but in most other countries as well. The impact of the sudden shortage upon the economy and way of life in affected countries illustrates the value of effective oil exploration in maintaining the equilibrium of our industrialized economy.

Hubbert[1] also predicted that peak production for the world as a whole would be reached during the decade between 1990 and 2000, the former year applying if the most pessimistic published estimates of total reserves were assumed and the latter year if the most optimistic estimates were used. Figure 1-1 shows these projections.

The hydrocarbons now being extracted from petroleum to meet the demand for energy need not come from conventional sources, as they can, in principle at least, be extracted from tar sands, oil shales, or coal. But the technological, economic, and environmental problems of changing over to such alternative sources are so

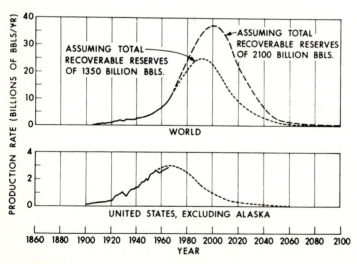

FIGURE 1-1
Crude oil production for the United States and world since 1900 and projected into the 21st century. (*From Hubbert.*[1])

formidable that we shall be almost entirely dependent on oil and gas in their present form for many years. Even when the capacity for large-scale conversion of other fossil fuels is achieved, there will be a demand for all the conventional petroleum that can be produced as long as the costs of exploration and production do not become so great that it is priced out of the market.

Present geophysical exploration techniques have reached the point of diminishing returns in most petroliferous areas and the only hope that presently exists for finding substantial new oil and gas reserves is in further technological development of geophysical methods. Many such deposits are located under water or in environments, e.g., desert or tundra, where conventional geological exploration is not very promising. Other undiscovered deposits are entrapped in such a way that existing geophysical techniques are incapable of finding them, and the only effective means of discovery is costly wildcat drilling. Many such oil deposits have never been discovered because geophysical methods that will locate them have not yet been developed. The limitation also applies to most oil located in stratigraphic (as opposed to structural) traps, although the appearance since 1970 of techniques for direct detection of gas by seismic reflection in marine areas should change the prospects to some extent. Other potential improvements are discussed by Dobrin.[2]

In consideration of these facts, the conclusion seems reasonable that no technical factor may be as important in governing the future supply of conventional oil as the development of improvements in geophysical techniques. Two areas for potential advances are particularly significant: (1) the attainment of capability for mapping productive structures in places where no usable data can be obtained by existing methods and (2) the development of more effective geophysical techniques for locating oil in stratigraphic traps.

Of the many important technical advances in geophysics since its introduction as a tool for oil exploration in the early 1920s, none has been responsible for any really significant increase in discovery rates. As pointed out by Lyons,[3] the ratio of discoveries in exploration wells located by geophysics remained constant through the 1950s at about one in six. Improvements in technology had thus done no more than keep pace with the increasing difficulty of finding oil as the easier-to-locate deposits were found. It is too early to assess the effect of the new seismic capabilities for direct detection of gas that appeared on the scene a year or so before this book went to press, but it is hoped that the ratio will show some improvement as a result. In any case, the current shortage of petroleum makes it more important than ever before that geophysical techniques be brought to the greatest degree of effectiveness that natural limitations allow.

Historical Development of Petroleum Geophysics

The earliest efforts to locate oil-bearing structures by geophysical tools involved gravity measurements. Shortly before the beginning of the present century, Baron Roland von Eötvös, of Hungary, completed development of the torsion balance that bears his name. This was a field device for measuring the distortions in the earth's gravitational field that would result from buried bodies with anomalous densities such as salt domes. In 1915 and 1916 the torsion balance was employed to detail the structure of what was then a one-well oil field at Egbell, now in Czechoslovakia. According to Eckhardt[4], this survey was highly successful. In 1917, Schweidar detailed a known salt dome at Hanigsen in Germany with a torsion balance, and the predicted structure was confirmed by subsequent drilling. Early in 1922, according to DeGolyer,[5] Shell surveyed the Horgada field in Egypt with this type of instrument, and later that year the Spindletop field in Texas was traversed by a torsion balance, yielding a striking anomaly over the known salt structure.

Early Gulf Coast surveys   In 1922 the torsion balance was used for the first time to explore for oil over areas where the structure was completely unknown. This was in the Gulf Coast of the United States, where the association of oil with salt domes had been well established, making the torsion balance particularly suitable as an exploration tool. Immediately afterwards, the instrument was introduced in Mexico.

At about the same time, seismic refraction equipment, very crude by modern standards, was brought from Germany to look for salt domes in the Gulf Coast.* In 1919, Ludger Mintrop had applied for a German patent on locating and measuring depths to subsurface features by refraction profiling. The earliest work, in 1923, was in Mexico, but later in the year a refraction (fan-shooting) survey was undertaken along the Mexia fault zone in the Texas Gulf Coast.

Both the torsion-balance and refraction campaigns were successful in locating salt domes as early as 1924. The gravity surveys led to the discovery of the productive Nash dome, and the seismic shooting was responsible for finding the Orchard dome, both in Texas. These successes led to more widespread application of the two techniques, and by 1929 virtually all the piercement-type domes in the Gulf Coast had been discovered.

Refraction in Iran   Meanwhile, in 1928, the seismic refraction method was introduced into the Middle East by the D'Arcy Exploration Company (which subsequently became British Petroleum). This technique turned out to be highly effective for finding the large oil-bearing limestone structures in Iran and was used successfully there for two decades.

Early reflection work   The earliest experiments with the seismic reflection method were carried out by J. C. Karcher from 1919 to 1921. To demonstrate the potential of the method for oil exploration he mapped a shallow reflecting bed in central Oklahoma early in 1921. On the fiftieth anniversary of this event, in April 1971, a monument (Fig. 1-2) was dedicated at the site where these tests had been

---

* Most of the material presented here on the early history of seismic prospecting is from Weatherby.[6]

FIGURE 1-2
Monument unveiled in 1971 at Belle Isle (Oklahoma City), Okla., to commemorate 50th anniversary of first reflection shooting by J. C. Karcher. (*SEG*).

conducted. Karcher was present at the ceremonies. Two of the first reflection records he shot there, as well as his interpretation of them, are shown in Fig. 1-3.

It was not until 1927, however, that the reflection method was put to work for routine exploration. In that year, the Geophysical Research Corporation used the technique to discover the Maud Field in Oklahoma. By the early 1930s, reflection

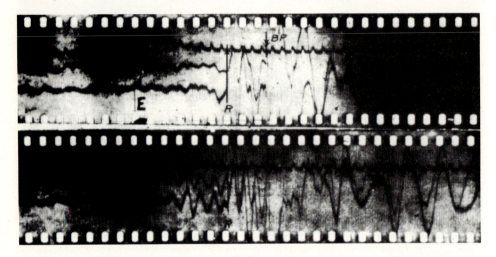

(*a*)

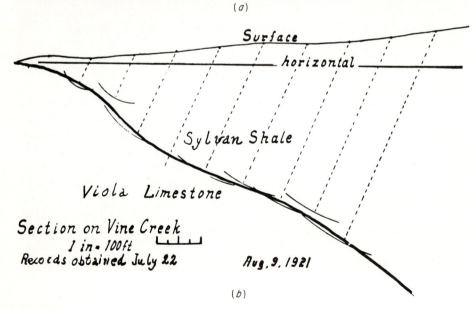

(*b*)

FIGURE 1-3
(*a*) Two of the earliest reflection records ever shot. Recorded by J. C. Karcher and colleagues at Belle Isle, Okla., in 1921. (*b*) Karcher's interpretation of his earliest reflection line, Belle Isle, Okla., 1921.

became the most widely used of all geophysical techniques, a status it has maintained ever since.

Gravity after the torsion balance   The torsion balance was slow and cumbersome to operate and more fieldworthy instruments were developed for measuring gravity by the early 1930s. A simple pendulum system was introduced by Gulf Research & Development Company in 1932, and it was used in the field until about 1935. In 1935, the first gravimeter (an instrument giving direct readings of gravity differences) was put to use in commercial prospecting, and in a short time gravimeters virtually displaced all other gravity-measuring tools in oil exploration work. Light-weight gravity meters appearing in the late 1940s increased the speed and economy of land surveys. Bottom meters were introduced for water work before World War II, and shipborne meters were in use by the early 1960s.

Early seismic work in water-covered areas   Ever since the later 1920s, when seismic surveys were conducted from houseboats and barges in the bays and bayous of the Gulf Coast, geophysics has been used to explore for oil in water-covered areas. By the end of World War II, seismic surveys were being undertaken in the open ocean, both off the Gulf Coast and off California. In the early 1950s marine surveys had been carried out in the Gulf of Mexico, on the California shelf, in the Persian Gulf, and in Lake Maracaibo. But it was not until the middle 1960s that marine exploration was extended to include almost all the shelf areas of the world, with widespread use of gravity, magnetic, and seismic instruments.

Advances after World War II   Since the end of World War II continual progress has been made by oil companies, geophysical companies, and equipment manufacturers in developing new and improved geophysical tools. Most of the advancements in the seismic reflection method were initiated with the primary objective of eliminating noise that interferes with reflections. A number of them were introduced to achieve greater safety, economy, or flexibility in field operations. Most of the new energy sources that have appeared since the 1950s are in this category.

Presentation of tape-recorded data on time-corrected record sections in variable-area or variable-density modes became common practice during the later 1950s. In 1963, digital recording equipment was first employed on a widespread basis, and digital computers were programmed to process the data thus acquired. The processing techniques have been greatly improved over the years, full advantage being taken of the increased storage capacity and speed of the computers as well as other advances made by communications engineers in signal-processing technology.

Other developments related to field operations have increased the economy and effectiveness of the seismic reflection method. Mechanical impactors and vibrators operating on the surface have supplemented dynamite in shot holes as sources of seismic energy although they have not supplanted it. The Thumper* and the Dinoseist† are impulsive sources which produce signals similar to those of dynamite,

---

* Trademark, Petty-Ray, Geophysical Inc.   † Trademark, Atlantic Richfield Oil Company.

but the Vibroseis* generates a continuous wave of slowly varying frequency, which gives data requiring special processing procedures before they can be used. Another highly significant development was common-depth-point recording, in which each geophone receives signals from different shot points, the individual recordings being subsequently composited for reduction of noise. This technique was invented by Harry Mayne of Petty Geophysical Engineering Company in the early 1950s,† but it did not come into widespread use until about a decade later, when processing technology had been developed that made it economically feasible.

The universal use of common-depth-point techniques in offshore seismic work led to the development of ship-towed streamer cables more than 2 mi long, with large numbers of pressure phones per channel. Such cables record reflections with less interference from noise than was usually possible with earlier types. The most striking innovation in marine geophysics, however, has been the replacement of dynamite by special energy sources such as sparkers, air guns, and propane-oxygen exploders for virtually all operations.

Gravity and magnetic instruments have also been adapted for work at sea. Shipborne gravity meters which give a continuous record of gravity variations from a moving ship make such surveys more economical than those with water-bottom meters of the type previously employed in offshore gravity work. Such shipborne meters are on gyroscopically stabilized platforms which compensate for accelerations due to ship motion. Proton and optical pumping magnetometers are often used for obtaining magnetic data on marine surveys undertaken primarily for seismic recording.

Direct Detection of Hydrocarbons (Bright Spots)

One of the most significant developments in the entire history of exploration geophysics is the recently developed capability for direct detection of gaseous hydrocarbons, usually in water-covered areas, by proper processing of seismic reflection data. Until this capability was established, it was generally taken for granted that geophysics could locate structures and in some cases stratigraphic features favorable for oil and gas accumulation but could not locate the hydrocarbons themselves. Its development can thus be looked upon as a revolutionary breakthrough in geophysical prospecting.

The detection technique is based on the principle that gas-saturated sands have a lower velocity than adjacent water- or oil-saturated sands, so that the velocity contrasts across surfaces bounding the gas zones above or below would give reflections of higher amplitude than would be observed from the same interface on either side of the gas zones. New processing techniques that allow true relative amplitudes to be observed on seismic record sections make it possible to observe the

---

* Trademark, Continental Oil Company.

† A technique for determining velocities using a similar approach in the field was introduced in the early 1930s by Cecil H. Green.

high reflection amplitudes directly on the sections in the form of *bright spots* (see Chap. 10).

At the time this book goes to press, the bright-spot technique is too new for adequate evaluation. Although there have been numerous verifications by drilling of gas accumulations predicted by bright spots, there is no way of knowing how many predictions have not been borne out or how many of the gas reservoirs actually found were noncommercial.

Statistics on Oil-Exploration Activity

Since 1933 the Society of Exploration Geophysicists (SEG) has published statistics on the level of geophysical activity expressed in terms of crew-months. Curves based on the statistics thus presented show how greatly such activity fluctuates because of economic and political factors.

Figure 1-4 illustrates the variation of seismic exploration activity in the United States since 1933 and that of gravity exploration activity since 1938 (expressed by the number of crews operating) along with corresponding statistics on exploratory drilling and on new oil discovered in the United States. It is evident that the level of seismic activity in the United States has declined rather steadily since the peak reached in 1952 and that gravity activity has decreased significantly since 1945. Yet the drop has not actually been as great as would appear from the curves. Marine activity off the coasts of the United States has increased greatly during recent years, but it is not evident from statistics based on number of crews alone that a marine seismic crew can cover 50 times as many miles of line per day as a typical land crew. Moreover, a modern land crew using common-depth-point shooting techniques and digital recording accounts for a vastly greater expenditure and volume of subsequent data-processing activity than a typical crew doing land work required as recently as the early 1960s.

The plots of oil discovery and drilling activity when properly smoothed show a correlation with geophysical activity if the time lags between exploration work and

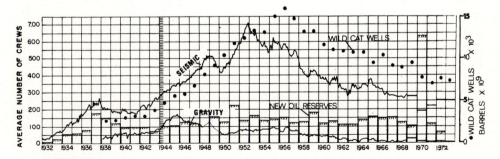

FIGURE 1-4
Statistics on crew activity in seismic and gravity prospecting along with number of wildcat wells drilled and new reserves discovered, 1933–1973. (*Data from SEG activity reports and American Petroleum Institute.*)

consequent drilling (sometimes as great as 6 years) are taken into account. The anomalously high level of new reserves indicated for 1970 reflects the discovery of Prudhoe Bay, the reserves estimates for which became official in that year.

It should be emphasized that the growth in foreign geophysical activity since the early 1950s has compensated to a large extent for the decline in domestic work. This is illustrated by Fig. 1-5. During this period, geophysical surveys have been carried out at an increasing rate in almost every part of the free world where there are known petroleum prospects. Nowhere has the increase been as pronounced as in offshore areas (see Fig. 1-6) although it is difficult to demonstrate this because published statistics on marine activity in various parts of the world were not reliable before 1967.

Summary

Geophysics has occupied a most important place in the exploration for oil and gas since its introduction in the early 1920s, but its effectiveness is limited by a number of factors. Although it now appears possible to detect gaseous hydrocarbons in offshore areas by seismic reflection, techniques are not available for locating offshore oil directly and the capability of geophysics for locating hydrocarbons in land areas has not been as widely demonstrated. If there are no source beds, or if rocks needed

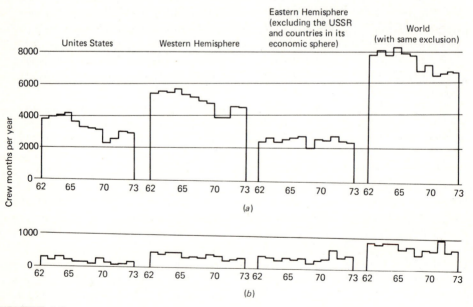

FIGURE 1-5
Distribution of geophysical activity in crew months per year for different parts of the world, 1962–1973: (*a*) Seismic; (*b*) gravity. (*Data from SEG activity reports.*)

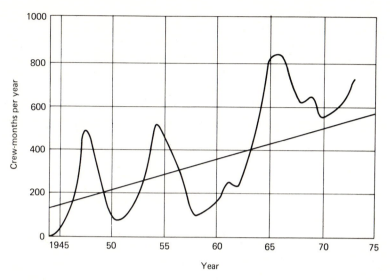

FIGURE 1-6
Worldwide seismic marine activity since 1944. Smooth line represents long-term average. (*Data from C. H. Johnson and D. M. Blue, Offshore Technology Conf. Proc., 1969 and SEG activity reports.*)

as reservoirs do not have the necessary porosity or permeability, structures which look highly favorable from a geometrical standpoint will inevitably be dry. The record of geophysics in locating stratigraphically trapped oil has been quite poor, although direct detection techniques promise to improve its performance as far as gas in marine areas is concerned. The likelihood that the record of success can be improved for other kinds of stratigraphic accumulations depends on the geological characteristics of the trap itself and on the amount of geologic information available. Reefs, usually classified as stratigraphic features, can generally be located by seismic reflection, and oil entrapped in reefs can be found as easily by geophysics as that in most structural traps. Pinchouts can often be located but seldom with the precision necessary for efficient location of hydrocarbons. The place of seismic prospecting in locating stratigraphic oil is discussed by Lyons and Dobrin.[7] The book in which this article appears contains numerous case histories illustrating all degrees of success with geophysics, the extent depending on the geologic nature of the accumulation.

   The only real limitations in finding structural traps with the seismograph have been inadequate resolution and interference by noise. Dramatic advances in seismic recording and data processing over the past decade have brought about great improvements in the capability of the seismic method for overcoming both limitations. Many structures with potential for oil accumulation can be located now that probably would not have been discoverable a few years ago. Further improvements should lend still greater effectiveness to finding structural traps favorable for oil accumulation.

As more oil is found, and as the amount of oil remaining to be discovered decreases, exploration targets will become deeper and the search will extend farther out to sea and into deeper water. Moreover, the exploration effort will be extended more and more into remote areas and into those with more hostile environments and more difficult logistics. It will also move into areas where new techniques show their ability to obtain useful data after earlier approaches have failed. This is the challenge which geophysics must face in meeting the world's needs for energy from hydrocarbons.

1-5 THE USE OF GEOPHYSICS IN MINING EXPLORATION

Geophysics and the world supply of minerals Just as our society is critically dependent on petroleum products for its energy requirements, it is equally dependent on mineral supplies for maintenance of the industrial economy which is the basis of our modern civilization. The rate at which such mineral supplies are being extracted to meet the increasing demands of an ever-expanding economy has created great concern among economic geologists lest we run out of critically important minerals sooner than is generally realized. In 1968 Charles F. Park, Jr.,[8] said, "whereas in the past many minerals were at times in troublesome surplus, it now appears that the world is about to enter a period when shortages of several minerals may well develop in the next decade." His statement is even more applicable now.

Of all metals in common use, three are already reaching the stage where they are in short supply in the light of present needs—copper, tin, and mercury. Although supplies of other common metals are not now in jeopardy, the world's annual consumption of them is so enormous that a high level of exploration activity is necessary to maintain adequate supplies. To the extent that geophysics can help in the discovery of such reserves, its place should become increasingly important as mineral shortages become more imminent.

Excellent surveys of geophysical techniques in mining can be found in Parasnis[9] and in Ref. 10. Geophysical methods have been most successful in locating two types of ores: (1) sulfides, both massive and disseminated, and (2) iron ores. Other minerals such as chromites and gold have been discovered by geophysical surveys but not nearly with the same degree of success.

The principal metals found in massive sulfide ore bodies are copper, nickel, lead, and zinc. The most common minerals in which they are found are chalcopyrite, pyrrhotite, galena, and sphalerite. Most such ore bodies are characterized by high conductivity, high density, and often, because of the frequent occurrence of magnetite as a "guest" mineral, high magnetic susceptibility. The electromagnetic, resistivity, and induced-polarization techniques appear best for detecting conductivity anomalies associated with such bodies. Gravity measurements are desirable for observing density anomalies where they are significant and magnetometer surveys for measuring any diagnostic magnetic anomalies.

Disseminated sulfide ores are favorable sources of porphyry copper and molybdenum. Among the minerals appearing most frequently in this form are chalcopyrite, chalcocite, bornite, molybdenite, and pyrite. The most effective geophysical tool for finding ores of this type is the induced-polarization technique.

Iron ores of greatest economic interest contain magnetite and hematite. Magnetite has the highest magnetic susceptibility of all minerals, and magnetic techniques are most suitable in the search for iron in this form. Hematite is not particularly magnetic, but it is often related genetically or stratigraphically to lithologic units which do contain magnetic minerals so that the magnetometer may be suitable in exploring for it as well as for magnetite. Since the density of magnetite and hematite is often greater than that of the host rocks containing these minerals, gravity surveys may be useful in the search for both kinds of ores.

History of mining geophysics   Geophysical tools were applied to mineral exploration almost three centuries before geophysics was used in the search for oil. The magnetic compass was employed in prospecting for iron ore as early as 1640, but it was not until about 100 years ago that a special type of instrument, the Swedish mining compass, was developed for such investigations. Its magnetic needle was so suspended that it could be rotated about both horizontal and vertical axes. American versions of this compass were used to explore for iron ore in New Jersey and Michigan during the last decades of the nineteenth century.

One of the earliest pioneers of exploration geophysics was Robert Fox, who in 1815 discovered that certain minerals exhibit spontaneous polarization. He proposed measurement of this effect as a means of prospecting for such ores. It was not until almost a century later, however, that a commercial discovery was made by this technique. In 1913, Conrad Schlumberger used it to locate a sulfide deposit at Bor. About the same time he developed practical field techniques for resistivity and equipotential-line prospecting, techniques which were based on experiments carried out by Osborn and others before the turn of the century in the Great Lakes mining area.

From 1915 to 1920, dip needles of various types were introduced for magnetic-mineral prospecting. The Schmidt magnetometer, still in occasional use, appeared about this time. Airborne magnetometers based on the flux-gate principle under development for exploration were used for submarine detection during World War II and applied to prospecting shortly after the war. Nuclear magnetometers, both for ground and airborne surveys, appeared around 1955, and the even more precise optical-pumping-type (cesium- and rubidium-vapor) magnetometers were introduced into exploration work around 1961. The airborne magnetic gradiometer came into use in the middle 1960s. All these magnetic instruments have been employed both in mining and petroleum exploration.

During the 1920s improved techniques were developed for resistivity prospecting involving multiple-electrode configurations. Electromagnetic methods were introduced by Hans Lundberg in the middle 1920s, and they were adapted for airborne surveys about 1947.

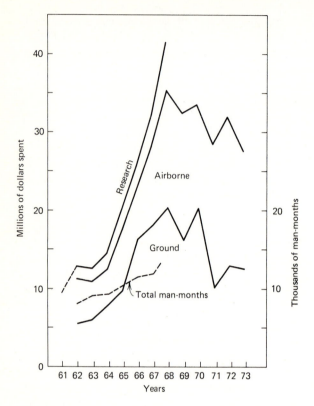

FIGURE 1-7
Worldwide activity in mining geophysics as separated by types of methods.
All curves for expenditure are cumulative with respect to those below them.
(*Data from SEG activity reports.*)

Before World War II, the theoretical basis for mining exploration was quite limited, and interpretation was often only qualitative. Since the war, there have been significant advances in the theory behind the interpretation methods used in mining geophysics, particularly with the magnetic and electromagnetic techniques.

A number of geophysical methods for mining exploration have come into widespread use since World War II. In 1948 the induced-polarization, or overvoltage, method was introduced commercially in the search for sulfide ores, and this is now the most widely used technique in mining geophysics. Magnetotelluric methods, including audio magnetotellurics, were also introduced after the war.

Statistics on activity in mining geophysics   While the use of geophysics in oil exploration may have reached a stage of maturity in the 1950s, with a leveling off or even a decrease in activity that was evident at least until the onset of the 1973 energy crisis, mining geophysics appeared to be entering a period of intensive growth as of the late 1960s. From the standpoint of dollar volume, worldwide geophysical

Table 1-1   PERCENTAGE OF MINING GEOPHYSICS EXPENDITURE BY METHOD: GROUND METHODS

Method	1973	1972	1971	1970	1969	1968	1967	1966	1965	1964	1963	1962	1961
Magnetic	8.5	6.2	5.5	7.3	10.1	8.0	9.3	20.9	11.6	9.9	13.2	16.3	14.8
Resistivity					1.9	5.3	7.7	4.8	10.5	13.3	16.8	15.9	21.7
Induced polarization	17.5*				33.1	18.3	31.4	15.8	25.2	23.8	23.0	13.0	14.2
Earth currents					0	0.2	0.2	8.9	0.1	0.1	0.5	0.3	1.9
Self-potential					0.3	1.8	1.0	0.1	2.5	2.3	1.5	1.4	3.1
Electromagnetic	6.4	4.3	4.3	5.6	9.7	7.3	11.1	26.1	14.6	14.6	8.8	14.5	14.5
Gravity	6.7	10.2	4.4	7.4	5.2	11.8	7.9	5.4	7.0	9.6	14.6	12.6	8.6
Seismic	2.5	2.6	3.8	8.9	0.8	6.5	15.0	8.2	8.9	12.3	5.4	15.7	5.6
Radioactive	1.3	2.4	5.0	8.2	1.4	4.3	2.2	2.5	1.1	0.7	0.5	0.2	1.4
Geochemical					6.9	32.4	8.9	1.3	12.1	8.8	13.4	7.1	11.8
Drill-hole logging	4.6	3.0	12.3	3.8	19.0	3.6	2.9	1.8	5.8	4.4	2.0	2.3	2.3
Geothermal					1.1	0.2	0.7	1.3	0.2				0.1
Miscellaneous					10.5	0.3	1.7	2.9	0.4	0.2	0.3	0.7	
Electrical		23.3	29.6	23.5									
Combined	52.5	48.0	35.1	35.3									
	100.0	100.0	100.0	100.0	100.0	100.0	100.0	100.0	100.0	100.0	100.0	100.0	100.0

SOURCE: Data from annual activity reports in *Geophysics*. (A difference in reporting procedures since 1969 makes it difficult to compare statistics for the electrical methods over the entire period.)

* Total for resistivity and induced polarization.

Table 1-2 PERCENTAGE OF MINING GEOPHYSICS EXPENDITURE BY METHOD: AIRBORNE METHODS

Method	1973	1972	1971	1970	1969	1968	1967	1966	1965	1964	1963	1962	1961
Magnetic	51.2	68.1	42.3	73.2	43.7	53.1	57.6	69.0	66.0	67.2	66.4	69.0	80.7
Electromagnetic	24.7	1.4	1.3	2.7	2.1	3.7	12.8	11.2	19.2	15.9	2.1	6.8	7.7
Combined magnetic and electromagnetic	23.0	17.0	30.3	19.3	39.0	17.2	24.8	16.4	14.5	16.2	27.7	24.0	
Radioactivity	1.0	13.5	26.1	4.8	15.1	25.4	4.8	3.4	0.3	0.7	3.6	0.2	10.8
Other	0.1	0			0.1	0.6					0.2		0.8
	100.0		100.0	100.0	100.0	100.0	100.0	100.0	100.0	100.0	100.0	100.0	100.0

SOURCE: Data from annual activity reports in *Geophysics.*

Table 1-3   SI PREFIXES FOR POWERS OF 10*

Multiple	SI prefix	Abbreviation	Example	
			Unit	Abbreviation
10^6	mega	M	megahertz	MHz
10^3	kilo	k	kilocalorie	kcal
10^{-1}	deci	d	decibel	dB
10^{-2}	centi	c	centimeter	cm
10^{-3}	milli	m	millisecond	ms
10^{-6}	micro	μ	microsecond	μs

* Abbreviations for powers of 10 not used in this text are omitted.

activity in mining exploration increased by a factor of more than 3 in the period from 1961 to 1968, having leveled off since that time. This is a much greater rise than can be accounted for by inflation alone. The statistics on expenditure for this period, with separation of the total into ground methods, airborne methods, and research, are presented in Fig. 1-7. On the same plot is shown the total number of man-months applied in the search for minerals from 1962 to 1968. Its growth was much less (about 50 percent) than that of the expenditures during this period, indicating that the cost of equipment per geophysical employee more than doubled between 1962 and 1968.

Mining exploration involves a greater number of geophysical techniques than petroleum prospecting, and there is a more even distribution of activity among the methods. Tables 1-1 and 1-2 show the relative expenditure for the different ground and airborne geophysical methods used in mining over the period 1961–1973. The growth of induced-polarization activity during the years from 1951 to 1969 is particularly noteworthy.

1-6   UNIT ABBREVIATIONS USED IN THIS BOOK

With the exception of gals (not abbreviated) and milligals (abbreviated mgal), abbreviations for units follow the style of the Système International (SI), now being adopted in most fields. In this system units named after people are given abbreviations that start with a capital letter, e.g., volt (V), joule (J), bel (B), and oersted (Oe). The name hertz (Hz) replaces cycles per second. Note that the capitalization applies only to the abbreviation for the unit; when it is spelled out, it is lowercase, as before. Units not named after people are given lowercase abbreviations, e.g., second (s), meter (m), and hour (h). To show multiples and submultiples of units, prefixes corresponding to certain powers of 10 are attached to the unit abbreviation (see Table 1-3). Many of these combinations, for example, kg (kilogram), mm (millimeter), and kΩ (kilohm), are probably familiar to most readers.

REFERENCES

1. Hubbert, M. King: "Resources and Man," National Academy of Sciences and the National Research Council, Freeman, San Francisco, 1969.
2. Dobrin, Milton B.: Geophysics and Our Future Oil Supply, in *Proc. Southwest. Leg. Found., Explor. Econ. Pet. Ind.*, vol. 8, Matthew Bender, 1970.
3. Lyons, Paul L.: Economics of Geophysics in Oil Exploration, *Geophysics*, vol. 27, pp. 121–127, 1962.
4. Eckhardt, E. A.: A Brief History of the Gravity Method of Prospecting for Oil, *Geophysics*, vol. 5, pp. 231–242, 1940.
5. DeGolyer, E.: Notes on the Early History of Applied Geophysics in the Petroleum Industry, pp. 245–254, in "Early Geophysical Papers," Society of Exploration Geophysicists, Tulsa, Okla., 1947.
6. Weatherby, B. B.: The History and Development of Seismic Prospecting, *Geophysics*, vol. 5, pp. 215–230, 1940.
7. Lyons, Paul L., and M. B. Dobrin: Seismic Exploration for Stratigraphic Traps, pp. 225–243, in "Stratigraphic Oil and Gas Fields," *Am. Assoc. Petrol. Geol. Mem. 16* and *Soc. Explor. Geophys. Spec. Pub. 10*, 1972.
8. Park, Charles F., Jr.: "Affluence in Jeopardy," Freeman, Cooper, San Francisco, 1968.
9. Parasnis, D.: "Mining Geophysics," Elsevier, New York, 1966.
10. Society of Exploration Geophysicists: "Mining Geophysics," vol. 1, Tulsa, Okla., 1966.

HOW SEISMIC WAVES PROPAGATE

The seismic wave is the basic measuring rod used in seismic prospecting. If we are to understand how it works and evaluate the information we get from it in geological terms, we must be familiar with the basic physical principles governing its propagation characteristics. These include its generation, transmission, absorption, and attenuation in earth materials and its reflection, refraction, and diffraction characteristics at discontinuities. Seismic waves are generally referred to as elastic waves because they cause deformation of the material in which they propagate like that in an elastic band when it is stretched. The deformation consists of alternating compressions and dilatations as the particles in the material move closer together and farther apart in response to forces associated with the traveling waves.

Many decades before the advent of seismic prospecting, the characteristics of elastic waves traveling in the earth were studied by seismologists concerned with signals from earthquakes. While the scale, both with respect to the wavelengths and distances involved, is much larger for natural earthquake waves than for those generated in seismic prospecting, both types of wave propagation are described by the same physical laws.

In introducing the basic concepts of elastic-wave propagation in the earth, we shall emphasize the physical and geological significance of the concepts rather than their formal mathematical expression.

2-1 ELASTIC CHARACTERISTICS OF SOLIDS

Stress and strain Seismic waves propagate in solids as patterns of particle deformation traveling through the materials with velocities that depend upon their elastic properties and densities. To show the nature of this dependence we shall describe these deformations in terms of the forces which cause them, defining two useful concepts stress and strain. The relation between them for a particular material enables us to describe the elastic properties of the material as well as the characteristics, such as velocity, of waves propagating therein.

Dilatational strain Let us consider the changes in position of two points A and B in the inside of a solid after a linear deformation (in this case tensional) (Fig. 2-1). A and B are a distance dx apart. A dilatational motion has shifted A to A' and B to B', the former having moved a distance u and the latter a distance $u + du$. Defining the *strain* as ratio of the change in separation to the original separation, we express the component of ϵ_x in the x direction as

$$\epsilon_x = \frac{u + du - u}{dx} = \frac{\partial u}{\partial x}\dagger \tag{2-1}$$

Taking dv as the differential elongation in the y direction and dw as that in the z direction, we similarly define

$$\epsilon_y = \frac{\partial v}{\partial y} \qquad \text{and} \qquad \epsilon_z = \frac{\partial w}{\partial z}$$

The strains observed when seismic waves pass through a material are much too small to be seen by the eye, falling in the neighborhood of 10^{-6} for most linear deformation.

Shear strain If a cubic block is cemented to an immovable surface along its bottom face, as shown in Fig. 2-2a, and a horizontal traction is exerted along the upper face, the initially vertical faces perpendicular to the paper will be inclined after distortion, the surface initially vertical now making an angle α with its original direction. For a more general analysis, we should relinquish the constraint we have introduced along one face and leave the block free to rotate under shear as well as

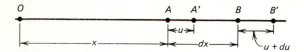

FIGURE 2-1
Quantities used in deriving expression for linear strain.

† The strain must be expressed as the partial derivative $\partial u / \partial x$ rather than as the total derivative du/dx because we must assume that y and z do not vary when x varies.

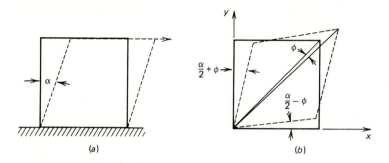

FIGURE 2-2
Two definitions of shear strain for cubical block: (*a*) one side of block in firm
contact with rigid surface; (*b*) block fixed at one point and free to rotate as
well as to deform under shear.

to deform elastically. This is because shearing deformation actually involves a
combination of lengthening along one diagonal, shortening along another, motion
along the diagonal, and rotation of the diagonal. Figure 2-2*b* shows the geometry.
The angle (very much exaggerated in the diagram) between the deformed surface
originally lying in the *yz* plane and the reference plane (*yz* coordinate plane),
which is $\alpha/2 + \phi$, where ϕ is the angle of rotation of the diagonal, is small enough
to be approximated by its tangent, which is the rate of increase of deformation in the
x direction with increasing *y*, or $\partial u/\partial y$. Similarly, $\alpha/2 - \phi$ can be approximated
by $\partial v/\partial x$. We then solve for α, the shear strain, and ϕ, the rotation as follows:

$$\frac{\alpha}{2} + \phi = \frac{\partial u}{\partial y} \tag{2-2}$$

$$\frac{\alpha}{2} - \phi = \frac{\partial v}{\partial x} \tag{2-3}$$

$$\alpha = \frac{\partial u}{\partial y} + \frac{\partial v}{\partial x} \tag{2-4}$$

$$\phi = \frac{1}{2}\left(\frac{\partial u}{\partial y} - \frac{\partial v}{\partial x}\right) \tag{2-5}$$

Dilatation When a three-dimensional body deforms in the same sense (either by
extension or contraction) along all three coordinate axes, there will be an expansion
of volume in the case of tensile deformation or a contraction of volume in the case
of compressive deformation. The ratio of the change in volume to the volume before
deformation is called *cubical dilatation*.
  Assume a rectangular element *dx dy dz* before strain. When there is strain, the
respective sides become $dx\,(1 + \epsilon_x)$, $dy\,(1 + \epsilon_y)$, and $dz\,(1 + \epsilon_z)$ and the resulting

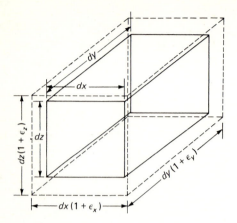

FIGURE 2-3
Expansion of rectangular parallelepiped under cubical dilatation.

volume, as indicated in Fig. 2-3, is

$$dx\,dy\,dz\,(1 + \epsilon_x)\,(1 + \epsilon_y)\,(1 + \epsilon_z) = dx\,dy\,dz\,[1 + (\epsilon_x + \epsilon_y + \epsilon_z)$$
$$+ (\epsilon_x\epsilon_y + \epsilon_x\epsilon_z + \epsilon_y\epsilon_z) + \epsilon_x\epsilon_y\epsilon_z] \quad (2\text{-}6)$$

The third and fourth terms on the right side are of second and third order, respectively, compared with the first term and may be neglected. Then the cubical dilatation θ becomes, by definition,

$$\theta = \frac{dx\,dy\,dz\,(1 + \epsilon_x + \epsilon_y + \epsilon_z) - dx\,dy\,dz}{dx\,dy\,dz} = \epsilon_x + \epsilon_y + \epsilon_z \quad (2\text{-}7)$$

Definition of stress   Let us consider a small plane area ΔA on the surface of an irregular solid as shown in Fig. 2-4. Such an area could be taken in its interior as well. Assume that an element of force ΔF is exerted uniformly over the area in a direction to make an angle ψ with the normal to the plane surface. This force can be resolved into components $\Delta \bar{F}_N$ normal to the plane and $\Delta \bar{F}_T$ tangential to the plane

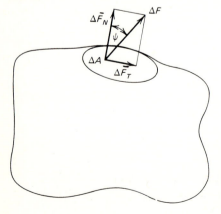

FIGURE 2-4
Resolution of force acting on a small area of a solid into its normal and tangential components.

in the directions indicated in the figure. The *stress S*, which is a vector since it has both magnitude and direction, is defined as the ratio of the force $\Delta \bar{F}$ to the area ΔA. As the area becomes infinitesimally small, S is expressible as the derivative

$$S = \frac{d\bar{F}}{dA} \tag{2-8}$$

The limiting ratio of the normal component of force $\bar{F}_N$ to the area is defined as the *dilatational stress* $d\bar{F}_N/dA$ and the ratio involving the tangential component as the *shear stress* $d\bar{F}_T/dA$.

To express stress in a more general way, we shall set up an *xyz* coordinate system and designate the orientation of the surface element upon which the force is exerted by the direction of its normal (Fig. 2-5). Let us consider a rectangular area with boundaries dy and dz with a normal in the x direction. A stress making an arbitrary angle ϕ with the normal would have one component X_x of dilatational stress and two components Y_x and Z_x of shear stress. In this notation, the capital letter represents the direction of the force component, and the subscript indicates the direction of the *normal* to the area on which it is acting. If we draw two other rectangles representing elementary areas in the xy and xz planes and then resolve the corresponding stress elements with respect to them, we have a total of nine stress components, X_x, Y_y, and Z_z being dilatational stresses and the other six (such as X_y, Z_x, etc.) being shear stresses. As stresses in pairs, such as X_y and Y_x, are equal to each other, the actual number of independent stress components is six rather than nine.

Relations between strain and stress   When an "ideal" elastic solid is subjected

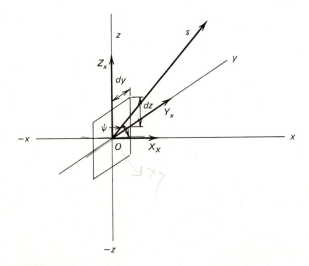

FIGURE 2-5
Resolution of stress acting on an element of area in the yz plane into one dilatational component (X_x) and two shear components (Y_x and Z_x).

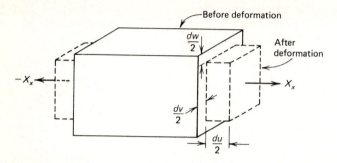

FIGURE 2-6
Deformation of rectangular parallelepiped by dilatational stress acting in one
direction.

to linear deformation in a single direction, the specification of the material as
elastic implies a direct proportionality between the compressional or dilatational
stress and the linear strain of the form

$$X_x = E \frac{\partial u}{\partial x} \tag{2-9}$$

where E, the proportionality constant, is *Young's modulus*. For most materials, E
is of the order of a megabar (10^{12} dyn/cm²).

For a three-dimensional body the relation between stress and strain is somewhat
more complex. A tensional stress X_x will cause an elongation du in the x direction,
as shown in Fig. 2-6, but at the same time there will be contractions in the y and z
direction of dv and dw, respectively. It is evident that dv and dw will be less than
du, both being related to du by the same proportionality constant σ, which is
Poisson's ratio.

The relationship can be expressed by the equations

$$E \frac{\partial u}{\partial x} = X_x \qquad E \frac{\partial v}{\partial y} = -\sigma X_x \qquad E \frac{\partial w}{\partial z} = -\sigma X_x \tag{2-10}$$

the minus signs representing contractions, so that σ can be expressed as $-\partial v/\partial u$
and $-\partial w/\partial u$.

If there is no volume change when a unidirectional stress is applied, σ becomes
0.5, the maximum value it can have. For highly consolidated, unweathered rocks
such as fine-grained limestones or deeply buried crystallines, σ ranges from 0.2
to 0.3, while for most nonindurated clastic sedimentary rocks it ranges from 0.05
to 0.02, depending on porosity and weathering.

Let us consider the case where tensile (or compressive) stresses act along all three
principal axes, which may be designated X_x, Y_y, and Z_z. Each strain component

can be written in terms of these stress components as

$$E\frac{\partial u}{\partial x} = X_x - \sigma Y_y - \sigma Z_z$$

$$E\frac{\partial v}{\partial y} = -\sigma X_x + Y_y - \sigma Z_z \qquad (2\text{-}11)$$

$$E\frac{\partial w}{\partial z} = -\sigma X_x - \sigma Y_y + Z_z$$

If the stress results from an excess hydrostatic pressure ΔP above the ambient pressure, all three of the stress components will be the same and each will be equal to ΔP. If the three equations of (2-11) are added, we have

$$E\left(\frac{\partial u}{\partial x} + \frac{\partial v}{\partial y} + \frac{\partial w}{\partial z}\right) = (1 - 2\sigma)(X_x + Y_y + Z_z)$$

$$= (1 - 2\sigma)3\Delta P \qquad (2\text{-}12)$$

or $$E\theta = (1 - 2\sigma)(3\Delta P)$$

where $$\theta = \frac{\Delta V}{V} \qquad (2\text{-}13)$$

Now define $(\Delta V/V)/\Delta P = \theta/\Delta P$ as the *compressibility* β of the material and $1/\beta = k$, the *bulk modulus*; then

$$k = \frac{\Delta P}{\theta} = \frac{E}{3(1 - 2\sigma)} \qquad (2\text{-}14)$$

This formula relates the constant for cubical dilatation resulting from pressure to the constants relating linear strain and linear stress.

The relationship between shear stress and shear strain is quite simple. For the small deformations involved in seismic-wave propagation, shear stress is proportional to shear strain, the proportionality constant being μ, as in the relation

$$X_y = Y_z = \mu\alpha = \mu\left(\frac{\partial u}{\partial y} + \frac{\partial v}{\partial x}\right) \qquad (2\text{-}15)$$

μ is called the *rigidity modulus*. For most rock materials it ranges from 0.1 to 0.7 Mbar.

The rigidity modulus can be expressed in terms of Young's modulus and Poisson's ratio as

$$\mu = \frac{E}{2(1 + \sigma)} \qquad (2\text{-}16)$$

A derivation of this relationship can be found in Dix[1] (pp. 300–303). Equations

(2-14) and (2-16) illustrate the interrelation between the four basic elastic constants E, σ, k, and μ.

By algebraic manipulation of Eq. (2-11), the nine components of stress defined in the previous section can be linearly related to corresponding strain components. Dix[1] (pp. 303–305) shows that a dilatational stress like X_x can be expressed in the form

$$X_x = 2\mu \frac{\partial u}{\partial x} + \lambda \left(\frac{\partial u}{\partial x} + \frac{\partial v}{\partial y} + \frac{\partial w}{\partial z} \right) \tag{2-17}$$

where λ, one of Lame's coefficients, is related to Young's modulus E and Poisson's ratio σ as follows:

$$\lambda = \frac{E\sigma}{(1 + \sigma)(1 - 2\sigma)}$$

Generalizing from Eqs. (2-15) and (2-17), we can write

$$X_x = 2\mu \frac{\partial u}{\partial x} + \lambda\theta \qquad X_y = Y_x = \mu \left(\frac{\partial u}{\partial y} + \frac{\partial v}{\partial x} \right)$$

$$Y_y = 2\mu \frac{\partial v}{\partial y} + \lambda\theta \qquad Z_x = X_z = \mu \left(\frac{\partial w}{\partial x} + \frac{\partial u}{\partial z} \right) \tag{2-18}$$

$$Z_z = 2\mu \frac{\partial w}{\partial z} + \lambda\theta \qquad Y_z = Z_y = \mu \left(\frac{\partial v}{\partial z} + \frac{\partial w}{\partial y} \right)$$

These are *Hooke's relations* for all stress components in terms of strains. They apply to "ideal" elastic solids (homogeneous and continuous) for deformations which are small enough to fall within the range of linearity implied by the equations.

2-2 PROPAGATION CHARACTERISTICS OF COMPRESSIONAL AND SHEAR WAVES

We shall next show how the propagation characteristics of seismic waves, particularly their velocities,* depend on the elastic constants defined in the previous section. To do this, we must show how one obtains the classical wave equation from Hooke's relations.

The general form of the wave equation, which is most applicable to the propagation of seismic waves through the earth, assumes deformation in three directions,

---

* The term *velocity* is, strictly speaking, a vector expressing both the magnitude, referred to as *speed*, and the direction of motion. In this book, velocity will be considered to be synonymous with speed.

each component of stress being associated with strain in more than one direction, as indicated in Eqs. (2-17) and (2-18).

The logic by which the more general equation is derived can be demonstrated very simply when applied to the case where the stress and strain are both confined to a single direction. This is what occurs when a thin rod is subjected to elastic deformation along its axis, as illustrated in Fig. 2-7. This rod, of area dA, has a Young's modulus E and density ρ. An element of the rod having a length of dx will be moved from the position bounded by the full lines to the position bounded by the dashed lines when subjected to an elastic stress $S(x)$ in the axial direction x. The force on any surface is the stress times the area. The net force on such an element will be the difference between the forces $S(x)\,dA$ and $S(x + dx)\,dA$, where $S(x + dx)$ is the stress at the position $x + dx$. This net elastic force will equal the mass $\rho\,dx\,dA$ (density multiplied by volume) of the element times the acceleration d^2u/dt^2 of a particle having the instantaneous deformation u. This relation is expressed in the form

$$[S(x + dx) - S(x)]\,dA = \rho \frac{d^2u}{dt^2}\,dx\,dA \qquad (2\text{-}19)$$

Now $S(x + dx) - S(x) = (dS/dx)\,dx$ and from Eq. (2-9), $S = E\,du/dx$ (the partial derivative not being needed because the deformation is only in the x direction). Differentiating the term for S we get

$$E \frac{d^2u}{dx^2}\,dx\,dA = \rho \frac{d^2u}{dt^2}\,dx\,dA \qquad (2\text{-}20)$$

and we can write the equation of motion

$$\frac{d^2u}{dx^2} = \frac{\rho}{E} \frac{d^2u}{dt^2} \qquad (2\text{-}21)$$

This is the form of the classical one-dimensional wave equation

$$\frac{d^2q}{dx^2} = \frac{1}{V^2} \frac{d^2q}{dt^2} \qquad (2\text{-}22)$$

where V is the velocity of propagation. A convenient solution is

$$q = A \sin k\,(Vt - x) \qquad (2\text{-}23)$$

and this can be verified by differentiation. In the case of the elastic wave in a rod,

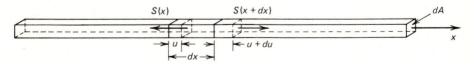

FIGURE 2-7
Elastic deformation in element of thin rod caused by longitudinal stress wave along axis.

comparison of Eqs. (2-21) and (2-22) shows that

$$V = \sqrt{\frac{E}{\rho}} \tag{2-24}$$

so that the velocity of this wave depends only on the elastic modulus and the density.

The three-dimensional wave equation, which can be derived in a similar way, is essentially analogous. For compressional deformation it is

$$\frac{\partial^2 \theta}{\partial x^2} + \frac{\partial^2 \theta}{\partial y^2} + \frac{\partial^2 \theta}{\partial z^2} = \frac{\rho}{\lambda + 2\mu} \frac{\partial^2 \theta}{\partial t^2} \tag{2-25}$$

where θ is the cubical dilatation. For shear deformation, it is

$$\frac{\partial^2 \alpha}{\partial x^2} + \frac{\partial^2 \alpha}{\partial y^2} + \frac{\partial^2 \alpha}{\partial z^2} = \frac{\rho}{\mu} \frac{\partial^2 \alpha}{\partial t^2} \tag{2-26}$$

where α is the shear strain.

By comparing Eq. (2-22) with Eqs. (2-25) and (2-26) it is easy to show that the velocity V_P for compressional waves is

$$V_P = \sqrt{\frac{\lambda + 2\mu}{\rho}} \tag{2-27}$$

and the velocity V_S for shear waves is

$$V_S = \sqrt{\frac{\mu}{\rho}} \tag{2-28}$$

The solution to the wave equation can be expressed as a displacement (or pressure) that maintains its form as distance x and time t change as long as x and t are so related that $Vt - x$ is constant. This is another way of saying that the wave propagates at a velocity equal to x/t. Such a solution has the general form $f(Vt - x)$. The significance of this function is illustrated in Fig. 2-8.

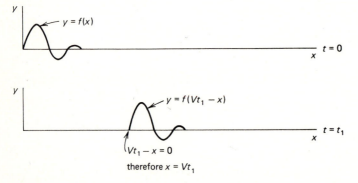

FIGURE 2-8
Movement along x axis of pulse having a waveform $y = f(x)$ and velocity v during time interval t_1.

A more complete and realistic form of the wave function, which holds for a pulse of any shape, is the Fourier series, which can be expressed as

$$f(Vt - x) = \sum_{n=1}^{n=\infty} [A_n \cos nk_1 (Vt - x) + B_n \sin nk_1 (Vt - x)] \quad (2\text{-}29)$$

where A_n and B_n are the Fourier coefficients for the nth harmonic term and k_1 is 2π divided by λ, the fundamental wavelength (normally the length of the initial pulse). The n's are successive integers (1, 2, 3, . . ., etc.). Another form of Eq. (2-29) is

$$f(Vt - x) = \sum_{n=1}^{n=\infty} \left[A_n \cos 2\pi n\left(\frac{t}{T_1} - \frac{x}{\lambda_1}\right) + B_n \sin 2\pi n\left(\frac{t}{T_1} - \frac{x}{\lambda_1}\right) \right] \quad (2\text{-}30)$$

where T_1 is the fundamental period. $T_1 = 1/f_1$ with f_1, the fundamental frequency, being equal to V/λ_1. This means that a seismic pulse can be looked upon as the summation of an infinite number of sine and cosine waves, each having a frequency that is an integral multiple of the fundamental frequency. The amplitudes of the frequency components can be determined by the conventional techniques of Fourier analysis if the initial waveform is known.

2-3 TYPES OF SEISMIC WAVES

In the previous section we summarized the relationship between the velocities of propagation for compressional and shear waves, generally referred to as *body waves*, and the elastic constants of the solid material in which they travel. In this section we shall endeavor to describe the characteristics of such waves in a way that is easier to visualize. We shall also consider the two types of surface waves, Rayleigh waves and Love waves, most widely observed in seismic prospecting and earthquake seismology.

Compressional waves   The particle motion associated with compressional waves consists of alternating condensations and rarefactions during which adjacent particles of the solid are closer together and farther apart during successive half cycles. The motion of the particles is always in the direction of wave propagation. It has been demonstrated by Dix[2] that a pressure pulse traveling as an expanding sphere through an elastic medium must have an oscillatory character and that the pulse passing any point involves at the very least an initial compression of the particles followed by a rarefaction and then a second compression before quiescence is restored. This is illustrated in Fig. 2-9.

If a pressure is suddenly applied, as by an impact, at a point inside a homogeneous elastic medium of infinite size, the region of compression will move outward from the disturbance as an expanding spherical shell, the increase of radius having the compressional wave velocity V_P as designated in Eq. (2-27). Behind this, we observe another expanding shell representing maximum rarefaction and later, at an

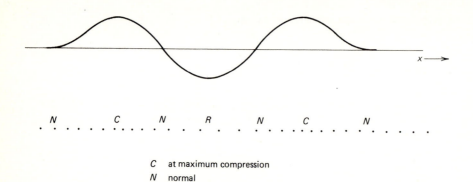

C at maximum compression
N normal
R at maximum rarefaction

FIGURE 2-9
Particle separations during passage of compressional pulse.

approximately equal distance, the second compressional pulse as shown in Fig. 2-10.

The relation between compressional velocity V_P and the elastic constants λ, μ, and ρ is given in Eq. (2-27). This velocity can be expressed in terms of other constants also, as indicated by the relations

$$V_P = \sqrt{\frac{k + \frac{4}{3}\mu}{\rho}} = \sqrt{\frac{E}{\rho}\left(1 + \frac{2\sigma^2}{1 - \sigma - 2\sigma^2}\right)}$$

$$= \sqrt{\frac{E}{\rho}\frac{1 - \sigma}{(1 - 2\sigma)(1 + \sigma)}} \tag{2-31}$$

These expressions were derived by applying Eqs. (2-14) and (2-16) to Eq. (2-27).

Shear waves When shear deformation propagates in an elastic solid, the motion of individual particles is always perpendicular to the direction of wave propagation. The velocity V_S of such waves was shown in Eq. (2-28) to be $\sqrt{\mu/\rho}$. An alternative expression is

$$V_S = \sqrt{\frac{E}{\rho}\frac{1}{2(1 + \sigma)}} \tag{2-32}$$

Comparing Eqs. (2-31) and (2-32), we see that the ratio of compressional to shear velocity is

$$\frac{V_P}{V_S} = \sqrt{\frac{k}{\mu} + \frac{4}{3}} = \sqrt{\frac{1 - \sigma}{\frac{1}{2} - \sigma}}$$

Either expression tells us that the compressional speed will always be greater than the shear speed in a given medium. Both radicals must be greater than 1, the first

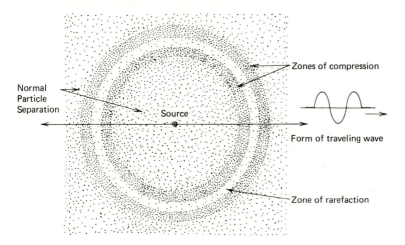

FIGURE 2-10
Spherical spreading of compressional pulse in plane through source at center of
expanding spheres. Particle separations indicated by density of dots.

because k and μ are always positive, the second because σ cannot be greater than
$\frac{1}{2}$ in an ideal solid. If, during the passage of a shear wave, the particles all move
in parallel lines the wave is said to be *polarized* in the direction of the lines. A
horizontally traveling shear wave so polarized that the particle motion is all
vertical is designated as an *SV wave*; when its motion is all in the horizontal plane,
it is called an *SH wave*. For most consolidated rock materials V_P/V_S is between 1.5
and 2.0. As shear deformation cannot be sustained in a liquid, shear waves will not
propagate in liquid materials at all. The outer portion of the earth's core is assumed
to be liquid (even though its density is approximately that of lead) because it does
not transmit shear waves from earthquakes.

Figure 2-11 shows the nature of the particle motion in an oscillatory shear pulse
passing through an elastic medium. Note that the actual movement in the material
is perpendicular to the direction of wave propagation.

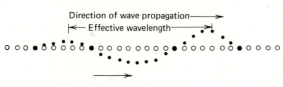

○ Normal position of particle

● Position during passage of
shear pulse

FIGURE 2-11
Particle deformation along line of wave travel during passage of shear pulse
through solid material.

Rayleigh waves Rayleigh waves travel only along the free surface of a solid material (Fig. 2-12a). The particle motion, always in a vertical plane, is elliptical and retrograde with respect to the direction of propagation. The amplitude of the motion decreases exponentially with depth below the surface. The speed of Rayleigh waves is slower than for any body waves, being about nine-tenths that of shear waves in the same medium. The mathematical relationships are derived by Richter.[3]

When a low-speed surface layer overlies a much thicker material in which the speed of elastic waves is higher, the Rayleigh-wave velocity varies with frequency. For wavelengths very short compared with the layer thickness, the speed is about nine-tenths of the shear velocity in the material comprising the surface layer. The properties of this layer govern the speed since such short waves will not penetrate to the underlying material.

For very long wavelengths the speed is nine-tenths the shear velocity in the substratum material since the effect of the surface layer is negligible when most of the wave travels in the zone below it. For intermediate wavelengths the velocity falls between these extremes. This variation of velocity with frequency or wavelength is known as *dispersion*. A dispersive wave, in which different wavelengths travel with different speeds, will appear as a train of events in which successive cycles have increasing or decreasing periods. When a low-speed surface layer is thin compared with wavelength, the longer wavelengths will have higher velocities as they penetrate farther into high-speed material. The periods will then decrease from the beginning of the train to the end. Some examples of such dispersion are illustrated in Dobrin.[4] By analysis of the dispersion of Rayleigh waves on earthquake records seismologists have been able to derive a great deal of useful information on the layering in the earth's crust and upper mantle.

Rayleigh waves are believed to be the principal component of *ground roll*, the common designation for low-velocity, low-frequency surface waves which often obscure reflections on seismic records obtained in oil exploration. They seem to be particularly troublesome in the Gulf Coast of the United States.

Love waves These are surface waves which are observed only when there is a low-speed layer overlying a higher-speed substratum. The wave motion is horizontal and transverse (Fig. 2-12b). The British mathematician A. E. H. Love demon-

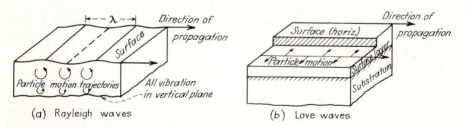

FIGURE 2-12
Characteristics of (a) Raleigh waves and (b) Love waves in traveling along surface of solid.

strated that these waves propagate by multiple reflection between the top and bottom surface of the low-speed layer.

All Love waves are dispersive, the velocity increasing with wavelength. The Love wave speed is equal to that of shear waves in the upper layer for very short wavelengths and to the speed of shear waves in the lower medium for very long wavelengths. Because their particle motion is always horizontal, Love waves are seldom recorded in the course of seismic prospecting operations, for which the detectors respond to vertical ground motion only. They are used extensively, however, in earthquake seismology to study the earth's near-surface layering. For a more advanced discussion of surface waves the reader is referred to Grant and West[5] (pp. 95–107).

2-4 ATTENUATION, REFLECTION, REFRACTION, AND DIFFRACTION OF ELASTIC WAVES

Falloff of energy with distance The energy of a wave in a given medium is proportional to the square of its amplitude (which may be expressed in terms either of pressure or displacement). As a spherical wave spreads out from its source, the energy must be distributed over the area of the sphere, which increases as the square of the sphere's radius. Thus the energy per unit area varies inversely as the square of the distance from the source; the amplitude, which is proportional to the square root of the energy per unit area, should be inversely proportional to the distance the wave has traveled. In addition to the loss of amplitude due to spreading out of the wave, there is also a certain loss from absorption, due to frictional dissipation of the elastic energy into heat. The loss from this source is exponential with distance and will be considered in more detail later in this chapter.

Combining both mechanisms of attenuation, we note that for a homogeneous material

$$I = I_0 \frac{r_0}{r} e^{-\alpha r} \tag{2-33}$$

where I = amplitude at distance r from source
         I_0 = amplitude of distance V_0
         α = absorption coefficient

The value of the absorption coefficient depends on the material.

Huygens' principle Waves in a homogeneous medium, as previously pointed out, spread out from a point source as expanding spheres. *Huygens' principle* states that every point on a wavefront is the source of a new wave that also travels out from it in spherical shells. If the spherical waves have a large enough radius, they can be treated as planes. Lines perpendicular to the wavefronts, called *wave paths* or *rays*, can often be used to describe the wave propagation more conveniently than wave-fronts.

Reflection   Let us apply Huygens' principle to a plane longitudinal wave imping-ing obliquely upon an interface between two elastic media having respective com-pressional velocities of V_{P1} and V_{P2}, shear velocities of V_{S1} and V_{S2}, and densities of ρ_1 and ρ_2 (Fig. 2-13). Consider the incident wavefront AB. The point A will become the center of a new disturbance, from which both longitudinal and transverse waves spread out hemispherically into each medium. Considering only the waves that return into the upper medium, we see that by the time the ray that passed through B reaches the interface at C, a distance x from B, the spherical com-pressional wave from A will also have traveled a distance x and the spherical shear wave a distance $(V_{S1}/V_{P1})x$. Drawing a tangent from C to the first sphere, we get the wavefront of the reflected compressional wave, which has an angle of reflection r_P (with the perpendicular to the interface) equal to the angle of incidence i. This is so because the incident and reflected compressional waves travel at the same speed. A tangent to the smaller circle represents the reflected wavefront for the shear wave, which will make an angle r_S with the interface, determined by the relation

$$\sin r_S = \frac{V_{S1}}{V_{P1}} \sin i \qquad (2\text{-}34)$$

In the case of normal incidence $(i = 0)$, the ratio of reflected *energy* in the com-pressional wave E_r to the incident energy E_i is

$$\left. \frac{E_r}{E_i} \right|_{i=0} = \frac{(\rho_2 V_{P2} - \rho_1 V_{P1})^2}{(\rho_2 V_{P2} + \rho_1 V_{P1})^2} \qquad (2\text{-}35)$$

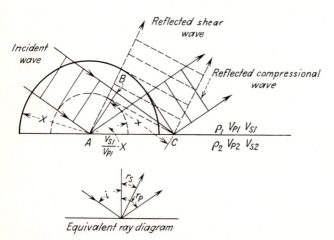

FIGURE 2-13
Reflection of plane compressional wave at interface.

The square root of this ratio, known as the *reflection coefficient R*, gives the relative *amplitudes* of the reflected and incident waves. This can be expressed in the form

$$R = \frac{A_r}{A_i} = \frac{\rho_2 V_{P2} - \rho_1 V_{P1}}{\rho_2 V_{P2} + \rho_1 V_{P1}} \tag{2-36}$$

The amount of energy reflected in this case is thus seen to depend on the contrast in the product of density by velocity (acoustic impedance) on opposite sides of the interface and is independent of the side from which the incident wave approaches. When the medium in which the incident wave travels has a smaller acoustic impedance than the medium across the interface from it, there is no phase change on reflection. When the incident wave is from the side of the interface having the higher acoustic impedance, the reflected wave shows a phase shift of 180°, as follows from the fact that the numerator of Eq. (2-35) becomes negative. Thus, a compression becomes a rarefaction upon reflection from a medium having a lower product of seismic velocity and density.

From a practical standpoint, the reflection coefficient depends mainly on the velocity contrast on opposite sides of the interface, since the variation in density among different kinds of rocks is usually small. It is theoretically possible for the velocity to increase and the density to decrease across an interface in such proportions that there will not be any contrast in acoustic impedance. Under such conditions, no reflection would be expected.

Refraction; Snell's law   When an incident wave strikes an interface, each point along the interface (according to Huygens' principle) becomes the center of a new hemispherical elastic wave that travels into the second medium with a speed of V_{P2} for compressional-wave propagation and with a speed of V_{S2} for the shear wave. From Fig. 2-14 one sees that the compressional wavefront in the lower medium travels a distance AD while the wavefront in the upper medium travels the distance x from C to B. The resulting refracted wave makes an angle R_P with the interface. From the diagram it is evident that

$$\sin i = \frac{BC}{AB} \qquad \text{and} \qquad \sin R_P = \frac{AD}{AB} = \frac{V_{P2}BC}{V_{P1}AB}$$

so that
$$\frac{\sin i}{\sin R_P} = \frac{V_{P1}}{V_{P2}} \tag{2-37}$$

This is *Snell's law*.

For the shear wave, the angle of refraction R_S is expressed by the relation

$$\frac{\sin i}{\sin R_S} = \frac{V_{P1}}{V_{S2}} \tag{2-38}$$

When $\sin i = V_{P1}/V_{P2}$, $\sin R_P$ becomes unity and R_P becomes 90°. This means that the refracted waves does not penetrate the medium but travels along the interface between the two materials. The angle $i_c = \sin^{-1}(V_{P1}/V_{P2})$ is known as the *critical*

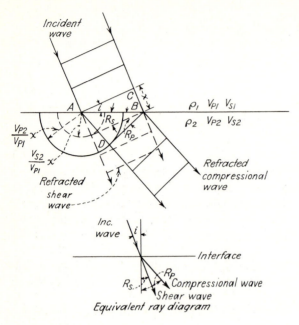

FIGURE 2-14
Refraction of plane compressional wave across interface.

angle for refraction of the compressional wave. For any value of i greater than this critical value, there is no refraction into the second medium and the wave is *totally reflected*. This concept of the critical angle is most important in seismic refraction work, since the ray used in refraction shooting (the *head wave*) is the one which impinges on the top surface of a high-speed bed at the critical angle, travels horizontally along this surface, and is refracted back to the earth's surface at the same angle.

Reflections and refractions at oblique incidence According to Eq. (2-35), the ratio of incident energy normal to the interface that is reflected upward is dependent only on the relation between the acoustic impedances on opposite sides of the interface. When the ray path makes any other angle than 90° with the interface, the reflected energy depends on the angle; also any compressional wave obliquely incident on the interface will be transformed into the four kinds of waves illustrated in Fig. 2-15, reflected compressional, reflected shear, refracted compressional, and refracted shear.

The partitioning of energy among the four types of waves depends on the angle of incidence and on the velocities (shear and compressional) and densities on each side of the interface. The relationships are expressed by Knott's or Zoeppritz's equations, which are rather involved and will not be presented here. Interested readers are referred to Richter[3] for their derivation. At vertical incidence no shear waves are generated at the interface. When the angle exceeds the critical angle, perceptible amounts of shear energy are reflected upward and also refracted into the medium below. Compressional reflections maintain an almost constant energy at small angles, but as the critical angle is approached, the percentage of compressional energy that is reflected increases sharply.

Energy of the refracted compressional wave remains almost constant with increasing angle of incidence until the critical angle is reached, at which point it is of course cut off. There is an increase in the conversion to shear energy, both reflected and refracted, as the critical angle is approached, the maximum shear amplitude being observed at an angle somewhat beyond. The curves of Fig. 2-15 were computed by Richards[6] for a typical interface between deep elastic formations and dense limestones. They are based on Knott's equation for partition of energy at an elastic interface. McCamy et al.[7] have published a more complete set of curves for the distribution of the energy among the four phases as a function of the angle of incidence and the velocities and densities on each side of the interface.

Diffraction   When seismic waves strike any irregularity along a surface such as a corner or a point where there is a sudden change of curvature, the irregular feature acts as a point source for radiating waves in all directions in accordance with Huygens' principle. Such radiation is known as *diffraction*. Figure 2-16 illustrates a buried corner at *A*, from which waves, excited by radiation downward from a source at the surface, spread out in all directions along paths which are rectilinear as long as the velocity is constant. Those waves shown in the drawing are returning to the surface along the indicated paths. A diffracted wave will reach the surface first at a point directly above the edge, because the path is shortest to this point. The event will be observed at successively later times as one moves along the surface away from the point. The amplitude of a diffracted wave falls off rapidly with distance from the nearest point to the source. Diffracted events are frequently observed on seismic records, but they are not always recognized as such.

Limits of applicability of elastic theory in earth materials   Up to this point, we have been studying the laws of elastic-wave propagation in ideal materials having properties not often found in the earth. In such materials there is microscopic and macroscopic homogeneity; stress and strain have a linear relationship; there is no volume change in shear deformation; and no energy is lost due to friction resulting from the wave motion. Yet any theoretical treatment that endeavors to take into account the deviations of the properties of earth materials from those of ideal solids can rapidly become too complex to handle, so that elastic theory (with all its inadequacies) provides the best groundwork for studying seismic properties of earth materials.

2-5   GENERATION OF SEISMIC WAVES FOR PROSPECTING APPLICATIONS

Until now our discussion of basic principles has been equally applicable to earthquake seismology and seismic prospecting. The two kinds of seismic waves are generated by processes that are quite different. Originally, explosive sources such as

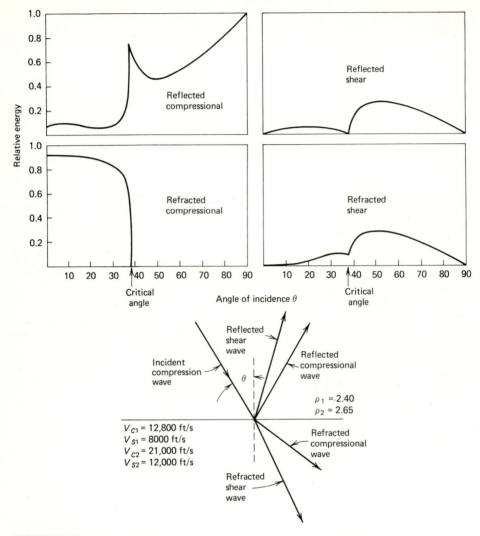

FIGURE 2-15
Partition of energy of incident compressional wave at boundary between materials having specified velocities and densities. Where the incident ray is not perpendicular to the boundary, four kinds of waves are generated and the energy in each type depends on the angle of incidence θ, as shown in the plots. (*After Richards.*[6])

dynamite were detonated in boreholes for virtually all seismic prospecting operations. For about two decades, however, impactive and vibratory mechanical sources of energy, which operate on the earth's surface, have also been employed; moreover, a large variety of nondynamite energy sources have virtually replaced dynamite and other explosives for generation of seismic waves in marine exploration.

The physical theory behind the generation of seismic energy has not been worked

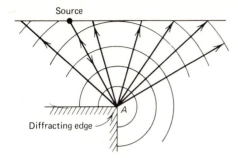

FIGURE 2-16
Diffraction from an edge. The source
A of diffracted radiation has been set
into oscillation by waves generated on
surface. Radial lines with arrows are
ray paths; circular arcs are wavefronts.

out as satisfactorily as that for other aspects of the seismic exploration process
(such as attenuation, reflection, and refraction) because materials in the immediate
vicinity of most energy sources are subjected to nonlinear deformation and the
physics of such deformation is much more complex than it is for elastic-wave
propagation.

The mechanics of generating traveling waves by underwater explosions was
investigated quite thoroughly during World War II, and we know more about the
physical processes involved in shooting underwater than those that occur when a
shot is fired in a borehole. The behavior of underwater energy sources such as
dynamite in generating seismic waves will be covered briefly in Chap. 5, while
vibratory or impactive surface sources used on land such as Vibroseis and Dinoseis
will be discussed in Chap. 4. In this section we shall consider the mechanics of
generating seismic waves only with explosives in boreholes. While such holes are
generally drilled below the weathered layer, the material is for the most part only
semiconsolidated, and when explosives are detonated at the bottom of the hole, a
more or less spherical or cylindrical cavity is formed in the rock surrounding the
explosion, as shown schematically in Fig. 2-17. Inside the wall of the cavity is a

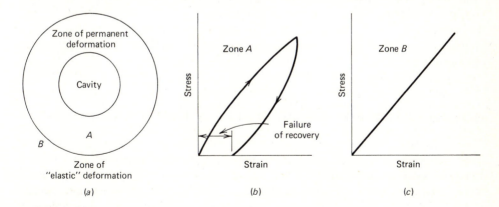

FIGURE 2-17
Deformation of earth when explosive is detonated underground: (a) locations
of zones representing different types of deformation around cavity left by shot;
(b) stress-strain relation in zone A; (c) stress-strain in zone B. (*After Dix.*[1])

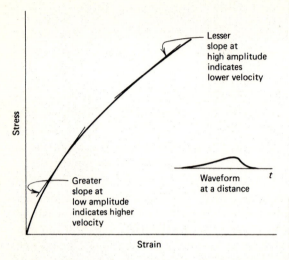

FIGURE 2-18
Spreading of pulse in soft material with nonlinear stress-strain curve. (*After Dix.*[1])

shell of rock material which has been compacted beyond the limit of elastic recovery. This is indicated on the stress-strain diagram for silty clay under very high stress. The curve shows a high degree of hysteresis (permanent displacement after removal of stress).

Dix[1] has used the form of the stress-strain curve in relatively unconsolidated materials (as in zone *A* of Fig. 2-17*a*) to explain the characteristics of the wave recorded from an explosion which takes place in such materials. The steeper the slope of the curve the greater the effective Young's modulus and (because of the relation between this constant and the velocity, at least over the linear parts of the curve) the higher the velocity of propagation of the seismic pulse. In Fig. 2-18 the slope is highest where the stress is least, and the smaller amplitudes thus travel at the higher speeds. As the energy of the explosion builds up, the slope decreases, so that the higher-amplitude part of the impact will generate waves with progressively lower velocities as the pressure continues to build up. This effect results in the spreading out of the wave and yields a waveform which shows a relatively slow buildup of energy with time rather than the near-vertical rise that one associates with shock waves.

2-6 ABSORPTION OF SEISMIC WAVES IN EARTH MATERIALS

The absorption of elastic waves in rocks has been the subject of extensive theoretical and experimental study. Attenuation constants have been measured for a variety of earth materials (as in Ref. 8, sec. 8), but the mechanism for attenuation in many types of rocks, particularly softer sedimentary rocks, is not very well understood.

The amplitude of a seismic wave falls off with distance r from the source in accordance with Eq. (2-33), which contains a term $1/r$ for spherical spreading and an exponential term e^{-ar} for absorption. The symbol α is referred to as the *absorp-*

tion coefficient. Experiments by Born[9] with samples of shale, sandstone, limestone, and cap rock indicated that α is proportional to the first power of frequency for the types of rocks which transmit seismic waves in the portion of the geologic section where oil is generally sought. This kind of dependence would suggest that the mechanism of absorption is solid friction associated with the particle motion in the wave.

The coefficient α can be easily related to δ, the logarithmic decrement (the logarithm of the ratio of amplitude of any cycle to that of the following one in a train of damped waves) and Q, which is π/δ. α can be expressed in terms of δ as $\delta f/v$ or in terms of Q as $\pi f/Qv$, where f is frequency and v is propagation velocity. Both δ and Q are frequently used in the literature to designate attenuation characteristics of materials.

In 1940, Ricker[10] published the first of a series of papers on the form of seismic pulses as governed by the absorption characteristics of the earth materials through which they propagate. He derived equations for the waveform that would be observed after an impulsive signal has traveled through absorbing material in terms of first-power, second-power, and fourth-power dependence of the absorption coefficient on frequency. The shape of the wave he computed for second-power dependence seemed to resemble observed waveforms rather closely. Such a dependence would indicate a viscoelastic frictional loss of the type usually associated with highly viscous liquids. Assuming this attenuation law, he developed equations predicting waveforms for both ground displacement and ground-motion velocity; computed waveshapes for the two at a number of distances from the source are shown in Fig. 2-19. The symmetrical wave representing the velocity at very large distances is referred to as a *Ricker wavelet.*

Waveforms having this appearance and some of the characteristics predicted for it by Ricker's theory have been observed in the Pierre shale of Colorado,[11] but subsequent experiments in the same formation reported by McDonal et al.,[12] which involved Fourier analysis of observed waveforms, indicated an absorption propor-

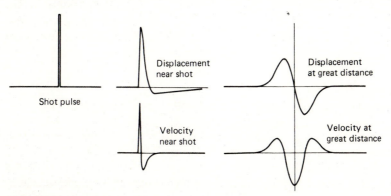

FIGURE 2-19
Change in waveforms of displacement and particle velocity at increasing distances from explosion in shothole. (*After Ricker.*[11])

tional to the *first power* of the frequency, implying solid friction of the type indicated in Born's laboratory experiments. The preponderance of evidence now appears to support the solid-friction hypothesis, and this is accepted by most geophysicists.

The Ricker wavelet has frequently been employed as a convenient representation of the basic seismic pulse in attenuating material. The equation for it is a useful mathematical expression for a seismic signal in the design of recording instruments as well as of programs for data processing.

Table 2-1 shows attenuation characteristics as observed in the laboratory or field for rock samples representing most major rock types of interest to exploration geophysicists. The values of the attenuation coefficient α have been computed for a frequency of 50 Hz from measurements tabulated in the literature of Q at various frequencies using the relation $\alpha = \pi f / Q v$, where f is frequency and v velocity. There is an appreciable overlap in the attenuation values, but it is evident that sedimentary rocks are generally more absorptive than igneous rocks. Actually, there is a large range of variation among different samples for the same type of rock, as is indicated in the tables in sec. 8 of Ref. 8. The great number of papers that have appeared in the literature on attenuation of seismic waves in rocks indicates the importance of this question to geophysicists.

But regardless of the physical mechanism within the rock fabric or the precise law of attenuation, the absorption of higher-frequency energy at a greater rate than

Table 2-1 ATTENUATION COEFFICIENTS FOR 50-Hz SEISMIC WAVES

Material and source of sample	Velocity, km/s	Attenuation α, km^{-1}
Granite:		
Quincy, Mass.	5.0	0.21–0.32
Rockport, Maine	5.1	0.237
Westerly, R.I.	5.0	0.384
Basalt:		
Painesdale, Mich.	5.5	0.414
Diorite	5.78	0.21
Limestone:		
Solenhofen, Bavaria	5.97	0.04
Hunton, Okla.	6.0	0.366
Sandstone:		
Amherst	4.3	0.71
Navajo	4.0	1.77
Shale:		
Pierre, Colo.	2.15	2.32
Sylvan, Okla.	3.3	0.68

SOURCE: Data from Sydney P. Clark, Jr. (ed.), Handbook of Physical Constants, rev. ed., *Geol. Soc. Am. Mem.* 97, 1966, Table 8-1.

lower-frequency energy is well established. This property of rocks causes a progressive lowering of apparent frequency of seismic events with increasing distance of travel through the earth.

2-7   VELOCITIES OF SEISMIC WAVES IN ROCKS

Most igneous and metamorphic rocks have little or no porosity, and the velocities of seismic waves depend mainly on the elastic properties of the minerals making up the rock material itself. This is also the case with massive limestones, dolomites, and evaporites. Sandstones, shales, and certain kinds of soft limestones, on the other hand, have more complex microstructures with pore spaces between grains which may contain fluids or softer types of solid material such as clay. For such rocks, velocity is very much dependent on the porosity and on the material filling the pores. Table 2-2 gives compressional and shear velocities for rocks of different types based mainly on laboratory measurements upon representative samples.

Igneous and metamorphic rocks   In general, igneous rocks have seismic velocities which show a narrower range of variation than sedimentary or metamorphic rocks. The average velocity for igneous rocks is higher than that for other types. The range for 15 samples of granite taken from the earth's surface which are listed in Ref. 8 is from 16,500 to 20,000 ft/s. For basalts from four locations the range is from 17,800 to 21,000 ft/s. The fastest rock is dunite, an ultrabasic rock that some believe may be an important constituent of the earth's mantle, for which the speeds measured for five samples range from 22,400 to 28,500 ft/s. Most types of metamorphic rocks show an even wider range of variation in velocities. Gneiss, for example, has speeds ranging from 11,600 to 24,800 ft/s, and marble velocities are listed from 12,400 to 23,000 ft/s.

Variation of velocity with depth, usually simulated in the laboratory by putting samples under high pressures, is rather small for most igneous rocks. As the pressure was raised from 10 bars (only slightly more than atmospheric) to 10,000 bars (corresponding to a 115,000-ft depth of burial) the compressional velocity of three granite specimens increased less than 15 percent. Sedimentary rocks generally exhibit a much greater percentage increase in velocity with overburden pressure for reasons which will be considered in the following paragraphs. Figure 2-20 shows the effect of depth of burial upon velocity for one sample of igneous rock (granite) and one sample of sedimentary rock (sandstone) as determined by laboratory measurements.

Sedimentary rocks   The velocity characteristics of sedimentary rocks are quite different for different types. Most evaporites such as rock salt and anhydrite have velocities which lie in the same range as igneous rocks. Rocks of this kind show little variation in speed even for different depths of burial. Dolomites exhibit a limited range of variation, and this is quite close to that for many types of evaporitic rocks. Velocities of limestones, sandstones, and shales vary over a much wider range. The

Table 2-2 COMPRESSIONAL AND SHEAR VELOCITIES IN ROCKS

Material and Source	Compressional velocity		Shear velocity	
	m/s	ft/s	m/s	ft/s
Granite:				
Barriefield, Ontario	5640	18,600	2870	9470
Quincy, Mass.	5880	19,400	2940	9700
Bear Mt., Tex.	5520	17,200	3040	10,000
Granodiorite, Weston, Mass.	4780	15,800	3100	10,200
Diorite, Salem, Mass.	5780	19,100	3060	10,100
Gabbro, Duluth, Minn.	6450	21,300	3420	11,200
Basalt, Germany	6400	21,100	3200	10,500
Dunite:				
Jackson City, N.C.	7400	24,400	3790	12,500
Twin Sisters, Wash.	8600	28,400	4370	14,400
Sandstone	1400–4300	4620–14,200		
Sandstone conglomerate, Australia	2400	7920		
Limestone:				
Soft	1700–4200	5610–13,900		
Solenhofen, Bavaria	5970	19,700	2880	9500
Argillaceous, Tex.	6030	19,900	3030	10,000
Rundle, Alberta	6060	20,000		
Anhydrite, U.S. Midcontinent, Gulf Coast	4100	13,530		
Clay	1100–2500	3630–8250		
Loose sand	1800	5940	500	1,650

SOURCE: Sydney P. Clark, Jr. (ed.), "Handbook of Physical Constants," rev. ed., *Geol. Soc. Am. Mem.* 97, 1966.

key to the variation appears to be the density and (a closely related quantity) porosity. Figure 2-21 shows how all sedimentary rocks except anhydrites exhibit a 0.25 power relationship between velocity and bulk density.

Such a correspondence between density and velocity is not confined to sedimentary rocks. Nafe and Drake[14] have plotted seismic velocity versus bulk density for a wide variety of materials ranging from muds at the bottom of the sea to ultra-

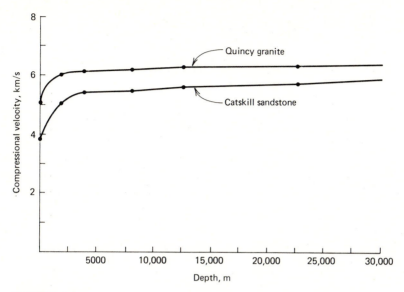

FIGURE 2-20

Increase of compressional velocity with depth for a typical granite and a typical sandstone. Leveling off of increase in velocity at shallow depths probably caused by closing of cracks in granite and maximum reduction of pore space in sandstone.

basic igneous rocks. Figure 2-22 illustrates their results. The best line through the points, which has surprisingly little scatter about it, indicates the same 0.25 power relation which shows up clearly in the logarithmic plot of Fig. 2-21. This relation makes it possible to estimate the velocity of rocks when only the bulk density is known and vice versa.

In most sedimentary rocks the actual velocity is dependent upon the intrinsic velocity in the minerals constituting the solid rock matrix, the porosity, the pressure, and the velocity in the fluid filling the pore spaces. It also depends on the composition of any solid cementing material between the grains of the primary rock constituents.

At shallow depths of burial the velocity of most sedimentary rocks increases rapidly with increasing pressure. For rocks consisting of grains that are approximately spherical, a theoretical relationship developed by Gassmann predicts that the velocity should be proportional to the pressure raised to the one-sixth power, the constant of proportionality being expressed in terms of the elastic constants and density of the rock material itself. For rocks such as quartzites that have almost no porosity, very small cracks are often present near the surface which tend to close under the weight of the overburden at depth, the result being a rapid increase in velocity in the first few thousand feet and a leveling off at greater depths.

Beyond the depth where such consolidation is reached, the influence of variation in pressure on velocity becomes small, and then porosity and mineral composition

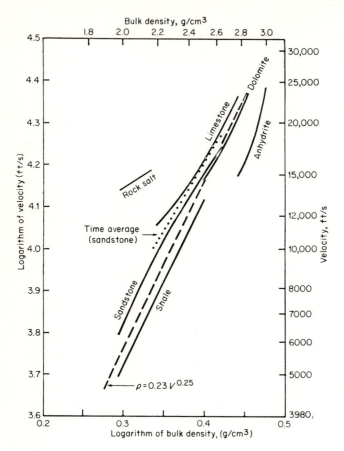

FIGURE 2-21
Velocity-versus-density relationships for different types of rocks. (Gardner, Gardner and Gregory.[13])

of the grains become dominant in governing velocity. A very simple linear relationship between the reciprocal of the velocity and porosity has been found by Wyllie et al.[15] to be valid for water-saturated sandstones at depths greater than a few thousand feet:

$$\frac{1}{V} = \frac{\phi}{V_F} + \frac{1-\phi}{V_M} \tag{2-39}$$

where V = velocity in saturated rock
ϕ = fractional porosity
V_F = velocity of fluid in pore space
V_M = velocity of solid material making up rock matrix

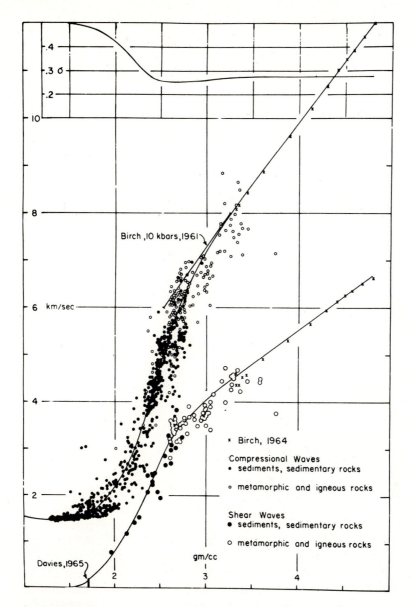

FIGURE 2-22
Velocity versus density for compressional and shear waves in all types of rocks.
Poisson's ratio versus density at top of figure. (*Data assembled by John E. Nafe,*
Lamont-Doherty Observatory. From Ludwig et al., "The Sea," vol. 4, part 1,
Wiley-Interscience, New York, 1970.)

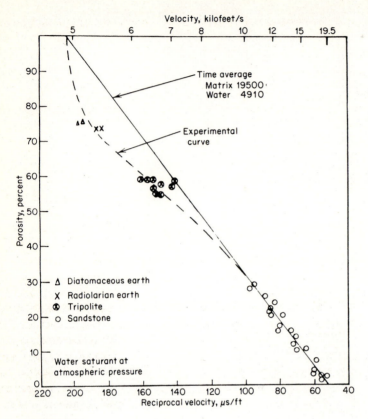

FIGURE 2-23
Velocity versus porosity for various silicic rocks. Straight line represents predicted time-average relationship. (*After Wyllie, et al., Geophysics,* 1958.)

Equation (2-39) is referred to as the *time-average relationship.* Where it holds, the
velocity V would be equal to V_M at zero porosity and V_F at 100 percent porosity,
its reciprocal $1/V$ being linearly related to ϕ for values in between. Figure 2-23 is a
plot of V versus ϕ for several types of sedimentary rocks. The observed velocities
for sandstone show a close adherence to the time-average relationship over the
porosity range between zero and 30 percent. The other rocks represented, exhibiting higher porosities, have a different matrix velocity, but there is too small a
porosity range to test the validity of the time-average law for these materials.

A similar equation developed from theoretical considerations by Pickett[16] is

$$\frac{1}{V} = A + B\phi \qquad (2\text{-}40)$$

where A and B depend on lithologic parameters and depth of burial. The equation
appears to be valid for a wider range of sedimentary rocks.

A large-scale statistical study of sedimentary-rock velocities has been made by Faust,[17] who showed that for sandstones and shales the velocity V (in feet per second) can be expressed empirically as

$$V = KZ^{1/6}T^{1/6} \qquad\qquad (2\text{-}41)$$

where $K = 125.3$ when Z is in feet, T in years, and V in feet per second

       Z = depth of burial

       T = age of formation

Although this relation is purely empirical, being based on conventional well-velocity surveys made before the days of velocity logs (so that velocities of sections hundreds of feet thick had to be used in compiling the averages), the correspondence between Faust's one-sixth power dependence on depth and Gassmann's theoretically determined one-sixth power relation with pressure is quite interesting even though it may be coincidental. The geologic age relationship cannot be so easily related to the physical mechanisms governing velocity, but increasing induration and grain cementation with time might be expected to lead to an increase in seismic velocity.

Figure 2-24 shows some curves based on Faust's law indicating velocities of sands and shales as a function of depth for rocks of different age.

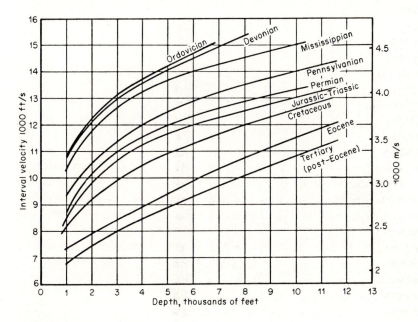

FIGURE 2-24

Compressional velocities as determined from borehole surveys in sandstone and shale. (*After Faust.*[17])

2-8 PRINCIPLES INVOLVED IN MEASURING SEISMIC-WAVE CHARACTERISTICS

The immediate objective of all seismic field measurements is to obtain a record in the most useful form possible of the ground motion resulting from the arrival at the surface of seismic waves reflected or refracted from subsurface formations. The record that is made in the field, whether on paper or on magnetic tape, will not be an exact representation of the actual ground motion because the characteristics of the measuring system introduce distortions that change the waveforms, often appreciably. Sometimes these distortions are inevitable, as it is generally impossible to observe the desired reflections or refracted events as well as might be required if the earth motion itself were recorded with really high fidelity.

One of the most important concepts entering into the interpretation of records showing ground motion is filtering, which is the process of changing the waveform of a signal. In Chap. 3 we shall consider electric filters used in connection with this process. In Chap. 6, we shall study the principle of filtering as carried out in processing data with electronic computers. Still another type of filtering occurs in the earth itself, and it is appropriate for us to consider it in connection with our investigations of the propagation characteristics of earth materials.

We have seen how earth material changes the shape of the pressure impulse from an explosion, which initially has the form of a spike with a negligible time duration, into an oscillatory pulse having a breadth that increases with distance of travel. This is one sense in which the earth acts as a filter, changing the form of the input much more drastically than most electric filters are capable of doing. If we could separate the filtering action of the earth from that of our recording systems, we could interpret the entire seismic record as an expression of the earth's filtering behavior. Such filtering action depends on the earth's structure and lithology. In Chap. 8, which is on reflection interpretation, we shall apply filter theory to deduce the layering characteristics of reflecting formations from the waveforms of the signals which they return to the surface.

REFERENCES

1. Dix, C. H.: "Seismic Prospecting for Oil," Harper, New York, 1952.
2. Dix, C. H.: On the Minimum Oscillatory Character of Seismic Pulses, *Geophysics*, vol. 14, pp. 17–20, 1949.
3. Richter, C. F.: "Elementary Seismology," Freeman, San Francisco, 1941.
4. Dobrin, Milton B.: Dispersion in Seismic Surface Waves, *Geophysics*, vol. 16, pp. 63–80, 1957.
5. Grant, F. S., and G. F. West: "Interpretation Theory in Applied Geophysics," McGraw-Hill, New York, 1965.
6. Richards, T. C.: Motion of the Ground on Arrival of Reflected Longitudinal and Transverse Waves at Wide-Angle Reflection Distances, *Geophysics*, vol. 26, pp. 277–297, 1961.

7. McCamy, Keith, R. P. Mayer, and Thomas J. Smith: Generally Applicable Solutions of Zoeppritz Amplitude Equations, *Bull. Seismol. Soc. Am.*, vol. 52, pp. 923–955, 1962.
8. Clark, Sydney P., Jr. (ed.): "Handbook of Physical Constants," rev. ed., *Geol. Soc. Am. Mem.* 97, New York, 1966.
9. Born, W. T., The Attenuation Constant of Earth Materials, *Geophysics*, vol. 6, pp. 132–148, 1941.
10. Ricker, Norman: The Form and Nature of Seismic Wavelets and the Structure of Seismograms, *Geophysics*, vol. 5, pp. 348–366, 1940.
11. Ricker, Norman: The Form and Laws of Propagation of Seismic Wavelets, *Geophysics*, vol. 18, pp. 10–40, 1953.
12. McDonal, F. J., F. A. Angona, R. L. Mills, R. L. Sengbush, R. G. Van Nostrand, and J. E. White: Attenuation of Shear and Compressional Waves in Pierre Shale, *Geophysics*, vol. 23, pp. 421–439, 1958.
13. Gardner, G. H. F., L. W. Gardner, and A. R. Gregory: Formation Velocity and Density: The Diagnostic Basis for Stratigraphic Traps, *Geophysics*, vol. 39, pp. 770–780, 1974.
14. Nafe, John E., and Charles L. Drake: Variation with Depth in Shallow and Deep Water Marine Sediments of Porosity, Density and the Velocities of Compressional and Shear Waves, *Geophysics*, vol. 22, pp. 523–552, 1957.
15. Wyllie, M. R., A. R. Gregory, and G. H. F. Gardner: An Experimental Investigation of Factors Affecting Elastic Wave Velocities in Porous Media, *Geophysics*, vol. 23, pp. 459–493, 1958.
16. Pickett, George R.: Principles for Application of Borehole Measurements in Petroleum Engineering, *Log Anal.*, May–June 1969, pp. 22–33.
17. Faust, L. Y.: Seismic Velocity as a Function of Depth and Geologic Time, *Geophysics*, vol. 16, pp. 192–206, 1951.

3

SEISMIC RECORDING INSTRUMENTS

Seismic records obtained in prospecting show the motion of the earth's surface, as generated by explosives or other energy sources, at different, usually closely spaced, observing positions. The motion is actually indicated in terms of particle velocity versus time rather than of particle displacement versus time. The proper translation of the signals thus obtained into geological information requires us to know as much as possible about the behavior of seismic waves as they propagate through earth materials, but we must also understand the characteristics and performance of the instruments that record the waves when they return to the surface. The properties of the seismic waves in the earth were taken up in Chap. 2; the operation of the recording instruments will be considered in this chapter.

It is important for the geologist to realize how the instruments with which reflection records are obtained can affect the interpretation of the data on the records. Unrecognized instrumentally generated distortions of the signals on seismic records could result in spurious geological conclusions.

The primary elements of modern instrumental systems used to record seismic ground motion are geophones, amplifiers, digital recorders (with associated hardware), and units such as galvanometric cameras for monitoring. In recent years, special-purpose digital computers have been put into the recording trucks to control the entire recording process in the field. They are generally used both for regulating and monitoring the field operations and for preliminary processing of the data more or less concurrently with the shooting.

3-1 GEOPHONES

The *geophone*, sometimes referred to as the detector or the seismometer, is the unit in direct contact with the earth which converts the motion of the earth resulting from the shot into electric signals. These signals constitute the input into an instrumental system the end product of which is the presentation of subsurface geological information in some visible form, usually as a record section comparable to a geological cross section.

Electromagnetic geophones   Nearly all geophones currently used for seismic recording on land are of the electromagnetic type. This kind of detector consists of a coil of wire and a magnet, one of the two elements being fixed as rigidly as possible to the earth's surface so that it will move along with the earth in response to seismic disturbances. The other element is inertial and is suspended by a spring from a support attached to the portion that moves with the earth. Relative motion between the two produces an electromotive force between the coil and magnet, the voltage being proportional to the *velocity* of the motion.

Figure 3-1 shows the principle of operation for a unit in which the magnet is inertial and the coil moves with the earth. In most geophones designed for exploration work, however, the coil is incorporated in the inertial element and the magnet is attached to the case, which of course moves as the earth in which it is planted moves. When the coil is the inertial element, it is ordinarily attached to a mass suspended by a spring. For the geophone shown in Fig. 3-2, the coil is wound about a bobbin, and the combination of the two acts as the inertial mass.

The sensitivity of an electromagnetic geophone depends on the strength of the magnet and the number of turns of wire in the coil as well as the geometry governing the interaction of the magnetic flux lines and the coil. Geophones have become steadily smaller as new magnetic materials of greater strength have become available. Some geophone elements used regularly in field operations are almost as small as golf balls (Fig. 3-3).

Every seismic detector, whether designed to record natural earthquakes or the artificial earthquakes generated in seismic prospecting, has a natural period which

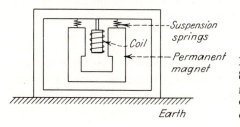

FIGURE 3-1
Schematic diagram of electromagnetic geophone. The magnet is the inertial element, and the case moves with the earth.

FIGURE 3-2
Cutaway of double-coil electromagnetic geophone. Width is 2 in. (*Mark Products, Inc.*)

depends on the mass and the restoring force of the spring suspension. In an electromagnetic geophone the natural period T depends on the mass m of the suspended inertial member (whether it is the magnet or the coil) and the stiffness coefficient k of the spring. The latter quantity is the proportionality constant between the force on the spring and the elongation attributable to the force. The period T is dependent on m and k as follows:

$$T = 2\pi \sqrt{\frac{m}{k}} \qquad (3\text{-}1)$$

For frequency, which is the reciprocal of the period, the relation is

$$f = \frac{1}{2\pi} \sqrt{\frac{k}{m}} \qquad (3\text{-}2)$$

If the damping in the geophone system is small, any seismic impulse setting the spring suspension into motion will generate an oscillatory or "ringy" output signal with a frequency that is the reciprocal of the natural period. This type of oscillation is generally undesirable in seismic recording. For one thing, the response to any component of the input signal that occurs at the natural period would be greatly accentuated compared with all other components.

By introducing proper damping, it is possible to make the geophone response approximately equal at all frequencies above the resonant frequency. When this is the case, the geophone output gives a high-fidelity representation of the ground motion. The damping which suppresses mechanical oscillation in the geophone originates from the eddy-current effect. The degree of damping is controlled by a

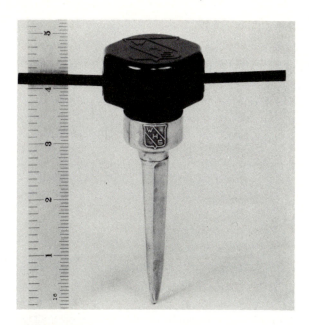

FIGURE 3-3
Electromagnetic geophone with spike for good coupling to earth. (*Walker-Hall-Sears, Inc.*)

resistor connected across the terminals and the effect of a conductive bobbin, which acts as a shorted turn.

While most geophones used in exploration are designed to have a relatively flat response, there are special circumstances where it is desirable for the response curve to be peaked at a predetermined frequency. If *detection* of a signal (such as the first break in refraction recording) is more important than precise registration of its waveform, and if the signal is immersed in a high level of background noise, sharp tuning of the detector to the frequency of the expected signal may be the only way to observe it at all.

The three variable parameters in a geophone system are thus the mass m of the suspended element, the stiffness coefficient k, and the damping, which is linearly dependent (the proportionality constant being R) upon the velocity of the moving element. The differential equation for the displacement from equilibrium u of a system where there is an external force F_0 (the ground motion) oscillating at frequency f is

$$m\frac{d^2u}{dt^2} + R\frac{du}{dt} + ku = F_0 \sin 2\pi f t \tag{3-3}$$

This equation for forced, damped oscillatory motion is a classic one in mathematical physics. The solution (see Morse,[1] p. 28) approaches the steady-state value

$$u = \frac{F_0}{2\pi f \sqrt{R^2 + (2\pi f m - k/2\pi f)^2}} \sin (2\pi f t - \phi) \tag{3-4}$$

where
$$\phi = \tan^{-1} \frac{2\pi f m - k/2\pi f}{R}$$

is the phase angle between input and output for frequency f.

The output of an electromagnetic geophone will be proportional to the velocity of the coil, which is the first derivative of Eq. (3-4). This is

$$v = \frac{du}{dt} = \frac{F_0}{\sqrt{R^2 + (2\pi f m - k/2\pi f)^2}} \cos (2\pi f t - \phi) \tag{3-5}$$

Figure 3-4, based on Eq. (3-5), shows a set of characteristic curves for geophone response as a function of the frequency of earth motion and of damping. Such curves can be obtained experimentally by putting the geophone on a shaking table that is set into oscillation at various frequencies.

Each curve of Fig. 3-4 corresponds to a different amount of damping. The ordinate is the voltage output divided by the output which would be obtained with an excitation having the same velocity amplitude and a frequency much higher than the natural frequency. The uppermost curve is for an undamped system. Theoretically the absence of damping results in an infinite response at resonance. As damping is introduced in increasing amounts, the amplitude and sharpness of the

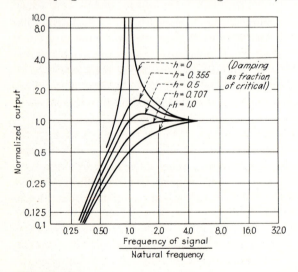

FIGURE 3-4
Normalized frequency-response curves for electromagnetic geophone with various values of damping (h is fraction of critical damping).

peak at the resonance frequency diminish. The maximum amount of damping that will just eliminate the oscillatory character of the response is referred to as *critical damping*, which is reached when $R = 2\sqrt{km}$. If the damping is half its critical value ($h = 0.5$), a maximum will still be observed but it will be at a somewhat higher frequency than the natural one. This is considered by many to be the most acceptable degree of damping for reflection recording. At a damping 0.707 times the critical value, the peak disappears and the output increases smoothly with increasing frequency, approaching its maximum value asymptotically. The curve for critical damping ($h = 1.0$) follows a similar pattern. If there is any substantial component of earth motion in the neighborhood of the detector's natural frequency, it is desirable to flatten this portion of the response curve by introducing enough damping resistance to remove the peak. Otherwise the effect of that component will be exaggerated in the output. If, for example, a geophone of natural frequency 6 Hz were used for reflection work in an area where the dominant frequencies to be recorded are in the neighborhood of 30 Hz, the damping would not have to be adjusted so carefully as in refraction work, where frequencies as low as 5 Hz are encountered.

The choice of natural frequency, which can be set by use of a spring with the proper k value, should be governed by the minimum frequency to be recorded. At one time it was considered desirable to restrict reflection frequencies to the higher portion of the range passed by the earth and to suppress lower frequencies (which might be associated with ground roll) by using the geophone itself as a high-pass filter. It was common then for natural frequencies of geophones to be set at 30 Hz or higher. Now that ground roll can be suppressed in the field by shot and geophone patterns and there is greater interest in reflection signals from deep formations that may have useful components far below 20 Hz, there has been a trend toward geophones with lower natural frequencies. It is simplest, as can be seen from Fig. 3-4, to set the natural frequency of the geophones at the lower limit of the range of frequencies it is desired to record. Refraction geophones, for example, should always have a natural period well below 10 Hz (actually below 5 Hz for exceptionally long shot-detector distances). Commercial geophones with natural frequencies ranging from 7.5 to 30 Hz are now available for reflection work. Variable shunts can be inserted to provide any desired degree of damping.

Pressure phones (hydrophones)   While electromagnetic geophones with moving coils are the standard detecting units for work on land, pressure geophones are more commonly used for receiving seismic signals in appreciable depths of water. Special waterproof cases have been designed (Fig. 3-5) to permit planting of moving-coil phones in marshy ground. Pressure-sensitive phones, often referred to as *hydrophones*, use piezoelectric crystals or comparable ceramic elements as pressure sensors. They generate a voltage proportional to the instantaneous water pressure associated with the seismic signal. It can be shown that this pressure is proportional to the velocity of the water particles set into motion by the signal.

One type of pressure phone is ordinarily placed inside a plastic hose filled with oil

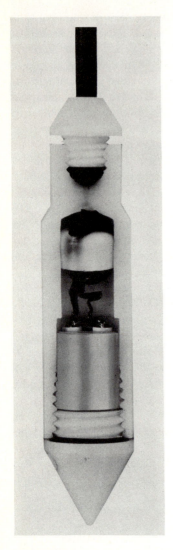

FIGURE 3-5
Geophone in special waterproof case
for planting in marshy ground.
(*Walker-Hall-Sears, Inc.*)

that transmits the pressure variations in the water to the sensitive element in the detecting unit. This type of cable is generally referred to as a *streamer*. Pressure phones of the type used in marine cables will be discussed further in Chap. 5.

3-2 ANALOG RECORDING

For the first 30-odd years of seismic reflection recording, all registration of signals from geophones or hydrophones was in continuous form, as was the case with the "wiggly lines" observed on paper oscillograms, or (since the early 1950s) with magnetic tape having a magnetization continuously varying with time. This type of storage, called *analog recording*, is to be contrasted with *digital recording*, which has been employed to an increasing extent since the early 1960s and is now found on nearly all seismic equipment operating in North America. In digital registration, signal amplitudes are recorded at discrete sampling intervals in a form that encodes a number for each sample, with no recording of the signal at times between the sampling instants. Figure 3-6 illustrates the basic differences between the two types of recording.

Although almost all seismic recording for oil exploration is now digital, we should review the techniques of analog recording for a number of reasons. Even on digital crews, the amplifiers into which the geophone signals are passed are almost always analog. Moreover, in many parts of the world analog equipment is still being used, the signals being digitized for processing subsequent to recording. Also, a large amount of seismic data still in the active files of oil companies has been recorded by analog techniques. Geologists and geophysicists involved in the processing or the interpretation of such data should understand how the original recording was carried out.

Analog amplifiers The primary problems encountered in designing amplifiers for seismic reflection work result from the exceptionally wide range of ground-motion amplitudes that the geophone picks up over the few seconds after the shot is fired. The ground motion in the vicinity of a shot point may have a million or more times the amplitude immediately after the explosion that it has at the end of the record, when the energy that is recorded has traveled many tens of thousands of feet to a deep reflector and back. The dynamic range of analog tape is between 40 and 45 dB (corresponding to an amplitude factor of 100 to 200), so that the range of 1 million in the signal amplitudes must be somehow compressed in the amplifier if it is to be

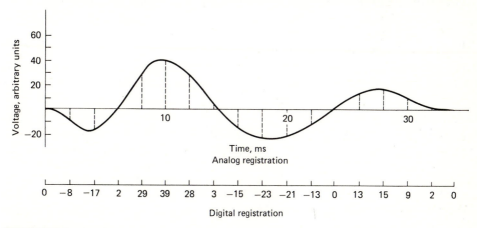

FIGURE 3-6
Analog and digital registration of the same signal. Sampling interval for the digital data is 2 ms.

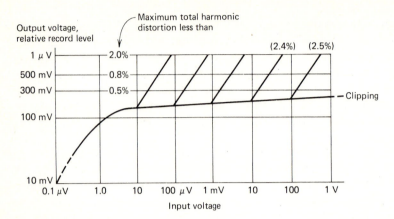

FIGURE 3-7

Characteristic curve for automatic gain control illustrating relative constancy of output for 10^6 to 1 range of input voltages. Diagonal lines represent response to and distortion from simulated noise burstouts with a 35-Hz input signal. (*SIE, Inc.*)

stored properly on analog tape. For direct registration on paper records of the type used for the first 25 years or so of seismic recording, only about 20 dB of dynamic range is available, making the requirements for signal compression even more stringent.

To compress the signal so that it can be recorded by media having such limitations, analog amplifiers have special circuitry for expanding or suppressing the gain level during recording in the form of automatic gain control. The characteristic curve of Fig. 3-7 shows how a 10^6 to 1 range of input voltages is reduced to a range of approximately 2 to 1 at the output of one type of analog system.

Early automatic gain control (AGC) systems operated with diode bridge circuits, which required balancing in the field to maintain stability. Such balancing is no longer necessary. The rate at which the AGC action is applied and relaxed after the signal has passed can be varied on some amplifiers. Special gain controls are used to regulate the amplitude levels at the beginning and end of the record.

Analog filters for recording and playback Filtering of seismic signals from geophones is often necessary before they can be recorded in a really useful form. Such filtering is done both in field recording and in subsequent processing of the recorded data. In the days before reproducible recording, filters in the field truck were used to remove undesired noise from seismic data before registration on the paper records used for the ultimate interpretation. Even after magnetic tape was introduced, it was sometimes desirable to remove higher-amplitude spurious signals, such as low-frequency surface waves, by prefiltering before registration on the tape so as to avoid saturation effects. Analog filtering is still carried out in conjunction with digital recording to avoid spurious signals on records.

Analog filters of the type that were used in the field trucks before the advent of digital systems had a choice of high- and low-cut frequency limits to be selected with knobs on the amplifier panel. The same types of filters were used for playing back tapes in the field to monitor the recorded data and for data processing in analog playback centers. The typical performance of filters available with field amplifiers is illustrated by the curves of Fig. 3-8. Slopes of the analog filter curves, ranging from 18 to 36 dB/octave, are generally much gentler than those obtained with digital processing filters.

In digital recording, analog filters must be employed in the field to prevent aliasing, a spurious effect associated with the sampling process which will be discussed in Chap. 6.

Magnetic analog recorders   Two basic types of tape-recording units have been employed in most analog systems, one using frequency modulation and the other using amplitude modulation. In the former type, the output from the amplifier modulates the frequency of a 3000-Hz carrier signal over the seismic range (10 to 300 Hz). The signal representing geophone motion is extracted upon playback by demodulation. In the other type, the signal is impressed directly upon the tape, as in sound recording, the magnetization of the tape being proportional to the strength of the signal up to the limit of the tape's dynamic range (usually about 45 dB). In either type of recording, the tape is stretched around a drum which revolves, during the time the recording is going on, past a bank of heads. There is one head for each trace plus one for timing and several more for auxiliary data channels.

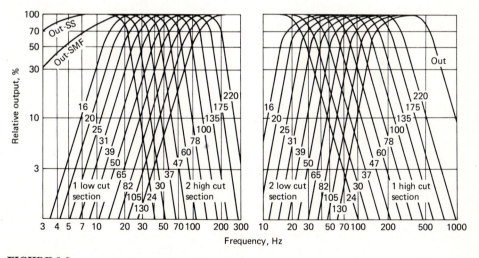

FIGURE 3-8
Characteristic filter curves for PT-100 analog amplifiers. The set of curves on the right is for a higher-frequency recording range than on the left although there is substantial overlap between the two ranges. (*SIE, Inc.*)

3-3 DIGITAL RECORDING EQUIPMENT

One of the most significant developments in seismic technology has been the introduction of digital recording in the field. The first use of digital equipment for actual exploration work was in 1963. The advantages of digital recording over analog recording had long been recognized, but it was not until computer technology reached the stage where speed and storage capacity were adequate for handling multichannel seismic signals in real time that the approach became feasible for seismic operations. There are a number of elementary explanations of seismic recording by digital means in the literature. An early paper on the subject was by Dobrin and Ward.[2] The reader is also referred to Silverman,[3] Evenden and Stone,[4] and Dohr[5] for further information.

Principles The analog signals constituting the amplified outputs of the geophone groups on the ground are digitized by analog-to-digital converters, the digital output being recorded on magnetic tape. In playback, the data on the tapes are introduced into digital computers, which, through filtering, time shifting, compositing, and other operations, put them into the form desired for presentation after digital-to-analog conversion. The description of the digitizing process that follows does not refer to any existing systems of digital recording; their operation is too complicated for presentation at an elementary level. What we shall consider is a schematic system chosen for pedagogic simplicity rather than for close correspondence to current practice.

All seismic signals that are recorded digitally are registered in the form of binary numbers. These differ from ordinary decimal numbers in that the digits in the binary numbers are the coefficients of 2 raised to successively lower integral powers while the decimal numbers are the coefficients of 10 raised to the corresponding powers. A decimal number like 1247 is actually

$$1 \times 10^3 + 2 \times 10^2 + 4 \times 10^1 \times 7 \times 10^0$$

while a binary number like 11001 is

$$1 \times 2^4 + 1 \times 2^3 + 0 \times 2^2 + 0 \times 2^1 + 1 \times 2^0$$

This adds up to 25 in decimal designation. It is evident from inspection of these two examples that the highest digit available for a number to the base n (such as 10 or 2 in the foregoing examples) is $n - 1$. Thus all binary numbers are expressed by combinations of digits which are either 0 or 1.

Figure 3-9 is a highly simplified demonstration of how seismic signals can be presented on tape in digital form. The output of the seismic amplifier, shown at the left, is sampled at uniform intervals, in this case 2 ms, and the amplitudes at each sampling position are converted into binary form as shown in the middle part of the figure.

The sample having an amplitude of 28, for example, is written in seven-digit binary form as 0011100, the zero at the beginning being the coefficient of 2^6, the

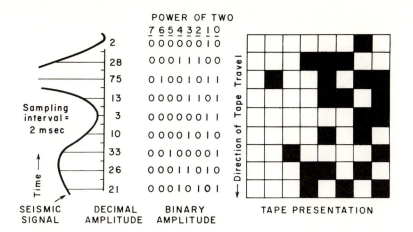

FIGURE 3-9

Representation of a seismic signal sampled at 2-ms intervals on digital magnetic tape. Black squares indicate binary digit 1, and white squares binary digit 0. This format is schematic and is not used in practice. (*Dobrin and Ward.*[2])

second of 2^5, and the two zeros at the end being the multipliers of 2^1 and 2^0, respectively. For tape recording of digital information, the binary system has the great advantage that any number can be represented on tape by a series of blocks, each of which when magnetized corresponds to the digit 1 or when not magnetized corresponds to the digit 0. This "yes" or "no" combination is all that is needed to represent any number on magnetic tape, regardless of its magnitude, if enough elements, or *bits*, are available. The crossword puzzle pattern on the right represents a strip of tape on which the binary numbers shown to the left of it are encoded. This arrangement would be used for a tape system having eight recording heads, each corresponding to a different binary digit. Actually no systems with as few as eight channels are used in modern seismic recording.

Figure 3-10 shows a block diagram of a system for converting analog signals into

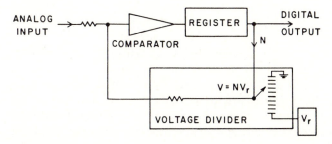

FIGURE 3-10

Schematic diagram illustrating operation of an analog-to-digital converter. (*Dobrin and Ward.*[2])

digital form. It is called an *analog-digital converter*. The input at the upper left could be the voltage output of a seismic amplifier in the field. Vr, at the lower right, is a constant reference voltage which is fed into a voltage divider where a variable fraction N of it is picked off for balance against the input voltage. By successive approximations the digitally controlled comparator locates at each instant the value of N that will null the voltage in the loop containing it and the voltage divider. This value is stored digitally in the register, and when the nulling voltage is found, a command is given for the contents of the register to spill out the digits for N in proper sequence, thus producing the output signal in digital form.

In Fig. 3-11 we see how the signals on 24 information channels corresponding to a conventional multitrace seismic system can be transferred to a single channel for storage and processing. For seismic frequencies, a sampling of the signal at short intervals such as 2 ms preserves all significant information. During this interval an electronic switch, called a *multiplexer*, sweeps each of the 24 input channels sequentially. The output, still an analog-type signal, consists of a series of chopped-up fragments corresponding to the respective samples, each continuous portion having a duration of 83 μs (2 ms divided by 24). This signal, shown below the multiplexer block, is next passed into the sample-and-hold unit. In it each 83-μs continuous signal is sampled, and the value thus obtained is fed into a capacitor which holds it in storage before digitization. The output of the sample-and-hold system now goes into an analog-to-digital converter, which operates as depicted in Fig. 3-10. The emerging signal is now in the form of discrete pulses or pips having uniform height. Each 83-μs interval includes one binary number consisting of eight digits which are either 1 or 0. Bits represented by pulses correspond to the binary digit 1; those without a pulse correspond to 0.

The final element through which the signals pass before being recorded on digital

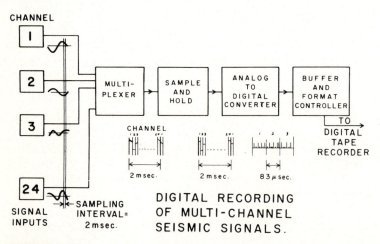

FIGURE 3-11
Digital recording of 24 seismic channels using multiplexer, analog-to-digital converter, and other elements. (*Dobrin and Ward.*[2])

tape is the buffer and format controller. This holds them in storage until they go into the tape recorder, and it also programs the sequence of the digits so that the samples from the different channels can be kept in order for redistribution upon playback. The tape recorder has a separate head for each bit. The seismic signals are recorded on the tape in real time, *i.e.*, the time that was actually required to register them initially. Later processing of the tape in a digital computer can be either in real time or at a faster rate.

Dynamic range in digital recording   A conspicuous advantage of digital recording is its dynamic range, which is the ratio of the maximum amplitude to the minimum amplitude that can be meaningfully stored. Being limited only by the number of bits in the recording system, the system can be designed to have a dynamic range that is as great as is necessary to recover the entire signal with fidelity. It is true of course that the dynamic range of the analog elements, such as geophones and amplifiers, that precede the digitizing stage sets a limit to the performance of the entire recording unit.

When digital systems were first introduced, they were designed to record "words" (amplitude values of 13 bits plus 1 for sign) each corresponding to a dynamic range of almost 78 dB. This should be compared with the 45-dB maximum range available in analog recording systems. Binary gain-ranging amplifiers, which automatically shift the gain by steps corresponding to a factor of 2, increase the effective dynamic range by another 78 dB. With such a capacity, it is possible to record the entire seismic signal on digital tape with preservation of true relative amplitudes over the entire length of the record. Moreover, the capability of recovering a weak reflection signal in the presence of a very strong noise is enhanced by the higher dynamic range of digital systems. A reflection superimposed on a stronger noise signal that might be irretrievably lost in an analog tape recording can still be recoverable in undistorted form with the dynamic ranges that are obtainable on digital recordings. The doubling of dynamic range made possible by binary-gain equipment is particularly valuable in recording true amplitudes for direct detections of hydrocarbons.

A more recent development designed to increase the dynamic range of the signal that is digitally recorded is the floating-point system. This records the gain level as a characteristic (comparable to that expressing the power of 10 in a logarithm), as does the binary gain. The number recorded, however, will be the coefficient for a higher power of 2, thus permitting a higher dynamic range. Also, there is no limit, of the type that exists in binary-gain systems, to the rate at which the gain can change between samples, so that the variation of recorded amplitude with time will be less smooth and will require special software for usable presentation.

Field recording systems   The complexity of recording seismic signals digitally in the field might lead one to expect digital field systems to be very large and bulky, requiring a greater truck capacity than was needed in analog recording. Yet thanks to miniaturization of electronic components, digital systems are exceptionally

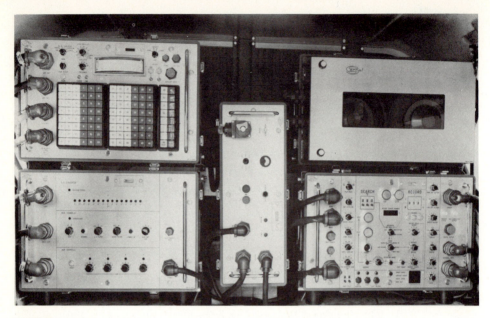

FIGURE 3-12
A 48-channel digital recording system mounted in the back of a Land Rover.
(*SERCEL*)

compact. Figure 3-12 shows a typical 48-channel recording system mounted in a Land Rover. The tape transport and various control units are shown in the pictures.

Figure 3-13 is a simplified block diagram illustrating the basic elements of a field system, from geophone input to tape transport, that is designed expressly for digital recording. The geophone output first enters an input module, which incorporates elements to test the performance of the individual geophone and to balance for variations in electrical characteristics of the cables leading from the phones. In addition, it is possible to introduce controlled oscillator signals into the system in place of geophone outputs as a means of checking the performance of the recording system.

In the amplifier module, there are 12 individual binary-gain channels, so that two modules are required for a conventional 24-channel system. The amplifiers are of the stepped-gain type with a change in gain by a factor of 2 between adjacent steps. The gain increases with decreasing input signal level and decreases with increasing input level. Before going into the binary amplifier, the signal passes through a preamplifier and various other filters (high-cut, notch, and low-cut). The antialiasing filter must be set to provide a high cut no higher than half the sampling frequency, this value being referred to as the *Nyquist frequency*, considered in more detail in Chap. 6.

The binary-gain value for each sampling instant (normally at 2- or 4-ms intervals) is computed and stored with a 4-bit binary word for permanent recording if absolute

1700 SYSTEM DIAGRAM

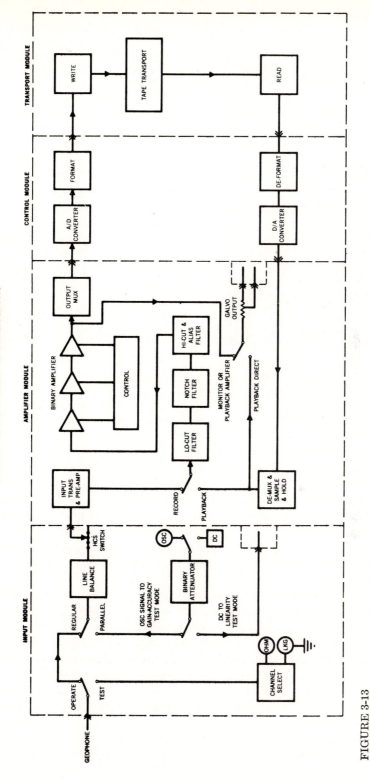

FIGURE 3-13
Block diagram of typical binary-gain digital recording system. (*Geo Space Corp.*)

FIGURE 3-14
Typical tape-transport unit for digital recording system mounted in cabinet with other components of system. (*SERCEL.*)

amplitude reconstruction should be desired at some later time. Amplifier gain ranging is provided in 15 gain steps for a total automatic gain change of 2^{15} or 90 dB. The output of the 24 amplifiers is multiplexed in this module, and the multiplexed signals, still in analog form, are passed into the analog-digital converter within the control-logic module. The digitized signal is here put into its proper format for input into the recorder itself. Tape transports are designed to register the signals on either 9- or 21-track magnetic tape, thus completing the actual recording operation.

The system is designed to sense the signals stored on the digital tape after the recording by a *read-after write element*, which is used as a monitor. This sensing is done by special reading heads, the outputs of which are converted into a format suitable for convenient digital-to-analog conversion. The analog output of this converter unit passes into the demultiplexer in the amplifier module, which un-scrambles the multiplexed signal into 24 separate channels comparable to those originally recorded. The outputs go into the same filters and amplifier units that were used at the original input stage and finally into a galvanometer, which makes a monitor record.

Figure 3-14 shows a typical nine-track recording transport unit employing $\frac{1}{2}$-in tape. A common tape format for such a tape recording is illustrated in Fig. 3-15.

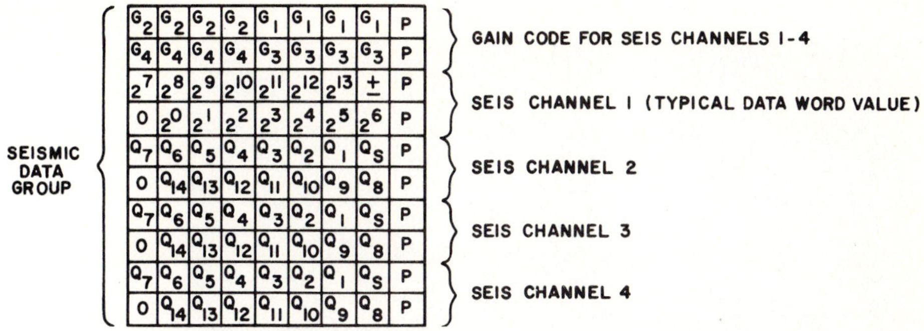

G_2	G_2	G_2	G_2	G_1	G_1	G_1	G_1	P	} GAIN CODE FOR SEIS CHANNELS 1-4
G_4	G_4	G_4	G_4	G_3	G_3	G_3	G_3	P	
2^7	2^8	2^9	2^{10}	2^{11}	2^{12}	2^{13}	$\pm$	P	} SEIS CHANNEL 1 (TYPICAL DATA WORD VALUE)
0	2^0	2^1	2^2	2^3	2^4	2^5	2^6	P	
Q_7	Q_6	Q_5	Q_4	Q_3	Q_2	Q_1	Q_s	P	} SEIS CHANNEL 2
0	Q_{14}	Q_{13}	Q_{12}	Q_{11}	Q_{10}	Q_9	Q_8	P	
Q_7	Q_6	Q_5	Q_4	Q_3	Q_2	Q_1	Q_s	P	} SEIS CHANNEL 3
0	Q_{14}	Q_{13}	Q_{12}	Q_{11}	Q_{10}	Q_9	Q_8	P	
Q_7	Q_6	Q_5	Q_4	Q_3	Q_2	Q_1	Q_s	P	} SEIS CHANNEL 4
0	Q_{14}	Q_{13}	Q_{12}	Q_{11}	Q_{10}	Q_9	Q_8	P	

SEISMIC DATA GROUP

FIGURE 3-15

Format (SEG B) for beginning information channels of a seismic record as registered on nine-track digital tape. (*Geo Space Corp.*)

The portion shown is for a typical scan of data over four of 24 channels being multiplexed at a particular sampling instant. The column at the right-hand side of each row is reserved for *parity* bits, which indicate whether all other bit positions in the row are registering. The two gain-code rows at the top of the figure define the respective gains (in 4 bits for each) corresponding to the first set of four seismic data words. Each of these words follows in subsequent pairs of rows. It consists of a sign bit, 14 bits expressing magnitude, and a blank bit (designated by 0). The gain code for channels 5 to 8 follows the information channels that are shown.

Before the gain and magnitude information for a scan as shown in the figure is recorded, some preliminary data (not indicated in the illustration) are registered. The start of the scan is signified by four rows of 0s in all bit positions except one (position 7 which contains 1s). There are then coded time identifications for the scan, the uphole time, and the time break. The eight rows following the portion of the format shown present the same kind of signal amplitude information for channels 5 to 8 as is indicated for channels 1 to 4. The sequence of gain and amplitude data is repeated until all 24 channels are covered, whereupon the *starting code* (all 0s but one) is repeated before starting the next scan.

Needless to say, the programming or software necessary to route such a complex array of information properly into the recording heads represents a remarkable engineering accomplishment. Equal competence in computer technology is required to extract the information from the tape with playback heads and put it into ultimately usable form. Fortunately, the geophysicist does not have to devise this software, but he must still work closely with the computer manufacturers and programmers to ensure that it will meet his needs efficiently.

3-4 OTHER FIELD INSTRUMENTATION

Monitoring cameras In digital recording, as in analog recording, it is desirable to obtain a visible record in the field of what has gone on the tape. Such a

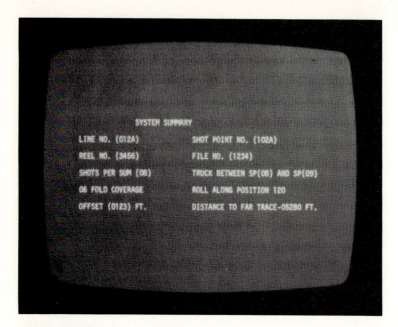

FIGURE 3-16
Visual display showing status of field operation as indicated by an oscilloscope
in GS-2000 computer-controlled recording system. (*Geo Space Corp.*)

monitor record tells the observer whether all the traces were recording and gives
some idea whether source conditions (charge size and depth, for example) were
appropriate or whether they should be changed.

For this purpose a galvanometer and camera are used to register the read-after
write signal taken from the digital tape immediately after recording. There is
usually one channel of the galvanometer for each geophone signal recorded by the
digital system. The galvanometers may register in variable-density, variable-area,
or wiggly-line modes. Cameras in the field use a "dry-write" process, and neither
a dark room nor a developing tank is needed to bring out the signal traces on the
recording paper.

Computers for control of recording   In recent years it has become increasingly
common to have special-purpose computers in the recording truck to handle many
functions in the data acquisition that were previously carried out by the field
observer. Tests of such items as continuity of the connections to each geophone
group on the ground can be performed automatically by the computer, which can
also guide the operator through the various steps in the initial field setup. At the
same time it can establish a sequence of operations which is followed automatically
until changed by the observer. Operational parameters are monitored with the aid
of a visual display unit (Fig. 3-16), which presents questions and indicates accept-

ance or rejection of the answers depending on whether they fall within predetermined limits of tolerance.

REFERENCES

1. Morse, P. M.: "Vibration and Sound," 2d ed., McGraw-Hill Book Co., New York, 1948.
2. Dobrin, Milton B., and Stanley H. Ward: Tools for Tomorrow's Geophysics, *Geophys. Prospec.*, vol. 10, pp. 433–452, 1962.
3. Silverman, Daniel: The Digital Processing of Seismic Data, *Geophysics*, vol. 32, pp. 988–1002, 1967.
4. Evenden, B. S., and D. R. Stone: "Seismic Prospecting Instruments," vol. 2, "Instrument Performance and Testing," Borntraeger, Berlin, 1971.
5. Dohr, Gerhard: "Applied Geophysics," Enke, Stuttgart, Germany, 1974.

4

ACQUIRING SEISMIC REFLECTION DATA ON LAND

In this chapter, we shall review the field procedures used in seismic reflection prospecting on land. This is a phase of the art which one might expect to change less than those involving the development of recording instrumentation and the processing and ultimate presentation of data. Yet advances in operational techniques, particularly since the middle 1950s, have brought about spectacular improvement in data quality as well as in operating efficiency. Many of the most fruitful innovations in field methods have been made possible because of recording instruments or processing capabilities that have been introduced in the past decade or so.

Since 1955 several new energy sources have come into use as alternatives to dynamite, and new arrangements of source and receiving elements have been introduced to improve data quality by minimizing noise. The new sources have many advantages in areas where conventional shooting is infeasible or uneconomical.

4-1 CONVENTIONAL FIELD PROCEDURES

We shall first consider the basic field techniques that have been used for conventional reflection recording since the earliest days of the art. In areas where only shallow information is needed and data quality is good, these techniques might still yield adequate geological information although they are seldom employed.

Need for multiple-channel recording of reflection signals   In recording ground motion from reflected seismic waves it is customary to receive the signals from each shot (underground explosion of dynamite or surface impact from a mechanical source) with a large number of geophones (actually geophone groups) spread out along a line extending from the shot point for a distance of thousands of feet. With such an arrangement each shot yields information on the structure of a subsurface interface at a large number of reflecting points distributed along the line, obviously a more economical approach than receiving a signal from only one point on each reflector for every shot. An even more important reason for multiple-channel recording is the need to identify the reflections as such and separate them from ground motion due to other sources. Much of the undesired ground motion, which we refer to in this context as *noise*, is associated with the shot and might result from waves that have traveled along the earth's surface or that have been scattered or diffracted by surface or subsurface irregularities. A large portion of such spurious ground disturbance can be removed by proper recording and data-processing procedures, many of which will be discussed later in this chapter and in subsequent chapters, but it is not likely that all undesired events can be eliminated.

There is generally no way of distinguishing the reflections from the noise on a single trace displaying the ground disturbance recorded by a detector group. But when several closely spaced groups are laid out along a line with the shot, the time (or phase) relations between corresponding events (such as peaks or troughs) in the signals on the respective traces from contiguous geophone channels make it possible to identify the reflectors on the basis of the patterns they show in their correlation from trace to trace.

Simple geophone spread arrangements   Figure 4-1 illustrates the ray-path geometry for a spread of 13 phone groups* distributed equally between two adjacent shot points *A* and *B*. We refer to *phone groups* rather than geophones because the individual phones are almost invariably planted in electrically connected patterns for cancellation of noise. Each group might consist of as few as three or as many as several hundred phones connected in series or parallel. The output, which goes into a single amplifier channel, represents the ground motion at the center of the group. The design of such geophone patterns will be considered later in this chapter. In practice, the phone group at the point from which the shot is actually fired is disconnected, so that in the simplified example of Fig. 4-1 only 12 groups record at any time.

---

* This small a number of phone groups is almost never used in current practice and is introduced here only for simplicity of description. In most present-day operations 48 or more phone groups are laid out at a time, the positioning of the shot points being more complex than the figure would suggest.

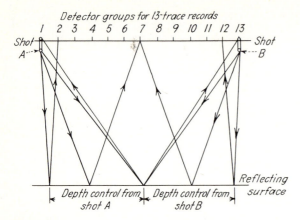

FIGURE 4-1
Obtaining continuous depth control along profile by recording shots from
opposite directions with same geophone spread.

In planning the layout of shots and geophones along a reflection profile, one must
remember that the subsurface depth computed from a reflection recorded by a
particular geophone group can be assumed to apply (for gentle dips) at a point
midway between the shot and the phone. If continuous depth control is desired along
a profile, the detector groups are uniformly spaced between adjacent shot points
and each group will record reflections from successive shots on opposite sides of the
spread. Each shot gives depths beneath the half of the spread nearest it, as shown in
Fig. 4-1. When the bed is inclined, the depth point will of course be displaced from
its position midway between shot and receiver; methods for correcting the error
that would be introduced by plotting it at the midpoint will be discussed in a later
chapter.

With the *split spread* (Fig. 4-2), by far the most common arrangement for
conventional coverage, an equal number of geophone groups is laid out* on each
side of the shothole. All 20 groups between shothole A and shothole C record the
shot from B simultaneously.

Sample monitor record for split spread Figure 4-3 illustrates a 36-channel
oscillographic (wiggly-line) record made from a split spread. This is the kind of
record on which all interpretation was carried out for the first quarter century of
reflection prospecting. Wiggly-line records like this have been superseded by
variable-area and combined-mode record sections corrected for variations in
reflection time not related to the structure of the reflecting surfaces. Yet the
greater detail that can generally be brought out on conventional oscillographic
traces and the opportunity such a presentation affords the interpreter to examine

---

* Here again the number of receiving groups does not correspond to what is now used in practice.
In present-day split-spread shooting, at least 24 groups are laid out on each side of the shothole.

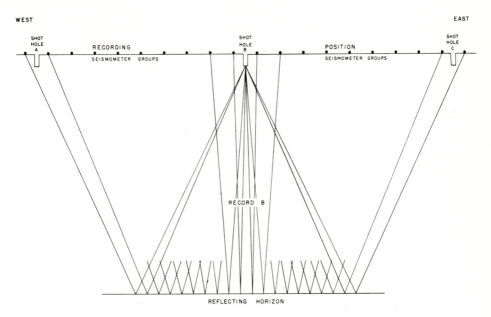

FIGURE 4-2
Split-spread arrangement for continuous correlation. Shot point at *B*. (*Geophysical Service, Inc.*)

raw uncorrected data could give old-fashioned records of this type certain advantages over more modern modes of presentation.

Each trace of the record represents the ground motion of a single group of geophones. The centers of adjacent groups are spaced 200 ft apart. All phones in each group are connected in series, and the resultant signal is transmitted by one pair of a multiconductor cable to an amplifier in the recording truck. The output of the amplifier is registered on magnetic tape, either analog or digital. Total distance from the shot (which is located between the phone groups corresponding to channels 18 and 19) to the farthest detector groups (corresponding to channels 1 and 36) is 3500 ft in each direction from the shot.

Times from the shot moment (labeled "time break") can be read by use of the vertical lines. The light lines are at 0.010-s intervals, the lines of intermediate weight at 0.050-s intervals, and the heavy lines at 0.100-s intervals. The interval of time represented by the portion of the record shown is about 2.5 s.

After the time break, we see that the two center traces stay quiescent until a time of about 0.025 s when "down kicks" are registered. The farthest traces are quiet for more than 0.4 s before motion of the ground is observed. For the traces in between, the times at which the first movement is observed appear to increase more or less linearly with distance from the shot to the geophone group for each trace. This lineup of first break times results from the near-horizontal paths followed by the rays responsible for these events on the record. The waves travel along the base of a thin low-velocity surface layer and reach the detectors before any other events.

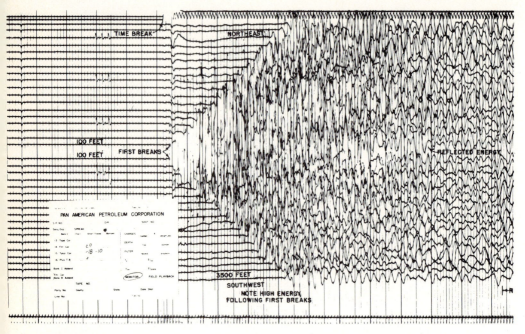

FIGURE 4-3
Typical oscillographic reflection record consisting, of 36 channels, as made in
the field recording truck to monitor the digital data. (*Amoco Production Co.*)

At a time of about 0.8 s, the first well-defined reflections are recognizable by the
way corresponding peaks or troughs on successive traces tend to fit into each other
even though there are generally systematic (if usually small) shifts in time between
one trace and the next. The troughs for some of these reflection events are marked.
The pattern of each reflection across the record is rather like an umbrella having a
horizontal shaft located between the center traces. Note that the curvature
decreases progressively with increasing time on the record. The increase in time
with distance from the shot results from the fact that the length of the travel
path of the reflected waves becomes greater at larger offset distances.

Progression of split spreads for single-coverage shooting   For the first three
decades of reflection prospecting, split spreads were used to obtain continuous
reflection-point coverage without the duplication that is characteristic of modern
shooting. It is instructive to begin our discussion of recording maneuvers in seismic
operations by describing split-spread shooting. Later in the chapter we shall cover
the more complicated operational logistics involved in common-depth-point
procedures.

   With single-coverage shooting, only half the split detector spread is shifted
between shots at successive shotholes. For example, after the shot has been fired
in *B* (Fig. 4-2), the detector groups between *B* and *C* are left for recording the shot
from *C* while the groups from *A* to *B* are moved to cover the interval from *C* to a new

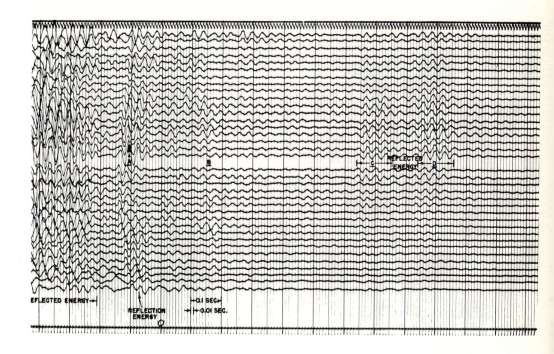

shot point *D*, located the same distance to the right of *C* that *B* lies to the left. Figure 4-4 shows a sample set of records made from spreads of this type before common-depth-point shooting came into universal use. Correlations of the reflections between adjacent end traces for continuous records are clearly defined.

Trends in multiplicity of recording channels   The number of channels used in seismic recording has continually increased since the earliest days of reflection prospecting. Originally seismic records contained as few as two traces. By 1936 the standard number was 6; by the early 1940s it was 12. From the end of World War II until the beginning of the 1970s 24-channel recording was standard. Now 48-channel systems are the most common, but larger numbers, up to 256, are sometimes used and there is every reason to expect that the number will become even greater as time goes on.

4-2 CHARACTERISTICS OF SEISMIC NOISE TO BE SUPPRESSED IN FIELD RECORDING

A major impetus for the development of new energy sources and new field techniques in reflection recording has been the necessity for eliminating or suppressing spurious seismic signals from ground motion not associated with reflections. Such signals are generally referred to as *noise*.

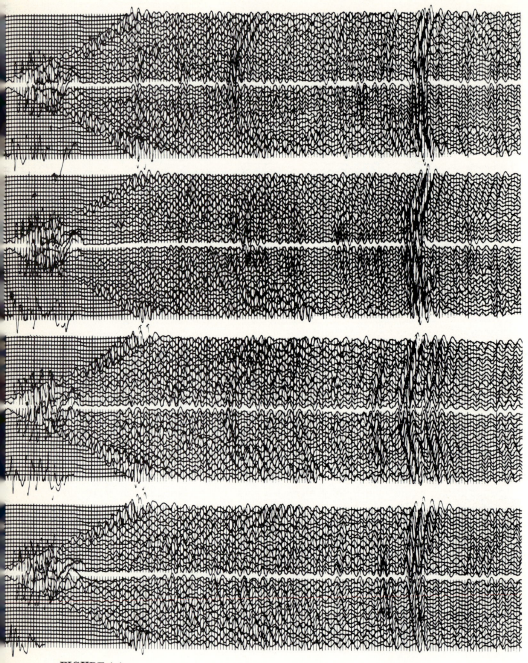

FIGURE 4-4
Oscillographic records from four adjacent shot points set up for 100 percent
coverage. Note the correlations of reflection events between records. (*Amoco
Production Co.*)

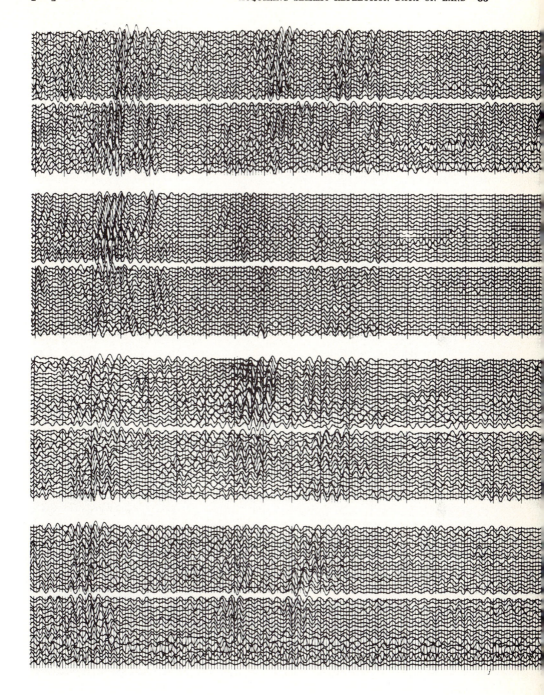

Noise has always been the most troublesome problem in seismic prospecting. In some areas it still poses more of a challenge than can be met at the present state of the art. And most of the improvement that has been achieved since the early 1930s in the effectiveness of the seismic reflection method has been brought about by the introduction of various techniques designed for suppressing noise on records.

Two primary approaches are used for reducing seismic noise to enhance desired reflections. One is to set up a recording arrangement that will cancel the unwanted signals before they are recorded in the field. Techniques for doing this will be considered in this chapter. The other is to process the data after they are recorded in a way that will suppress the noise, generally by appropriate filtering. In Chap. 6 we shall review techniques for enhancing reflection quality by the latter means. Both approaches to noise reduction are generally pursued simultaneously.

The principal types of noise associated with land shooting are surface and near-surface waves, scattered or incoherent noise, and multiple reflections.

Surface and near-surface waves   In the early days of reflection shooting in the Gulf Coast, low-velocity, low-frequency surface waves with an amplitude level that is high compared to reflections would often override useful reflection information. These events were referred to as ground roll. Because the frequencies of the waves were usually much lower than those for reflection, low-cut filters were introduced into the amplifier circuits to eliminate interference from this source. Somewhat later, groups of series-connected geophones were laid out over distances corresponding to one or more wavelengths of the ground roll. Such an arrangement would result in suppression of the horizontally traveling ground roll and enhancement of vertically traveling reflections.

In the early days of seismic prospecting, it was assumed that ground roll consisted primarily of Rayleigh waves (see page 38). Subsequent research, however, has shown that Rayleigh waves account for only a part of such interference. In some areas, *e.g.*, one in east Texas studied by Dobrin, Simon, and Lawrence,[1] dispersive Rayleigh waves are readily recognizable. In other areas, e.g., one in west Texas described by Dobrin, Lawrence, and Sengbush,[2] more complex physical mechanisms may be responsible for much of the interference. Among these mechanisms are refracted waves multiply reflected in a surface layer, shear refractions, and guided waves trapped in a near-surface layer with a lower speed than that in the material above and below.

A typical reflection record on which ground roll causes perceptible inference with reflections is illustrated in Fig. 4-5. Along with it is another record showing how the ground roll can be removed by low-cut filtering alone, leaving the reflection in a more easily interpretable form. Where frequency filters alone do not yield such good discrimination, multiple geophones or multiple shots must be used with the number of units and spacings selected to give optimum cancellation of the wavelengths observable in the ground roll. The procedures involved in determining the geophone and/or shot patterns that will be most effective in removing such

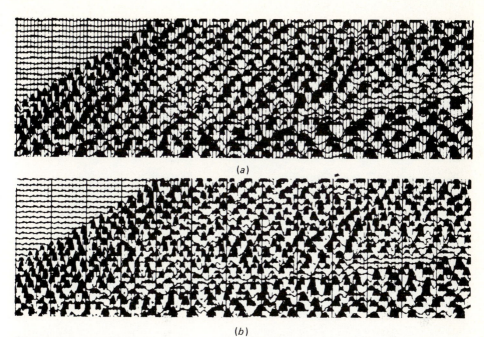

FIGURE 4-5
Ground roll from single impact on ground: as recorded in variable-area mode
(*a*) without filtering and (*b*) with low-cut frequency filter. (*Petty-Ray Geophysical Division, Geosource Inc.*)

interference from surface waves will be considered later in this chapter.

Scattered and other incoherent energy   The surface and near-surface waves discussed in the previous paragraphs are coherent, traveling events which can be followed for substantial distances along the receiving profile. Cancellation can be accomplished by applying principles of directivity similar to those used in designing antennas for receiving radio waves. A different type of interference which is generally more difficult to cancel comes from *incoherent noise*, sometimes referred to as *random noise*. This is usually associated with scattering from near-surface irregularities.

Incoherent noise is particularly common when the shot point overlies or is close to gravel, boulders, or vuggy limestone, all of which can cause scattering of waves. The strength of the waves thus scattered depend inversely upon the distance of such materials from the shot. Sometimes the scattering occurs where stream banks and other topographic irregularities diffract energy from the shot and return it to the recording line in the form of incoherent noise. Figure 4-6 illustrates a record section on which most of the useful reflections appear to have been obscured by incoherent-noise signals.

Incoherent noise observed at one point on the surface should by its very nature

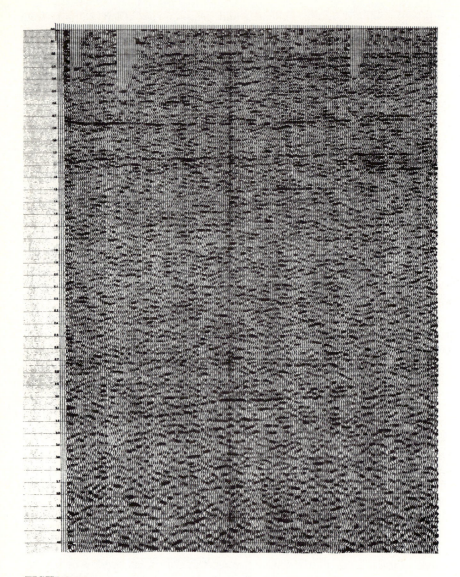

FIGURE 4-6
A poor reflection record section, in which nearly all desired information between 1 and 3 s is obscured by incoherent noise. (*Sun Oil Co.*)

be entirely unrelated to that at another point only a short distance away. This would not be the case with coherent noise, where there would be a predictable relationship between the two signals. Addition of signals containing incoherent noise should result in some cancellation of the noise, and the more signals are added the more complete the cancellation to be expected. The multiplicity factor N is the product of the number of source elements s and the number of receiver units r from which signals are added. The cancellation effect is proportional to $\sqrt{N}$. To obtain the greatest suppression of incoherent noise one would therefore use as many shots and as many receivers (geophones per trace) as possible. Yet addition of signals from 100 receiving elements per trace results in only 3 times the noise cancellation that would be obtained if only 10 elements were used per trace. A major advantage of most surface sources such as dropped weights lies in the fact that it is a simple matter to generate a great number of individual signals within a limited area by moving the source a short distance between impacts, whereas it would be economically prohibitive to drill an equivalent number of holes for shooting dynamite.

Multiple reflections A common and particularly troublesome type of interference is that from multiple reflections, which can look so much like primary reflections that the geophysicists may not always be able to identify the multiples as such. Multiple reflections can be of many kinds, some of which are shown in Fig. 4-7, but the simplest and probably the most common is the surface multiple (Fig. 4-7b), which arrives at twice the time of the primary reflection from the same bed. Other types of multiple are the interbed reflection (Fig. 4-7c) and the more complex kind shown in Fig. 4-7d. Elimination of the surface multiples is best accomplished by proper application of common-depth-point shooting, discussed in detail later in this chapter. Often there is a distinct frequency change between the primary and multiple reflections because of differences in the materials through which the respective waves travel as well as differences in path lengths. In such

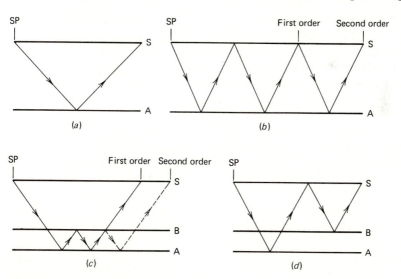

FIGURE 4-7
Several types of multiple reflections: (a) primary; (b) surface multiple; (c) interbed multiple; (d) combination multiple. S is earth's surface. A and B are reflecting interfaces. (*Amoco Production Co.*)

cases it may be possible to discriminate between the two by means of frequency filtering alone.

4-3 ENERGY SOURCES FOR REFLECTION SHOOTING ON LAND

Although a number of nonexplosive energy sources have come into use for reflection prospecting since the middle 1950s, the source employed for more than 60 percent of all work on land is still dynamite exploded in shotholes. During the first 25 years of reflection activity, this was the only source which provided sufficient energy to yield satisfactory reflection data. The introduction of magnetic-tape recording and compositing systems in the 1950s made it possible to build up usable signals from mechanical impactors and similar low-energy sources by adding properly synchronized returns from a multiplicity of individual impacts. These newer sources offered economic and operational advantages over dynamite in many areas, and the percentage of land crews which use them has risen steadily since their introduction. In 1972, 46 percent of the land shooting in the United States and 33 percent of all such work outside the U.S.S.R. made use of nondynamite sources. Such sources are now employed for virtually all seismic work at sea; some of them will be discussed in Chap. 5.

Dynamite*

The mechanism for generation of energy by exploding dynamite in a shothole was discussed in Chap. 2 (pages 43–46). Because of the impulsive nature of the seismic signal it creates and the convenient storage and mobility it provides for energy that can be converted into ground motion, dynamite should constitute an ideal seismic source from many points of view. There are, however, a number of drawbacks to its use: (1) In seismic operations, the dynamite is planted in sticks or cans in boreholes (usually about 4 in in diameter) that may range from 30 ft to several hundred feet in depth; this requires drilling the holes, which is difficult and expensive in many areas, particularly in hilly terrain into which heavy drilling equipment cannot be easily moved or in desert areas where water is not readily available. (2) Dynamite can be dangerous. Although the safety record of the geophysical industry has been remarkably good, the hazard involved in using dynamite has led to legal restrictions, some reasonable and others perhaps not, upon its transportation, storage, and handling, which often cause considerable inconvenience and expense.

Dynamite ordinarily comes in cylindrical cardboard-covered sticks 20 in long with a diameter of about 3 in, and a weight of 5 lb. The dynamite itself is implanted in a matrix of inert materials so that it constitutes only about 60 percent of the

---

* Although dynamite as invented by Alfred Nobel used nitroglycerin as the active constituent, it is customary to designate other explosives used for seismic-energy generation, such as ammonium nitrate or TNT, as "dynamite" also.

weight of the sticks themselves. The principal constituent of dynamite, nitroglycerin, is mixed with a sufficient amount of inert material to be stable at ordinary temperatures. The heat and mechanical impact of a detonator explosion convert the chemically bonded energy of the active material (such as the nitroglycerin) into kinetic energy of molecular motion in the gases formed when the explosion takes place. It is this energy conversion that is responsible for the violent impact against the earth that generates the seismic signal.

The amount of dynamite needed for a reflection shot varies from less than 1 lb to several hundred pounds, depending on the nature of the material in which the shot is fired, the lithologic characteristics of the geological section below the shot, and the depth of penetration desired.

To maximize the amount of explosive energy transmitted into the earth, it is customary to fill the hole with a heavy drilling mud, which prevents the pressure generated by the explosion from being released too easily into the air. The tamping effect of the mud increases the efficiency of the shooting and improves the signal quality. Sometimes a delayed expulsion of the mud into the air produces a geyser that can be quite spectacular (Fig. 4-8).

In areas where interference from noise is severe, shots are often fired in linear or areal patterns for noise reduction. The same considerations that govern the geometry

FIGURE 4-8
A "geyser" created by expulsion of mud from shothole after dynamite explosion. (*Petty-Ray Geophysical Division, Geosource Inc.*)

of geophone patterns apply in determining the geometry of shothole patterns as well. The design of such patterns will be discussed later in this chapter.

Buried Primacord

We have just seen how noise can be reduced by exploding dynamite in arrays of shotholes, achieving the same directivity effects as those obtained with multiple geophones. The greatest possible enhancement of downward-traveling energy over horizontally traveling energy can be obtained with a continuous source horizontally elongated. Such downward directivity can be achieved by burying a proper length of Primacord* (an explosive extruded into a ropelike form), detonating it at one end (or at its center), and letting the explosive disturbance propagate along it at a speed (22,000 ft/s) much greater than the speed of seismic propagation in the near-surface material within which it is buried.

The Geoflex† source system operates on this principle. It is used in lengths of several hundred feet, being plowed into the ground to a depth of 2 or 3 ft. Burial is necessary to suppress noise and increase the efficiency of energy transfer into the earth. If the velocity of detonation of the cord were infinite, the energy would propagate vertically downward in phase and there would be maximum vertical directivity. Because the velocity is finite, the sharpest and highest-amplitude pulse will be directed forward at an angle whose sine is the velocity in the earth divided by that in the cord. The ray path will generally be at an angle of 20 to 30° with the vertical, as shown in Fig. 4-9. It is evident that the greatest seismic effect is observed in the direction of in-phase wave propagation shown in the figure. The variation with angle in the time duration of the signal gives rise to a filtering effect such that the highest frequencies are observed at the angle θ and the lowest in the horizontal direction.

Nondynamite Surface Sources

The advantages of distributed generation of energy for attenuating horizontally traveling noise have led to the development of nondynamite sources which allow an almost unlimited number of effectively simultaneous impulses to introduce energy over any desired distance along the surface without the cost of drilling or plowing and without the hazards involved in using dynamite. These sources involve mechanical impact upon the earth's surface or shaking of the surface with a mechanical vibrator. Multiplicity of the impulses necessary for noise suppression is obtained by a combination of multiple source units and by storage and subsequent compositing of reflected signals sequentially generated at locations close to one another along the surface.

* Trademark of Ensign Bickford Co.
† Trademark of Imperial Chemical Industries, Ltd.

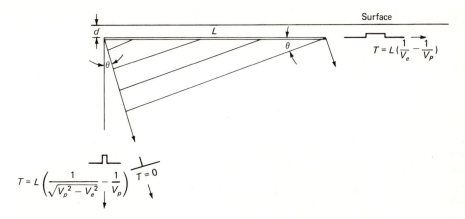

FIGURE 4-9
Use of Primacord planted in plowed trench as seismic source. L = length of Primacord, V_e = velocity of seismic waves in earth, V_p = velocity of detonation in Primacord, θ = direction of maximum amplitude with vertical, T = duration of pulses in different directions.

All sources of this type are so disposed in the field that signals received from impacts (or sweeps in the case of vibratory source systems) are applied to the earth over a linear distance comparable to what would be used for a line of shotholes at a shot point or (where necessary and feasible) over an area that would be used for a two-dimensional array of shotholes. The signals so generated are composited to simulate the effect conventionally obtained from a single extended shot point.

Weight dropping Although experiments with dropped weights as energy sources for reflection work were carried out as early as 1924, it was not until magnetic recording made it possible to composite impulses from a multiplicity of drops in close proximity that this type of impact could be used as an energy source for recording deep reflections. The first commercial system of this type was invented by McCullom[3] and put to use for oil exploration in 1953. The source he developed is called the Thumper.

A high level of undesired noise is generated by weight dropping, mostly in the form of surface waves. The characteristics of such noise have been studied by Mason[4] and Neitzel.[5]

The weight, which is a 3-ton slab of iron, is attached by chains to a crane on a special truck (Fig. 4-10) and when dropped is released to fall 9 ft. It is hoisted from the ground immediately after the impact so that it can be dropped again within a few seconds at another spot the order of 10 ft away to which the truck has meanwhile moved. Waves from each drop are picked up by a detector spread and recorded on magnetic tape for later compositing. The distance from the source to the nearest detector is considerably greater than with conventional dynamite

FIGURE 4-10
Truck for dropping weight used in reflection recording. Weight is falling.
(*Petty-Ray Geophysical Division, Geosource Inc.*)

shooting because of the necessity for receiving the reflections before the slow-moving high-amplitude surface waves arrive from the drop.

During the first decade the weight dropper was in use it was customary to lay a large pattern of geophones at the position that would ordinarily be occupied by the shot point and to move the source itself for drops at successive positions along the line. The more recent practice (see Peacock and Nash[6]) is to use the same geophone arrangement as with dynamite shooting and to move the weight-dropping truck to positions along the line that would correspond to shotholes in conventional arrangements. Various drop patterns may be used, the configuration depending on the level and type of noise to be canceled as well as the space available. Figure 4-11 illustrates a few source patterns that are common for weight-dropping surveys. As many as 100 drops have been used per shot point in particularly noisy areas.

The time interval between drops generally ranges from 10 to 12 s. Often two trucks are used in tandem, the interval between drops then being half as great, with signals from both trucks being composited, often in the recording truck, to obtain a single output tape.

The Dinoseis   In 1962, the Sinclair Research Laboratories introduced a source system designated as Dinoseis, in which an explosion of gas (propane and oxygen) is detonated inside a closed chamber in contact with the ground (see Johnson[7]). The chamber is expandable so that the bottom plate, which rests on the ground during detonation, is free to move with respect to the bell-shaped mass enclosing the chamber at the top.

In the earlier models, the weight of the truck carrying the system was supported by the chamber itself. In a more recent version, the detonating chamber is lowered to the ground from the undercarriage of the truck body; it is so suspended that at the time of firing it is not in rigid contact with the truck itself. Upon detonation it quickly springs away from the earth. At the top of its bounce the chamber is caught hydraulically and locked into place to prevent its falling back to earth and causing

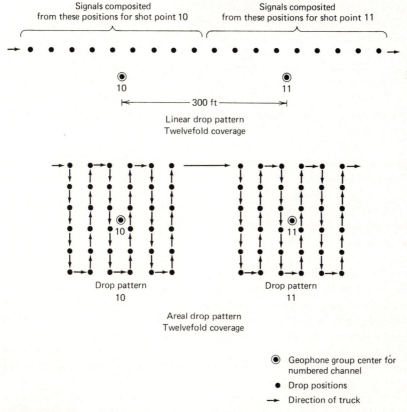

FIGURE 4-11
Typical linear and areal patterns for dropping weights in seismic reflection operations. Linear pattern involves nine impacts per shot point, areal pattern 36.

FIGURE 4-12
Dinoseis source element mounted on truck. (*Atlantic Richfield Co., Geo Space, Inc., and Rogers Explorations, Inc.*)

a second impact that would complicate the waveform of the down-going seismic pulse. Figure 4-12 shows how a unit of this type is suspended.

In normal operation, a number of Dinoseis units, generally three or four, are used at the same time. Signals actuating the firing system in each chamber are sent by radio so that the detonation for all trucks will be simultaneous. The fact that multiple units can be synchronized gives the Dinoseis an advantage over the weight dropper in developing a higher energy level than that obtainable from a single unit. With weight dropping there is no way to synchronize impacts from multiple units within a few milliseconds.

After each detonation the group of trucks is moved 20 ft or so, detonating chambers are lowered to the ground, and the cycle is repeated. Patterns of shooting and receiving are similar to those described for weight dropping. Here also the current practice is to use 48 or more geophone groups and to shoot in an arrangement suitable for subsequent compositing. As with the weight dropping, offset distances (to the nearest phone) are often as great as several thousand feet because it is easier to avoid interference from surface-generated ground roll by having the reflections arrive before the surface waves than to cancel superimposed high-energy surface waves by subsequent filtering.

Vibroseis The two mechanical impactor systems we have discussed, like dynamite, introduce an impulsive seismic signal into the earth. The wave put into the earth by a Vibroseis source is oscillatory rather than impulsive and persists for many seconds, the frequency changing slowly over the duration of the signal. The returned signals recorded in the field cannot be interpreted directly, as is generally possible with the other sources. The recorded data must be processed by *cross-correlation* of the signals received by the geophones with the oscillatory source signal itself. The technique involves a comparison of the two signals with progressively increasing delay times. Reflections and other seismic events related to the source signal give a greater degree of correlation with the generated waveform than random noise.

OPERATIONAL PROCEDURES   The Vibroseis technique was developed during the 1950s by Continental Oil Company after a long period of experimentation. The earliest account of this system is by Crawford et al.[8] The present source consists of a 2-ton mass with a hydraulic vibrator controlled by a preprogrammed sinusoidal wave signal of *continuously varying frequency* activated by a starting "beep" sent by the recording truck. The individual wave cycles generated by the vibrator travel into the earth like pulses from conventional sources and are reflected in the same way. The reflection signals returned to the surface are all superimposed, and the signal received by the geophones must be processed in such a way that the complex pattern of energy so recorded is converted into reflection traces of the type obtained from impulsive sources.

Vibroseis operations, like Dinoseis, involve the use of several trucks, usually four simultaneously, as shown in Fig. 4-13. In open fields the trucks travel along parallel trajectories; when confined to roads, they operate in tandem. They stop for vibratory sweeps at intervals determined by the total number of sweeps needed per shot point, all four units sweeping simultaneously for times lasting 7 to 21 s. Records are made from the combined outputs of the series-connected geophone groups, and data from all sweep positions within the distance range designated as the shot point are composited into a single channel, often by an analog or digital summing unit in the field truck.

Like the weight dropper or the Dinoseis, the Vibroseis is employed with 24, 48, or more geophone groups, and sweep patterns may be placed to coincide with every second, third, or fourth patch of phones, depending on the multiplicity (sixfold, twelvefold, etc.) of subsequent common-depth-point compositing. A common field arrangement for a linear array is shown in Fig. 4-14.

PROCESSING OF VIBROSEIS DATA   Because the signal put into the ground persists for a long time (generally many seconds), the reflection signals actually recorded in the field are entirely incoherent to the eye and special processing is necessary, as has been pointed out, to convert the data into usable form.

A reflection record obtained from a Vibroseis source consists of superimposed signals from each of the reflecting surfaces, as illustrated in Fig. 4-15. Each reflection has approximately the same waveform as the source signal, but the wave train corresponding to each reflector is delayed by the time required for it to travel from the source to the reflecting interface and back. The combination of all the individual

FIGURE 4-13
Four Vibroseis trucks operating along parallel lines in a west Texas survey. (*Continental Oil Co.*)

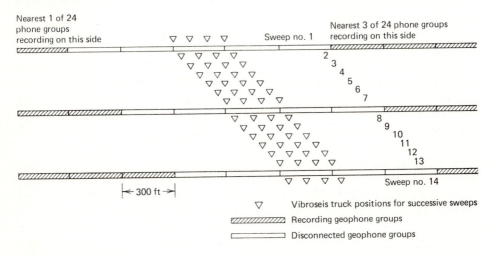

FIGURE 4-14
Progression of Vibroseis sweep positions along geophone line in four-truck operation. Note change in recording connections as trucks move.

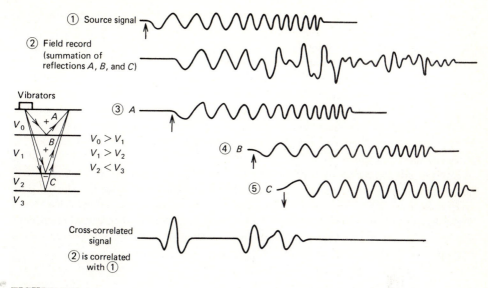

FIGURE 4-15
Recording and analysis of a Vibroseis source signal reflected from three interfaces. (*Continental Oil Co.*)

reflections is so spread out that it would not be expected to give an interpretable
record.

Qualitatively, it might be instructive to look on the cross-correlation process as
a test of the fit of the source signal (1 in Fig. 4-15) and the recorded signal (2 in
Fig. 4-15) containing the reflections at successive relative displacements of the two
signals along the time scale. The initial fit is tried with the beginnings of the two
signals coincident, and then the fits are determined with the respective zero
positions progressively shifted by constant increments such as 2 ms.

The degree of fit is determined at each juxtaposition by multiplication of the two
signals over their entire length at closely sampled ordinate positions and addition
of the products. The greater the sum the greater the cross-correlation coefficient
obtained in this way will be. If the two signals are random, the cross-correlation
is zero.

At a time shift equivalent to the two-way travel time for a reflection, the returned
signal displays a partial coincidence with the source signal (only partial because
other reflections and noise will be superimposed). When such a coincidence occurs,
the cross-correlation value (degree of fit) will be a maximum. Elsewhere the
relation between the two signals is random and the cross-correlation value is low.
If the cross-correlation so determined is plotted against the time shifts, the reflections will show up as high-amplitude events that look exactly like reflections on
conventional recordings from impulsive sources, the portions of each trace in
between being more or less quiescent depending on the noise level.

ADVANTAGES OF VIBROSEIS Energy from a Vibroseis unit can be introduced into the earth over the entire range of seismic frequencies although the efficiency of transmission by the earth may vary with the frequency. The frequency content of a signal from an impulsive source is not subject to control and may, in the case of dynamite, be influenced primarily by the material in which the explosion occurs. In many places the best signal-to-noise ratio is observed over a limited range of frequencies which can be specifically programmed into the Vibroseis source signal. Shallow reflections, for example, would call for a sweep over a range of frequencies at the high end of the usual seismic spectrum while deep reflections would call for a sweep, such as from 8 to 25 Hz, at the low end.

Another advantage of Vibroseis lies in the fact that a signal from it, being spread out over many seconds, will have a much lower amplitude level at the source than an impulse in which all the energy is injected into the earth within a few milliseconds. This feature makes it possible to use Vibroseis in populated areas where explosions would not be acceptable. Vibroseis surveys have been carried out near the centers of Chicago and Los Angeles, to cite two examples.

4-4 SHOT AND GEOPHONE ARRAYS

In the earliest days of seismic reflection prospecting it was customary to use a single shothole and one geophone for each trace. By the later 1930s, groups of geophones spread out over tens to hundreds of feet were connected in series or in series-parallel arrangements, and the combined outputs of the geophones were fed into a single amplifier channel corresponding to the group as a whole. The purpose of the grouping, which at first involved three to six phones, was for cancellation of ground roll and other horizontally traveling noise. In areas such as west Texas, where noise was particularly severe, patterns consisting of 100 or more geophones became common during the late 1940s. At the same time it became customary to drill shotholes in patterns over areas where it appeared desirable to reinforce the noise attenuation obtainable with the geophone groups alone. In west Texas, Florida, and New Mexico, where usable reflections are hard to get, it has been customary to use patterns involving as many as 36 shotholes per shot point and 96 geophones per trace. Figure 4-16 shows records obtained with shot and geophone patterns typical of those employed in Florida during the early 1950s. It is from a paper by McKay[9] that illustrates the effectiveness of many types of patterns, some of which are exceptionally large. Nondynamite sources make it possible to achieve the advantages of shooting a large number of distributed shots without the disadvantage of drilling many shotholes.

Principles The theory of noise attenuation by patterns is the same whether applied to the shot or the geophone end of the wave trajectory. The basic idea is to design both groupings so that waves traveling vertically or nearly vertically are reinforced while those traveling in the horizontal direction are reduced.

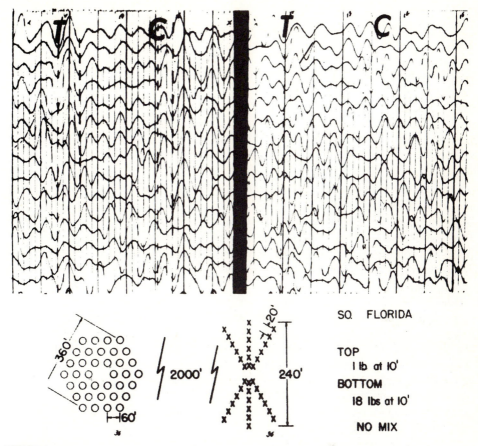

FIGURE 4-16

Comparison of record obtained using single hole and 36 phones per trace (*right*) with record obtained along same spread with pattern of 36 holes and 36 phones per trace (*left*). The respective shot and geophone patterns are shown in the lower part of the figure. (*From McKay.*[9])

Figure 4-17 shows a four-geophone group which covers a horizontal distance equal to a wavelength of the surface wave to be canceled. With this arrangement at any given time the horizontal wave will cause upward motion in two detectors of each group and downward motion in the other two. If all four are connected in series, the net signal from this source will be very nearly zero because of cancellation. Reflected waves, on the other hand, are very nearly vertical when they reach the surface, and all four detectors of each group will respond to them by moving in the same direction at the same time. The outputs for the reflection signals should therefore be additive. Proper grouping of geophones should thus reinforce reflected events and cancel horizontally traveling noise.

The optimum number of elements in a shot or geophone pattern and the best spacing of the units within the pattern are determined by applying the same

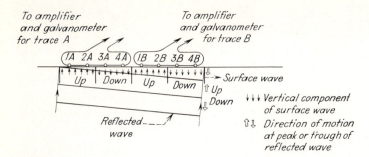

FIGURE 4-17
Eliminating effects of ground roll by use of multiple geophones in series.

principles as those used in designing radio antennas. The theory is reviewed by Lombardi[10] and Parr and Mayne,[11] the basic concept being illustrated by Fig. 4-18. Here we have an array of five uniformly spaced geophones, the separation between adjacent phones being D. We can calculate the response of the phones to waves of various wavelengths most expeditiously at the instant a peak of the traveling wave coincides with the center phone of the group. If the wavelength λ equals the spacing D ($\lambda/D = 1$), the signals picked up by the phones are in phase and the output when all phones are connected together is 5 times the peak value for the center phone. If $\lambda/D = 2$, the first, third, and fifth phones record maximum *positive* (upward) motion, while the second and fourth phones record the maximum *negative* (downward) motion. The sum is thus the peak output for a single phone or one-fifth the output for $\lambda = D$.

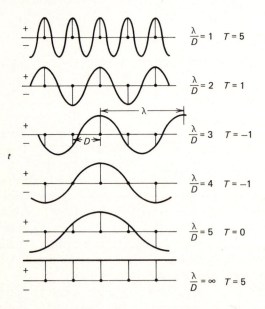

FIGURE 4-18
Cancellation of waves of different wavelengths using five geophones separated by a distance D. Numbers in right-hand column represent relative amplitudes of outputs for the array.

Similarly it is evident that both for $\lambda/D = 3$ and $\lambda/D = 4$ the downward motion exceeds the upward motion by one unit. The sign is not of significance in evaluating attenuation characteristics, and we can consider the response for these to be effectively equivalent to that for $\lambda/D = 2$. When $\lambda/D = 5$, the positive and negative contributions are equal and cancel one another, leaving zero response. As the wavelength gets very large compared with D, it approaches the limit of $\lambda/D = \infty$ at which value the peak output for five units is observed again, just as with $\lambda/D = 1$.

Figure 4-19, based on the model just considered, shows how the respective outputs from groups of five and eight phones vary as λ/D changes. For the five-unit group, one sees that there are three values of λ/D in addition to $\lambda/D = 5$ at which there is complete cancellation. There are three λ/D values at which transmission is a maximum. The corresponding attenuation curve for an array with eight units shows seven positions along the λ/D axis at which cancellation is complete and six positions for maximum transmission. From these two examples we can make several generali-

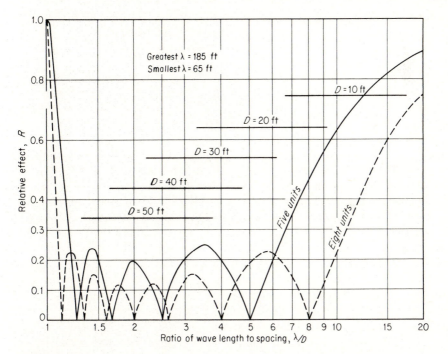

FIGURE 4-19

Noise-cancellation curves for five and eight geophones in groups. Relative response referred to amplitude of a wave recorded from a single geophone. To illustrate use of curve, suppose eight phones of group were spaced 20 ft apart. For wavelength of 100 ft $\lambda/D = 5$; the chart shows that only 20 percent of the original wave amplitude is recorded by the system. The bars illustrate the range of λ/D values from an area where the noise wavelengths lie between 65 and 185 ft. (*Adapted from Parr and Mayne.*[11])

zations. The number of zeros (positions where the geophones have no output) on the curve is 1 less than the number of phones, and the number of peaks or lobes between the zeros is 2 less than the number of phones. The greater the number of phones the lower one can draw the envelope (line tangent to the peaks) of the attenuation curve, indicating less transmission of horizontally traveling waves. Also, the greater the number of phones the wider the zone along the λ/D axis within which significant reduction of horizontal noise takes place. This means that the range of noise wavelengths that can be effectively reduced increases with the number of phones per group.

The equation for the fractional transmission T by an array of N geophones a distance D apart for horizontally traveling waves having a wavelength λ is

$$T = \frac{1}{N} \frac{\sin 2\pi\ (N\lambda/D)}{\sin 2\pi\ (\lambda/D)}$$

where T is the ratio of the signal observed with the N phones a distance D apart to that which would be recorded if they were all in a huddle (at zero distance apart).

Determination of optimum phone spacing   To determine the optimum geophone spacing for a given number of phones per trace in an area where the characteristics of the noise are not known, it is advisable to carry out special measurements in the field. These tests are designed to establish the nature of the noise (whether coherent or incoherent) and (for the coherent noise) to determine the range of wavelengths encompassed. A series of records is made with geophones (generally single) spaced 5 to 30 ft apart, the spread being moved (or, alternatively, if the geophone spread is held fixed, the source being moved) so that there is continuous geophone coverage from the shot point itself out to distances as great as several thousand feet. Figure 4-20 illustrates a typical *noise section* recorded during a test of this kind.

All events on the section having a lineup at a low enough velocity to be associated with horizontally traveling surface and near-surface waves are readily identifiable on such sections, and their velocities are determined simply by dividing the distance between the nearest and farthest phone receiving the event by the differential time required to traverse this distance. The wavelength is the velocity multiplied by the period (time between successive peaks or troughs) of the traveling wave. A range of wavelengths is determined for all noise events of this type, and a spacing is selected that will optimize the noise attenuation.

Above the curves in Fig. 4-19 are drawn a series of bars indicating the range of λ/D values corresponding to noise events from an area where tests of the type described show the noise wavelengths to lie between 65 and 185 ft. The respective bars correspond to trial values of D ranging from 10 to 50 ft. Taking the envelope of the respective attenuation curves for five and eight units, we see that the 50-ft spacing gives optimum cancellation for the five-phone group while the 40-ft spacing appears best for the eight-phone group.

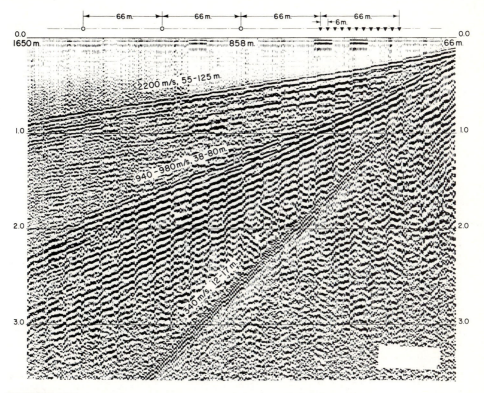

FIGURE 4-20
Traveling waves recorded over spread about 1600 m long from surface source.
Velocities of waves range from 2200 m/s (probably first arrivals) to 340 m/s
(air waves). 940–980 m/s events are probably Rayleigh waves. (*Petty-Ray
Geophysical Division, Geosource Inc.*)

Parr and Mayne[11] have shown that an improvement in noise attenuation is
obtained if the sensitivity of the geophones in the patterns is tapered in such a way
that the greatest response is at the center of the group, with the response of the
phones decreasing in both directions as one goes toward the outermost geophones.
The taper could be obtained by individual controls, such as potentiometers, on each
geophone, but since geophones are not equipped with such devices, the same effect
can be achieved more easily by placing two or more phones side by side or by
changing the spacing between adjacent units. For example, a five-unit array with
equally spaced phones could yield as much as double the noise cancellation if extra
phones were added so that three phones would stand in a huddle at the center
position, two phones together at the second and fourth positions, and single phones
at the first and last positions.

Areal arrays   Noise often travels in directions that are not coincident with the
direction of the radial line extending outward from the shot. This happens when the

noise which initially travels in a direction different from that of the geophone spread is reflected back to the line of phones by some feature such as a vertical escarpment or river bank or even a hidden lateral irregularity below the earth's surface. Where this occurs, it is necessary to have geophones in areal, i.e., two-dimensional, rather than in linear patterns. A properly designed areal pattern should yield adequate attenuation regardless of the direction in which the noise approaches.

Areal patterns may be rectangular (for example, 24 phones may be in four rows x ft apart with 6 phones in each row spaced y ft apart) or along concentric circles, depending on the nature of the noise and on the space available for the array. In a wooded area, it may not be feasible to bulldoze clearings that would allow circular symmetry of phones around the center position for each group. Yet it might be quite feasible to lay out the phones in an elongated rectangular array.

In determining the attenuation for an areal array one must specify the direction of the wave travel. The attenuation curves (transmission versus λ/D) for a star pattern consisting of 13 geophones will be different for different directions of propagation, as shown in Fig. 4-21. To use conventional attenuation curves designed

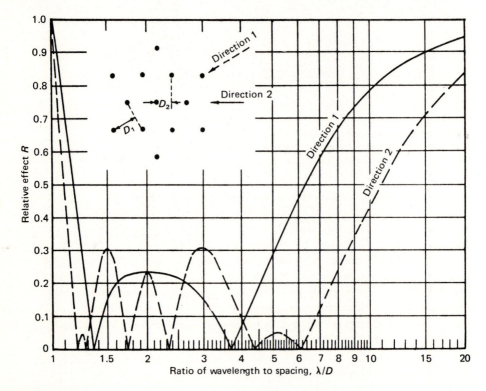

FIGURE 4-21
Cancellation curves for an areal pattern of geophones consisting of 13 units. Note the differences in cancellation for the two directions of approach of the wave to be attenuated. (*Parr and Mayne.*[11])

for linear arrays with areal patterns the position of each phone would have to be projected on a line in the direction of wave travel, as shown in the figure.

The detailed procedures followed in designing areal shot and detector patterns will not be discussed here. Reynolds[12] describes a spoke-and-wheel pattern which has worked successfully in many areas. McKay[9] has published a large collection of sample records showing the improvement in reflection quality that is possible with elaborate two-dimensional shooting and receiving patterns.

Where the noise is incoherent and traveling events cannot be followed between geophones only a few feet apart, cancellation techniques of the type described will not be effective. Statistically incoherent noise can be reduced by a factor proportional to $\sqrt{n}$, where n is the number of detectors in the group irrespective of spacing. Where m shots are used along with the n phones per trace, the improvement in signal-to-noise ratio is proportional to $\sqrt{mn}$. This may be a difficult way to obtain the necessary degree of cancellation. Yet in some areas, e.g., the Sahara Desert or in west Texas, where there is a high level of incoherent noise the only way to obtain usable reflections is with very large shothole patterns and an even larger number of phones per trace. Patterns comprising 48 shots and 36 phones in a star configuration for each trace are not unusual in west Texas, and much larger patterns have been used for work in the Sahara Desert (see Pieuchot and Richard[13]).

4-5 COMMON-DEPTH-POINT SHOOTING

The use of simple arrays of shots or geophones for canceling noise is practical only where the noise has wavelengths that are not appreciably greater than the lengths of the arrays themselves. The length of a linear array of n phones a distance D apart is $(n-1)D$. Moreover, the length of most arrays is limited to a distance that is not substantially greater than the separation between group centers along the spread. Thus noise with wavelengths of more than a few hundred feet cannot be handled by conventional arrays.

Occasionally high-speed refracted or diffracted events are observed with apparent velocities of 15,000 ft/s or more corresponding to wavelengths as great as 500 to 1000 ft. Such wavelengths will not be adequately canceled by the geophone groups having a much shorter length which are customarily employed in reflection work. A more common and even more troublesome type of long-wavelength noise is that associated with multiple reflections. The multiple events ordinarily have a different moveout from the primaries arriving at the same time because of the difference in average velocities for the respective paths. Thus when primary reflections are corrected to have no normal moveout, multiple events at appreciable distances from the shot will appear to make a small angle on the corrected record with the primaries, cutting across them in roughly the same way as very high velocity long-wavelength noise events.

The common-depth-point technique is designed to cancel noise of large apparent wavelength, regardless of its origin. As with conventional cancellation, outputs of phone groups distributed over a distance comparable to a wavelength are summed.

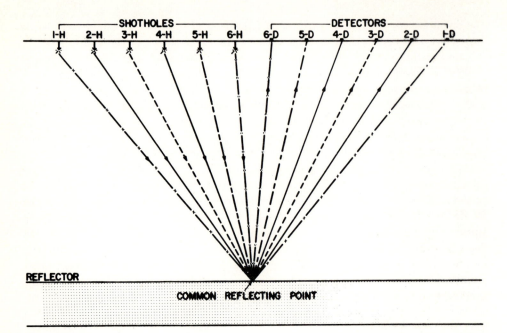

FIGURE 4-22
Ray paths for reflections from a single point in sixfold common-depth-point shooting.

The loss of definition which averaging over such an extensive baseline would otherwise cause is averted by a special arrangement of shots and geophones that combines only those signals reflected from the same region of the subsurface.

The method for doing this has been described by Mayne,[14] the inventor of the technique. Basically, signals associated with a given reflection point but recorded at a number of different shot and geophone positions are composited either with special field processors or in a playback center after appropriate time corrections are applied to compensate for the increasing length of ray path as the shot-geophone distances are increased. Figure 4-22 illustrates the recording arrangement when six signals are composited.

In an actual field setup, such common-depth-point shooting (or multiple coverage) involves a greater number of shot points per unit distance along the line than conventional split-spread shooting does. With 24 recording stations threefold coverage will be obtained if the shots are separated by four geophone-group intervals, fourfold if by three intervals, and sixfold if by two intervals. If the shooting is twelvefold there is a shot for every geophone group center. Twenty-four-fold shooting on land, which is becoming increasingly common, requires 48 recorded geophone positions. Figure 4-23 shows the successive shot positions for a set of single-end shooting spreads giving sixfold multiple coverage. Consider the reflecting

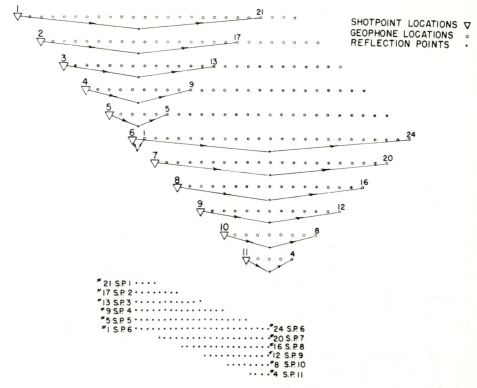

FIGURE 4-23
Shot and geophone combinations giving sixfold multiplicity for two subsurface reflecting points. Pattern at bottom shows reflecting points for successive spreads. (*Mayne.*[14])

point halfway between geophone positions 10 and 11. Reflection takes place at this position when the shot is at shot position 1 and the phone at geophone position 21, also when the shot is at position 2 and the geophone at 17. Such a reflection also occurs with shots and geophones at the four other combinations of positions shown on the left side of the figure.

To composite, or *stack*, the data for this reflecting point, the playback is so programmed that after all signals are corrected for normal moveout, the information on channel 21 of the tape from shot point 1 is added to that from channel 17 of the tape made from shot point 2, with similar contributions from the respective tapes corresponding to the four shot points 3 through 6.

At the bottom of the figure the dots show reflection points which correspond to all recorded wave paths. It is seen that there are six such paths for each subsurface position. Adding the proper signals for coincident subsurface reflection positions as indicated on the diagram and plotting the output traces thus obtained on a record section is referred to as *stacking*. Shooting of this type is carried out continuously

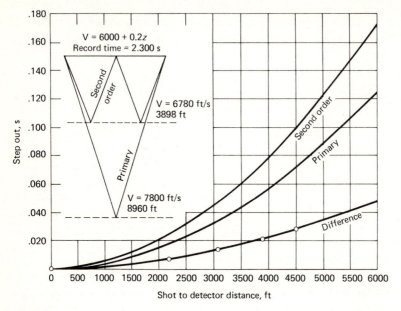

FIGURE 4-24
Cancellation of multiple reflections by common-depth-point processing for
five shot-geophone distances. Difference in arrival times for the multiple at
these distances covers a wave period, resulting in cancellation upon addition
of the signals. (*Mayne.*[14])

along a line, the phones at the rear being picked up in groups of four or six and
moved to the front of the spread as the shooting progresses in a kind of leap-frog
maneuver. This is referred to as *rollalong*.

Figure 4-24 illustrates how multiple reflections are attenuated by common-depth-
point stacking. The primary reflection and the multiple shown in the diagram at the
upper left arrive at the geophone at the same time. The average velocity for the
primary must be higher than that for the multiple when the velocity increases with
depth; thus the multiple travels through formations having an average speed slower
than the average along the path taken by the primary. The moveout times for the
multiple and primary are plotted against horizontal receiving distance. The multiple
shows a greater moveout time than the primary event. If a moveout correction is
made using the velocity for the primary and the traces thus corrected are then
added together, all primary reflections should have phase coincidence and the
signal after summation should have accentuated amplitudes for such events. The
multiples, however, will be out of phase, and addition of such signals for equally
distributed distances from 0 to 6000 ft will tend to result in cancellation. In this
example, the total range of differences in time between the two corrected events
over a 6000-ft distance is more than 40 ms, which is greater than the periods of most
seismic reflection events.

It is evident from the diagram that the deeper the primary event to be brought out over a multiple that arrives at the same time the smaller the moveout differential over a given spread length. Yet the deeper a reflection the greater the likelihood that there will be interference from multiples. The only way to increase the time differential from moveout is to lengthen the spread. For this reason, spreads are much longer with common-depth-point shooting than they were with conventional single coverage. Spreads allowing recording as much as 2 mi from the shot point are not uncommon.

Actually, the cancellation of multiples would be more complete if the shot-geophone distances were such as to allow equal time intervals between successive samples along the difference curve. The practical difficulties of laying out spreads in which the distances between phones are nonuniform would make such an arrangement infeasible. The irregularity of time intervals can be minimized, however, if there is an appreciable offset distance between the shot and the nearest, phone, as is customary (to reduce surface-wave interference) when nondynamite sources are employed. With the nearest phone at a large offset distance, all the phones lie along the portion of the difference curve that approaches linearity. The nearer the differential time-distance curve is to being linear the more nearly equal the phase differences for successive intervals and the better the cancellation to be expected.

Whether the shooting should be programmed for threefold, fourfold, sixfold, twelvefold, or twenty-four-fold coverage depends on the quality of the data and the relative amplitudes of the multiple reflections to be canceled. Now that common-depth-point data are often used to compute interval as well as average-velocity information, it may be desirable to shoot with as great a degree of multiplicity as is operationally feasible (twenty-four-fold or more) so as to attain the greatest possible precision in the resultant velocity. Current techniques for determining velocities from seismic records shot in common-depth-point configuration will be described in a later chapter.

REFERENCES

1. Dobrin, M. B., R. F. Simon, and P. L. Lawrence: Rayleigh Waves from Small Explosions, *Trans. Am. Geophys. Union*, vol. 32, pp. 822–832, 1951.
2. Dobrin, M. B., P. L. Lawrence, and Raymond Sengbush: Surface and Near-Surface Waves in the Delaware Basin, *Geophysics*, vol. 19, pp. 695–715, 1954.
3. McCullom, Burton: Seismic Exploration Apparatus, U.S. Patent 2,767,389, 1956.
4. Mason, R. G.: A Small-Scale Field Investigation of Motion near the Source, *Geophys. Prospec.*, vol. 5, pp. 121–134, 1957.
5. Neitzel, E. B.: Seismic Reflection Records Obtained by Dropping a Weight, *Geophysics*, vol. 23, pp. 55–80, 1958.
6. Peacock, R. B., and D. M. Nash, Jr.: Thumping Techniques Using Full Spread of Geophones, *Geophysics*, vol. 27, pp. 952–965, 1962.
7. Johnson, James F.: Surface Sources of Seismic Energy, *Proc. 7th World Petrol. Cong., Mexico City*, vol. 2, pp. 709–716, 1967.

8. Crawford, John M., W. E. N. Doty, and M. R. Lee: Continuous Signal Seismograph, *Geophysics*, vol. 25, pp. 95–105, 1960.
9. McKay, A. E.: Review of Pattern Shooting, *Geophysics*, vol. 19, pp. 420–437, 1954.
10. Lombardi, L. V.: Notes on the Use of Multiple Geophones, *Geophysics*, vol. 20, pp. 215–226, 1955.
11. Parr, J. O., Jr., and W. H. Mayne: A New Method of Pattern Shooting, *Geophysics*, vol. 20, pp. 539–565, 1955.
12. Reynolds, F. F.: Design Factors for Multiple Arrays of Geophones and Shot Holes, *Oil Gas J.*, Apr. 19, 1954, pp. 145–146, 1954.
13. Pieuchot, M., and H. Richard: Some Technical Aspects of Reflection Seismic Prospecting in the Sahara, *Geophysics*, vol. 23, pp. 557–573, 1958.
14. Mayne, W. Harry: Common Reflection Point Horizontal Data Stacking Techniques, *Geophysics*, vol. 27, pp. 952–965, 1962.

ACQUISITION OF SEISMIC DATA
IN WATER-COVERED AREAS

In Chap. 4 we considered techniques for seismic recording on land. Some of the topics discussed there, e.g., use of multiple receivers and common-depth-point techniques, are equally applicable to seismic work at sea. Other aspects of seismic-data acquisition, however, are uniquely associated with marine applications. For example, the physical processes by which seismic energy is generated in the water are quite different from those involved when the generation is in solid earth materials. Special energy sources can thus be used at sea which are not generally applicable on land. Receiving transducers and cables are designed quite differently, and position location for marine exploration may require methods which have been specifically developed for solving the special problems encountered at sea.

While the seismic method has been used in offshore oil exploration since before World War II, most marine activity was confined to the Gulf Coast of the United States until the early 1960s, when the amount of offshore seismic exploration in other parts of the world began to increase at a rapid rate. Concurrent with this expansion of activity was a rapid development of new and improved equipment and techniques which have led to greatly increased efficiency and economy in marine operations as well as to much better data quality.

Because of the vast amount of data recorded each day in a seismic operation at sea, digital recording was put to use for nearly all marine shooting within a few years after its introduction. From the middle 1960s on, virtually all marine reflection work has employed multiple (common-depth-point) coverage. Since late 1967, explosives have only rarely been used as sources of seismic energy except for special detonating systems in which small ($\frac{1}{2}$ lb or less) charges confined in cans or plastic jackets are exploded.

5-1   GENERATION OF SEISMIC ENERGY UNDERWATER

The function of any underwater source of seismic energy, whether or not it involves the detonation of dynamite or similar explosives, is to introduce a sudden positive (or sometimes negative) pressure impulse into the water. This impulse involves a compression (or rarefaction) of the water particles, creating a shock wave that spreads out spherically into the water and then into the earth. A delayed effect of the shock wave is an oscillatory flow of water in the area around the explosion, which gives rise to subsequent pressure pulses designated as *bubble oscillations*. A simple description of these phenomena, published by Kramer et al.,[1] will be summarized here.

Formation and properties of the gas bubble in water shooting   The properties of the seismic signals generated by all marine energy sources are strongly dependent upon the bubble oscillation, which in the case of dynamite and certain nonexplosive sources such as air guns is associated with an actual bubble in the water. With other types of sources where no bubble exists, periodic pressure pulses are generated which have characteristics similar to those from the bubble oscillations. To understand how these pulses are generated for all types of sources, let us consider what happens to a gas bubble created underwater by an explosion of dynamite.

If we assume a spherical charge of dynamite 1 ft in diameter to be detonated at its center, we shall have a very rapid conversion of the solid material constituting the explosive into high-temperature, high-density gas, which spreads outward into the still undetonated portions of the sphere at a velocity as high as 22,000 ft/s, causing combustion of the solid explosive material with which it makes contact. The pressure in the interior of the solid shell has a peak value of 2 million pounds per square inch. When the outward traveling detonation front reaches the surface of the spherical charge 23 μs after the detonation, the water immediately outside the sphere becomes highly compressed and a strong shock wave is initiated in it. At a time of 30 μs from the detonation, the peak pressure of the shock wave, now traveling in the water, has been reduced to 550,000 lb/in². The falloff in pressure on either side of the shock front is sharp, as is demonstrated in Fig. 5-1.

From this point on the process of generation is the same whether the source is dynamite, an air gun, an electric sparker creating a steam bubble in the water, or any other system that suddenly injects a bubble of gas in the water. The important consideration from the standpoint of putting seismic energy in the earth

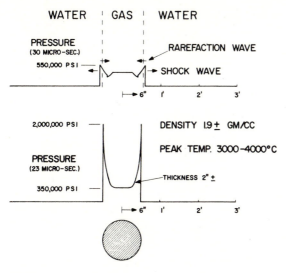

FIGURE 5-1
Pressure fields in vicinity of spherical dynamite charge 23 μs (*below*) and
30 μs (*above*) after detonation at center. (*From Kramer, Peterson, and Walter.*[1])

is the creation of a shock wave by sudden compression of the water that occupies
the space adjacent to the bubble immediately after it is generated.

As the shock front progresses outward, its pressure and particle velocity continue
to decrease. Figure 5-2 shows both quantities as a function of distance from the
point of detonation at a time of 630 μs. At this time (still less than 1 ms after the
explosion) the gas bubble is 2 ft in diameter, and the shock front is a spherical
surface having a radius of 6 ft. The water pressure at the outer edge of the front is
now 16,000 lb/in². Just inside the shock front the pressure decreases in the direction
of the source, reaching a minimum at a distance of 2 ft from the detonation point.

The particle velocity in the water consists of two components: (1) the outward-
directed *compressive* flow of water required to fill the rarefaction left behind the
shock front which transports water under compression away from the source, and
(2) the *afterflow* which supplies water to accommodate the tangential expansion
that occurs as the shock front travels. The afterflow represents a production of
kinetic energy which becomes converted into a pressure wave when the outward
flow of water is reversed.

Because of momentum, the gas bubble continues to expand until 200 ms after
the shot, at which time its radius is about 10 ft. The pressure inside the bubble is
now only 2 lb/in², which is 35 lb/in² *below* the ambient hydrostatic pressure. Here
the expansion stops and contraction begins. The rapid shrinking of the bubble
causes an increasing inward velocity of the water and a rapidly increasing pressure
in the contracting bubble. At 400 ms the bubble has collapsed to its smallest
diameter, and expansion starts again. If it were not for frictional losses and the rise

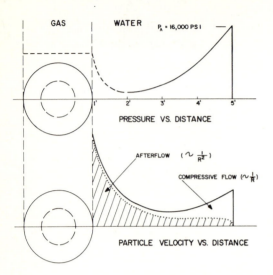

FIGURE 5-2
Pressure and particle velocity versus distance from source 630 μs after underwater explosion. (*From Kramer, Peterson and Walter*[1].)

of the bubble to the surface, the process would continue indefinitely. Figure 5-3 demonstrates this cycle of bubble oscillation. The depth of the bubble stays almost constant while its diameter is large because the resistance of the water above it inhibits upward motion. When the bubble diameter is smallest, the water resistance is the least and the bubble rises at its greatest rate of speed.

Relation between oscillation period of gas bubble and energy of source The period of the bubble oscillation is of great practical importance because each oscillation generates a new seismic impulse. The seismic signal associated with the initial pressure injection is thus repeated at intervals equivalent to this period. Such multiple repetitions in the down-traveling source signal are likely to cause reverberation effects in the reflection signals that can obscure desired information. Fortunately, the magnitude of the bubble oscillation can be minimized by proper design of the source and by the use of appropriate shooting procedures. In many cases the entire bubble train can be used as an effective energy source by recording the original bubble sequence and using an appropriate mathematical filter (of a type to be discussed in Chap. 6) for simplifying the desired reflection signal.

The relation between the bubble-oscillation period T and the potential energy associated with the oscillation was developed by Willis[2] on the basis of an equation previously published by Lord Rayleigh[3] for the dependence of the period of the bubble upon its radius and its ambient hydrostatic pressure. Lord Rayleigh's equation, originally developed to explain the sounds made by steam bubbles in a teakettle, is

$$T = 1.83 A_m \sqrt{\rho/P_0} \qquad (5\text{-}1)$$

where
T = period of bubble oscillation, in seconds
A_m = maximum radius of the bubble, in centimeters

ρ = density of the fluid, in grams per square centimeter

P_0 = ambient pressure, in dynes per square centimeter

Willis' equation introduced the relation between Q, the potential energy in the bubble, and its pressure and volume:

$$Q = \tfrac{4}{3}\pi A_m{}^3 P_0 \qquad\qquad (5\text{-}2)$$

Eliminating A_m, we obtain the following relationship between T and Q:

$$T = 1.14\,\rho^{1/2}\,P_0^{-5/6}\,(KQ)^{1/3} \qquad\qquad (5\text{-}3)$$

where K is a constant depending on the units of Q. It is 1.0 if Q is in ergs, 1.00×10^{10} if in kilojoules, 1.36×10^7 if in foot-pounds, and 4.18×10^{10} if in kilocalories. If we assume a density of 1.024 g/cm³ for seawater and replace P_0 by $d + 33$, where d is the depth in feet of the center of the bubble below the water surface, this equation becomes, for T in seconds,

$$T = \frac{0.000209\,(KQ)^{1/3}}{(d + 33)^{5/6}} \qquad\qquad (5\text{-}4)$$

If we operate on the assumption that the potential energy of the bubble Q is proportional, for any particular type of source, to the intrinsic energy in the source, we have a means of determining the relative efficiency of different sources and of comparing the effective performance of the sources on the basis of the oscillation period, which is often easily measurable. A simple method for doing this will be discussed later in this chapter (see page 129).

The mechanism for pressure generation by bubble-pulse oscillation is somewhat different for sources which do not create an actual bubble in the water. In the Esso

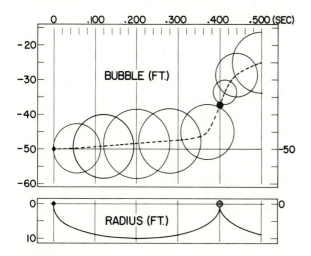

FIGURE 5-3
Variation of bubble radius and position in water with time after dynamite explosion. (*From Kramer, Peterson, and Walter.*[1])

Sleeve Exploder or Aquapulse* source, for example, a propane-oxygen explosion takes place in enclosed chambers separated from the water by more or less flexible walls. Even here, however, there is an oscillation effect caused by the buildup of potential energy in the afterflow, which is generated whenever there is a shock wave. As with explosives, the period of the oscillation depends upon the potential energy associated with the source. For sources where there is no actual bubble, the signal will still exhibit all the effects of a bubble oscillation because of the afterflow mechanism, which is inextricably associated with the shock wave. The amplitude of the pressure signal associated with the oscillation will, however, be lower.

5-2 MARINE ENERGY SOURCES

It is not feasible to describe all the energy sources developed for marine seismic work in recent years, but the number of basic types is relatively small, and examples of each will be discussed. In addition to explosive sources conventionally used for open-water detonation, we shall investigate types involving specially encapsulated charges fired under controlled conditions, sudden release of air and steam into the water, underwater electric discharges, plates displaced by eddy-current effects, and confined propane-oxygen explosions.

Dynamite

Referring to dynamite and other high explosives, Kramer et al.[1] write:

Probably no other type of source provides such a compact package of concentrated energy and no other source provides such a simple means for rapid and almost instantaneous release of energy. However, for various economic, technological, and political reasons the last few years have seen a rapid increase in the use of alternative types of energy sources, particularly in marine seismic surveying operations.

Dynamite was used as an explosive for early seismic work at sea until the early 1950s, but since then nitrocarbonitrate (NCN) has accounted for all marine shooting with explosive sources. In the middle 1960s the percentage of work making use of NCN decreased almost to zero until specially controlled NCN sources such as Flexotir† and Maxipulse,‡ described later in this chapter, came into extensive use. In 1974 20 percent of all marine shooting employed small NCN charges.

In shooting explosives at sea, it had been customary to detonate the charge at such a shallow depth that the bubble would break through the surface of the water and not oscillate. Lay[4] found that the maximum depth d (in feet) at which the bubble will break is related to the charge weight w (in pounds) by the formula

$$d = 3.8w^{1/3} \tag{5-5}$$

* Trademark of Western Geophysical Company of America.
† Trademark of Institut Français du Pétrole.
‡ Trademark of Western Geophysical Co. of America.

Worzel and Ewing[5] reach a similar conclusion in a memoir on shallow-water seismic-wave propagation.

The explosion of NCN at such shallow depths substantially lowers the efficiency of the operation because there is always a pressure node at the free surface of the water at which any excitation should theoretically have no effect. The most efficient input of energy is actually at an antinode, which occurs at depths equal to any odd number of quarter wavelengths. One-quarter of a wavelength for a typical seismic reflection signal in water is 35 to 40 ft.

The bubble-oscillation period in water for a NCN charge of 1 lb detonated at 30 ft is about 120 ms, and for a charge of 50 lb it is about 700 ms. Such long intervals cause repetitions of reflection events on records, which can make proper interpretation very difficult. Thus large charges must always be fired at shallow enough depths to destroy the bubble in spite of the loss in efficiency that results when a large part of the energy of the explosion is released into the air.

There are many reasons for the great decrease in the use of explosives for marine seismic work during the later 1960s. Most of them relate directly or indirectly to the potential hazards, both to life and to property, associated with explosives as conventionally used. For example, it was generally not considered safe to handle explosives after dark for any seismic crew, land or marine. This restriction limited offshore operations to daylight hours and, depending on latitude and time of year, could cut down on productivity by 50 percent or more. Also, the explosive was seldom fired from the ship that contained the recording equipment and towed the receiving cable because any misconception about location of the charge under the water surface could result in the loss of a cable costing more than $100,000 or even the destruction of the recording ship. The need for two ships in such operations (one for shooting and one for receiving) involved a much greater initial investment and operating expense than the single ship, which could suffice for both purposes if a less hazardous source were used.

Another important reason for abandoning explosives detonated just below the surface was their very low efficiency. A large part of the energy of the detonation went into the geyser each shot created instead of into seismic events. The cost of using many charges having such poor effectiveness became prohibitive with the introduction of common-depth-point shooting.

Also the logistic difficulties in shipping, storing, and handling large quantities of explosives, particularly in parts of the world that are far from factories where it is manufactured or in politically sensitive areas, added complications and delays that were not encountered with other energy sources.

Finally, there was the danger of destroying fish by dynamite which has led to restrictions on seismic activity in many areas where fishing is important. Such restrictions could prohibit such activity altogether at certain times of year. In areas such as the Cook Inlet of Alaska and the Bay of Biscay, the severity of regulations against underwater explosions made it particularly urgent that some seismic-energy source be found which would be acceptable to the agencies charged with protecting the fisheries industry.

Although the devices which have largely replaced conventional high explosives for offshore seismic work are ordinarily referred to as *nonexplosive sources*, this terminology has euphemistically come to include systems which involve the detonation of NCN charges under such conditions that safety and economy are not compromised.

Sources Using Controlled Charges of Dynamite

Flexotir   In one such system, developed by the Institut Français du Pétrole and designated Flexotir, the source itself is a small pellet of dynamite weighing about 2 ounces embedded in a plastic cartridge. The charge is detonated at the center of a multiply perforated cast-iron spherical shell about 2 ft in diameter, which is towed behind the ship at a depth of about 40 ft. The perforations (130 of them) are 2 in across. A flexible hose conveys the charge from the fantail of the ship into the sphere, the driving force being water pumped down the hose under high pressure. The sphere must be replaced after about 2000 explosions.

The mesh made by the perforations in the spherical enclosure has the effect of breaking up the bubble, in accordance with a principle first established by Knudsen,[6] so that oscillation is aborted and the undesired effects of bubble pulsing on the signal are suppressed. Because the shooting depth is about a quarter wavelength for typical seismic reflection waves, the efficiency of the explosion for generating seismic energy is much greater than for detonation just below the surface and, as Lavergne[7] has shown, the effective seismic signal is as strong as that for a much larger charge fired at conventional depths.

Flexotir systems had their greatest use from the mid-1960s to the early 1970s.

Maxipulse   Another source containing dynamite, the Maxipulse, is also designed for detonation under conditions that combine safety, efficiency, and elimination of bubble-pulse effects. The charge, about $\frac{1}{2}$ lb, is packed in a can which is injected into the water at a depth of about 40 ft by a delivery device trailed from the ship. The detonation takes place about 1 s after the injection, the delay making it possible for the can to explode far enough from the injector to avoid damaging it. The arrangement is illustrated in Fig. 5-4a.

Upon detonation, a bubble is formed, and it expands and collapses with a period of approximately 100 ms. A detecting hydrophone on the injector device picks up the signal generated by the source, including the bubble oscillations (Fig. 5-4b). In the final processing of the data, this signal is compared with the reflection signal, and its effects are minimized by use of a filter program that yields an output equivalent to what would have been obtained if the source had generated a single sharp pulse. This system appears to yield reflections from exceptionally great depths.

Nonexplosive Sources

Sparker   The sparker was originally developed for fundamental studies of underwater sound propagation (Anderson[8]) and of subbottom geology (Knott and

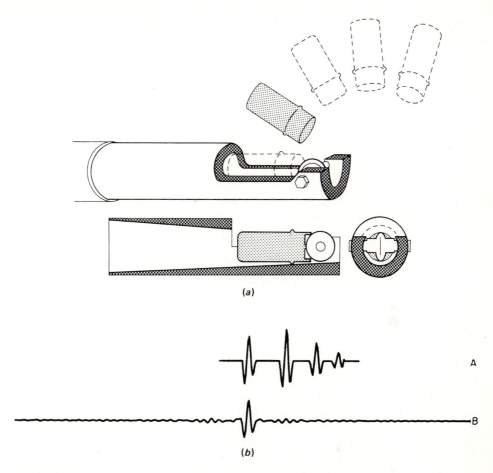

(a)

(b)

FIGURE 5-4
Maxipulse system: (a) gun and charge, showing how charge strikes firing wheel and how it is subsequently ejected; (b) collapse of bubble pulse by computer processing. Waveform A includes bubble pulses as generated. Waveform B shows collapsed waveshape after the processing. (*After C. H. Savit, Copyright Offshore Technol. Conf., Houston, 1970, vol. I, p. 604.*)

Hersey[9]). It has been used to investigate the structure of the sediments below the deep ocean basins as well as under continental shelf areas.[10] It was employed to establish the continuity of the limestone under the English Channel with the object of evaluating the feasibility of a tunnel between England and France. It is now used primarily for engineering studies.

In the sparker, seismic waves are generated by the sudden discharge of current between electrodes in the water. The electrode assembly, towed behind a ship, is connected to capacitor banks on board the ship which are charged by an electric generator. A specially designed switching system suddenly closes the circuit, discharging the capacitors through the saltwater gap between the electrodes at the instant the shot is to be fired. The discharge through the water between the

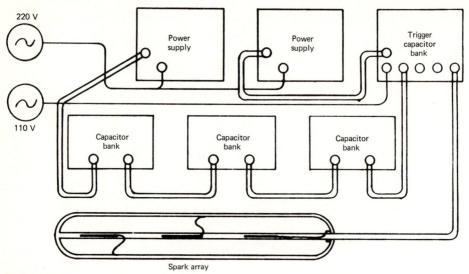

FIGURE 5-5
Three-electrode Sparkarray unit with associated shipborne electrical system.
(*E.G. & G. International.*)

electrodes generates such intense heat that the water is suddenly vaporized, creating a steam bubble which starts a shock wave just as a dynamite explosion does. The signal returned from the bottom and from subbottom reflectors is recorded by a series of pressure hydrophones on a cable, all the phones being connected together for a single-channel recording. The recording system is a graphic recorder, somewhat similar to a fathometer, with simultaneous recording on magnetic tape.

Figure 5-5 illustrates the principle of the sparker system (Sparkarray) manufactured by E.G.&G. The electrodes here are mounted in groups of three on a tubular frame. Each three-electrode source unit uses 8 kJ of energy. The energy is increased by adding units together, the only limitation on the possible number of units being generator capacity and space available for the capacitor banks. The Teledyne sparker uses larger electrode units of 40 kJ each. The maximum energy available from sparker sources has increased steadily over the years. Single-channel surveys have been conducted with sparkers having more than 200 kJ total energy, and multichannel sparker recording has recently come into use for shallow-penetration surveys.

Boomer   The Boomer, manufactured by E.G.&G. is similar to the sparker in that it also makes use of a sudden discharge of electricity stored in capacitor banks on board ship. Like the sparker, it is mainly used for bottom studies in connection with engineering projects. The current does not pass through the water as it does with the sparker; instead it passes through a coil of wire embedded in a slab of epoxy. An aluminum disk is attached to the slab, as shown in Fig. 5-6. The eddy-current

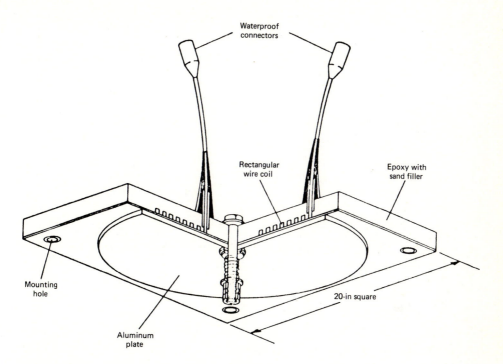

FIGURE 5-6
Design of a Boomer source. (*E. G. & G. International.*)

effect causes a sudden strong movement of the plate against the water that generates
a sharp pressure pulse. Following the pressure pulse is an implosive pulse which,
although negative, corresponds to the bubble pulse associated with explosive
sources. It is caused by the rush of water into the void left behind the plate when it
is suddenly set into motion. A spring returns the plate to its position of contact with
the slab. The pulsing can be repeated every few seconds.

The energy of the seismic impulse is concentrated at a higher range of frequencies
than with most other sources. For this reason, the Boomer is employed where only
a limited penetration is needed but where the highest possible resolution is desired.
It has been most commonly used in bottom surveys for engineering purposes.

Air gun The most widely used of all nonexplosive sources is the air gun. Like the
sparker, it was originally employed for academic studies of subbottom geological
structure in marine areas. It was also used for research on sound transmission in
the ocean.

The Bolt PAR*, employed more extensively than any other type of air gun, is

* Trademark of Bolt Associates.

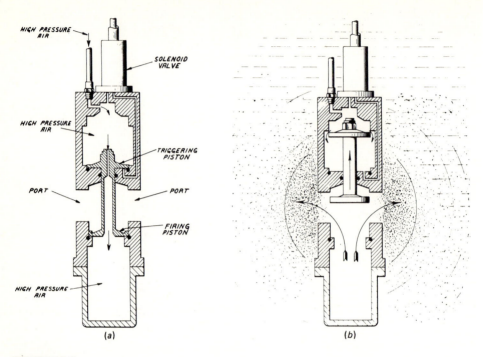

FIGURE 5-7
Bolt PAR air gun, showing two stages of the firing cycle: (a) armed; (b) fired.
(*Bolt Associates, Inc.*)

manufactured in a number of models with capacities ranging from 1 to 2000 in³ of air and makes use of compressors that operate at pressures up to 2000 lb/in².

Figure 5-7 shows how the PAR source works. High-pressure air, which passes through a hose from the compressor to the towed submerged unit, enters through the connection at the upper left. It flows into the upper chamber, across which is fitted the top piston of a shuttle consisting of a shaft with a triggering piston at the upper end and a firing piston at the lower. There is a hole in the shaft through which the air from the upper chamber enters the lower one.

Although the same pressure is developed in each chamber, the area of the triggering piston above is somewhat greater than that of the firing piston below, and the net downward force on the shuttle causes it to move down until it is stopped by the base of the upper chamber. At the instant the gun is to be fired, a solenoid opens a valve that injects high-pressure air between the triggering piston and the base of the upper chamber through the opening on the right side of this chamber. The sudden introduction of the air through the solenoid-controlled valve upsets the equilibrium of the system, and the shuttle moves upward at a high velocity. As the firing piston passes the four large ports (two of which are shown in Fig. 5-7), all the high-pressure air from the lower chamber is suddenly spilled out into the water, creating an air bubble quite similar to that from a dynamite explosion and giving rise to repetitive

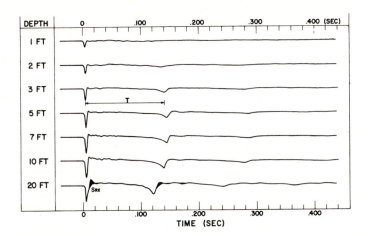

FIGURE 5-8
Air-gun pressure signature at various depths. (*Kramer, Peterson, and Walter.*[1])

bubble pulses at a rate determined by the oscillation period of the air mass thus generated. The larger the volume of the air, the longer the period. Figure 5-8 shows the waveform obtained from an air gun at different depths. The attenuation between the initial impulse and the first bubble pulse is quite significant, and the second bubble-oscillation signal is just perceptible at the shallower depths. The decrease in oscillation period with depth conforms to the Rayleigh-Willis prediction that the period should fall off as the five-sixth power of source depth.

The effect of the bubble-pulse repetition is to give an oscillatory and hence unsatisfactory reflection record. Special measures are therefore taken in the shooting or in the processing center to eliminate the bubble oscillations. The most effective way of doing this in the field is to use an array of guns, all fired in synchronism, having a variety of air-chamber capacities. The intervals between the initial pulse and the first bubble pulse will be different for each gun having a different air capacity. The pressure signal actually recorded from the array will consist of an impulse representing the sum of the initial pulses from all the guns followed by a train of much weaker bubble pulses spread out over a period of time and partially canceling one another. With one arrangement, 25 guns ranging in capacity from 10 to 40 in³ are towed in an array. Another procedure is to use a smaller number of larger air guns (say three guns with capacities of 300, 160, and 80 in³).

A different technique for bubble-pulse suppression is to add air to the bubble at the stage where it begins to contract, thus weakening the force of the contraction and hence the pressure in the pulse generated when the bubble reaches minimum diameter.

Aquapulse* The Aquapulse, one of the more extensively used marine sources,

* Trademark of Western Geophysical Co. of America.

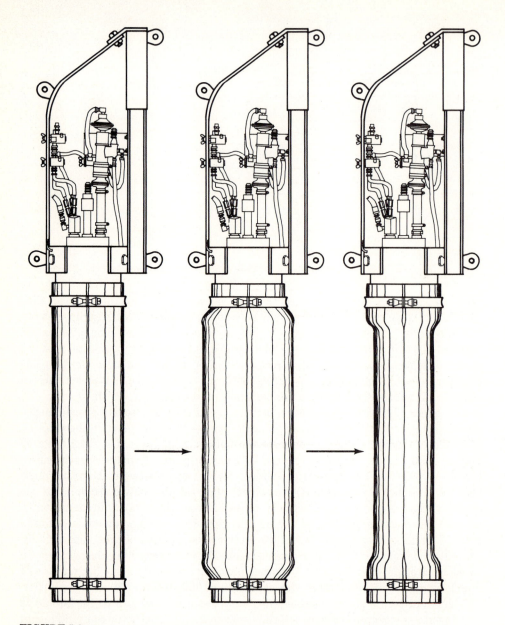

FIGURE 5-9
Three stages in the operating cycle of the Aquapulse gun: (*left*) normal shape
of sleeve in air before it is filled with gas; (*center*) sleeve at maximum expanded
position approximately 3 ms after detonation; (*right*) contraction of sleeve
after products of combustion are removed by vacuum pump on ship. (*Western
Geophysical Co. of America.*)

generates a pressure pulse by detonation of a propane-oxygen mixture in an enclosed underwater chamber with flexible walls. The chamber in this case consists of an elongated heavy-rubber cylinder, which is necked inward over most of its length except at the ends. The cylinder is filled with an explosive mixture of propane and oxygen which is detonated by an electric spark. The explosion causes ballooning of the cylinder (Fig. 5-9), which introduces a pressure pulse into the water. The products of combustion are vented immediately afterwards to prevent oscillation of gases inside the enclosure. Figure 5-10 shows two guns in tandem, a common arrangement, being lowered into the water from the seismic boat. The afterflow in the water surrounding the cylinder causes oscillation pulses which are attenuated in strength because of the venting of the gaseous combustion products, and the resultant waveform (Fig. 5-11) indicates that the amplitude of the initial pulse is exceptionally high compared with that of subsequent oscillations.

Other nonexplosive sources   The extent to which various marine sources have been put to use has changed with time. Some that were once used quite extensively are no longer used on a large scale, often for operational or economic reasons. Among

FIGURE 5-10
Two-gun Aquapulse being lowered into the water. (*Western Geophysical Co. of America.*)

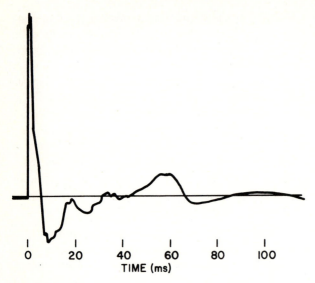

FIGURE 5-11
Pressure signature from Aquapulse source. (*Western Geophysical Co. of America.*)

sources in this category are marine versions of the Vibroseis and Dinoseis. Other available sources have been less widely used than those discussed above for reasons which are not always evident.

The Vaporchoc*[12] injects a bubble of high-pressure steam into the water after transmission through an insulated hose from a generator on the ship's deck. The condensation of the steam in the cold water causes collapse of the bubble before oscillatory motion can progress very far, resulting in a relatively simple pressure (mainly implosive) pulse. It is said to work best at depths less than 20 ft. The principle of its operation is illustrated in Fig. 5-12.

In some sources the energy is generated along a line rather than at a point. An example is the Aquaseis,† which employs a 100-ft length of towed explosive in ribbon form such as Primacord‡. The detonation travels along the cord at a speed of about 20,000 ft/s, the effect of the travel being to separate the bubble into small segments with weaker pressure pulses from each so that oscillation effects are minimized.

The Hydrosein,§ in which a large steel plate is pushed outward into the water by a quick-acting air piston, is implosive and generates a negative pulse. The acceleration of the plate is so high that a vacuum is created behind the plate, which is filled by water rushing into the void.

* Trademark of Compagnie Générale Géophysique.
† Trademark of Imperial Chemical Industries.
‡ Trademark of Ensign Bickford Co.
§ Trademark of Western Geophysical Co. of America.

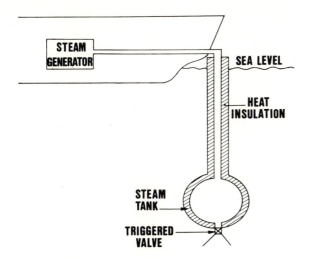

FIGURE 5-12
Schematic diagram showing operation of Vaporchoc steam source. (*After Staron*.[12] *Copyright, Offshore Technol. Conf.*)

A recently introduced source is the Seismovac,[13] developed by EG&G, Inc., in which a vacuum introduced into an underwater chamber sets a piston into inward motion. Because of inertia the piston continues to move until it has compressed the gas remaining in the chamber to the point where its pressure is much higher than the external water pressure. The piston is then driven outward with high velocity, generating a shock pulse in the water. It is restrained from returning at the end of its outward stroke so that bubble pulsing will be inhibited.

Comparison of Marine Sources

From these descriptions it is evident that there is a large variety of sources which are available for a marine survey. Until recently it was quite difficult to compare the performance of the different sources in any objective way, but Kramer et al.[1] have demonstrated a simple method for determining one important property, the relative potential energy available for seismic signal generation. This is done by observing the bubble-oscillation interval for each source and, from a chart based on the Rayleigh-Willis relation, measuring the potential energy actually generated. Figure 5-13, which is called the *Rayleigh-Willis curve* and is based on Eq. (5-4), shows how the energy from different sources can be compared from measurements of bubble-oscillation period under various conditions. The energy, expressed in foot-pounds and in equivalent pounds of dynamite, is read off the abscissa scale.

The observed data points for dynamite and the PAR air guns lie quite close to the theoretical Rayleigh-Willis curve, indicating that a large percentage of the potential

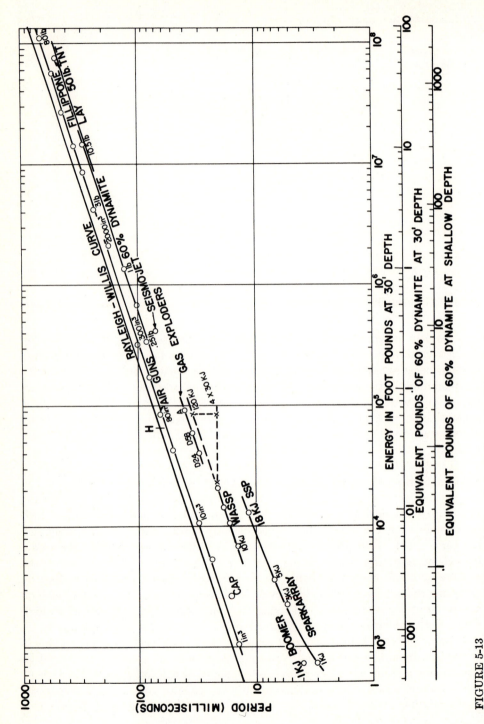

FIGURE 5-13
Rayleigh-Willis diagram relating bubble-pulse period to potential energy for
representative sources at 30-ft depth. (*Kramer, Peterson, and Walter.*[1])

energy is available for developing the bubble. Points for the sparker and gas-exploder types of sources fall farther below the theoretical curve, and we would therefore assume that their efficiency (in terms of percentage of energy available for bubble formation) is lower.

In addition, one can evaluate relative frequency characteristics for the different sources from this plot. Assuming that the low-frequency cutoff point varies as the reciprocal of the bubble period, one could predict the spectrum of the source with reasonable reliability from this type of study. The inverse relation between energy and frequency should be taken into consideration in the light of desired resolution and penetration, bearing in mind that the penetration becomes greater with decreasing frequencies while the resolution becomes poorer.

5-3 CABLES USED IN MARINE SHOOTING

In the early years of seismic exploration in water-covered areas, gimbal-mounted geophones were attached to cables designed to keep the phones 6 to 15 ft below the surface. In those days, cables were seldom more than 1000 ft long. Cables in use today are considerably longer, more expensive, and more sophisticated in design.

Multiple-channel streamer cables The streamer cable, originally developed for antitorpedo warfare in World War II, is the most widely used type for modern seismic recording. This cable, a plastic tube $2\frac{1}{2}$ to 3 in in diameter, is neutrally buoyant and filled with oil. If the density of the water changes because of temperature or salinity variations, the overall density of the cable can be changed to maintain desired buoyancy by adding or removing thin lead sheets wrapped around the cable. The hydrophone elements, wires, and transformers are inside the plastic, generally transparent, along with strain members. In general, the cable consists of detachable and interchangeable sections, about 200 to 300 ft in length.

Most marine recording makes use of cables 3200 m long, which contain 48 recording segments, each feeding a separate channel. Individual "live" segments are $66\frac{2}{3}$ m long, and each may contain 20 to 32 hydrophones. Some cables are designed for only 24 channels, while others have as many as 96. In the latter case, there is often a 50 percent overlap of the sensors for each channel so that the total length of each group and that of the cable itself are the same as for the 48-channel type.

Figure 5-14 shows the configuration of such a 96-channel cable as it is towed in actual use. There is a heavily weighted faired lead-in some 200 m long, made of nylon-reinforced neoprene, the weights being necessary to depress the front end of the cable to the desired operating depth.

A tail buoy is connected to the far end of the cable. Sighting this from the ship or tracking it with radar makes it possible to tell how closely the cable is trailing the ship along its line of motion. Currents often cause feathering of the cable away from the ship's path.

At suitable intervals along the cable, eight or ten pressure-sensitive depth

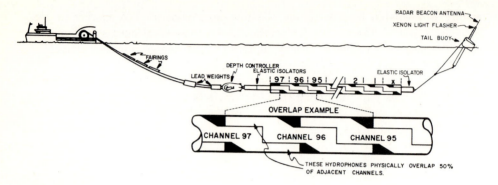

FIGURE 5-14
Schematic diagram of streamer cable being towed from ship. (*Western Geophysical Co. of America.*)

controllers of a type developed by Continental Oil Company keep the cable at the optimum depth. Each unit contains wings that lift or lower the cable, depending on the angle it makes with the horizontal. The controller is set for the desired depth, and a pressure gauge actuates the wings when the actual depth of the cable begins to deviate from that for which the setting was made.

Streamer cables perform satisfactorily in most areas where marine shooting is carried out, but they are not suitable in shallow water where the depth is not much greater than (or sometimes not as great as) the level corresponding to quarter wavelength optimum submergence. For shallow-water exploration, bottom-reference cables, for which sensors maintain a constant elevation above the water bottom, or cables which lie on the bottom are generally used.

Single-channel streamers In shallow reconnaissance exploration as well as in engineering surveys, it is more expedient and more economical to record from a single channel than from a long multichannel spread. On marine reflection surveys carried out for academic rather than exploration studies, single-channel recording has been more frequently employed than multichannel cables, which have been only recently introduced for this type of research. The cable used for single-channel surveys is of the streamer type. Because only one channel is used, it may be economically feasible to introduce many more hydrophone elements than would be possible for each element of a multichannel cable. As many as 100 elements may be built into a 300-ft length of cable when it is to be used in this way. This number results in a better response to high frequencies than is usual for 30 elements per channel. Figure 5-15 shows noise as a function of water speed, illustrating how the increase of noise with speed is similar for a single phone or an array.

5-4 REFLECTION PROCEDURES AT SEA

Almost every marine reflection survey today is carried out as a single-ship operation. With explosives in conventional form no longer being used, the same ship tows

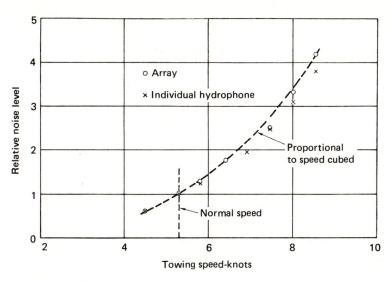

FIGURE 5-15

Noise level versus towing speed for streamer cable. (*From M. Schoenberger and J. F. Mifsud, Geophysics, vol. 39, p. 788, 1974.*)

both the energy source and the recording cable. Such an arrangement has obvious economic advantages over any requiring two ships.

Nearly all energy sources in common use are towed at a distance far enough from the stern of the ship to avoid the possibility of damage to the ship's hull. As noted in our discussion of marine cables, there is generally a distance of about 1000 ft between the source and the center of the nearest receiver group on the cable. With the recording portion of the cable generally 3200 m in length, the total distance from the ship to the end of the cable can thus be 2 mi or more. It is important to know the actual position of the cable as it is towed through the water. When, because of cross currents, the cable drifts away from the line of motion of the ship and the positions of the geophones are different from those assumed in making time corrections for common-depth-point stacking, appreciable deterioration of the processed data could result. With a neutrally buoyant streamer cable, visual or radar contact can only be made with the tail buoy, which is hard to see. The buoy is therefore designed to be observable on the radar screen and it generally contains a light for visibility at night so that its position will always be known.

The adverse effects of excessive *feathering*, as this type of deviation is called, upon data quality depend to a large extent on the dip and structural relief of the reflecting formations. The greater the relief the less such deviation can be tolerated. Many oil companies require that operations be shut down when the current across the shooting line becomes strong enough for the deviation of the cable due to feathering to go beyond specified limits.

Virtually all marine reflection work is carried out with common-depth-point shooting. The time interval between the shots depends on the degree of multiplicity. In sixfold coverage, for example, the shot is fired every two geophone intervals, which is about 200 m (or once a minute at 6 knots). Twelvefold shooting would involve shots every 100 m (about once every 30 s with a 6-knot speed). Now that more rapid repetition rates are feasible with most sources, there is an increasing trend toward twenty-four-fold coverage (one shot every 50 m or 15 s) and even forty-eight-fold coverage. The timing may be determined from the ship's position as indicated on a plotter associated with the electronic navigation (discussed later in this chapter) or the shots may be fired at constant intervals of time determined from the ship's speed.

With sources such as the Aquapulse allowing fast repetition of energy impulses, several shots can be fired at uniform intervals between the predetermined common-depth-point shooting positions. Where this is done, the signals from all shots nearest to each position (between one midpoint and the next) are composited, the sum constituting the input for that position. The compositing may either be done on the ship, using a special summing system, or during subsequent playback. This procedure, designated as *vertical stacking*, increases multiplicity and should therefore reduce noise. It had been common to composite two to four shots for each input although the practice has largely been discontinued.

5-5 MARINE REFRACTION

Until the later 1960s, all refraction work at sea was carried out with dynamite as the energy source, making use of the separate ships for sending and receiving. Such work involved considerably greater expense than reflection because of the cost of the explosives as well as the cost of operating two ships instead of one. These costs tended to price refraction shooting out of the market, in spite of some unquestionable advantages it has demonstrated over reflection for reconnaissance in previously unexplored offshore areas. Since 1968, a sonobuoy receiver, developed by the U.S. Navy for submarine detection from aircraft, has been used for refraction exploration with nondynamite energy sources. This development, which allows single-ship refraction operations, has made marine refraction so economical that the volume of activity of this type has shown a significant increase since it became available for exploration work. Explosives are generally employed only when the desired range from source to detector is too great for sonobuoys to operate.

Marine refraction with explosives In refraction shooting at sea with dynamite as the energy source two ships are employed. The charges are usually dropped to the water bottom from the fantail of the shooting ship. Detonation at the bottom will generally result in more efficient transmission of refracted energy than firing just below the water surface, as is necessary in reflection work. The charge is fired from a switching unit on the shooting ship, which is armed by the shooter and triggered via a radio signal actuated by the recording drum on the receiving ship. The shot moment is transmitted to the recording ship by radio.

The same 24- to 96-channel cable designed for receiving reflection signals can generally be used for refraction work as well. The spacing of receiving phone groups, usually 67 m, is quite suitable for refraction recording. In addition to the hydrophone groups and associated amplifiers designed for detecting the low-frequency refraction arrivals, it is common to introduce special hydrophones with amplifiers set for receiving high frequencies every 1500 ft or so along the cable to determine water-wave arrival times with the greatest possible precision. Knowing these times at points well distributed along the cable makes it possible to measure shot-detector distances precisely without depending on electronic survey systems, which give the distances between the shooting and recording ships but not the actual distance from the shot to the individual phones on the receiving cable.

Sonobuoy refraction system   Sonobuoys are self-contained systems for receiving sound waves in the water and transmitting them to a distant receiving point by radio. A hydrophone or array of hydrophones is suspended from the buoy, which contains sonic amplifiers and a radio transmitter with an antenna projecting upward from the floating buoy. Sonobuoys have been employed for several decades in academically oriented marine refraction surveys, using dynamite as a source. One such survey was carried out in the English Channel to determine the subbottom geology there.[11] It was only in the late 1960s that refraction techniques involving sonobuoy receivers became practical for petroleum exploration. These techniques involve the use of air gun or other nonexplosive sources and expendable sonobuoy receivers.

The sonobuoys were initially developed for submarine detection. When the buoy is thrown into the water from a ship, the hydrophones drop from the bottom of the floating buoy to a depth of about 60 ft and a 3-ft antenna for transmission of the hydrophone signals by radio springs into the air from its top.

Figure 5-16 shows a disassembled sonobuoy of this type. Originally four hydrophone elements were suspended 18 inches apart along a cable. Now a single phone is used along with a preamplifier. The main amplifier and the circuitry for the radio transmitter are on cards inside the buoy. Power is from a battery activated by seawater.

Figure 5-17 illustrates the shooting-receiving configuration for a sonobuoy refraction operation. The shooting ship, equipped with an air gun and a recording unit, drops the buoy from the stern and starts to move away from it along the profile. A single-channel recording drum giving a visible stylus-written record triggers the energy source each time the recording stylus reaches zero time. The pulse from the source travels through the subbottom as a refracted signal and also through the water as a direct wave, both events being picked up by the hydrophones on the sonobuoy for broadcast to the recording ship through the sonobuoy antenna. The signal received by the ship's antenna is registered on a drum-type direct-writing recorder for monitoring as well as on a digital tape recorder for ultimate processing and presentation. The water-wave arrival time provides a direct measure of the shot-receiver distance, and refracted events are generally easy to identify as first

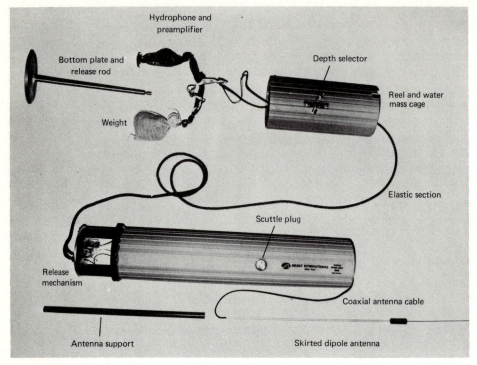

FIGURE 5-16
Refraction sonobuoy disassembled. (*Select International, Inc.*)

and often later arrivals of coherent energy. As the sonobuoy drifts during the recording of a profile, the use of the water-wave arrival times makes it possible to maintain desired precision in the shot-receiver distances for each shot along the profile. Figure 9-17 shows a typical refraction record made with an air-gun source and a sonobuoy receiver.

The earliest use of this type of sonobuoy system for refraction surveys in shallow-water areas was by the Lamont-Doherty Geological Observatory off the coast of Brazil.[14] Its first application to oil exploration was in late 1968 off the coast of Java.

5-6 NOISE PROBLEMS IN MARINE SEISMIC WORK

In the previous chapter we considered the various types of noise encountered in reflection work on land. We pointed out that most of the major improvements that have been made in seismic recording and data-processing techniques have been devised with the objective of eliminating noise and enhancing the quality of desired reflections. Some types of noise which are recorded on land are also observed on marine records. There is one type, however, which while sometimes encountered in land work is predominantly associated with marine shooting. This is *surface-layer*

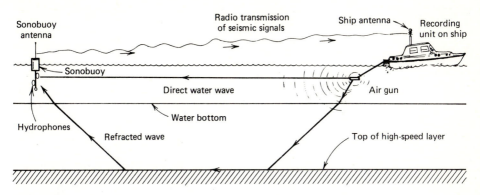

FIGURE 5-17
Sonobuoy refraction recording operation.

reverberation, also referred to as *ringing* or *singing.* It is caused by multiple reflection of waves, both at the source and receiving end of the reflection path, that bounce back and forth between the top and bottom of the water layer.

Figure 5-18 shows the frequency spectrum computed by Backus[15] for reverberation in a water layer 100 ft deep with a "hard" bottom, i.e., a bottom having a sound speed higher than that of the water. In this case, the fundamental frequency is one-fourth of the reciprocal of the one-way time through the bed, and frequencies for higher harmonics are odd multiples of this frequency, as expressed by

$$f_n = \frac{(2n-1)V_w}{4d_w} \tag{5-6}$$

where f_n = frequency of nth harmonic (n = any integer)
   V_w = velocity of sound in water = 5000 ft/s
   d_w = water depth

For the example given, the fundamental frequency would be $12\frac{1}{2}$ Hz with harmonics at $37\frac{1}{2}$, $62\frac{1}{2}$, $87\frac{1}{2}$, ... Hz.

The net effect of the multiple reflection in the water is to give all reflections the same frequency characteristics as the water reverberation exhibits itself. Each peak in the spectrum [corresponding to the various harmonics of Eq. (5-5)] can be related to repetitive events in the reverberating signal, but as the individual trains are superimposed, it is not generally possible to resolve them on the recorded signal. On a variable-area record section these events will often have the appearance of uniformly spaced stripes which entirely obscure the reflections from subbottom interfaces which are desired. Figure 6-29 (p. 192) illustrates such a ringing record.

Reverberation effects can be eliminated by proper processing of the records, making use of digital filter programs that effectively cancel the undesired filtering effect of the water layer. This process, known as *inverse filtering* or *deconvolution,* will be discussed in the next chapter.

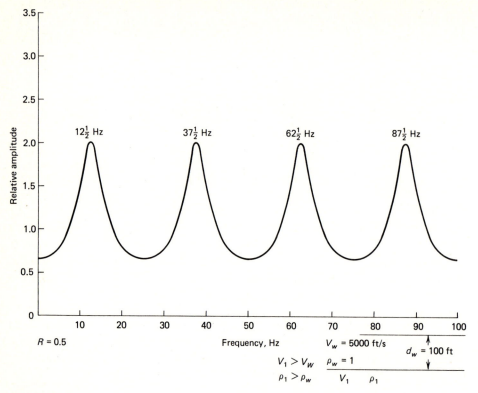

FIGURE 5-18
Frequency spectrum for reverberation of pressure pulse in water layer 100 ft
deep. (*After Backus.*[15])

5-7 POSITION LOCATION FOR MARINE SURVEYS

The precise determination of position coordinates is an important aspect of all
geophysical surveys, whether on land or in marine areas. In most land work,
long-established surveying procedures have generally been adequate, providing all
the precision needed to meet geophysical requirements. For marine operations,
however, no position-location techniques usable over areas out of sight of land were
in existence until World War II. To meet wartime and other requirements for
guidance of ships and aircraft, a number of radio navigation systems were developed
in the 1940s. Among these were reflection radar, Loran, Shoran and continuous-wave
phase-comparison methods such as Decca and Raydist.

The earliest techniques to be employed in radio position location were reflection
radar and Shoran. The continuous-wave methods were first applied in petroleum
exploration during the later 1940s, and for two decades they were used more
extensively than any other positioning systems in offshore geophysical surveys,
mainly because of their great range and high accuracy. In 1968, the Transit satellite
system, initially developed by the U.S. Navy and operated by it since 1964, was

first put to use in geophysical exploration work and it is being employed to an increasing extent in seismic surveying.

Line-of-Sight Methods

Radar reflection and Shoran both involve the direct propagation of pulsed microwaves along a line-of-sight path. For this reason, they are applicable only at relatively short distances from shore. The range limitation may not be too restrictive in areas where transmitters can be located on high mountains close to the coastline.

Radar can be used in two ways: (1) A transmitter on board the ship can send out radar waves that are returned by reflection from passive targets at known locations near shore or from buoys at predetermined positions in the water. Measurement of the times required for waves to make the round trip from the same source to a number of targets make it possible to determine the ship's location quite precisely, as the velocity of the radar waves, like that of all electromagnetic radiation $(3 \times 10^{10}$ cm/s), is well known. (2) The other approach is to set up two radar transmitters at known positions on shore and to time the reflections from the ship, feeding the times by radio into a shipborne plotter, which produces a continuous track of the ship's position.

Shoran makes use of two transmitting stations on land, but there is a third station located on the ship. Pulses are sent out from the ship's unit which are received by the land stations. After a brief but precisely known delay, a transmitter at the land station broadcasts the signal back to the ship. The time required for each round trip less the delay determines the distance from the ship or plane to the corresponding fixed point on land. The intersection on a chart of the radii representing ranges from the two stations on a chart gives the location of the ship.

Each pulse is transmitted at a frequency between 210 and 320 MHz and has a duration of less than 1 μs. With such high frequencies, an accuracy of ± 25 ft is possible under favorable conditions. For a transmitter 1000 ft above sea level (as on a mountain near shore), the range is about 40 mi.

Continuous-Wave Systems

Continuous-wave systems give results almost as accurate as those of Shoran, but they are usable over much greater distances from shore. The primary continuous-wave systems are Decca, Raydist, and Lorac, all of which are quite similar in design and performance. Because the systems involve the use of continuous waves, it is not possible to measure absolute distances, only differences in distance between one position and another. These differences are determined by counting the number of wave cycles, or lanes, in the standing-wave pattern surrounding each shore station that are crossed as the ship moves from one position to another.

Hyperbolic systems The hyperbolic technique involves the use of three shore stations. *Differences* (usually expressed as integers) in lane count between two adjacent pairs of stations are plotted on a map as a family of hyperbolas for each pair. The differences as measured for the respective pairs are located on the map. The actual position is plotted at the intersection point of the respective hyperbolas.

Figure 5-19 shows a family of curves (the heavy lines) representing such locations, one lane apart. Each of these lines is a hyperbola, the locus of a point whose distances to two other fixed points, called foci, differ by a constant. In this case the foci are the two transmitter locations A and B. If the output of a receiver tuned to the frequency of the two transmitters is fed to a phasemeter, the reading will be zero along any of the hyperbolas, but elsewhere it should have some value between 0 and 360°. The phase angle will be proportional to the distance of the point from the nearest line of zero phase difference. If the phase can be read to the nearest degree, the position can be established to $\frac{1}{360}$ times the distance between adjacent hyperbolas in the vicinity of the point. At 1000 kHz, a wavelength corresponds to about 1000 ft, and the distance between adjacent hyperbolas, which is a half wavelength, will be about 500 ft. In this case the interpolation could be made with a theoretical accuracy of the order of 1 ft. Because the propagation paths of the radio signals are not actually straight lines, and because the propagation velocity of radio waves is affected by air temperature and humidity, accuracies are generally of the order of 300 to 500 ft, but a given position will be found to have the same hyperbolic coordinates within about 100 ft on repeated readings on different days.

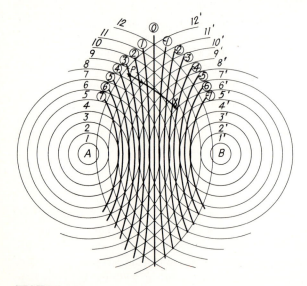

FIGURE 5-19

Contour map showing lines of zero phase difference in vicinity of two sources (A and B) of synchronized radio waves of same frequency. Light lines are successive waves spreading out from source at any time. Heavy lines are hyperbolas representing path-length differences of integral wavelengths as labeled. Ship moving from P to X must cross six such hyperbolas.

Lane identification Although a phasemeter will enable one to interpolate precisely between adjacent lines of zero phase difference, it cannot distinguish between the lines themselves and thus cannot locate positions absolutely. With continuous waves generated at the source, there is no way that the phasemeter can identify individual waves. If, however, the receiver is at a known location at one point along one of the hyperbolas, it is possible to *count* cycles as the shipboard receiver moves away from the known position. Suppose the receiving ship starts at point P (Figure 5-19), known to be on hyperbola 3 (which represents a difference of three cycles between the wave from B and that from A), and moves eastward as shown. Before reaching point X it crosses six lanes, which means that the phasemeter would make six revolutions (up to hyperbola -3) and then a fractional part of a revolution equal to the distance from line -3 to point Z divided by the distance between lines -3 and -4. If the phasemeter were linked to a counter, such as the mileage indicator on a speedometer, which is set at 3 when the receiver is at position P, it will shift six digits in the negative direction and read -3 at the end of the run. The fractional distance between -3 and -4 can be read directly from the phasemeter dial.

This does not give the position in itself but only one locus. Another pair of transmitters (one of which is usually common to both pairs) will give the position on a second hyperbola in the same way. The two hyperbolas can intersect in two points, but one of these is usually so located that it can be readily ruled out, leaving the position specified uniquely within the limits of accuracy discussed. Figure 5-20 illustrates how three transmitters give two sets of hyperbolas from which actual positions are located on a specially designed chart. The common transmitting station M is called the *master*, and the other two, A and B, are called *slaves*. Assume

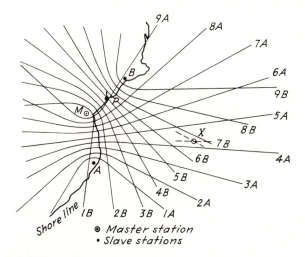

FIGURE 5-20
Use of three transmitting stations (one master, two slaves) in continuous-wave position location.

● TRACKING STATION
◉ TRACKING AND INJECTION STATION
◈TRACKING, INJECTION AND COMPUTATION STATION

FIGURE 5-21
Navigation satellite monitoring and update stations.

that the ship containing the receivers starts from a dock at point P, located at the intersection of lane 9 for station pair MB, the respective meter dials being initially set at these values. At position X on the map the B counter will read 6 and the B phasemeter 300°, while the A counter will read 5 and the A phasemeter 270°.

Satellite Navigation

The U.S. Navy Navigation Satellite System makes use of a number of navigation (or Transit) satellites operating in polar orbit (as of the writing of this book, the number is 5) along with tracking stations and monitoring or injection stations to check and maintain the accuracy of the system by frequent correction of the information each satellite transmits. The configuration of a typical orbit is shown in Fig. 5-21. The satellites are in trajectories that are measured from the tracking stations whenever the satellite comes within range. The tracking information is transmitted to a computing center, where it is used for calculating constants, updated as frequently as twice a day, which are injected into the satellite by radio. The satellite continuously retransmits this orbital information on two carrier frequencies, 400 and 150 MHz. A review article by Dobrin[16] presents an elementary discussion of the theory and operation of satellite navigation systems.

Principles of position location from satellite data   The receiver on board the ship, which is completely passive, picks up signals from the satellite which provide the information necessary to determine the height, latitude, and longitude of the satellite at any time of day. Once the position of the satellite is established as a function of time, it is only necessary to determine the distance or *range* of the satellite at the instant of closest approach to determine the ship's position. This is found by observing the doppler shift in the frequency of the radio signals broadcast

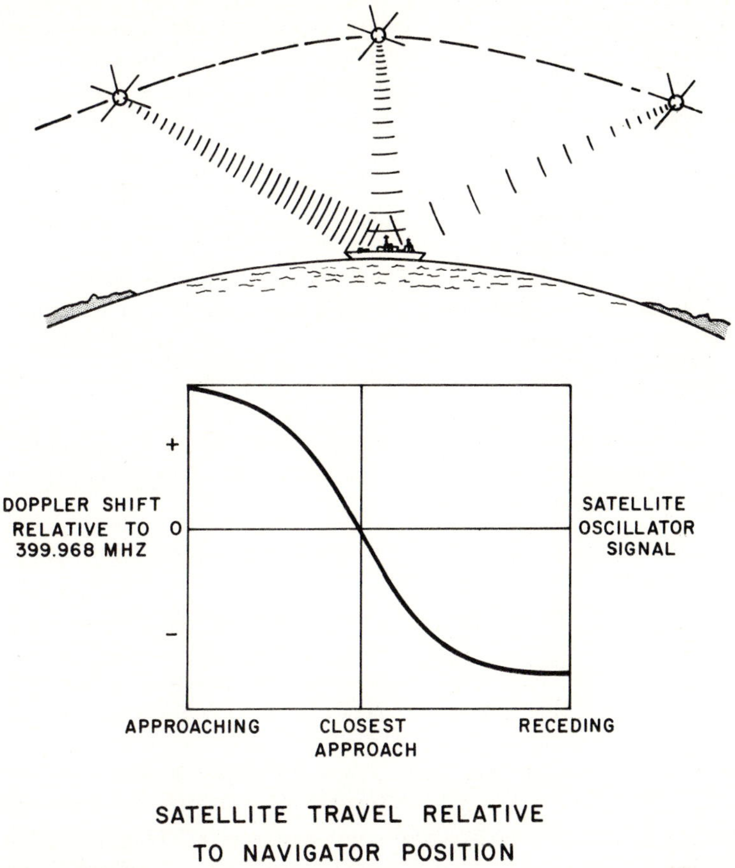

SATELLITE TRAVEL RELATIVE
TO NAVIGATOR POSITION

FIGURE 5-22
Use of doppler-shift characteristics to determine range of satellite at closest
approach. Frequency shift is plotted versus time.

from the satellite. The actual shift varies with time as shown in Fig. 5-22. The
instant at which the doppler shift changes sign from positive to negative gives the
time of the satellite's closest approach to the ship. The sharpness of the slope of this
curve is a function of the *distance of closest approach*, so that the distance can be
computed from the variation of the doppler shift with time. The trajectory constants
contained in the beamed signal are stored in a small shipboard computer (such as
PDP8-I) which calculates the distance of the satellite at its closest approach by use
of the measured frequency-time relationships, which are also stored, and then
determines the position coordinates of the ship by applying the updated satellite
trajectory constants. The technique is described by Talwani et al.[17] and Guier.[18] The
latitude and longitude of the ship are printed out by a special typewriter connected
to the computer output.

Frequency and precision of fixes For accurate position data, information should be used only from satellite passes in which the elevation of the satellite at closest approach is within the range from about 20 to 70° above the horizon. If six satellites are in orbit, there should be about 24 passes per day falling within this range at middle latitudes. This means that there should be an average of one recording every hour. The intervals between usable passes will vary depending on the relation, which tends to be random, of the trajectories for the different satellites that are in orbit at the same time. Intervals could range from a few minutes to several hours. An accuracy of readings within about 300 ft can generally be obtained provided the ship's velocity is accurately known. This is usually measured in shallow water (less than 1000 ft) by doppler Sonar and in deep water by inertial guidance, both discussed later in this chapter.

Advantages and disadvantages of satellite positioning There are a number of advantages of satellite navigation over other positioning techniques such as those making use of continuous waves:

1 There is no limitation in the applicability of the system because of distance from shore.
2 Atmospheric disturbances do not restrict operations to daylight hours.
3 No shore installation is needed.
4 Because only a satellite receiver and small computer are required by the user, the overall expense is less than for other positioning systems employed in operations far from shore.

The principal disadvantages of satellite navigation are:

1 Within 50 to 100 mi from the coast the uncertainty in positioning is usually greater with satellite navigation than with range-measurement techniques. The difference decreases, however, with increasing distance from shore.
2 As readings can be obtained only at intervals of minutes to hours, additional hardware is required for interpolating positions between readings to track the ship's course continuously.

Methods used for such interpolation will be discussed in the following paragraphs.

Interpolation between satellite fixes Interpolation between satellite positions was initially carried out by dead-reckoning procedures. Starting at a satellite fix, the navigator plotted the ship's position on the map from the best velocity and heading information he could get. When the next satellite fix was plotted, there would generally be a discrepancy between the position of the fix on the map and the position along the trajectory plot for the time the fix was made. On the basis of such differences a least-squares iterative adjustment was made in the trajectory plot that would correct the velocities to those giving the best agreement at all fix positions.

In some geophysical surveys on which satellites were used for position location, velocities were determined by devices that measure the ship's speed with respect to

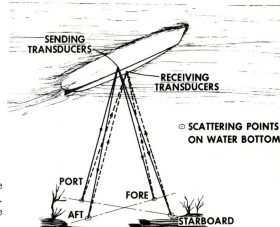

FIGURE 5-23
Use of doppler Sonar to determine
velocity of ship with respect to bottom.
View from below hull of ship. (*The
Marquardt Company.*)

the water. When there are water currents, the velocities thus measured will deviate
from the actual ship speed by the amount of the current. Among the current meters
used for measuring water speed were the *pit log* (incorporating a pitot tube), the
spinner log, and the Electrolog,* which injects electric current into the water
between electrodes trailed from the ship and then measures the ship's speed through
the water by determining the rate at which the magnetic field generated by the
current is cut by a detecting coil trailed along with the electrode assembly.

More accurate position information can be obtained between fixes if velocities
can be measured with respect to the bottom itself. If such an arrangement is used,
water currents do not affect the measurement. The only system which measures
absolute speed referred to the bottom is doppler Sonar, which beams very high
frequency (150- to 300-kHz) continuous-wave sound signals to the bottom at an
angle with the vertical and measures the doppler shift in frequency (due to the
ship's motion) in the waves scattered back to the ship. Four beams are sent to the
bottom, as illustrated in Fig. 5-23. The fore and aft shifts are averaged to determine
the component of velocity along the ship's axis, and the port and starboard shifts
are averaged to obtain the athwartships component of velocity.

Because of attenuation of the high-frequency sound waves, the doppler Sonar
system will not operate reliably in water deeper than about 1000 ft. Scattering layers
in the water resulting from a high density of marine life, such as plankton, sometimes
limit the depth range even further. The accuracy of distance measurement is often
rated as 0.2 percent. This means that in water less than 1000 ft deep a ship moving
at 6 knots should not go more than about 100 ft off the course plotted from the
output of a doppler Sonar system during a typical 90-min interval between satellite
fixes. In water between 600 and 1000 ft deep, the error may be somewhat greater
than it is at depths less than 600 ft. A system employing the satellite receiver and
ocean-bottom doppler Sonar in relatively shallow waters (1000 ft) has an accuracy
of 200 to 500 ft and is considered at least comparable to the best radio-positioning
systems that can be used beyond line-of-sight ranges.

* Trademark of Chesapeake Instrument Company.

Where the water is too deep for doppler Sonar to operate, inertial-guidance systems can maintain positions between satellite fixes with an accuracy of about 500 to 650 ft. Such systems, originally developed for aircraft navigation and adapted by Litton Industries for geophysical surveys, are highly complex and require sophisticated computer software to meet exploration requirements.

The integration of satellite-fix information with that from the interpolation system (such as doppler Sonar or inertial guidance) is generally carried out by a computer program with feedback elements that enable the inaccuracies in the data from the satellite to be reduced by appropriate application of the interpolation data and vice versa. The most probable trajectory is plotted directly from the computer output. Upon return to port, statistical error-correction and interpolation routines will generally produce track plots to an absolute accuracy of 100 to 200 ft.

REFERENCES

1. Kramer, F. S., R. A. Peterson, and W. C. Walter: "Seismic Energy Sources, 1968 Handbook," Bendix United Geophysical Corporation, Pasadena, Calif., 1968.
2. Willis, H. F.: Underwater Explosions: Time Interval between Successive Explosions, *Br. Admir. Rep.* WA-47-21, 1941.
3. Rayleigh, Lord: On the Pressure Developed in a Liquid during the Collapse of a Spherical Cavity, *Phil. Mag.*, vol. 34, pp. 94–98 1917.
4. Lay, Roy L.: Repeated P-Waves in Seismic Exploration of Water Covered Areas, *Geophysics*, vol. 10, pp. 467–471, 1945.
5. Worzel, J. L., and Maurice Ewing: Explosion Sounds in Shallow Water, in "Propagation of Sound in the Ocean," *Geol. Soc. Am. Mem.* 27, 1948.
6. Knudsen, W. C.: Elimination of Secondary Pressure Pulses in Offshore Exploration, *Geophysics*, vol. 26, pp. 425–436, 1961.
7. Lavergne, M.: Emission by Underwater Explosions, *Geophysics*, vol. 35, pp. 419–435, 1970.
8. Anderson, V. C.: "Wide Band Sound Scattering in the Deep Scattering Layer," *Scripps Inst. Oceanogr. Publ.* 53–36, 1953.
9. Knott, S. T., and J. B. Hersey: High Resolution Echo Sounding Techniques and Their Use in Bathymetry, *Deep Sea Res.*, vol. 4, pp. 36–44, 1956.
10. Hersey, J. B.: Continuous Reflection Profiling, pp. 39–46, in "The Sea," vol. 3, Wiley, New York, 1963.
11. Giles, Ben I.: Pneumatic Acoustic Energy Source, *Geophys. Prospec.* vol. 16, pp. 21–53, 1968.
12. Staron, Philippe: Vaporchoc: Utilization of Steam as a New Seismic Source, *Offshore Technol. Conf., Houston, 1973, Prepr.* vol. I, pp. 487–497.
13. Goldberg, Seymour: The Seismovac Monopulse Pneumatic Rebound Source, *Geophysics*, vol. 37, pp. 174–177, 1972.
14. Ewing, John, R. Leyden, and M. Ewing: Refraction Shooting with Expendable Sonobuoys, *Bull. Am. Assoc. Petrol. Geol.*, vol. 53, pp. 174–181, 1969.
15. Backus, M. M.: Water Reverberations: Their Nature and Elimination, *Geophysics*, vol. 24, pp. 233–261, 1959.

16. Dobrin, Milton B.:   Satellite Navigation and Position Determination for Offshore Petroleum Exploration and Drilling, *Proc. 8th World Petrol. Congr.*, vol. 3, pp. 391–402, Applied Science, London, 1971.

17. Talwani, M., J. Dorman, J. L. Worzel, and G. M. Bryan: Navigation at Sea by Satellite, *J. Geophys. Res.*, vol. 71, pp. 5891–5902, 1966.

18. Guier, W. H.:   Satellite Navigation Using Integral Doppler Data: The AN/SRN-9 Equipment, *J. Geophys. Res.*, vol. 71, pp. 5903–5910, 1966.

6

ENHANCEMENT OF SEISMIC REFLECTION DATA IN PROCESSING CENTERS

The greatest impact of the electronic computer upon seismic prospecting has been in the processing of the seismic data recorded in the field. Such processing had its beginnings when field systems were introduced in the early 1950s for reproducible recording on analog magnetic tape, but it was not until modern high-speed digital computers became available that the full potential of processing techniques could be realized for enhancing the quality and usefulness of seismic field data.

The basic objective of all seismic processing is to convert the information recorded in the field into a form that most greatly facilitates geological interpretation. The data initially recorded on magnetic tape (digital or analog) are transformed in the processing center into a corrected record section comparable in many ways to a geological structure section. One object of the processing is to eliminate or at least suppress all noise (defined here as signals not associated with primary reflections, particularly those which might obscure or be confused with such reflections). Another is to present the reflections on the record sections with the greatest possible resolution.

Reproducibly recorded data and playback facilities have made it economically feasible to implement new field recording systems designed for suppressing noise at its source. Many of these systems require processing of the data as an integral part of the data acquisition. Among these are common-depth-point recording, Vibroseis, and Dinoseis, as well as many marine systems, all described in the preceding two chapters. Most nondynamite sources require summing of signals, which can be carried out either on special digital or analog compositing units in the field or, at a later stage, in a playback center.

Data processing covers filtering, trace correction, stacking, compositing, velocity analysis, true-amplitude registration, migration, and plotting. Our primary concern in the present chapter will be with data enhancement by various types of filtering. The principle of stacking was introduced in Chap. 4 in connection with common-depth-point recording. Velocity determination and migration will be considered in Chap. 7, which deals with conversion of reflection times to structures. Processing of reflection data to preserve proper amplitude relations is discussed in Chap. 10, which is on direct hydrocarbon detection.

In this chapter we shall concentrate on digital filtering techniques although we shall begin with a brief discussion of analog processing operations.

6-1   DATA PROCESSING WITH ANALOG SYSTEMS

Most of the processing operations now carried out on digital computers were done, if only on an experimental basis, with analog systems before suitable digital approaches became available. Analog equipment is still used for initial processing of data recorded in analog form, particularly where there would be long delays in transporting the data to digital playback centers. Shifting of times and compositing of signals, particularly for common-depth-point stacking, were routinely accomplished by mechanical, electrical, and optical means long before digital systems were available for such operations. Movable pickup heads on drums holding analog tape were widely used for time shifting. These mechanisms allowed not only static corrections (compensating for weathering and elevation differences) but also corrections for normal moveout due to angularity of the ray path. The moveout corrections required a continuous shifting of the heads with respect to one another as the time along the trace increased. To allow a longer time interval over which shifting could be carried out, magnetic delay lines with 800 ms or more excursion of the recorded signals were introduced.

Compositing   Signals can be composited on analog playback systems in either of two ways: (1) A pair of drums is used, and the signals are transformed successively from one to the other as the summation proceeds. (2) An analog compositing system widely used in the 1960s makes use of both narrow heads and wide heads, the width of the latter being 10 or 12 times that of the former. The signals to be composited are recorded side by side on the tape by the narrow heads. The compositing is done by sweeping the 10 or 12 narrow tracks recorded on the tape by the narrow heads with a single wide head covering them all. The output of the wide head is impressed on another tape, usually wound around the same drum as the tape registering the recorded signals.

Filtering   During the period between the introduction of reproducible seismic recording on analog magnetic tape and the advent of digital techniques, it was customary to play the recorded data back through analog filters both in the field and in playback centers. Most analog equipment used for such filtering consisted of conventional electrical elements for high-pass or low-pass frequency cutoffs. Filters of this type are discussed in Chap. 3 (pages 66 to 67).

A more flexible type of analog filter, which was used on a production basis in several large playback installations, operated in the time domain. Although equipment of this type has been superseded by digital systems, the principles involved in its operation are the same as those that enter into time-domain digital processing. Since use of the time domain in analog filtering is probably easier to visualize than in digital processing, it appears worthwhile to introduce the concepts involved by describing how they can be applied in analog systems.

To express the action of a filter in the time domain, we must transform its frequency-response curve into a function of time called its *impulse response*. This is the output signal from the filter when the input is a spike having a time duration too short to observe on conventional recording modes. The output signal on the right side of Fig. 6-1a is the impulse response for the filter shown at the center. It is evident that the filter has the effect of spreading out the spike and converting it into an oscillating pulse of finite duration.

Even if the input signal is extended in time, we can still use the impulse response of the filter to predict the output. To do this, as shown in Fig. 6-1b, we approximate the input as a succession of uniformly spaced spikes, each having a height corresponding to the signal amplitude at the instant when it occurs. Every spike can be looked upon as independent of all the others so that the output it generates is simply the impulse-response curve with an amplitude proportional to its height and a time for its beginning equivalent to the time when that spike occurs on the signal. The amplitude of the input thus acts as a weighting factor to be applied to the impulse-response function. If the spike is negative (pointing downward), the output signal corresponding to it will be negative also. The actual output is closely approximated by the sum of the output for the individual spikes. As the interval between the sampled spikes gets smaller, the sum of the weighted impulse responses approaches the true output more and more closely until in the limit they coincide.

It is easy to demonstrate that the same output signal would be obtained if the impulse-response function were sampled instead of the input signal and the samples were used as weighting factors for the respective amplitudes of the individual input functions to be shifted in time and added.

Figure 6-2 illustrates how such weighting and shifting of the impulse response might be done with analog playback equipment. A single-channel magnetic tape is stretched around a drum having a fixed recording head which impresses the input signal on the tape. A series of reading heads is spaced at uniform intervals around the drum to pick up the signal thus recorded. The output of each of these heads is passed into a potentiometer with an attenuation set at the value proportional to the impulse response of the desired filter at the time corresponding to the head's

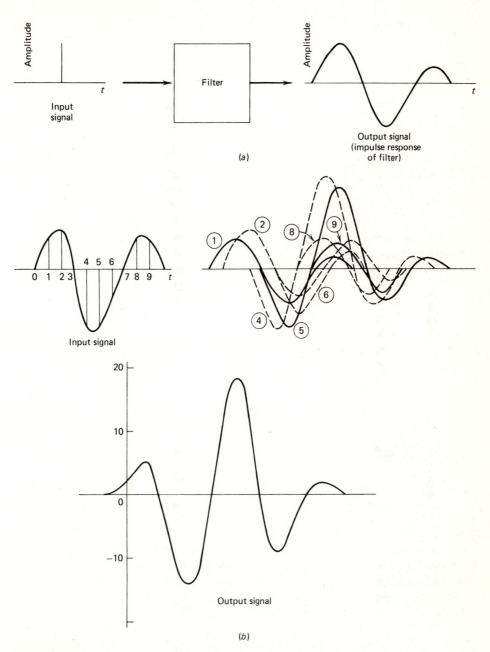

FIGURE 6-1

(*a*) The impulse response of a filter as the output signal when the input is a spike. (*b*) Use of impulse-response characteristic for filtering input signal of finite duration. Numbers on curves in diagram with superimposed signals represent sequence of corresponding spikes on left.

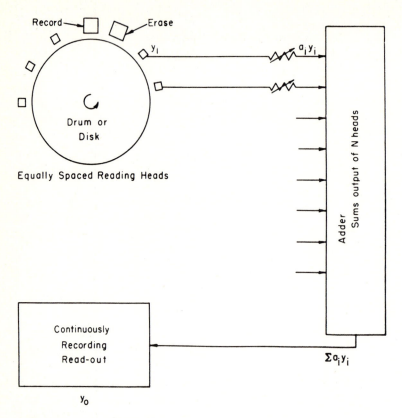

FIGURE 6-2
Filtering of signal on analog tape with delay line making use of equally spaced
reading heads with adjustable sensitivities. (*Jones et al.*[1])

position on the drum. If the weighting function is negative for a pickup head, a
switch is introduced that will reverse the sign of the output. The weighted signals,
properly shifted in time by amounts corresponding to the head separations on the
drum, are added together, the sum being the output signal.

Figure 6-3 illustrates how the input signal can be used as a weighting function
in filtering by progressive shifting of the impulse response. The output is obtained
by summation in a mixing unit of the weighted and shifted signals.

To change the filter characteristics it is only necessary to shift the weighting
factors to different values (corresponding to a different impulse response) by adjustment of the potentiometers which govern the amplitudes and the switches
which govern the signs for each delayed signal.

To filter a conventional 24-channel analog tape by such a means would require
an impractically large number of heads. It is therefore most expeditious to transfer
the channels sequentially to a single-channel drum with multiple reading heads.
Such sequential handling of the data channel by channel is intrinsically slow and

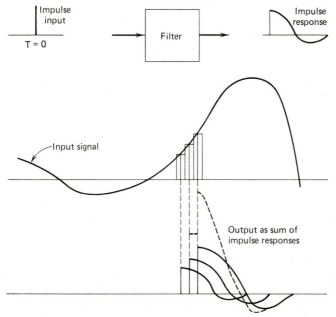

FIGURE 6-3
Superposition of impulse response weighted by input signal values and time
shifted as means of time-domain filtering. (*Jones et al.*[1])

awkward when large quantities of data have to be processed. Actually the elaborate
multiple-drum installations used for such processing in the early 1960s were aban-
doned as soon as digital systems that would perform the same function became
available.

6-2 OPERATING CHARACTERISTICS OF DIGITAL COMPUTERS USED IN SEISMIC-DATA PROCESSING

Because of low speeds and storage capacities, the earliest computers used for processing seismic data were limited to such relatively simple operations as time shifting
and compositing of sampled signals. The higher-speed computers, which became
available in the early 1960s, e.g., the IBM 360 series, the CDC 3200, and the TIAC,
besides handling all operations for which the earlier computers were used, could
perform filtering operations in real time, allowing much greater flexibility and
effectiveness than the conventional electric and time-domain analog filters previously employed.

Only data in digital form can be processed with digital computers. If the initial
recording is digital, the signals on the tape are demultiplexed and fed into computer
storage directly. If the initial recording is on analog tape, the analog signals must

be digitized by a suitable converter before being put into storage. Once the data are introduced into the computer, the processing procedures are the same regardless of whether the recording in the field was analog or digital.

The arithmetic operations computers actually perform are limited to addition and subtraction. Transfer of numbers from one storage position to another is easily programmable. Operations such as multiplication and division are carried out by a combination of addition and shifting of storage locations just as is in manual multiplication.

Time correction can be carried out by simple shifting of the number representing the amplitude of a signal at a sampling instant from a storage position representing the time of the sample as actually recorded to a position representing the corrected time. Compositing, like stacking, is accomplished by adding all signals in each storage position corresponding to the same time on the output record. Such operations are easily programmed for all electronic computers.

Seismic signals are generally sampled at either 2- or 4-ms intervals, the latter being more common with marine data. The samples for each trace are extracted from the multiplexed signal, stored on the digital tape by a process called *demultiplexing*, and put into computer storage as a series of *words** having the sequence of the original recording. If, for example, 2-ms sampling were being used, the 743d word put into storage for a particular channel would represent the amplitude of the signal at a record time of 1.484 s. (The first sample is for zero time, the second for 0.002 s, etc.) An instruction to shift the information stored for that time to another time 8 ms later because of a time correction, e.g., one needed to compensate for a change in elevation, would be translated by the computer into a shift of the stored number to the storage position for the 747th sample on the trace.

But it is in the filtering of the seismic signals that digital computers display their most unusual capabilities. In such filtering, demands are placed upon the electronic computer with regard to speed of operation and storage capacity that are exceeded in few if any other areas of technology. This processing almost always involves such a rapid rate of addition and high-speed multiplication (which is actually multiple addition) that the capabilities of general-purpose computers are often overtaxed and special-purpose computers are generally needed.

The conceptual basis for digital filtering of seismic data was brought to its present state of usefulness during and after World War II by mathematicians and electrical engineers who, as Silverman[2] has pointed out, were faced with the problem of improving the signal-to-noise ratio for radar, radio, and telephone systems. In all these applications, a signal of known form is introduced into the system at the input end and must be extracted from noise at the output end. In seismic prospecting, the problem is more complex. The input signal, a spike at the shot end of the seismic wave path, is so changed by the earth materials through which it travels that one

---

* A word in this sense is the (usually binary) number for the amplitude of the seismic signal at a particular instant of sampling.

can seldom if ever identify the input signal in the waveform of the output. Actually, the changes in the signal resulting from the earth's own transmission and reflection characteristics (both associated with the geology of the subsurface) are what we are trying to observe. The earth thus acts as a filter, and we can look upon a seismic trace as the output of this earth filter, the input being the signal (a simple pulse except in the case of Vibroseis) that is introduced by the energy source. In addition to filtering, computers carry out many other operations in seismic data processing, among them velocity and relative-amplitude determination, frequency analysis, and migration (presentation of the reflections in their true positions on the cross section). All these applications will be discussed in subsequent chapters.

6-3   PRINCIPLES OF DIGITAL FILTERING

The main objective of filtering in seismic reflection work is to remove undesired signals (collectively referred to as noise) from the record, leaving, ideally, only primary reflections having geological meaning. Two properties of the noise can be used as a basis for separating it from the signal. One is frequency; the other is apparent velocity. Only frequency filtering can be carried out with data on a single channel. Filtering based on apparent velocity requires the use of multiple channels.

The earliest reflection surveys required frequency filtering (then carried out in conjunction with the field recording) to remove low-frequency ground roll (surface waves) from the records so that reflections, generally of higher frequency, would not be obscured. This is still necessary, whether done in the field or in a playback center subsequent to the recording. Other types of noise can also be rejected by frequency discrimination, but because overlap is common between the frequency ranges covered by the noise and the reflection signals, the process often requires intricate control of the discrimination characteristics. Only digital filtering gives the flexibility necessary for optimizing filter performance in such cases. The need for sharp discrimination can be demonstrated where there is a large amount of scattered incoherent noise mixed with a reflection signal, as illustrated in Fig. 6-4. Analog filters generally have such gentle slopes that it would be difficult to extract the reflection from the noise by frequency filtering, but the nearly vertical slope which can be obtained from a properly programmed digital filter does allow such discrimination.

The techniques of manipulating the waveforms of seismic signals by digital computers to maximize the desired information and minimize noise were developed from information theory, a rather new and mathematically sophisticated branch of electrical engineering. To appreciate how digital filters work, it is not necessary to be familiar with all aspects of this theory; yet one must understand its basic concepts. Most of them can be explained intuitively without the use of advanced mathematics. Several elementary discussions of digital-filtering principles are available in the literature which cover the subject in more detail than is possible here.

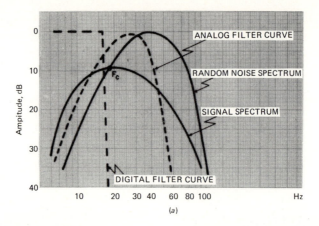

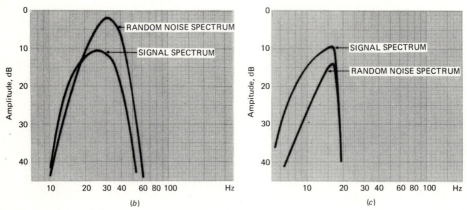

FIGURE 6-4
Comparison of performance of digital and electric analog filters designed for
extracting reflections from high-frequency scattered noise: (*a*) before filtering,
(*b*) after optimum analog filtering, and (*c*) after optimum digital filtering.

Elementary treatments by Silverman,[2] Robinson and Treitel,[3] Peterson and
Dobrin,[4] Anstey,[5] and Lindseth[6] are particularly recommended for those who wish
to explore digital processing further.

In our discussion earlier in this chapter of time-domain filtering by analog methods we showed how the action of a filter can be described by its impulse response as
well as by its frequency-response curve. The impulse response is in the *time domain*,
as is the input signal itself. The frequency-response curve conventionally used by
engineers is in the *frequency domain*. Both modes of expression are functions of
each other; i.e., if one is known, the other can be derived from it.

In digital filtering, either domain can be employed, but both the seismic signal
and the characteristics of the filter must generally be converted into the same form.
If the operation is to be in the time domain and only the frequency characteristic

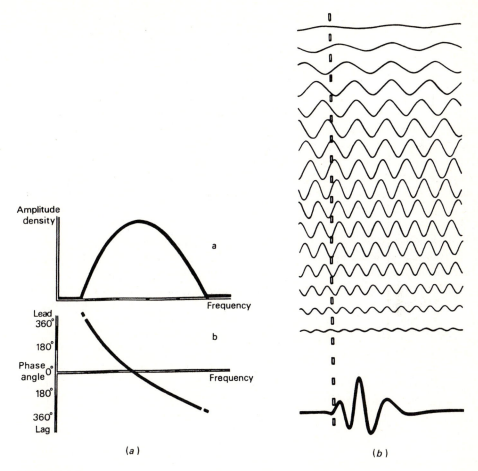

FIGURE 6-5
Fourier representation of seismic waveform: (*a*) amplitude and phase spectra;
(*b*) synthesis of the waveform (at bottom) by summation of sinusoidal waves
having amplitudes and time shifts corresponding to these spectra. (*From
Anstey.*[5])

of the filter is specified, the frequency curve must be expressed as an impulse response (in the time domain) before the operation can be carried out. *Fourier transformation* provides the physical basis for such conversions from one domain to the
other.

The Fourier Transform

The concept of Fourier transformation is based on the Fourier series, which expresses any function such as a time signal of limited length as a summation of an
infinitely long series of sine waves and cosine waves (see page 35). The wavelength

of the longest wave in the series (the fundamental) is the same as the length of the initial function, and subsequent terms have wavelengths one-half, one-quarter, one-eighth, etc., as great as the fundamental, the number of such terms being theoretically infinite. The component with one-half the wavelength of the fundamental is called the second harmonic, that with one-third the wavelength is called the third harmonic, etc., the amplitude of each term being governed by the waveform of the function to be represented in this way. The Fourier series is ultimately expressed in terms of the amplitude coefficients for each of the harmonic terms as well as phase terms, which give a measure of the displacement, in fraction of a cycle, of the starting point (zero crossing) for each sine or cosine wave in the series. The amplitude coefficients and phase shifts for each term are determined by Fourier analysis, the theory of which is covered in most textbooks on engineering mathematics.

The *Fourier transform* is an integral expression for a Fourier series applied to an infinitely long signal. As the wavelength of the fundamental approaches infinity in this case, the fundamental frequency must approach zero. All harmonics of this fundamental are separated by infinitesimal increments, and the Fourier series, consisting of amplitudes and phases for the successive harmonics, becomes a continuous function. Fourier-transform techniques are used to express a time function as a continuous function of frequency and also to synthesize a function expressed in terms of amplitude versus frequency (a frequency spectrum) into the function of time to which it corresponds.

Transformation of frequency spectrum to time signal Let us now consider how a continuous frequency spectrum can be transformed into a time signal by Fourier methods. Figure 6-5a illustrates an amplitude and phase spectrum, the frequency coverage being limited to a rather narrow range. The individual sinusoidal waves shown in Fig. 6-5b have different frequencies, increments between the successive frequencies being small. Amplitudes for each frequency are least at the top and bottom of the frequency range and greatest in the middle, as would be expected from the spectrum of Fig. 6-5a. At the same time the positions of the peaks for each frequency, when measured with respect to the reference marks indicating a constant time, shift systematically in accordance with the phase spectrum.

The sum of all the sinusoidal waves gives the time signal shown at the bottom of Fig. 6-5b. The process constitutes a Fourier transformation from the frequency function (amplitude and phase) to the corresponding time function.

Transformation of time signal into frequency spectra Let us now apply the principle of Fourier transformation to determine the frequency spectrum of a seismic signal. In the limit we want a continuous spectrum, but we shall start by obtaining components of it at discrete frequencies 2 Hz apart. Figure 6-6a and b illustrates the logic by which the analysis is carried out for one typical frequency, namely 22 Hz. Figure 6-6c to f shows the amplitudes of all the discrete cosine and sine terms calculated over the range of frequencies generally used in seismic reflec-

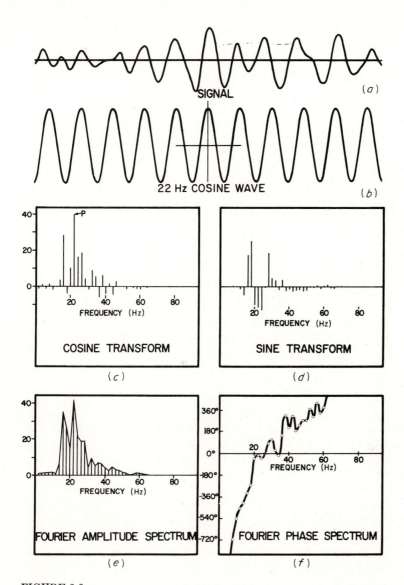

FIGURE 6-6
Fourier analysis of seismic signal by use of cosine and sine waves of different frequencies to derive cosine and sine portions of Fourier spectra: (*a*) signal, (*b*) 22-Hz cosine wave, (*c*) cosine transform, (*d*) sine transform, (*e*) Fourier amplitude spectrum, (*f*) Fourier phase spectrum.

tion work, as well as the resultant amplitude and phase spectra determined from the sine and cosine coefficients plotted in Fig. 6-6*c* and *d*.

To determine the cosine transform of the seismic signal at our 22-Hz frequency, we align it with a 22-Hz cosine wave and determine the degree of fit between the two curves. A window 500 ms long is used, and the origin is taken at the center. The cosine wave should be symmetrically placed with respect to the origin. Corresponding amplitude samples of the signal and the cosine wave are taken every 2 ms and multiplied at each instant of sampling. The sum of these products is plotted in normalized form, each at a sampling point such as P, where the spike observed on the cosine plot gives a measure of the amplitude component at this frequency.

If the seismic signal is itself a 22-Hz cosine wave, all products will be positive and the summation will yield a maximum value at this frequency. If the signal has *no* component of frequency at 22 Hz, there will be a random relation between it and the 22-Hz wave, so that there will be as many negative products (plus times minus or minus times plus) as positive ones (plus times plus or minus times minus). The net contribution to the spectrum will then be zero. In between, the relative contribution at each frequency is indicated by the net positive or negative value after the summation.

The process shown for the 22-Hz reference signal is repeated with cosine signals having all other frequencies between 2 and 100 Hz and then with sine signals over the same range of frequencies. Each sine wave is different from the cosine curve having the same frequency only because its origin comes at zero amplitude (midway between a trough and a peak) instead of at its peak value. The cosine and sine components of the spectrum are plotted as a series of discrete spikes. Conventional cosine and sine transforms can be obtained if the tops of the spikes are connected to produce smooth curves.

The spectrum is represented by two functions, in Fig. 6-6e and f. One is expressed as amplitude versus frequency and the other as phase versus frequency. The amplitude term for each frequency is the square root of the sum of the squares of the sine and cosine terms for that frequency, and the phase term is the angle whose tangent is the sine term divided by the cosine term.

Mathematical expression of Fourier transforms The integral expression for a Fourier transform is in two parts, one for the frequency spectrum $F(n)$ in terms of the time function $f(t)$ and the other for the time function in terms of the frequency spectrum. The frequency function $F(n)$ can be obtained from the time function $f(t)$ by use of the integral

$$F(n) = \int_{-\infty}^{\infty} f(t) \cos 2\pi nt \, dt - i \int_{-\infty}^{\infty} f(t) \sin 2\pi nt \, dt \qquad (6\text{-}1)$$

Time here is designated by t and frequency by n.

Similarly, the transform $f(t)$ of the frequency function $F(n)$ can be expressed by the integral

$$f(t) = \int_{-\infty}^{\infty} F(n) \cos 2\pi nt \, dn + i \int_{-\infty}^{\infty} F(n) \sin 2\pi nt \, dn \qquad (6\text{-}2)$$

The integral containing the cosine term is designated as real, while that containing the sine term is designated as imaginary; hence the use of the coefficient i. This complex representation is a mathematical convenience, as the two terms cannot be added directly but must be added vectorially, the sum having a magnitude and phase angle (proportional time delay) for each frequency which can be determined by following the rules for addition of complex quantities.

Designating the cosine integral in Eq. (6-1) as $C(n)$ and the sine integral as $S(n)$, we can show that the amplitude $A(n)$ of the spectrum at frequency n is

$$A(n) = \sqrt{C(n)^2 + S(n)^2} \tag{6-3}$$

and the phase angle, also a function of frequency, is

$$\phi(n) = \tan^{-1}\frac{S(n)}{C(n)} \tag{6-4}$$

The computation of Fourier transforms is simplified for functions that are symmetrical about the respective axes for which time or frequency are zero. With such functions, the imaginary sine term of the transform disappears because in the integration between $-\infty$ and ∞ each positive product on the right side of the axis is canceled by an equal negative product the same distance on the left side. In this case the sine wave, being an odd function, conforms to the relation

$$f(t)\ \sin 2\pi nt = -f(t)\ \sin 2\pi n(-t) \tag{6-5}$$

An important property of transforms is their complementary nature, which can be expressed as follows:

1   If $F(n)$ is the Fourier transform of $f(t)$, then $f(t)$ is the Fourier transform of $F(n)$.

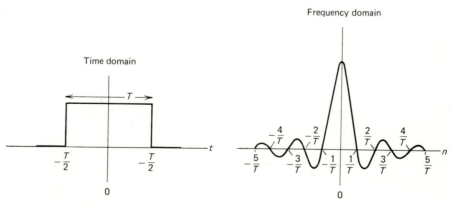

FIGURE 6-7
Fourier transform of a square wave of width T in the time domain.

2   If $f(t)$ is the Fourier transform of $F(n)$, then $F(n)$ is the Fourier transform of $f(t)$.

This reciprocal relationship has widespread application in seismic filtering.

Square wave   To illustrate how a transform is determined for a simple function of time, we shall demonstrate the derivation of the frequency spectrum of a square wave of amplitude A and breadth T which is symmetrical about the zero time axis. Its waveform is expressed by the relations

$$f(t) = \begin{cases} A & \text{when } t \geq -\dfrac{T}{2} \leq \dfrac{T}{2} \\[4mm] 0 & \text{when } t \leq -\dfrac{T}{2} \geq \dfrac{T}{2} \end{cases} \tag{6-6}$$

as illustrated in Fig. 6-7.

As the sine term is always zero for a symmetrical function such as this one and the cosine integral is zero for times outside the range $-T/2$ to $T/2$ bounded by the sides of the square wave, we can use only the cosine term in our integration and change the effective limits in Eq. (6-1) to encompass this range alone, obtaining the frequency spectrum

$$F(n) = \int_{-T/2}^{T/2} A \cos 2\pi n t \, dt = \left[\frac{A}{2\pi n} \sin 2\pi n t \right]_{-T/2}^{T/2} \tag{6-7}$$

$$= \frac{A}{2\pi n} 2 \sin \frac{2\pi n T}{2} = \frac{A}{\pi n} \sin \pi n T \tag{6-8}$$

$$= A T \frac{\sin \pi n T}{\pi n T} = A T \frac{\sin u}{u} = A T \text{ sinc } u \tag{6-9}$$

where $u = \pi n T$ and sinc x is a shorthand notation for $(\sin x)/x$.

Plotting the spectrum $F(n)$ versus n, as we do on the right side of Fig. 6-7, we see that the function becomes zero at values of frequency equal to $1/T$, $2/T$, $3/T$, etc., with the maximum amplitude at zero frequency. The frequency at which the first zero crossing occurs and the frequency intervals between successive crossings are inversely proportional to the breadth of the square wave. The width of the central peak is twice as great as the width of the lobes alongside it (generally referred to as *side bands*), and both widths are inversely proportional to the width of the square wave constituting the original time signal. Thus the longer the duration of the square wave, the narrower the lobes and the greater the concentration of the energy of the wave at low frequencies. Because the frequency is in the denominator of the expression, the peak amplitude for each half cycle of the sinc function decreases with increasing frequency, approaching zero as a limit.

The appearance in a spectrum of amplitude values at negative frequencies equivalent to the amplitudes observed at the corresponding positive values of the frequency may be somewhat confusing to the reader who is accustomed to thinking of frequencies from a practical engineering standpoint. The significance of a negative frequency can best be visualized by considering a wheel which can rotate in either direction, the rates of clockwise rotation representing positive frequencies and those of counterclockwise rotation negative frequencies. Mathematically it can be shown that the spectra are valid for both positive and negative values. It is therefore necessary to plot both, in symmetrical patterns, for proper representation.

The reciprocal relation between the transforms is illustrated for this case in Fig. 6-8, which demonstrates that a time function in the form of a square wave will have a frequency spectrum which is a sinc function, while a time signal in the form of a sinc function has a frequency spectrum which is a square

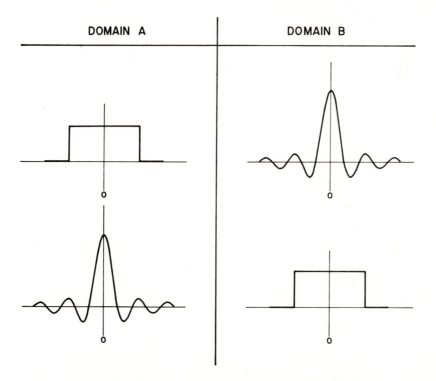

IF DOMAIN A IS TIME, DOMAIN B IS FREQUENCY
IF DOMAIN A IS FREQUENCY, DOMAIN B IS TIME

FIGURE 6-8
Reciprocity of Fourier-transform pairs.   The transformation is the same regardless of the domain of the signal to be transformed.   (*From Peterson and Dobrin.*[4])

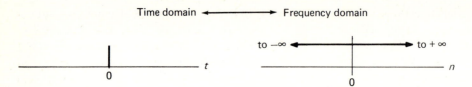

FIGURE 6-9
Fourier transform of a spike or delta function at zero time.

wave. As the sinc function theoretically extends from minus infinity to plus infinity, it is necessary for practical reasons to cut it off at predetermined positive and negative values, whether it represents a time signal or a frequency spectrum. Such cutoff should be at distances from the origin where the amplitude of the function is negligibly small. Later we shall see how the truncation of the sinc function affects the performance of a filter with a frequency characteristic that can be represented by a square wave.

Spike impulse   The Fourier transform of a sharp impulsive signal which has the appearance of a spike on a seismic recording is exceedingly simple. On a plot of amplitude versus frequency it is a horizontal line raised a constant distance above the time axis, indicating a uniform amplitude level for all frequencies (Fig. 6-9). To obtain this transform mathematically is rather difficult. Yet it is intuitively evident that a spike represents a square wave with a width of zero, so that the breadth of the central peak (the distance between the zero crossings nearest the central axis) on the corresponding transform becomes infinite, approaching a horizontal line. If, on the other hand, the square wave has infinite breadth, the sinc function contracts to a spike representing zero frequency.

Cosine wave   When the signal is a wave having the form $A \cos (2\pi t/T)$ (amplitude A, period T), as shown in Fig. 6-10, the transform, or frequency spectrum, is particularly evident. It consists of two spikes symmetrically placed around the zero-frequency axis, one at frequency n equal to $1/T$ and the other at frequency n equal to $-1/T$. As the function is symmetrical about zero time, there is no phase term in the transform.

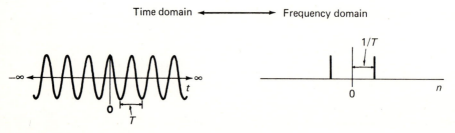

FIGURE 6-10
Fourier transform of an infinitely long cosine wave.

Although the derivation of this transform is somewhat complicated, it is easy to see intuitively that the frequency spectrum of an infinitely long cosine wave having a period of T would have to be confined to the frequencies $n = 1/T$ and $n = -1/T$ (the negative term being required because the cosine wave could be generated by either clockwise or counterclockwise rotation). There is obviously no component of the signal at any other frequency.

Infinite series of equally spaced spikes   Let us now consider a time signal consisting of an infinitely long succession of spikes separated by equal time intervals having a value of T. The transform of this array (as shown in Fig. 6-11) is also a series of spikes separated by a uniform frequency interval which is equal to the reciprocal of the time interval between the spikes in the time function. For symmetrical arrays of spikes that include a spike at zero time, there is no phase term in the transform. The mathematical derivation of this transform is beyond the scope of this chapter.

Let us investigate the physical significance of the frequency function for this case. The spike at zero frequency represents the dc component of the signal which exists because all spikes have positive amplitudes, so that the net flow of the current for a spike representing an electric voltage would be unidirectional and uniform. The two spikes nearest to, and on opposite sides of, the one at zero frequency correspond to cosine waves at the fundamental frequency n_1 which is the reciprocal of the repetition period of the time signal (n_1 being equal to $1/T$). The pair of spikes at frequency n_2, which is twice the frequency of n_1, represents cosine waves having the second harmonic frequency ($n_2 = 2/T$). By the same token, the frequency n_3, which is $3/T$, is the third harmonic of n_1, and so on, there being an infinite number of harmonic-frequency components, all having the same amplitude. We shall consider later what happens when the pulses that are periodically repeated consist not of spikes but of square waves having a finite width. We shall also take up the case of a finite number of uniformly spaced pulses, which has more practical significance than an infinite number.

Phase as a function of frequency   Whenever there are both sine and cosine terms in the transform, as is the case for time functions that are not symmetrical,

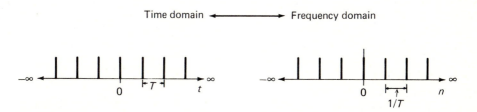

FIGURE 6-11
Fourier transform of a uniformly spaced series of spike signals extending to infinity in both directions.

it follows from Eq. (6-4) that phase as well as amplitude will have to be taken into account in specifying the frequency spectrum. Physically, we may picture the phase as the time (expressed as a fraction of the wave period multiplied by 360° or 2π rad) that a sinusoidal Fourier component of the signal must be shifted to bring its zero crossing to zero on the time axis. This shift in time will be different for different frequencies. For most signals of the type encountered in seismic work a plot of phase angle versus frequency turns out to be a smooth, continuous curve.

To illustrate the nature of the phase-frequency relationship, let us consider a square wave that is shifted in time away from a position of symmetry about the $t = 0$ axis, as in Fig. 6-12. The amplitude term of the transform is unchanged, but the phase curve shows that the phase angle (the tangent of which is the sine transform divided by the cosine transform) increases linearly with increasing frequency. A linear phase curve of this kind means that all frequency components of the signal are shifted by an equal amount of time, as would be necessary if the signal itself were shifted with no change in waveform.

The need for considering phase as well as amplitude is seen in Fig. 6-13. It shows two signals, each made up by adding two infinitely long sinusoidal waves of equal amplitude. One of the two waves has a frequency double that of the other. The only difference is in the phase of the constituent sine waves. The higher-frequency wave has a zero crossing at a time displaced from the zero time axis by a quarter of the wave period in Fig. 6-13a, whereas the crossing is at zero time in Fig. 6-13b. The lower-frequency wave has its zero crossing at zero time in both Fig. 6-13a and b. The difference in the appearance of the waves made up by adding respective Fourier components which have the same waveform but different phase relations is quite striking.

Sine wave Let us now consider the transform of an unsymmetrical time signal, one having the form $A \sin (2\pi t/T)$. It differs in phase from that of the cosine wave previously discussed because the sine wave, as can be seen in Fig. 6-14, is not symmetrical about $t = 0$. It is evident that the phase of the sine wave is shifted 90° with respect to the cosine wave. For a sine wave the spike in the transform at

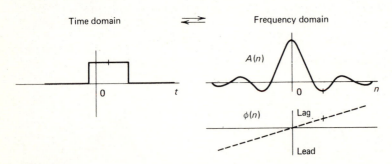

FIGURE 6-12
Phase shift introduced when square wave is displaced from its symmetrical position. (*From Peterson and Dobrin.*[4])

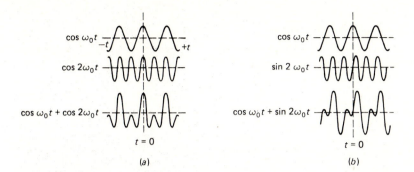

FIGURE 6-13
Effect of phase relation upon signal obtained by summation of two sinusoidal waves: (a) addition of two cosine waves with one having double the frequency of the other; (b) addition of the same waves after 90° phase shift of the higher-frequency event. Note the difference in the synthesized waveforms. (*From Anstey.*[5])

frequency $n = -1/T$ has a sign opposite to that of the spike in the transform at $n = 1/T$.

Significance of Fourier transforms To those whose primary concern with Fourier transforms is in their application to seismic data processing the physical significance of the concepts involved is likely to be of more interest than their mathematical expression. From the standpoint of geophysical applications, the following aspects of Fourier transforms are most important:

1 The Fourier transform of a function of time such as a seismic signal is the frequency spectrum of the signal.
2 The Fourier transform of a frequency function such as a filter response curve is a time function. An example of such a time function is the impulse response of a filter, as shown in Fig. 6-1. It is often more expeditious to program a digital filtering operation when the filter specifications are expressed in terms of time than when they are given in the form of a frequency spectrum. A Fourier transformation makes the necessary conversion to the time domain where the filter characteristics are initially specified in terms of frequency.

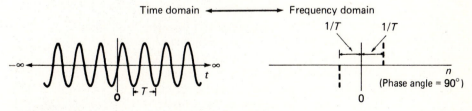

FIGURE 6-14
Fourier transform (amplitude and phase) of infinitely long sine wave.

3 It follows from items 1 and 2 that any function of time can be expressed as a corresponding function of frequency and vice versa.

4 Reciprocity exists between the appearance of a function and its transform regardless of whether the initial function is in the time or frequency domain. The appearance of a frequency spectrum associated with a time signal of a particular shape will be the same as that of the time function associated with a frequency spectrum having the same shape.

Convolution

Earlier in this chapter we introduced the concept of impulse response as the output of a filter whose input is a simple spike. We then showed that the impulse-response function can be used to predict the output of this filter for any input wave. This is done by representing the wave as a series of closely spaced spike samples, each with an amplitude determined by the waveform of the input signal. The output is then built up by superimposing the individually weighted response curves for each spike after shifting the time appropriately for each.

A convenient variation upon the process of superposition thus described is a mathematical operation designated as *convolution*. This operation was originally used to study the effect of an instrument's characteristics (such as its limitations in resolution) upon the measurements it carried out. It is a particularly useful concept when the instrument scans the feature being measured. One example might be the registration, using a scanning photometer, of the density variation along a line crossing a transparent film emulsion on which a picture is impressed. The width of the scanning window would affect the sharpness with which density variations on the original film can be resolved on a recording of the photometer output versus time. The relation between the density function on the film (the input), the response characteristics of the meter, and the photometer output during a scan can be determined by convolution.

Jennison[7] has described convolution in two different ways:

> A convolution may be simply considered as the operation whereby a structure is smeared or spread out by the response or resolution of an instrument or mathematical operation.
>
> The convolution of two functions is the result we would obtain by scanning one function with each element of the other (from minus infinity to infinity) and summing the products at each position of the scan.

The scanning of the film by the photometer in the spatial domain is equivalent to the operation of a filter on a signal flowing through it in the time domain. For this equivalence to be applicable, the frequency characteristic of the filter must be expressed in the time domain (by use of the impulse response), just as the photometer response had to be expressed as a variation of output with *position* of each input element being scanned with respect to the center of the scanning window. To get the output of the filter it is necessary to *convolve* the input signal with the impulse-response characteristic of the filter.

Mathematical expression   Let us express this concept mathematically. Consider two functions of time $f(t)$ and $g(t)$. The first might be a seismic signal. The second might be the impulse response of a filter through which the signal is passed. The output of the filter, which we call the *convolution product* of the two functions $f(t)$ and $g(t)$, can be expressed in the form

$$h(t) = \int_{-\infty}^{\infty} f(\tau)g(t - \tau) \, d\tau \tag{6-10}$$

or

$$h(t) = f(t) * g(t)$$

the symbol $*$ meaning "convolved with."

The physical significance of this mathematical operation is illustrated by Fig. 6-15. The input signal $f(t)$ is plotted in Fig. 6-15a and the impulse response of the

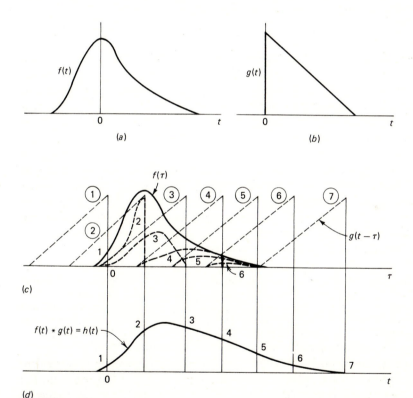

FIGURE 6-15
Convolution of two functions $f(t)$ and $g(t)$ to obtain convolution product $h(t)$. If $f(t)$ is an input signal and $g(t)$ the impulse response of a filter, $h(t)$ is the output signal. (*a*) Plot of $f(t)$ versus t; (*b*) plot of $g(t)$ versus t; (*c*) multiplication of $f(\tau)$ and $g(t - \tau)$ versus τ, products shown by dashes; (*d*) plot of $h(t)$ versus t.

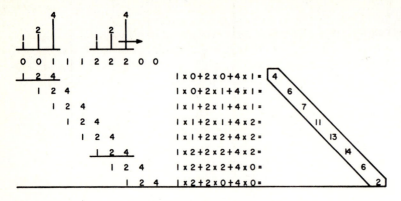

FIGURE 6-16
Convolution of sampled signals (1, 1, 1, 2, 2, 2) and (4, 2, 1). The first time
series represents the input signal, the second the impulse response, and the
eight-term series shown as a diagonal the output. Note reversal of second term
(*From Peterson and Dobrin.*[4])

filter $g(t)$ in Fig. 6-15b. To demonstrate how $f(t)$ is convolved with $g(t)$, we must
reverse one of the functions and multiply the reversed function by the unreversed
function ordinate by ordinate at various values of the time shift between them. In
doing this we change the designation of the time variable t to τ so that $f(t)$ becomes
$f(\tau)$ and $g(t)$, because of its reversal, becomes $g(t - \tau)$. The argument of g indicates that its original zero-time value has been shifted an amount t.

When the two functions are lined up so that their respective zero-time positions
coincide along the time axis, t is zero, as shown for position 2 in Fig. 6-15c. The
separation t of the two zero positions can be either positive or negative. For position
1 for $g(t)$ in Fig. 6-15c, t is negative; for positions 3 to 7 it is positive. For each of
the six time shifts t (including zero) the product $f(\tau)$ and $g(t - \tau)$ is plotted versus
τ using the heavy dashed lines in Fig. 6-15c. The areas under these lines, which correspond to the integral in Eq. (6-10), are plotted versus t after changing the time
variable back to t, as shown in Fig. 6-15d. The value of the integral is the convolution product $h(t)$, which is the output signal.

Convolution of sampled data The approach applied to continuous data in
Fig. 6-15 can be used to convolve sampled data. Figure 6-16 illustrates how a signal
represented by six samples equally spaced in time is convolved with the impulse
response of a filter which consists of three samples. The series representing the impulse response is swept after reversal across the input signal in the steps shown.

The need to reverse the impulse response with respect to the input can be visualized more easily if it is remembered that both functions are conventionally plotted
with time increasing from left to right. The output signal has its beginning when
the earliest part of the input (the left end) interacts with the beginning (also the
left end) of the impulse-response function. The termination of the output signal is
associated with the interaction of the latest part (right end) of the input signal

with the corresponding end of the impulse response. This interaction can be visualized most readily if one of the functions is reversed so that the contacts representing successive increments of time in the outputs can be looked upon as the sweeping (from left to right) of the reversed and unreversed signals, as shown in Figs. 6-15 and 6-16.

At each position of the sweep the products obtained by multiplying each input sample by the appropriate weighting function associated with the impulse response are added to give the output at the instant corresponding to that position. The successive shift, multiply, and add operations provide the basic sequence of steps involved in filtering by convolution, the type most widely used in digital processing of seismic signals.

Another example of convolution   Let us apply the mechanics of the convolution process to a type of measurement well removed from seismic prospecting since it may be easier to visualize than seismic signals passing through a filter. Assume that there is a series of randomly separated lights having different intensities along a line on the ground and that we want to record a profile of light intensity versus distance along the line. We must carry out the measurement with a photometer on an airplane which flies above the line at a height of 1000 ft. The response characteristics of the photometer are such that it measures only light *ahead* of the plane with a sensitivity that varies linearly from zero at a distance 500 ft in front of the plane to a maximum directly below. The layout of the lights and the sensitivity characteristic of the meter as a function of horizontal position of the plane are shown in Fig. 6-17.

As the plane moves, each light on the ground that lies within the cone of sensitivity will contribute an output voltage (from the photometer) that increases linearly with time until the plane reaches a point directly above it; afterwards there is no contribution. The reversal associated with the convolution process can be understood if one considers the contribution of each light on the ground as a function of time on the recording chart. The total reading on the chart at any time is the sum of the individual outputs associated with each light.

To get better resolution of the individual lights on the recording chart that registers the photometer output we would need an instrument with a narrower beam of sensitivity, one that is no wider at the earth's surface than the minimum distance between the lights if all of them are to be resolved individually on the photometer record.

Applications   It is easy to see from the foregoing discussion that the product of convolution, the output signal, will have a length that is the sum of the lengths of the input and the impulse response. This means that the effect of filtering is to lengthen the input signal by a time equal to the duration of the filter operator with which the input is convolved. If the signal is convolved with a spike, it follows from Fig. 6-16 that the signal is shifted without change of waveform by the amount that the spike is displaced from the zero-time reference of the filter operator.

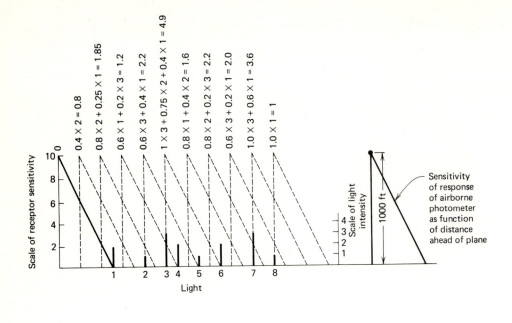

FIGURE 6-17

Recording of light intensities of individual sources on ground from photometer in airplane having triangular sensitivity function illustrated in upper right. Output of chart recorder in plane is the convolution of the light-intensity distribution on the ground with the response function of the measuring system. The numbers above the vertical dashed lines represent the summation of the contributions of each light within photometer range when the plane is at the position of the vertical line.

Relation between Fourier Transformation and Convolution (The Convolution Theorem)

One of the most important and most useful relationships in data processing is expressed by the *convolution theorem*. This establishes the equivalence of filtering by convolution and filtering by multiplication of the signal spectrum by the spectrum of the filter. This interchangeability of filtering approaches is taken advantage of extensively in programming digital computers for filtering seismic data.

If $F(n)$ is the Fourier transform of the function $f(t)$, $G(n)$ is the Fourier transform of $g(t)$, and $H(n)$ the Fourier transform of $h(t)$, the convolution product of

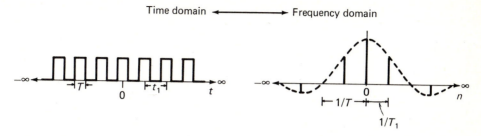

FIGURE 6-18
Fourier transform of an infinite series of square waves of width T with centers separated by interval t_1. Convolution theorem used to obtain frequency function. Dashed line shows how heights of spikes were derived.

$f(t)$ and $g(t)$, we can write

$$H(n) = F(n)G(n) \qquad (6\text{-}11)$$

This means that the function obtained by convolution of any two variables can also be obtained by taking the Fourier transform of each variable, multiplying the transforms, and then taking the transform of the product.

If, instead of time functions, we want to convolve the two frequency functions $F(n)$ and $G(n)$, the reciprocal relation holds. Defining

$$Q(n) = F(n) * G(n) \qquad (6\text{-}12)$$

gives

$$q(t) = f(t)g(t) \qquad (6\text{-}13)$$

where $q(t)$ is the Fourier transform of $Q(n)$.

The convolution theorem is often applied to compute Fourier transforms of functions for which the determination by integration would involve much more labor. The calculation, for example, of the transform for an infinitely long equally spaced series of square waves (shown in Fig. 6-18) would be very tedious by direct calculation. But if we were to build up the series of square waves by convolving an infinite succession of equally spaced spikes with a single square wave of the desired width, we could obtain the transform simply by multiplying the transform for the spikes by that for the square wave (the sinc function shown as a dashed line), the results being indicated by the suite of vertical lines on the right side. The effect of going from the spikes to the square waves is to reduce the amplitudes of the higher-harmonic terms for the spikes (see page 165) in such a way that those beyond a certain order become negligibly small.

Suppose, as another example, that we replace the infinite series of square waves with a finite succession of such events. How is the Fourier transform affected? We can find out, as shown in Fig. 6-19, by multiplying the original infinitely long train of pulses by a square wave with a breadth equal to the length desired for the finite series of square waves. To get the transform of this product we simply convolve the respective transforms of the two time functions that were multiplied. In the result-

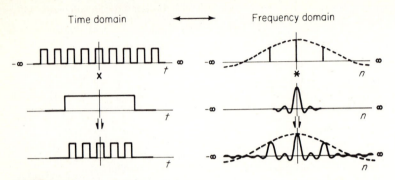

FIGURE 6-19
Development using convolution theorem of Fourier transform of truncated
series of uniformly spaced square waves. (*From Peterson and Dobrin.*[4])

ing transform obtained by this convolution, the spikes for the infinite succession of
square pulses have been spread out into a series of sinc functions having distances
between zero crossings that are inversely proportional to the number of pulses contained in the truncated series.

By the same procedure, we can show how the truncation of an infinitely long
cosine function will broaden the simple two-spike spectrum shown for this type of
signal in Fig. 6-11. When this infinitely long function is multiplied by a square
wave of appropriate length, as in Fig. 6-20, the transform of the truncated train of
cosine waves now consists of two sinc functions instead of two spikes, as each spike
has been convolved with a sinc function corresponding to the transform of the
square wave. The width of each sinc function decreases as the number of cosine
cycles increases. This is because the breadth of the square wave used in the multiplication is proportional to the number of cycles involved.

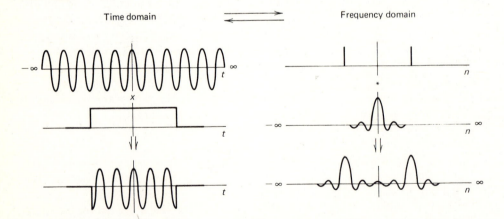

FIGURE 6-20
Development using convolution theorem of Fourier transform of truncated
cosine wave. (*From Peterson and Dobrin.*[4])

Other applications of the convolution theorem will be considered when we take up the design of filter operators for digital processing of seismic signals.

Autocorrelation

Fourier transforms provide one approach to the frequency analysis of seismic signals. Another way of obtaining the frequency spectrum which is used even more widely involves autocorrelation. This requires a series of mathematical operations similar to those in convolution except that the two functions operating on each other are the same and one of them is not reversed with respect to the other.

Initially the two identical functions are aligned with each other as shown in Fig. 6-21. The ordinates of the respective curves are picked off at equal time intervals. The samples are multiplied at each of the ordinate positions, and the products are added to give what is defined as the *autocorrelation function* at zero time lag. The function has its highest value at this lag because all ordinate values are coincident.

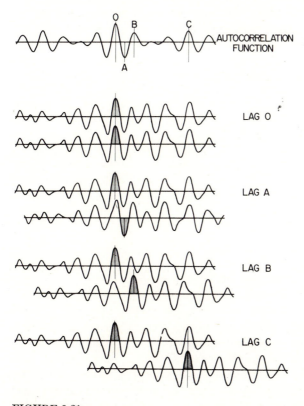

FIGURE 6-21
Determination of autocorrelation function illustrated by matching curve with itself at zero time lag and three other values of lag. (*From Peterson and Dobrin.*[4])

Now consider the case of a half-cycle shift A between the identical functions. Because of the shift, the products of the ordinate values are mostly negative, so that the sum shows a peak negative value when plotted against the time lag, designated here as A. Shifting the lower curve still farther to B, we obtain a positive autocorrelation value. Another peak is shown when the lower curve lags by an amount C, which apparently represents another periodicity in the original function.

The equation for the autocorrelation function of $f(t)$ is

$$\theta_{11}(t) = \int_{-\infty}^{\infty} f(\tau)f(\tau - t) \, d\tau \tag{6-14}$$

where τ is the time shift between the two identical functions. Note its resemblance to the expression for convolution [Eq. (6-10)], even though there is no reversal.

The autocorrelation function will have peaks at values of t corresponding to periodicities or repetition times in the signal that is correlated with itself. The reciprocals of such times will represent dominant frequencies.

If we take the Fourier transform of the autocorrelation function, we get the *square* of the frequency spectrum of the signal. A spectrum obtained in this way contains amplitude information but none on the phases of the various frequency components. A spectrum giving the square of the amplitude as a function of frequency but containing no phase information is referred to as a *power spectrum*.

Cross-Correlation

The same mathematical or graphical procedure that is used in autocorrelation is known as *cross-correlation* when it is applied to analyze the relationship between two different functions. The process involved was shown in Fig. 4-15 along with our discussion of Vibroseis recording. The equation for the cross-correlation of the time function $f(t)$ and $g(t)$ is

$$\theta_{12}(t) = \int_{-\infty}^{\infty} f(\tau)g(\tau - t) \, d\tau \tag{6-15}$$

At a lag τ of zero, the two functions are juxtaposed so that their individual zero time axes are coincident. They are then progressively moved apart, and at each value of time lag τ ordinate values sampled at closely spaced intervals are multiplied and the individual products added. For any values of the lag τ at which the functions tend to have the same shape and thus seem to fit into each other or correlate, the cross-correlation function will show a peak value.

A widely used application of cross-correlation in seismic exploration is the reduction of Vibroseis data, discussed in Chap. 4. To recapitulate, a varying-frequency signal of many seconds' duration reflected from a bed at a depth corresponding to a two-way travel time of t is superimposed on reflection signals of similar duration from other beds, so that it will probably not be identifiable on the record made in the field. If, however, a cross-correlation is run between the signal as initially generated and the ground motion as actually recorded, a high value of the cross-correlation coefficient at a lag of time τ indicates the presence of a reflection at that time. When the shift is of this amount, the source signal and the portion of the recorded signal made up of reflected energy from the bed in question will coincide, resulting in a maximum cross-correlation value for this time lag. Each reflection event will show a cross-correlation maximum at a time shift equivalent to its two-way reflection time so that the cross-correlation of a Vibroseis recording should be equivalent in appearance to a reflection record.

Deconvolution

We have shown that the output $h(t)$ of a filter having a given impulse response $g(t)$ when the input signal $f(t)$ is known can be computed by convolving the input with the impulse response, i.e.,

$$h(t) = g(t) * f(t) \tag{6-16}$$

Suppose, however, that the output $h(t)$ is known and we would like to recover the form of $f(t)$ as it was before it was modified by the filter. To do this we need to find another filter with a response $e(t)$ through which $h(t)$ might be passed to recover $f(t)$; $e(t)$ would then be an *inverse filter* with respect to $g(t)$. The process of canceling the effect of a filter with a second filter designed to be its inverse is called *deconvolution*.

The equation for the deconvolution operation is

$$f(t) = h(t) * \frac{1}{g(t)} = h(t) * e(t) \tag{6-17}$$

or in the frequency domain

$$F(n) = \frac{H(n)}{G(n)} = H(n)E(n) \tag{6-18}$$

where $F(n)$, $H(n)$, $G(n)$, and $E(n)$ are the frequency spectra or Fourier transforms of the respective time functions $f(t)$, $h(t)$, $g(t)$, and $e(t)$.

In practice we deconvolve seismic signals to convert the waveform of a reflection complicated by the filtering that takes place in the earth (such as by reverberation within a water layer) into a simple pulse representing the reflection waveform before the filtering took place. The signal generated by the shot initially had the form of a very short pulse which was broadened by the filtering action of the earth with a consequence decrease in resolution. Any reverberations of the pulse, as in a water layer, alter the waveform still more, interfering further with the desired definition of reflection events. Any process that will compensate for such distortions of the initial pulse should improve reflection quality. In deconvolution we endeavor to compensate for the undesired filter characteristics of the earth itself by creating a new filter that is the inverse of the unwanted earth filter.

The first step in doing this is to measure the spectrum of the signal and to manipulate this spectrum in a way that makes it approach the uniform character of the transform associated with a single spike. Computer programs are available for determining the spectrum (often by autocorrelation), designing the inverse filter, and applying it to the input data without human intercession. Practical aspects of the process will be discussed later in this chapter.

Principles of Sampling in Seismic Processing

Up to now we have considered only those aspects of processing principles which would apply, at least conceptually, to analog as well as digital operations. Although the theory of sampling enters into analog time-domain filtering, as pointed out by Swartz and Sokoloff[8] in an important paper published some years before the digital revolution, some aspects of it are unique to digital technology. The basic difference between digital and analog methods, as pointed out in Chap. 3, is that the analog data represent a continuous flow of (usually varying) voltage whereas digitally recorded data represent a discontinuous series of voltage values, expressed as numbers sampled at discrete time intervals, such as every 2 or 4 ms. Between sampling instants there is no registration of the voltage.

The parameters associated with digitizing of seismic data which are of greatest concern in processing are the sampling interval and the number of bits (or binary digits) for each sample. The former quantity governs the frequency range over which the processing system gives usable results while the latter governs the dynamic range, resolution, and precision of the processed information.

Effect of number of bits   As pointed out in Chap. 3 (68), seismic signals are stored digitally in binary code. The number of bits available for encoding the analog voltages at the sampling points sets a limit on the precision with which the data can be registered. Suppose that a peak voltage of 123.53 mV is to be encoded on analog tape with only 7 bits available. We could express 123 mV by the digits 1111011 and 124 mV by the digits 1111100, but there would be no way of designating any values in between. With 7 bits our final result can be read only to the nearest millivolt, and this is not precise enough for most seismic registration. A real digital recording system requires 14 bits for encoding the input voltage with the degree of precision inherent in seismic data, which allows resolution to the order of microvolts.

Amplitude relations   The lower the sampling rate (the smaller the number of samples per second) the lower the cost of digital recording and processing. What factors govern the lowest sampling rate that will yield acceptable data quality? The answer to this question depends on the upper limit of seismic frequencies that are of interest.

Let us consider first the effect of sampling rate on the amplitude measurement. The reconstruction in analog form of a digitized signal is accomplished by a digital-to-analog converter that passes the voltage corresponding to each sampling instant into a storage element. This value is held in storage until the next sampling instant, at which point the voltage in the hold element is switched abruptly to its next sampled value. The result is the staircase pattern shown in Fig. 6-22. The original signal, indicated by a dashed line, is a 62.5-Hz sine wave with an amplitude of 100 units and the sampling is every 2 ms. The staircase signal can be transformed by filtering to a sinusoidal wave having a frequency of 62.5 Hz which has an amplitude of 97 units plus a series of higher-order harmonics which are superimposed on the 62.5-Hz signal as a ripple. The ripple components are removed by a proper high-cut filter, but the reconstructed signal has now been reduced 3 percent in amplitude.

If the sampling positions are shifted somewhat to fall at points along the sine wave different from those shown in Fig. 6-22, the distortion in amplitude will probably have a different value; it will depend on the amplitude of ripple components which are observed with the new distribution of sampling points.

Frequency relations   Sampling has no adverse effect on the fidelity with which desired frequency components can be reproduced so long as the sampling frequency is more than twice the highest-frequency component of interest in the signal. If the sampling frequency is less than this, we can expect a serious distortion of output frequency which is generally referred to as *aliasing*.

Figure 6-23 illustrates the principle involved. The sine wave at the top has a frequency of 417 Hz and is sampled at a frequency of 500 Hz. The sampling fre-

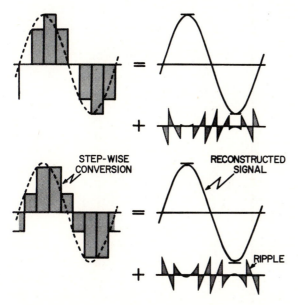

FIGURE 6-22
Reconstruction of a sampled signal in digital-to-analog conversion.   Ripple removed by filtering.   (*From Peterson and Dobrin.*[4])

quency is thus considerably less than twice the signal frequency. The output frequency is obtained by drawing the best smooth curve through the sampling points; this turns out to be 83 Hz.

The lower part of the figure illustrates the relation between input and output frequencies as the input frequency increases. Here the sampling frequency is 500 Hz. For inputs up to 250 Hz the output has the same frequency as the input. This limit of half the sampling frequency is referred to as the *Nyquist frequency*.

Above 250 Hz the output frequency decreases linearly with increasing input frequency until at 500 Hz the output frequency is zero. As the frequency increases beyond 500 Hz, the output repeats the pattern that began at zero frequency, reaching a peak of 250 Hz at 750 Hz input and reaching zero again at 1000 Hz.

z transforms   The z transform provides a mathematically convenient means for carrying out convolution and deconvolution of sampled signals with a digital computer. Using it, one expresses the amplitudes of successive samples taken at uniform intervals as a polynomial of z, an arbitrarily designated variable; its significance

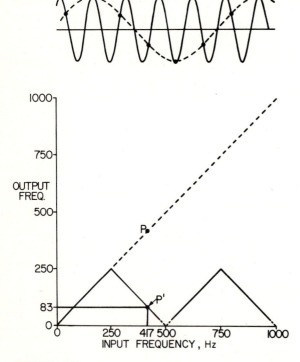

FIGURE 6-23
Frequency folding, or aliasing, in case where sampling frequency is greater than half-signal frequency. As input frequency increases beyond half of the sampling frequency (Nyquist value), output no longer follows input. The dashed output signal at 83 Hz is spurious. (*From Peterson and Dobrin.*[4])

lies only in its exponent, which indicates the time of the sample for which the amplitude constitutes the coefficient. Specifically, the power of z is the ordinal number of the sample, and the coefficient for each term is the value of the sample. For example, suppose the seismic signal of Fig. 6-24 is sampled so that at zero time the amplitude is 1, at the end of the first sampling interval (say 2 ms) it is 4, at the end of the second interval it is 3, and at the end of the third interval it is 2; then we could express the entire sampled signal in the form

$$f(z) = 1 + 4z + 3z^2 + 2z^3 \qquad\qquad (6\text{-}19)$$

If this signal is convolved with a filter having a response characteristic that can be represented by samples with a corresponding time series 2, 4, 3, the filter operator can be expressed in the form $g(z) = 2 + 4z + 3z^2$ and we can obtain the output

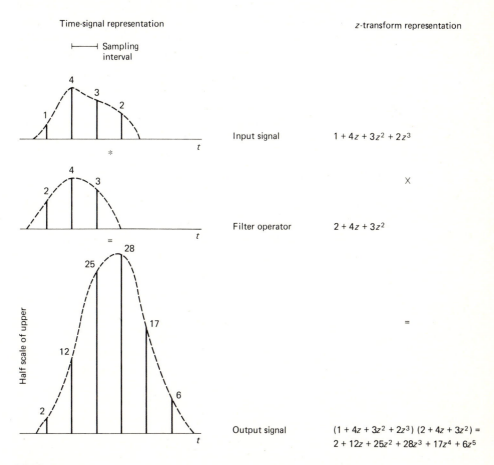

FIGURE 6-24
Filtering by multiplying z transform of input signal by z transform of filter operator.

$h(z)$ of the filter by multiplying the polynomials as follows:

$$h(z) = (1 + 4z + 3z^2 + 2z^3)(2 + 4z + 3z^2)$$

$$= 2 + 12z + 25z^2 + 28z^3 + 17z^4 + 6z^5 \qquad (6\text{-}20)$$

The time series for the output is obtained by listing the successive coefficients corresponding to the increasing powers of z as follows:

$$2, 12, 25, 28, 17, 6$$

This representation follows from the fact that each exponent of z designates the time at which the sample with the amplitude equivalent to its coefficient occurs in the output signal.

Multiplication of the terms corresponding to $f(z)$ with those corresponding to $g(z)$ makes use of the same shift-and-add operations that are carried out in conventional convolution processes.

The z transform provides a handy mathematical approach to *deconvolution* also. If the output signal is to be a spike at time zero, its time-series representation $1, 0, 0, 0, 0, \ldots$ becomes 1 in z-transform notation. If the input signal is $1, 0.5$ (in z-transform notation, $1 + 0.5z$), we can determine the inverse filter function $f(z)$ which when operating on the input will yield a spike as its output. This is a deconvolution operation which can be carried out by simple polynomial division as follows. Starting with

$$(1 + 0.5z) f(z) = 1,$$

$$f(z) = \frac{1}{1 + 0.5z} \qquad (6\text{-}21)$$

the function can be approximated by binomial expansion as the polynomial

$$f(z) = 1 - 0.5z + (0.5)^2 z^2 - (0.5)^3 z^3 + (0.5)^4 z^4 + \cdots \qquad (6\text{-}22)$$

so that the desired inverse filter operator can be specified by the converging time series

$$1, -0.5, 0.25, -0.125, 0.0625, \ldots$$

The convergence indicates that only a limited number of terms is needed to construct a usable filter operator as later terms in the infinite series become too small to be significant.

6-4   DIGITAL PROCESSING OF SEISMIC DATA: PRACTICAL ASPECTS

In this section we shall consider the principles governing the technique employed in digital playback operations. Digital hardware will be touched upon only very briefly, and no actual computer programs will be developed. Our object is to show the logic of digital processing with emphasis on the geophysical problems that can be solved if proper digital hardware and software are available.

Digitized seismic signals, whether recorded in the field or converted into digital form from analog tape, are sampled at intervals of 1 to 4 ms (generally 2 ms in land recording and 4 in marine recording), and the samples are expressed as words consisting of 14 or 15 binary numbers, or bits. In binary-gain systems 4 more bits are used to specify the amplitude range within which the samples fall. A single 24-channel record 6 s long sampled at 2-ms intervals would thus require storage of a 1.25 million bits. For common-depth-point processing as many as 48 signals must be composited to obtain an output trace, so that the storage capacity must be many times as great as the core memory of any conventional computer. Thus a large auxiliary storage is required for seismic processing applications. In order to handle filtering operations in real time, each multiply-and-add step in a convolution must be executed in about 0.5 μs. Although the normal time for such a sequence in conventional third-generation computers may not be as great as 0.5 μs, special high-speed convolvers are used for this operation to avoid tying up the main computer.

The first step in digital processing is for the reading heads on the input transport to take off the multiplexed signals from the digital tape and to route them into core storage after demultiplexing. When the signals are transferred into storage, it is necessary for each sample, consisting of a series of binary digits that express the signal amplitude for a particular instant of time, to go into a *cell* with a location that is indexed in terms of the channel number and the time (as well as the record number if more than a single record is in core storage at the same time).

Time Shifting

Once such an address is established for each data word, time-shifting operations become just a matter of readdressing storage positions, an operation easy to program. The format of the input tape (as illustrated in Fig. 3-15) is taken into account in designing the software, and the time on the record corresponding to any bit picked up by a head can be identified simply by counting. An elaborate program must be built into the computer to keep track of each bit and to ensure the maintenance of the correct time sequence as the respective bits are routed into core storage.

For time shifting, it is only necessary for the address of a stored number to be changed a specified amount. If, for example, an elevation or other static correction requires that an entire trace be shifted toward greater times by 10 ms, the computer simply moves the location of each sample for that trace by five positions (assuming 2-ms sampling) in the direction of increasing time. If a dynamic correction, in which the amount of time shift varies with time on the trace, is to be made, a more complex readdressing program is necessary, but the operation is still an easy one for the computer.

Compositing

Compositing, or addition, of two samples stored digitally (a vital step in filtering by convolution) is accomplished simply by execution of a command to add what is in the storage cell containing the first number to the contents of the cell containing the second.

Filtering

As digital filtering generally requires the use of filter operators which are actually impulse-response functions for the desired filters expressed as time series, it is necessary to put each filter function into storage (along with instructions as to how the function is to change with record time in the case of time-varying filters) so that it can be called from memory during subsequent filtering of the seismic signals. The position in storage of each term in the filter operation must be indexed for instantaneous access during the operation. The filtering itself, which generally involves the convolution of the input signal with the filter operator, breaks down into a programmed succession of multiplications of predetermined samples, additions, and time shifts.

Filter operators used in seismic processing vary greatly in length, but a typical one would be 200 ms long. With 2-ms sampling such an operator would consist of 101 points. The conversions between the time domain and the frequency domain which are necessary in digital filtering require the calculation of Fourier transforms. A technique introduced about 1965 by Cooley and Tukey,[9] called the *fast Fourier transform*, has greatly reduced the time necessary for carrying out this process.

Designing Digital Frequency Filters

Let us illustrate the principles governing the design of filter operators by applying them to the relatively simple type of frequency filtering required in passing all signals above one specified frequency or removing all below another such frequency or both. Such filters are generally designated as *high-pass*, *low-pass*, or *bandpass*. In designing digital filters that operate in the time domain we express the desired filter characteristic as a function of frequency and then compute the Fourier transform of this function so as to set up a filter operator as a function of time.

Suppose we want to design a low-pass filter in the time domain that transmits all frequencies from 0 to 40 Hz without attenuation but allows no frequencies above 40 Hz to pass. Remembering that frequency-characteristic curves must be symmetrical about zero, as they apply equally to positive and negative frequencies, we can draw the characteristic curve for the desired filter in the form of the square wave shown at the upper right of Fig. 6-25. This has uniform transmission between frequencies of -40 and 40 Hz and zero transmission outside this range. The time representation of this rectangular frequency function, which is its Fourier transform, is a sinc function having zero crossings at ± 0.0125, ± 0.025, ± 0.0375, . . . s.

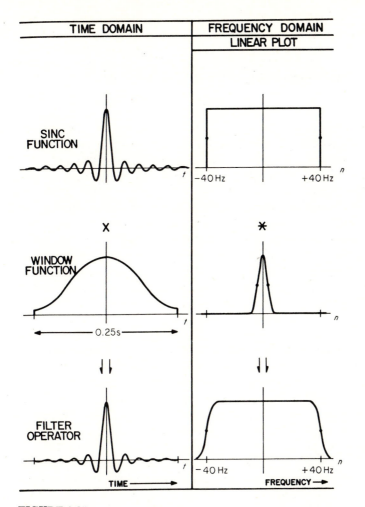

FIGURE 6-25

Design of low-pass filter operator based on convolution theorem. Desired cutoff at 40 Hz cannot be achieved exactly because the operator must be of finite length and the frequency function in the form of the square wave shown at the upper right would require an infinitely long filter function. (*From Peterson and Dobrin.*[4])

The times of these crossings are, as we saw on page 162, equal respectively to 1, 2, 3, . . . divided by the width, 80 Hz, of the square wave as measured on the frequency scale. Even though the amplitudes of the lobes thus obtained attenuate rapidly as time increases in the positive and negative directions, the sinc function must be infinitely long to represent an accurate transform of the square wave. An infinitely long filter operator is, of course, not physically realizable, and we must consider what effect truncation of such an operator will have on its filter characteristics. One way of truncating the time operator would be to multiply it by a rectangular (square-wave) function having a time duration of proper length, but the sharp corner at the cutoff point would leave undesired side-band effects. A common solution of this problem is to truncate the operator more gradually by multiplying it by a *window function* of the type shown at the left center of Fig. 6-25. This function consists of one cycle of a cosine wave with its trough raised slightly above zero; it is called a *Hamming window*.

The frequency transform of the window is shown at the left of the function itself. It consists of a peak at zero with a fairly sharp drop on either side. The purpose of the window function is to reduce the side bands to the point where they have no perceptible effect on the output.

When the infinitely long sinc function is multiplied by this window function, we get a time operator having a length of 0.250 s. What effect does the truncation have on the frequency characteristic of the filter? We can find out by convolving the original square filter curve with the transform of the window in accordance with the convolution theorem. Instead of a vertical cutoff, we have an inclined cutoff curve centered at 40 Hz, the slope depending on the width of the truncation window. The broader the window the steeper the slope. Thus the longer the filter operator the sharper the cutoff at the desired frequency.

The convolution of what is effectively a single-peak spectrum for the window function with the desired square-wave frequency function yields a filter with a frequency characteristic that has sloping rather than vertical sides. Even so, the slope is almost always steeper than what would be observed with the sharpest electric analog filters ordinarily used.

Once they are programmed, the operations involved in this type of filtering are carried out internally by the computer. The only variables to be included in the programming are the value of cutoff frequency and the width (in time units) of the truncation window. Once the filter operator is computed, it must be sampled at appropriate intervals and the samples put into storage along with the signal to be convolved.

Deconvolution Filtering

The purpose of deconvolution, as previously pointed out, is to compensate for filtering effects in the earth, e.g., those due to water-layer reverberation, that complicate and inevitably broaden the simple reflection pulses from subsurface interfaces. The end product of such an operation should be reflection signals that are simple wavelets with the narrowest breadth that the earth's absorption characteristics allow. Mathematically, the ideal reflection signal is a spike of zero time duration.

The frequency spectrum of a time signal in the form of a spike having infinitesimal width is "white"; i.e., components at all frequencies from plus to minus infinity have equal amplitudes. If the actual frequency spectrum can be converted to "whiteness" by introduction of a suitable inverse filter, the principle of Fourier transformation tells us that all reflection events will then become spikes and the greatest conceivable resolution in reflection character will be achieved.

Actually, the attenuation caused by absorption in the earth of the less-than-millisecond-long pulse emanating from the shot broadens the pulse to the point where a simple reflection signal will have a measurable width after it has traveled only a short distance from its source. Its frequency spectrum is then no longer white but approximates a square wave with a high cutoff at a frequency inversely proportional to the width of the pulse, which might be looked upon as a crude approximation to a sinc function. Once the high frequencies are attenuated by absorption, they cannot of course be restored by any kind of inverse filtering. However, it is possible to obtain a spectrum by such filtering that is approximately white over the frequency band (say 10 to 90 Hz) which can normally be transmitted through substantial thickness of sedimentary rocks. The wider the zone of flatness in the spectrum the narrower the corresponding wave pulse and the better the overall resolution.

Application to reducing reverberation   Although many characteristics of the earth's transmission can give rise to irregularities in the spectrum, the most troublesome type of anomaly is a peaking of the frequency response due to reverberation of the signal between two surfaces such as the top and bottom of a water layer (see Fig. 5-18). As indicated in Fig. 6-10, a frequency spectrum consisting of equal spikes at frequencies f and $-f$ corresponds to a cosine wave of infinite extent with a period of $1/f$. If each spike is broadened until the spectrum resembles the lower right-hand curve of Figure 6-20, the infinite cosine signal becomes truncated. The number of cycles will be inversely proportional to the width of the two spectral peaks. If the broadening extends to the point where the peak disappears and the spectrum is flat out to infinite frequencies (both positive and negative), the signal contracts to a spike.

As the range of reflections with which we work is seldom much below 10 Hz or above 90 Hz, the frequency spectrum representing the best resolution we can normally expect consists of a pair of rectangles lying between these limits on the positive and negative sides of the frequency axis, as shown in Fig. 6-26. This spectrum can be obtained by convolving the transform of an infinitely long cosine wave having a frequency of 50 Hz with a rectangular frequency function 80 Hz in breadth. The time signal corresponding to the rectangular frequency function is a sinc function with a central peak 0.025 s wide having side bands 0.0125 s in breadth. Multiplying this by a cosine wave with a period of 0.020 s (the reciprocal of 50 Hz) results in a pulse 0.010 s wide at its zero crossing. This represents the best resolution that can be expected from a reflection signal traveling in a medium which passes frequencies from 10 to 90 Hz.

To illustrate how the principle operates in an actual case, let us consider the spectrum observed from shallow-water shooting shown in Fig. 6-27 (along with the

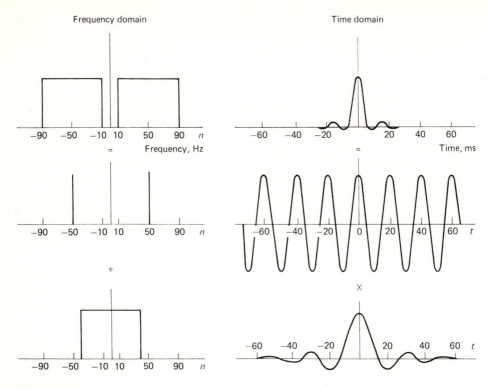

FIGURE 6-26
Determination using convolution theorem of waveform corresponding to a symmetrical frequency spectrum with flat response from 10 to 90 Hz and sharp cutoffs at both limits. Narrowness of resultant pulse (upper right) should result in good resolution.

waveform on which it is based). The sharp peak at about 38 Hz is about 25 Hz broad at its base. This peak frequency corresponds to an oscillation period of 0.026 s, which is in the neighborhood of what is observed on the record. After deconvolution, the peak is removed, and an essentially flat spectrum is observed from 15 to 80 Hz. Assuming vertical cutoffs at these frequencies, we would expect that seismic pulses would have a breadth approximately equal to the zero-to-zero width of the central peak of the sinc function corresponding to a rectangular frequency curve 65 cycles wide. The expected breadth, which is twice the reciprocal of this value, is about 0.030 s. The portion of the seismic record obtained after deconvolution shows discrete pulses in the general vicinity of this predicted period.

For water less than 100 ft deep analog methods of eliminating reverbration such as notch filters or three-point time-domain filter operators might suffice. Actually, such analog filtering was fairly successful in shallow-water work before digital processing became available. For deep water, on the other hand, only a very flexible approach such as digital deconvolution is likely to be effective for eliminating ringing effects. The complexity of the layering in most bottom materials, with the frequent occurrence of mud layers above hard rock, makes it difficult to predict ringing characteristics on the assumption of a simple lithologic model.

Peaking in the frequency spectrum due to repetition of signals by multiple reflection is not confined to water layers. Any other layering with strong reflection coefficients at the top and bottom can give rise to pronounced reverberation effects. Permafrost, which generally has a velocity much higher than that of the underlying unfrozen rock material, is responsible for strong ringing effects observed on reflection records shot on the North Slope of Alaska. The physical mechanism is very similar to that for water-layer reverberation in the case of a soft bottom.

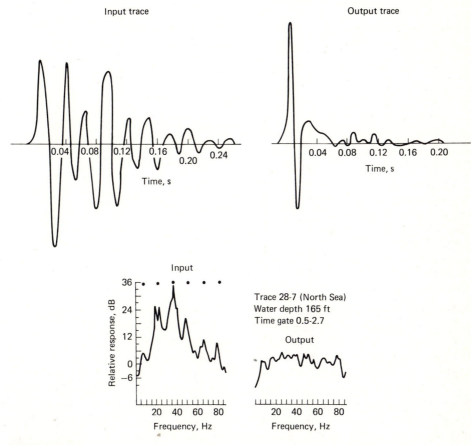

FIGURE 6-27
Waveforms and spectra of reflection signal from North Sea before (input) and after deconvolution (output). (*Spectra from D. W. Rockwell, Geophysics, vol. 32, p. 262, 1967. Waveforms computed from spectra assuming minimum phase.*)

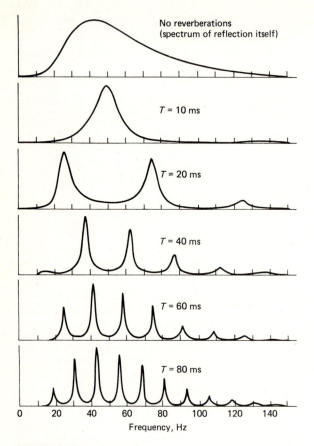

FIGURE 6-28
Spectra of reverberating pulses in water layer. T is two-way "thickness" (in time) of layer. (*From F. J. McDonal and R. L. Sengbush, Proc. 7th World Petrol. Congr. Mexico, vol. 2, p. 591, 1967.*)

Filtering of repetitive signals It should be emphasized that reverberation in near-surface layers and ghosting modify the waveforms of down-traveling or emerging seismic impulses and hence must be looked upon as filtering mechanisms within the earth. In the case, for example, of a reflection pulse returning from a subsurface, the spectrum of the wave reaching the layer from below must be multiplied by the spectrum associated with the reverberation in the water layer. Figure 6-28 shows how a wavelet spectrum would be changed by passing through water layers overlying a hard bottom in a system with the frequency characteristic illustrated in Fig. 5-20. This means that a reflection signal emerging from a deep interface must be convolved as shown with an impulse-response function associated with the individual reflections encountered by the wave during reverberation. Most deconvolution of marine data is carried out to cancel irregularities superimposed on the spectrum of the reflection signal approaching the water bottom from below by multiplication with the spectrum associated with the water layer itself. This is accomplished in a digital computer by convolving the input with a filter operator (expressed as a time function) associated with the inverse of the water-layer spectrum.

Enhancing resolution by deconvolution Another important application of deconvolution is in the improvement of resolution, usually by building up higher frequencies as the spectrum is whitened. Deconvolution carried out for this purpose must be applied with caution. Many types of noise, particularly scattered energy and wind noise, have substantial energy at the high-frequency end of the seismic recording range. Any deconvolution program that raises the response at this end of the output spectrum could magnify existing noise in a way that could cause deterioration of record quality. The fact that absorption of seismic waves is an exponential function both of frequency and distance traveled means that noise originating from near-surface irregularities is likely to contain more high-frequency energy than reflections from considerably deeper sources.

Mechanics of deconvolution A conventional form of deconvolution involves the automatic design by the computer of a filter operator which when convolved with the output yields a time series that is the closest possible approach to a single spike. As the time series for a spike is simply 1, 0, 0, 0, . . . , we can get the desired operator by dividing the z transform for the spike by the z transform of the signal to be deconvolved. The time series for the filter operator needed to convert the recorded signal into a spike would be infinitely long and hence would not be practical for use in actual filtering operations. The need is to determine an *approximate* inverse filter which can be represented by a time series of *finite* length. The greater the length of operator the less the error in the truncation that must be carried out for this purpose.

The logic behind this type of deconvolution process is discussed by Robinson[10] (chap. 7, pp. 142–193) and will be summarized very briefly here. The problem is to find an operator containing a specified number of terms which when convolved with the input wave will give a time series having the least possible deviation (on a least-squares basis) from the time series 1, 0, 0, 0, . . . corresponding to a spikelike signal. This minimum deviation is determined by squaring the error terms for each coefficient of the operator, adding the squares, differentiating the sum with respect to each coefficient, equating the derivatives to zero, and solving for the coefficients. These operations are carried out separately for each coefficient of the time series.

The calculation is facilitated by using the autocorrelation function of the input signal to obtain the spectrum needed for computing the appropriate inverse operator. It is a simple matter for the computer to determine the autocorrelation of the signal and to express it as a time series. Because the autocorrelation function does not contain phase information, and because different input functions could have the same autocorrelation function, we set up the specification that the filter func-

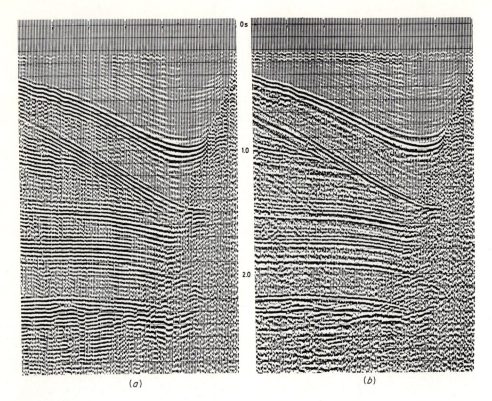

FIGURE 6-29
Marine record section: (*a*) before deconvolution; (*b*) after deconvolution.
(*Prakla-Seismos.*)

tion represented by the time series actually used has the smallest overall phase lag of all the possible functions that correspond to the same autocorrelation function. The inverse filter calculated by least squares will then have a minimum phase delay. When the resultant inverse-filter operator is used to convolve the input signal, the energy in the output will be concentrated at an earlier time in the signal for the minimum-phase-lag filter than for any other filter operator corresponding to the same autocorrelation function. This should yield an output having the best possible approximation to the ideal spike which we are endeavoring to approach.

There are other deconvolution techniques used in seismic processing, e.g., those involving prediction or Hilbert transforms, but the approach that has been outlined here illustrates the basic principles involved. Figure 6-29 shows the effect of a typical deconvolution operation on a record in which reflection information is seriously obscured by marine reverberation. Figure 6-30 shows how deconvolution can enhance the resolution of reflections recorded on land under conditions giving rise to observable ringing effects.

Wiener Filtering

During World War II, Norbert Wiener of MIT, one of the foremost mathematicians of this century, was asked to develop a technique for extracting radar signals from noise. The method of filtering he worked out turned out to have many other applications to information processing, including enhancement of seismic reflection data.

In formulating his filtering technique Wiener assumed that an information trace, e.g., in a seismogram, consists of a desired signal immersed in noise. Criteria were developed for deriving a filter operator that when convolved with the recorded trace would yield an output as close as possible to the desired signal. Starting with the observed trace one would only need to specify the desired signal in order to determine this operator.

The mathematical theory of Wiener filtering is too advanced for presentation here in more than rudimentary form. Its application to seismic processing is well explained by Robinson and Treitel[11] and Foster et al.[12] The earliest work on adapting Wiener's pioneering studies to seismic data was reported by Wadsworth et al.,[13] who were associated with the Geophysical Analysis group at MIT.

Figure 6-31 illustrates the principle of a simple Wiener filter. Three signals are shown: the input $x(t)$, the desired output $z(t)$, and the actual output $y(t)$. The

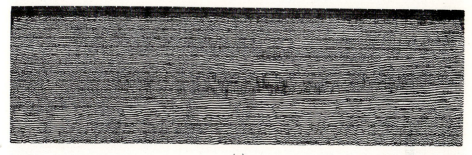

(a)

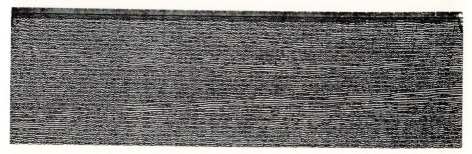

(b)

FIGURE 6-30
Section recorded on land: (a) before deconvolution; (b) after deconvolution.
(*Prakla-Seismos.*)

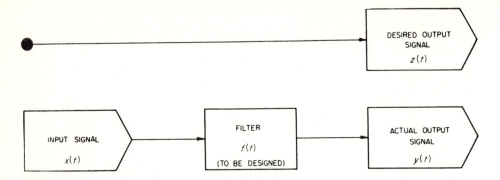

FIGURE 6-31
Principle of Wiener filtering. Desired output signal provides basis for
designing filter. (*From Robinson and Treitel.*[11])

problem is to determine the filter operator $f(t)$ that brings $y(t)$ as close as possible
to $z(t)$ by minimizing the differences (on a least-squares basis) between $y(t)$ and
$z(t)$. The agreement shown between the actual output and the desired output in
Fig. 6-32 is remarkably good considering the very small number of terms used in
the time series for the filter.

An approach like this is so tedious and so difficult to apply that it is not generally
employed in processing operations. An equation developed by Wiener and Hopf

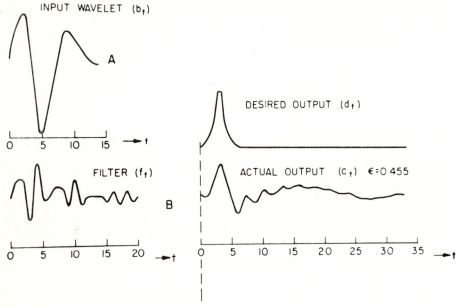

FIGURE 6-32
Comparison of desired output and actual output from Wiener filter. (*From
Robinson and Treitel.*[11])

provides a convenient basis for determining the filter operators. This equation makes it possible to determine the filter coefficients from the autocorrelation function of the input signal and the cross-correlation function between the desired output and the input signal. The two correlation operations can usually be determined by standard subroutines. As with the least-squares approach to filtering previously considered, simultaneous equations are set up which are solved by the computer to obtain the respective coefficients of the desired filter operators. The derivation and application of the Wiener-Hopf equation to seismic filtering are presented by Robinson and Treitel.[11]

The output function obtained by convolving the input with the optimum filter operator will not coincide exactly with the desired output, but the differences can be minimized by proper choice of parameters, as Fig. 6-32 demonstrates.

Weiner filters can be used either in single-channel or multichannel processing. In a sense, any deconvolution filtering designed to give an output which has the form of a spike is closely related to Wiener filtering.

To illustrate practical problems for which Wiener filtering has been applied, we shall consider two examples from the literature, ghost suppression and velocity filtering.

Ghost suppression   The problem often arises of removing undesired repetitions, or ghosts, of primary reflection signals caused by downward reflection of energy from the shot at interfaces a short distance above, as illustrated in Fig. 6-33. The use of Wiener filters to remove such spurious signals has been discussed by Foster et al.[12] and Schneider et al.[14]

One approach to the problem involves coordination of both shooting and processing procedures. Two shots are fired at different depths in the same shothole and recorded separately. The signals from the respective shots are filtered and added. The object of the filtering is to enhance the direct pulse from the shot and to attenuate the ghost. The difference between the uphole times for the respective signals is used in designing the filter operator that will optimize the output signal in this respect.

Velocity filtering   Another important application of multichannel Wiener filters is to the problem of removing coherent noise on reflection records that has a different moveout time or apparent velocity than the desired reflection events. In this case the time differential for a particular event between adjacent traces is the criterion on which the filtering operation is based. A technique for this type of filtering using Wiener operators has been described by Embree et al.[15]

Figure 6-34 shows the range of moveouts for reflections and low-velocity traveling noise on a plot of wave number (reciprocal of wavelength) versus frequency, which is effectively a conventional time-distance plot rotated by 90°. This is referred to as an *fk plot* (frequency *f* versus wave number *k*), and the slopes on it are proportional to the apparent velocities of the events on the record rather than to the reciprocals of the apparent velocities, as on a time-distance plot. A reflection

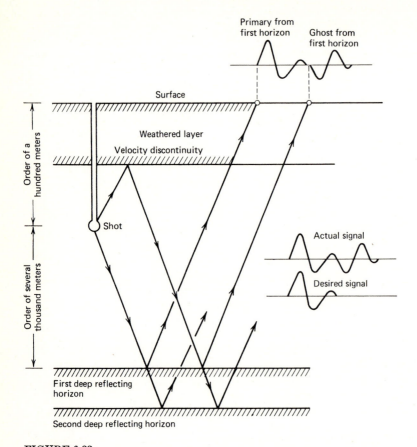

FIGURE 6-33
Primary and ghost reflections from deep horizon. Wiener filter is designed to
convert actual signal with ghost reflection to desired signal without it. (*After
E. A. Robinson, Geophysics, vol. 31, p. 487, 1966.*)

will normally have a small moveout across the record, and even when there is steep
dip, the moveout will be relatively small compared with most traveling noise such
as surface waves, which usually have low velocities. Reflections will fall within a
narrow wedge on the plot that is centered on its vertical axis. Low-velocity noise,
such as ground roll, falls within wedges encompassing a range of relatively gentle
slopes; high-velocity noise falls within wedges of intermediate slope.

Ground roll is conventionally suppressed by a combination of low-cut frequency
filtering and wave-number filtering, which is simply the type of discrimination accomplished by using multiple shots and/or geophones. High-velocity noise, on the
other hand, cannot be rejected in this way without risking deterioration of the
reflections, particularly those of higher velocity associated with steeply dipping
beds. The object of velocity filtering is to suppress high-velocity noise without
detriment to the reflection quality.

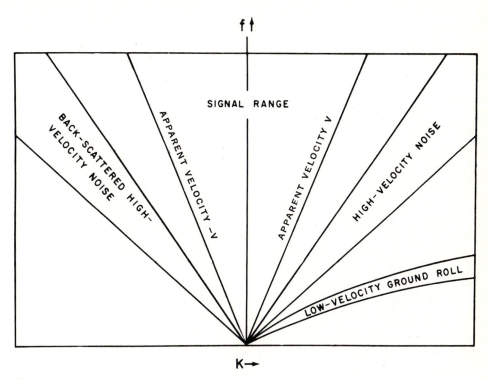

FIGURE 6-34
Ranges of noise events in frequency–wave-number (*fk*) representation.
(*Embree et al.*[15])

Referring again to Fig. 6-34, we set up a condition that all signals be passed without modification in the range of apparent velocities from V to $-V$. The filter response is specified not as a function of frequency alone but as a two-dimensional function of frequency and wave number that can be represented on an *fk* plot of the type shown in this figure. A transmission of unity (no filtering effect) is desired in the range where k has values between $-|f|/|V|$ and $|f|/|V|$ (the vertical bars represent absolute values irrespective of the sign of f). For all k values outside this range, the transmission is specified to be zero.

The determination of the optimum filter for each channel to realize these conditions is discussed by Embree et al.,[15] and their derivation of the appropriate operator is presented in an appendix to their paper.

Figure 6-35 shows the form the filter operator takes for six pairs of traces, each member of the respective pairs, such as 6, 7; 5, 8, etc., representing the same distance x_m from the source. Input signals corresponding to increasing distances from the output position at the center are fed into low-pass filters with progressively lower cutoff frequencies. Looking at the filtering action in a qualitative way, we might say that the highest-frequency components of the desired signal are passed only by the two channels nearest the center of the array, but the relative contribu-

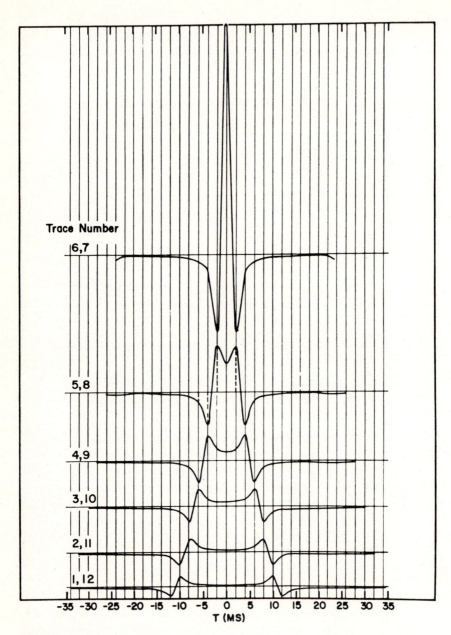

FIGURE 6-35
Filter operators used with pairs of traces at progressively greater distances
from shot point in velocity filtering by the pie-slice process. (*Embree et al.*[15])

tion of these components to the output signal is raised to its proper level by the high amplitude of the weighting function. Successively lower frequencies can be composited over a successively greater number of traces without attenuation.

6-5   PLOTTING OF SEISMIC DATA ON RECORD SECTIONS

The plotting of processed data on record sections can be done either with analog or digital equipment. The most widely used analog system for presentation of digitally processed data, the Geo Space plotter, operates by photography of the seismic signals displayed on a cathode-ray screen. This device can be used for registering data from analog tape or from multiplexed digital tape. The motion of the spot on the scope corresponding to the output from analog or digital storage of the seismic signal to be plotted is along a horizontal line, the time axis being provided by the rotation of the drum which moves the film or paper wrapped around it in the vertical direction. A digital-to-analog converter is of course necessary if digitally stored signals are to be plotted in this way.

All the usual modes of presentation are available with this system, the selection depending on the setting of the beam on the scope. For a wiggly-line trace, the beam is focused on the scope face at a point. For variable-area registration, the length of an elongated line on the scope is modulated by the signal amplitude, and for variable density the intensity of the beam is modulated and the light is focused on the screen as a line. The electron optics of the scope can be readily adjusted for combined-mode, e.g., wiggly-line plus variable-density, presentation.

The plotting may be carried out with 6, 12, 18, or 24 channels at a time, depending on the number of multiplexed signals that can be accommodated on the oscilloscope screen. A precision mechanism advances the recording drum along its axis as the plotting proceeds.

Two kinds of plotters are in current use which register digital signals directly without prior conversion to analog form. One type is electrostatic, as exemplified by the Gould and Varian plotters. The other transforms the digital signal into a dot matrix on a cathode-ray tube, which is photographed in the same way as with the Geo Space analog plotter. The Petty-Ray Photo-dot system is of this type.

Electrostatic plotters are generally used for processing in the field with minicomputers such as the Phoenix or ComMand systems or for preliminary stages of plotting in a playback center. The definition is not generally sharp enough for final, permanent presentation of the data, and other types of plotters such as the Geo Space or Photo-dot are generally employed for this purpose.

REFERENCES

1. Jones, Hal J., John A. Morrison, G. P. Sarrafian, and L. J. Spieker: Magnetic Delay Line Filtering Techniques, *Geophysics*, vol. 20, pp. 745–765, 1955.

2. Silverman, Daniel: The Digital Processing of Seismic Data, *Geophysics*, vol. 32, pp. 988–1002, 1967.

3. Robinson, E. A., and S. Treitel: Principles of Digital Filtering, *Geophysics*, vol. 29, pp. 395–404, 1964.

4. Peterson, Raymond A., and Milton B. Dobrin: "A Pictorial Digital Atlas," United Geophysical Corp., Pasadena, Calif., 1966.

5. Anstey, N. A.: "Signal Characteristics and Instrument Specifications," vol. 1 of "Seismic Prospecting Instruments," Borntraeger, Berlin, 1970.

6. Lindseth, Roy O.: "Recent Advances in Digital Processing of Geophysical Data: A Review," Teknika Ltd., Calgary, Alberta, Canada, 1970.

7. Jennison, R. C.: "Fourier Transforms and Convolutions for the Experimentalist," Pergamon, London, 1961.

8. Swartz, C. A., and V. M. Sokoloff: Filtering Associated with Selective Sampling of Geophysical Data, *Geophysics*, vol. 19, pp. 402–419, 1954.

9. Cooley, J. W., and J. W. Tukey: An Algorithm for the Machine Calculation of Complex Fourier Series, *Math. Comput.*, vol. 19, pp. 297–301, 1965.

10. Robinson, Enders A.: "Statistical Communication and Detection: With Special Reference to Digital Data Processing of Radar and Seismic Signals," Hafner, New York, 1967.

11. Robinson, Enders A., and Sven Treitel: Principles of Digital Wiener Filtering, *Geophys. Prospec.*, vol. 15, pp. 311–333, 1967.

12. Foster, M. R., R. L. Sengbush, and R. J. Watson: Design of Sub-Optimum Filter Systems for Multi-Trace Seismic Data Processing, *Geophys. Prospec.*, vol. 12, pp. 173–191, 1964.

13. Wadsworth, G. P., E. A. Robinson, J. G. Bryan, and P. N. Hurley: Detection of Reflections on Seismic Records by Linear Operators, *Geophysics*, vol. 18, pp. 539–586, 1953.

14. Schneider, William A., Kenneth L. Larner, J. P. Burg, and Milo M. Backus: A New Data-Processing Technique for the Elimination of Ghost Arrivals on Reflection Seismograms, *Geophysics*, vol. 29, pp. 783–805, 1964.

15. Embree, Peter, John P. Burg, and Milo M. Backus: Wide-Band Velocity Filtering: The Pie-Slice Process, *Geophysics*, vol. 28, pp. 948–974, 1963.

TRANSFORMATION OF REFLECTION TIMES
INTO GEOLOGICAL STRUCTURE

In the previous chapter we reviewed the techniques employed to convert seismic reflection data recorded in the field into corrected record sections on which all signals other than primary reflections from subsurface lithologic boundaries are suppressed to the greatest possible extent. The sections thus produced often appear to be equivalent to geological structure sections, so much so that geologists and geophysicists may feel strongly inclined to look upon them as if they were actual pictures of the subsurface structure. Great as the resemblance may be, however, it is neither proper nor safe to assume that seismic record sections show structure directly. The relationships which must be established between the times of seismic reflections and the actual positions of their sources in the earth can be much more complex than an oversimplified model of the subsurface might lead us to expect.

The conversion of observed reflection times to a correct geometrical picture of the subsurface involves a number of steps. First, the distribution of seismic velocities must be determined in the subsurface, and changes, both vertical and lateral, must be noted. Second, all reflection times must be corrected for variations in elevation and changes in the thickness of the low-speed layer just below the surface. Corrections must also be made to take account of the fact that the reflection rays do not travel along vertical paths when the geophone is displaced from the shot position. Where the reflecting beds dip, the point from which each reflection originates is not located vertically below the midpoint between shot and geophone, and an adjustment called *migration* is made to put this point closer to its true position in space. Finally the presentation of such adjusted information on time or depth sections and on maps can involve some complex manipulation of the data.

All these steps will be considered in the current chapter, but before we go into them we shall review the geometrical principles governing the conversions with which we are concerned.

7-1 GEOMETRY OF REFLECTION PATHS

Geometrical versus physical optics in seismic analysis In Chap. 2 we considered the physical processes involved in the reflection of seismic waves at an interface between rocks having different elastic constants and presumably different lithologies. The percentage of the seismic energy reflected at such an interface was shown to depend on the acoustic impedances (products of velocity and density) of the materials on respective sides of the interface and the angle of incidence of the down-going wave. At normal incidence the amplitude of the reflection is simply proportional to the difference between these acoustic impedances over their sum, the relation becoming more complex at oblique angles of incidence as shown in Fig. 2-15.

When there is a series of interfaces separating individual formations having different velocities and the distances between the interfaces are large compared with the seismic wavelengths employed, a separate reflection should theoretically be observed from each interface, although this does not always occur for a number of reasons to be discussed later.

The times required for the waves to travel from a near-surface source to the reflectors and back to receivers on the surface are used, along with all available information on seismic velocities, to determine the structure of the reflecting surface. This process forms the geometrical basis for the reductions used with the reflection method.

Actually, the separations of individual reflecting interfaces in the earth are often much smaller than a seismic wavelength, so that resolution of reflections from every lithological boundary in the geologic section is seldom if ever possible (nor would it be generally desirable). In considering the ray-path geometry of seismic reflections, we shall initially work with models of discrete interfaces that are separated from adjacent boundaries by distances large compared with seismic wavelengths. In the next chapter we shall consider the more usual case of boundaries that are closer together than a wavelength. Both physical optics and geometrical optics must be taken into account in the interpretation of seismic reflections from this more complex kind of layering.

Reflection from horizontal surfaces Let us assume a horizontal reflecting interface, as shown in Fig. 7-1, at a depth z below the earth's surface. The seismic

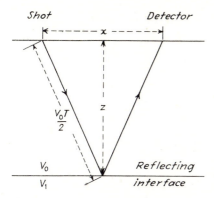

FIGURE 7-1
Wave reflected from single interface.
Speed constant at V_0 between source
and reflecting surface.

velocity above the interface is V_0, and that below it is V_1. The path of a reflected wave generated at the shot point and received by a detector a distance x away from it consists, as indicated in the diagram, of two rectilinear segments which have traveled from the surface to the reflecting interface and back to the surface. The total path length L is related to x and z by the formula

$$L = 2\sqrt{z^2 + \left(\frac{x}{2}\right)^2} = V_0 T \tag{7-1}$$

where T is the total travel time. For a horizontal reflector (z = constant) it is evident that the relation of T to x is hyperbolic. Solving for T, we have

$$T = \frac{2}{V_0}\sqrt{z^2 + \left(\frac{x}{2}\right)^2} \tag{7-2}$$

Similarly,
$$z = \tfrac{1}{2}\sqrt{(V_0 T)^2 - x^2} \tag{7-3}$$

Figure 7-2 shows the relation between travel time and horizontal distance for a reflected ray. The curve is actually a symmetrical one, as it holds for negative as well as positive values of x, the portion corresponding to the negative values not being shown. The axis of symmetry of the hyperbola representing the relationship is the line $x = 0$. The inset shows the linearity between T^2 and x^2 that results from squaring Eq. (7-1). The velocity V_0 can readily be determined from the slope of the line thus obtained.

As the inclination of the down-going ray decreases (the angle with the vertical increases) the down-traveling ray eventually approaches the boundary at the critical angle, $\sin^{-1}(V_0/V_1)$. At angles smaller than critical, a large proportion of the energy in the wave is transmitted (refracted) downward into the layer below the interface, as shown in Fig. 2-14. At critical-angle incidence, the refracted wave travels horizontally along the boundary at the speed of the underlying medium. At any greater angle of incidence, there is total reflection, and the wave does not penetrate into the lower material at all. Reflection continues to take place at angles greater than the critical angle, but it is evident from Fig. 7-3 that the ray refracted

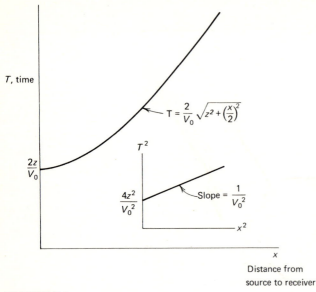

FIGURE 7-2
Time-distance relation for a wave reflected from a horizontal surface at depth
z in a medium of constant velocity V_0. Inset shows linear relation between
square of time and square of distance.

horizontally along the interface that is returned to the surface at the critical angle
will reach a distant point on the surface before the reflected wave reaches the same
point. Up to the distance X_{cross} which is shown on the diagram, the direct waves
that have taken a horizontal path along the earth's surface will be the first events
to arrive.

The depth z can be determined from the time-distance relation for the waves
returned to the surface after refraction along the interface. This is the basis of the
seismic refraction method, which will be covered in Chap. 9. Refraction theory
applied to first arrival times is frequently used for correcting reflection times for
delays in the low-speed weathered zone just below the earth's surface. The techniques used for such corrections will be discussed later in the present chapter.

Continuous increase of velocity with depth   In a model that is more realistic
in many areas than one involving discrete layers, the velocity of the medium is
assumed to increase continuously rather than stepwise with depth. The time-
distance curves for this case will depend, of course, on the functional relationship
between velocity and depth. A linear increase of "instantaneous" velocity* $v(z)$
with depth z is not only simple to handle mathematically but turns out to be a good
approximation to actual velocity functions in basin areas, such as the Gulf Coast

---

* This is the velocity across an infinitesimally thin layer of the material within which the velocity
V, although continually changing with depth, can be considered to be constant. For further
discussion of different kinds of velocity, see pp. 223–226.

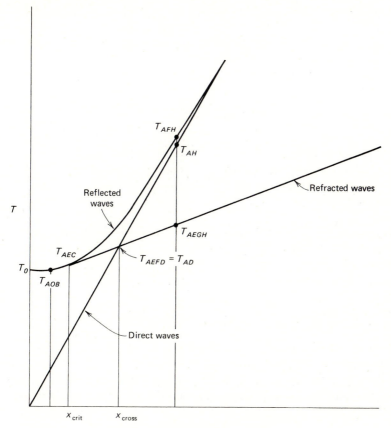

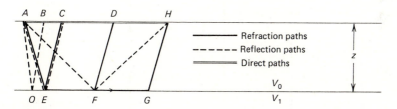

FIGURE 7-3

Relation between times for waves reflected and refracted from a horizontal interface. Times for direct wave also shown.

of the United States, where the sediments are predominantly clastic. The relation is expressed by the equation

$$V(z) = V_0 + kz \qquad (7\text{-}4)$$

where V_0 is the velocity at the surface ($z = 0$) and k is the velocity gradient, or the acceleration factor. It will be shown in Chap. 9 that a ray traveling downward in this kind of medium follows a trajectory that is a circular arc like that shown in Fig. 7-4. This arc has its center along a horizontal line a distance V_0/k above the surface and a radius such that the ray will reach the reflector at a point having an x coordinate midway between that of the source and that of the receiver. But when the shot and receiver are coincident and the ray paths are vertical, the time-depth relation is quite simple. It is

$$z = \frac{V_0}{k}\left(e^{kt/2} - 1\right) \tag{7-5}$$

where t is the two-way reflection time. This relation is plotted in Fig. 7-5. It is sometimes most useful to express the velocity as a linear function of vertical travel time rather than depth, although the relations between time and depth become more complex for this case.

Angle of emergence The angle at which a seismic wave returned by reflection or refraction reaches the earth's surface, generally referred to as the *angle of emergence* α, can be determined from the difference Δt between the arrival times at two nearby receivers a distance Δx apart along the surface (Fig. 7-6):

$$\sin \alpha = \frac{V\,\Delta t}{\Delta x} = V\frac{dt}{dx} \qquad \text{as } \Delta x \to 0 \tag{7-6}$$

where V is the velocity at the surface. A knowledge of this angle makes it possible to trace the ray toward its source, allowing calculation of the dip of the reflecting or refracting boundary from which it comes to the surface. The value of dt/dx can be determined by taking the slope of the time-distance curve at the receiving point.

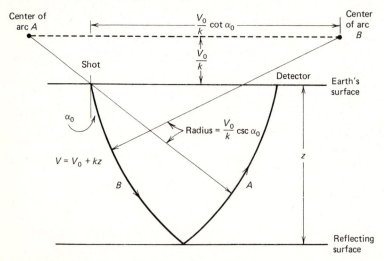

FIGURE 7-4

Ray-path trajectories for reflected wave traveling through material in which velocity increases linearly with depth.

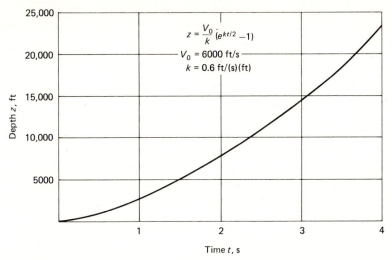

FIGURE 7-5
Time-depth relation for linear increase of velocity with depth.

Reflections from dipping beds   Let us consider the case of a reflecting surface, as shown in Fig. 7-7, dipping at an angle ϕ and underlying a homogeneous medium having a velocity V_0. The source is at O, and receivers are located on both sides of it as shown. The shot and geophones are laid out in a split-spread arrangement, as described in Chap. 4.

Above the cross section showing the wave paths is a plot of reflection times as a function of distance along the line. The path lengths for the reflection are best determined by locating the mirror-image point O' along the perpendicular to the reflecting surface a distance below the surface equal to the distance H.

Consider the path OQR, which is equal to $O'QR$, where R is a distance x from the source O, the origin of our coordinate system. From the law of cosines we can write

$$(V_0T)^2 = (2H)^2 + x^2 - 2(2H)x \cos (90° + \phi)$$

$$= 4H^2 + x^2 + 4Hx \sin \phi \tag{7-7}$$

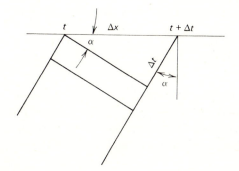

FIGURE 7-6
Determination of emergence angle α.

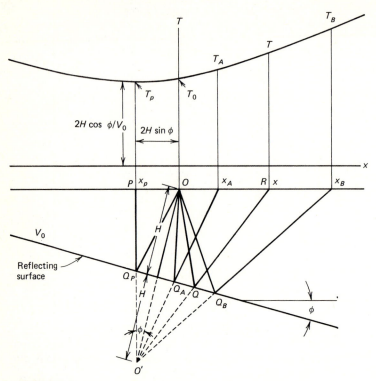

FIGURE 7-7
Ray-path trajectories and time-distance relation for dipping bed.

This can be shown to be the equation of a hyperbola whose vertical axis of symmetry passes through the point P (shown on the upper part of Fig. 7-7) having the coordinates

$$x_p = -2H \sin \phi \quad \text{and} \quad T_p = \frac{2H \cos \phi}{V_0} \tag{7-8}$$

This displacement of the minimum-time position on the x scale as well as the minimum time values can be used to determine the dip angle ϕ. Dividing the first part of Eq. (7-8) by the second, we get

$$\tan \phi = -V_0 \frac{x_p}{T_p} \tag{7-9}$$

The tangent is negative only because of the sign convention chosen. If the dip were in the opposite direction, it would be positive.

Another way of measuring dip from these data is to observe the reflection times T_A and T_B at distances x_A and x_B, respectively, as shown in Fig. 7-7. Substituting these times in Eq. (7-7), we get

$$(V_0 T_A)^2 = (2H)^2 + x_A^2 + 4H x_A \sin \phi$$

$$(V_0 T_B)^2 = (2H)^2 + x_B^2 + 4H x_B \sin \phi \tag{7-10}$$

We then subtract the first of these equations from the second, getting

$$V_0^2 (T_B^2 - T_A^2) = (x_B^2 - x_A^2) + 4H(x_B - x_A) \sin \phi \tag{7-11}$$

so that

$$\sin \phi = V_0^2 \frac{(T_B^2 - T_A^2)}{4H(x_B - x_A)} - \frac{x_B^2 - x_A^2}{4H(x_B - x_A)}$$

$$= \frac{V_0^2 (T_B^2 - T_A^2)}{4H(x_B - x_A)} - \frac{x_B + x_A}{4H} \tag{7-12}$$

If $(T_A + T_B)/2$, the average of T_A and T_B, is designated as T_{av}, the time difference $T_B - T_A$ as ΔT, and $2H$ as $V_0 T_0$, where T_0 is the time of a reflection received at the shot point ($x = 0$), we can write Eq. (7-12) in the form

$$\sin \phi = \frac{V_0 T_{av} \Delta T}{2 T_0 (x_B - x_A)} - \frac{x_B + x_A}{2 V_0 T_0} \tag{7-13}$$

If receiver A is located at the shot point ($x_A = 0$),

$$\sin \phi = \frac{V_0 T_{av} \Delta T}{2 X_B T_0} - \frac{x_B}{2 V_0 T_0} \tag{7-14}$$

If the reflecting bed is flat ($\phi = 0$), $\sin \phi = 0$ and the time differential at x_B with respect to a vertical reflection from the same interface will be

$$\Delta T = \frac{2 x_B^2}{V_0^2 (T_0 + T_B)} \tag{7-15}$$

For a split spread with $x_A = -x_B$ we have

$$\sin \phi = \frac{V_0^2 (T_B^2 - T_A^2)}{8 H x_B} = \frac{2 V_0^2 (T_{av}) \Delta T}{4 V_0 T_0 x_B} = \frac{V_0 T_{av}}{2 T_0} \frac{\Delta T}{x_B} \tag{7-16}$$

These relations hold only for the plane of the wave paths. If the shooting line is not in the direction of dip, the angle ϕ is the *apparent dip*, which is related to the angle of true dip by the equation

$$\sin \phi = \cos \psi \sin \delta \tag{7-17}$$

where ψ is the angle between the shooting line and the direction in which the reflecting surface actually dips.

If the direction of the true dip is not obtainable from other information, one must shoot in two different directions (not necessarily perpendicular to each other) and measure apparent dips in these directions. A simple geometrical construction can then be used for obtaining the magnitude and direction of the true dip of the reflecting interface from the apparent dips in two respective directions. Figure 7-8 illustrates the use of a chart designed for solving this problem. The mathematical justification for this construction can be found in Slotnick[2] (p. 164). The time differential for a specified distance along line A is ΔT_A, and that for line B is ΔT_B. These times are converted to distances, using a convenient scale factor, which are plotted from the center of the chart in the directions of the respective shooting lines. Perpendiculars are drawn from the end of each line. The direction of true dip is obtained by drawing a line from the center to the point of intersection of these perpendiculars. The magnitude of the true dip is proportional to the length of the line. A chart of this type is exact only to the extent that the sines of the different dip angles are small enough to be approximated by their tangents.

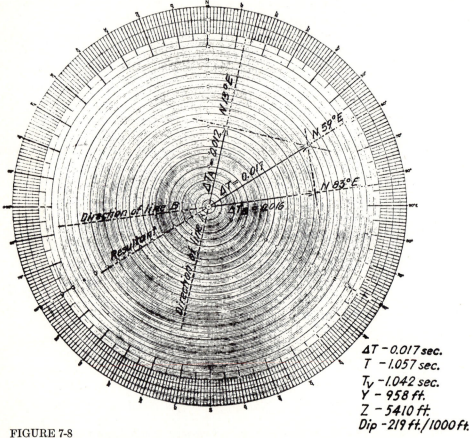

$\Delta T - 0.017\ sec.$
$T - 1.057\ sec.$
$T_V - 1.042\ sec.$
$Y - 958\ ft.$
$Z - 5410\ ft.$
$Dip - 219\ ft./1000\ ft.$

FIGURE 7-8
Use of chart to determine true dip from two dip components making an arbitrary angle with each other. (*Geophysical Service, Inc.*)

7-2   CORRECTIONS USED IN REDUCTION OF REFLECTION RECORDS

To obtain the greatest possible accuracy in mapping subsurface structures from reflection times it is necessary to correct reflection times for predictable irregularities not associated with the actual structures. An obvious source of such irregularity is surface elevation. If reflection waves from a flat subsurface interface were received by geophones spread over a hill, valley, or other topographic feature, the reflection times would indicate a structure that could be associated with the elevations at the earth's surface rather than with those of the subsurface formations being mapped. From a knowledge of the elevations and near-surface velocities, one can compute the variations in reflection time at points along the surface that are attributable to such topographic irregularities. Observed reflection times can be corrected for the effects of these irregularities by proper subtraction or addition of the time increments thus determined. Such a correction leaves time variations associated only with the subsurface structure under investigation.

Another source of irregularity in reflection times is the *weathered layer* just below the earth's surface. This generally consists of rock materials, usually above the water table, that are quite unconsolidated and hence have a very low seismic velocity, often 2000 ft/s or less. Because of this anomalous speed, any change in the thickness of the layer has an unduly large effect on the travel time of a reflected signal passing through it on its way to the surface. A number of techniques have been developed for determining the thickness of this layer at each detector station. A knowledge of the thickness and velocity of the low-speed layer makes it possible to correct for its effect on reflection times.

Before corrected record sections came into use, the elevation and weathering corrections were made *after* reflection times were read from the records. Now the corrections are introduced into computer storage for automatic application to the signals recorded in the field when they are played back in processing centers. These *static corrections*, as they are called, are applied to the data along with the *normal moveout*, or *dynamic*, *corrections* for angularity of ray path, which are computed from shot-geophone separations, reflection times, and average velocities.

Elevation Correction

Since the paths of reflected waves are usually close to vertical near the surface,* the correction for elevation difference can be made simply by subtracting (or adding) the time required for the wave to travel the vertical distance between a reference elevation and that of the earth's surface at the point in question. Two procedures, both illustrated in Fig. 7-9, are commonly used. One is to adjust the

---

* Even when the ray path makes a large angle with the vertical at the reflecting surface, refraction into progressively slower materials as the wave travels upward toward the earth's surface almost always assures a near-vertical ray path in the vicinity of the surface.

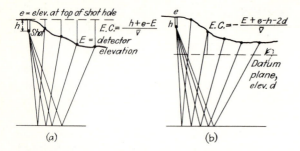

FIGURE 7-9
Elevation corrections (a) by putting all geophones at elevation of top of shot-hole and (b) by putting both shot and geophones on datum plane. Weathered layer removed.

reflection times to what they would be if the shot and all the detectors were at the same elevation as the top of the shothole. Then depths to reflecting horizons are computed with respect to the ground elevations at each shot location, and depths with respect to sea level are computed with reference to such elevations. The other method is to make a correction which puts both the shot and detectors on an arbitrarily chosen datum plane. The times required for the wave to travel down to the datum plane from the shot and up from the datum plane to the detector are then removed. The excess times are computed simply by dividing the differences in elevation between shot and datum and also between detector and datum by the near-surface velocity. Depths below datum are calculated from the corrected reflection times.

Weathering Corrections

The object of the weathering corrections, as has been noted, is to eliminate the effect on travel times of variations in the thickness of the low-speed zone. The effect of such variations on the apparent structure of reflectors is illustrated in Fig. 7-10. The velocity in the surface layer is so much lower than the average speed of the material between the base of the layer and the reflecting surface that any change in its thickness is observed on the record as a much greater change in the depth of the reflector. The increase in the weathering thickness of 20 ft gives an anomaly in one-way time of 7.5 ms, and when this time variation is multiplied by half the average velocity to the flat reflecting bed of 8000 ft/s, the reflector has an apparent relief of 30 ft. Unless a correction is made for the change in thickness of the low-speed zone, a spurious depression 30 ft deep will be indicated on the final map.

Many techniques are used for making such a correction. With some methods one "removes" the weathered layer altogether, putting all detectors effectively on a datum below its base; with others one "replaces" the layer with high-speed material of the kind that is found below the base of the weathered zone.

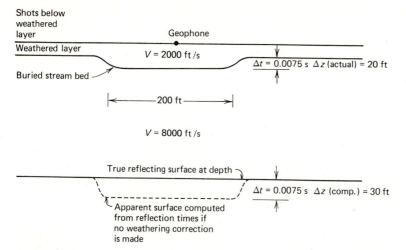

FIGURE 7-10
Spurious structure introduced at depth by failure to take local thickening of
weathered zone into account. Note magnification of relief at deep reflection
because of increase in velocity.

Weathering corrections using first arrivals   The first events to arrive at the
geophones on a reflection profile have traveled by refraction along the top of the
high-speed zone just below the weathered layer. Using the time-distance curves of
the first arrivals for shots from opposite directions, one can calculate the thickness
of the weathered layer by conventional refraction methods of the type that will be
taken up in more detail in Chap. 9.

SHOT BELOW WEATHERED ZONE   Figure 7-11 shows the wave paths of the first
energy to be received at points C_1 and C_2 from a shot exploded in a borehole at A
just below the base of the weathered layer. The weathered zone with a thickness
z_A from 0 to C_1 increasing to z_B at C_2 is assumed to have a constant velocity V_0,
and the speed of the consolidated material below it is V_1, which is greater than V_0.
   Consider first the time required for this wave to travel from A to C_1. From A to
B_1 its velocity is V_1, and it travels from B_1 to C_1 at a speed of V_0 along a path making
the critical angle α_c, which is $\sin^{-1}(V_0/V_1)$, with the vertical.
   The total time from A to C_1 is

$$T_{AC_1} = \frac{AB_1}{V_1} + \frac{B_1C_1}{V_0} \tag{7-18}$$

$$= \frac{OC_1 - z_A \tan \alpha_c}{V_1} + \frac{z_A}{\cos \alpha_c} \frac{1}{V_0} \tag{7-19}$$

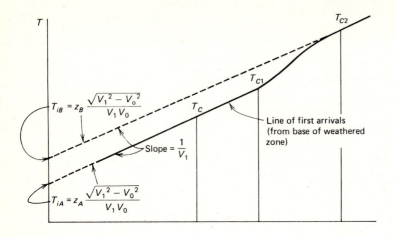

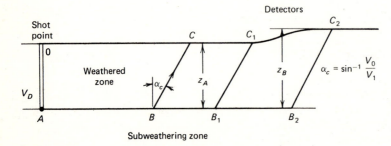

FIGURE 7-11
Changes of intercept times with thickness of weathered layer in determining weathering delay by plotting first arrivals.

$$T_{AC_1} = \frac{OC_1}{V_1} + z_A \left(\frac{1}{\cos \alpha_c V_0} - \frac{\tan \alpha_c}{V_1} \right) \qquad (7\text{-}20)$$

$$= \frac{OC}{V_1} + \frac{z_A}{V_0 \cos \alpha_c} (1 - \sin^2 \alpha_c)$$

$$= \frac{OC_1}{V_1} + \frac{z_A \cos \alpha_c}{V_0} \qquad (7\text{-}21)$$

$$= \frac{OC_1}{V_1} + \frac{z_A}{V_0} \sqrt{1 - \frac{V_0^2}{V_1^2}} \qquad (7\text{-}22)$$

$$= \frac{OC_1}{V_1} + \frac{z_A \sqrt{V_1^2 - V_0^2}}{V_1 V_0} \qquad (7\text{-}23)$$

Also
$$T_{AC_1} - \frac{OC_1}{V_1} = \frac{z_A\sqrt{V_1^2 - V_0^2}}{V_1 V_0} \tag{7-24}$$

Now if the time from A to C_1 is plotted against the distance OC_1, the term on the right will be the travel time at zero distance, which is designated the *intercept time* T_i.

Thus if the arrival times are plotted versus the shot-detector distance, as at the top of the figure, the time where the best line through the points intercepts the time axis is proportional to the layer thickness, as follows:

$$T_{iA} = z_A \frac{\sqrt{V_1^2 - V_0^2}}{V_1 V_0} \tag{7-25}$$

or
$$z_A = \frac{V_1 V_0 T_{iA}}{\sqrt{V_1^2 - V_0^2}} \tag{7-26a}$$

Similarly,
$$z_B = \frac{V_1 V_0 T_{iB}}{\sqrt{V_1^2 - V_0^2}} \tag{7-26b}$$

The excess time that must be removed to replace a surface layer of thickness z and speed V_0 with material having a speed V_1 is

$$\Delta T = \frac{Z}{V_0} - \frac{Z}{V_1} = z\left(\frac{1}{V_0} - \frac{1}{V_1}\right) \tag{7-27}$$

$$= T_i \frac{V_1 V_0}{\sqrt{V_1^2 - V_0^2}}\left(\frac{1}{V_0} - \frac{1}{V_1}\right) \tag{7-28}$$

$$= T_i \left(\frac{V_1}{\sqrt{V_1^2 - V_0^2}} - \frac{V_0}{\sqrt{V_1^2 - V_0^2}}\right) \tag{7-29}$$

$$= T_i \frac{V_1 - V_0}{\sqrt{V_1^2 - V_0^2}} = T_i \sqrt{\frac{V_1 - V_0}{V_1 + V_0}} \tag{7-30}$$

The value of the constant under the radical must always be less than 1, and for the velocities usually observed above and below the base of the weathered layer (such as 2000 and 7000 ft/s, respectively) it is generally in the neighborhood of 0.7.

The velocities V_0 and V_1 can be determined by shooting small charges at a number of depths in a shothole and receiving the signals from them with a detector at the top of the hole. Times for such events are plotted against depth, as in Fig. 7-12. Note that the break in apparent speed in the example shown in the figure does not coincide with any of the lithological boundaries indicated on the drillers' log. It is likely, therefore, that the break in velocities is associated with the water table, which would not show up on the log. The V_1 velocity can also be obtained from the

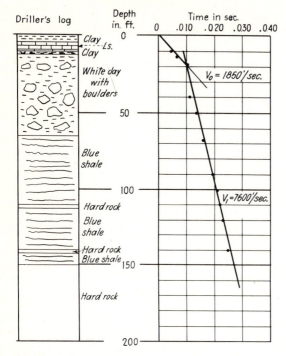

FIGURE 7-12
Vertical velocity distribution near the surface determined by "shooting up the hole." (*Geophysical Service, Inc.*)

inverse slope of the time-distance curve for first arrivals on the reflection record as observed in Fig. 4-3.

Figure 7-13 shows a sample set of time-distance curves from which intercept times can be determined at each detector position. For geophone 5, for example, the intercept time T_i of the shot from A is 0.015 s, and of that from B is 0.018 s. Both times were obtained by adding or subtracting the intervals between the points corresponding to the positions of geophones on the curve and respective average lines. V_0 is known to be 2000 ft/s. With an average V_1 velocity of 8150 ft/s and an average T_i taken as 0.016 s, the excess time T_{ex}, obtained by substituting these numbers in Eq. (7-30), is

$$T_{ex} = \sqrt{\frac{8150 - 2000}{8150 + 2000}} \times 0.016 = 0.012 \text{ s}$$

This correction is subtracted from the observed arrival time of the reflections received at this location.

SHOT ABOVE BASE OF WEATHERING Where the shothole has not penetrated the weathered layer, the refraction method is difficult to apply because there is no way

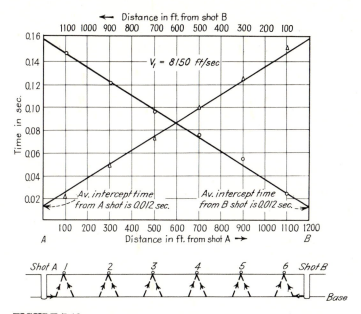

FIGURE 7-13
Plot of arrival times from two adjacent reflection shots received by same geophones. Average intercept time at each geophone used to calculate weathering correction. (*Geophysical Service, Inc.*)

of separating the portions of the intercept time corresponding to the paths through the low-speed zone at the respective ends of the trajectory. For an individual shot, this complication would not be serious, since the total delay in the weathered zone, wherever it might originate, would be taken care of in the correction. It is not possible, however, to average intercept times correctly for shots from opposite directions (as shown in Fig. 7-13) unless the contributions from each end can be separated. For this reason, a method which does not use intercept times at all is preferred when low-speed material appears to lie below one or both of the shots on opposite sides of the detector station. Figure 7-14 shows the ray paths for the first arrivals in this case.

The times of refracted waves received at D from shots A and B are

$$T_{AD} = T_{AS} + T_{SP} + T_{PD} \tag{7-31}$$

and

$$T_{BD} = T_{BT} + T_{TQ} + T_{QD} \tag{7-32}$$

respectively, the subscripts representing the legs along the paths indicated in the ray diagram of Fig. 7-14. At detector E, the refracted wave from A has a travel time

$$T_{AE} = T_{AS} + T_{SM} + T_{ME} \tag{7-33}$$

where

$$T_{ME} \approx T_{BT} + T_{uhB}$$

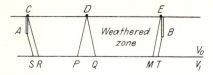

FIGURE 7-14
Shallow refraction paths where shot-holes do not penetrate below weathered zone.

T_{uhB} being the uphole time for shot B,

and
$$T_{SM} \cong T_{SP} + T_{PQ} + T_{TQ} - T_{MT}$$

From these relations it is easy to show that

$$T_{PD} + T_{QD} = T_{AD} + T_{BD} - T_{AE} + T_{uhB} - T_{PQ} + T_{MT} \qquad (7\text{-}34)$$

When, as is usually the case with the large velocity contrasts across the base of the weathered zone, the slant paths through the low-speed zone are nearly vertical, the last two terms on the right-hand side of Eq. (7-34) become negligible. In addition, times T_{PD} and T_{QD} are theoretically equal, so that each can be designed as T_w, which can be expressed in the form

$$T_w \cong \tfrac{1}{2}(T_{AD} + T_{BD} - T_{AE} + T_{uhB}) \qquad (7\text{-}35)$$

or alternatively,

$$T_w \cong \tfrac{1}{2}(T_{AD} + T_{BD} - T_{BC} + T_{uhA}) \qquad (7\text{-}36)$$

T_w closely approximates the time required for a reflected wave to pass through the weathered layer on its way to detector D. When it is subtracted from the time of all reflections received at D, the detector is effectively put at the base of the weathering.

T_{AD}, T_{BD}, T_{AE}, and T_{BC} are first-arrival times which are read directly from the appropriate traces on the records. Table 7-1 gives a sample calculation of corrections made with this procedure. The bottom line represents the weathering thickness z_w obtained by multiplying T_w and V_0, taken here to be 2000 ft/s.

Weathering corrections from uphole times   Where the weathering correction does not require high precision at each detector position, or where the precision that is desired is not obtainable from first-arrival data, considerable time can be saved if the correction is made directly from uphole times rather than from the first arrivals on the individual reflection traces. This approach gives the corrections only at the shot points, but with multifold coverage shot points may be located at each detector position (for twelvefold shooting) or at every second position (for sixfold).

The time necessary for the wave to travel vertically from the shot to an uphole detector on the surface is the same as would be required for a reflection to travel vertically upward from the level of the shot to the same detector. Where the shot is below the weathered layer, subtraction of this uphole time from the reflection time effectively puts the detector in the usually consolidated material at the level of the shot. By subtracting twice the time corresponding to vertical travel (at the speed of the subweathering material) between the shot position and the datum, one can effectively place both shot and detector on the datum plane, as shown in Fig. 7-15.

Table 7-1   SAMPLE WEATHERING CALCULATION GIVING TIMES IN SECONDS WHEN SHOTS DO NOT PENETRATE WEATHERED LAYER*

	A 0	100	200	300	400	500	600	700	800	B 900
				Distance from A, ft						
T_{AD}	0.012 ($= T_{uhA}$)	0.038	0.065	0.089	0.101	0.128	0.138	0.159	0.175	0.188 ($= T_{AE}$)
T_{BD}	0.174 ($= T_{BC}$)	0.159	0.148	0.129	0.114	0.097	0.078	0.078	0.048	0.010 ($= T_{uhB}$)
$Q = T_{AD} + T_{BD}$		0.197	0.213	0.218	0.215	0.225	0.216	0.237	0.223	
$R = T_{AE} - T_{uhB}$		0.178	0.178	0.178	0.178	0.178	0.178	0.178	0.178	
$2T_w = Q - R$		0.019	0.035	0.040	0.037	0.047	0.038	0.059	0.045	
T_w		0.010	0.018	0.020	0.019	0.023	0.019	0.030	0.023	
z_w, ft		20	36	40	38	46	38	60	46	

* See Fig. 7-15 for definitions of times.

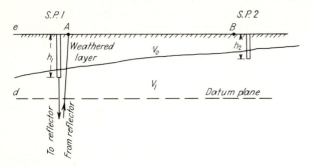

FIGURE 7-15
Use of uphole times for weathering corrections. T_{uh1} = time uphole at shot point 1; T_{uh2} = time uphole at shot point 2; e = surface elevation above sea level (assumed the same for shot and detector); h_1 = depth of shot; d = datum elevation above sea level; C_A = correction to datum in time of reflection from shot point 1 received at A; C_B = correction to datum in time of reflection from shot point 1 received at B; $C_A = T_{uh1} + 2(e - h_1 - d)/V_1$; $C_B = T_{uh2} + [(e - h_1 - d) + (e - h_2 - d)]/V_1$. Corrections at points between A and B are interpolated.

If the center of a group of geophones corresponding to a particular trace is not located at a shot point but between shot points, it will be necessary to distribute the correction uniformly between the adjacent shots.

Automatic Determination of Elevation and Weathering Correction

The widespread use of common-depth-point techniques in reflection shooting has made the requirements for precision in trace corrections much more stringent than with conventional single-coverage shooting. A reflection pulse with a period of say 20 ms can be badly distorted upon stacking or even lost if the random error due to imperfect corrections in the times of the individual events to be stacked is as little as ±5 ms. For this reason, the matter of trace corrections has become much more important in seismic reflection work than it was when there was only single coverage and when corrections were computed only for traces nearest the shot.

The tedious effort necessary to make manual trace corrections by techniques like those we have described, as well as the impossibility of making manual corrections when the anomalies in travel time result from velocity variations deeper than the weathered layer, has led to the development of special techniques for obtaining the corrections automatically with a computer.

In most such procedures, the computer is programmed to obtain corrections that will best align the reflections on traces from a common depth point so that primary reflections will be most likely to add up exactly in phase during the subsequent stacking. All automatic correction techniques make use of the redundancy that results when signals from many different shots are recorded at the same geophone positions and when signals from the same shot are received at a multiplicity of geophone positions. The static correction for each trace is of course made up of a component associated with the shot position and one associated with the detector position. By assuming that all traces recorded from a particular shot will have the same shot correction and that all those recorded by the same detector group will have the same detector correction, it is possible to set up a matrix for which the computer can solve for all the individual elements. One problem is to separate changes in time due to structural relief on the reflector to changes caused by near-surface irregularities.

Approximate trace corrections are initially computed by conventional methods like those considered earlier in this section. The residual values necessary to optimize alignment would then be determined by computer operations such as cross-correlation or matrix solutions.

One technique of this type has been described by Hileman et al.[3] Cross-correlation functions are used to measure the relative time shifts of each trace contributing to a single output trace from a common depth point. An arbitrary reference trace is used for correlating the individual signals for which static shifts are desired. Figure 7-16 shows the computation array these authors use for their correction technique. Time shifts obtained by cross-correlation are averaged over the rows to obtain the correction associated with the shot corresponding to each row. The value for the receivers can be obtained by averaging the time shifts in the columns representing particular detecting groups. The assumption is made that there are no systematic reflection effects as from dips.

Another technique for automatic trace corrections, described by Disher and Naquin,[4] involves the correlation of traces in pairs. One pair represents traces

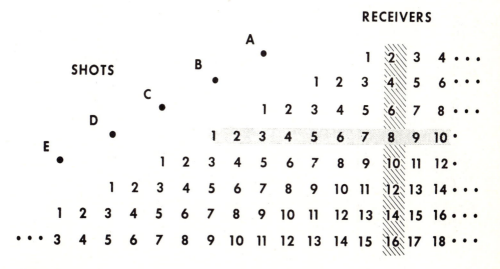

FIGURE 7-16
Array for computing static corrections with sixfold shooting geometry. Time shifts common to the same shots lie along the rows, and time shifts common to the same geophones lie along the columns. (*From Hileman et al.[3]*)

recorded by the same receiver from two consecutive shots. These give static corrections for the shots. The other type of pair is obtained when two adjacent detectors receive signals from the same shot. Correlation between these yields corrections for the detector positions. Differences that are consistent over a long distance are assumed to be associated with dip rather than trace irregularities, and these differences are subtracted from the observed differentials to give the desired residual static corrections. Figure 7-17 illustrates the improvement in reflection quality that such techniques can sometimes bring about.

Blondeau-Swartz Method for Weathering Corrections

In many areas, e.g., those where the surface material consists of glacial drift, the low-speed surface-layer velocity is not constant, as has been assumed in the correction procedures we have considered, but increases rapidly with depth below the surface. A velocity function that closely approximates what is observed from uphole surveys in such areas is

$$v = Az^{1/n} \tag{7-37}$$

where v = instantaneous velocity
$\quad z$ = depth
$\quad A$ = constant
$\quad n$ = number characteristic of surface material in area under consideration

In the late 1930s E. E. Blondeau developed a method of making weathering corrections for surface layers having this type of velocity increase with depth. In 1950 C. A. Swartz made some improvements on Blondeau's method, but the details of the technique were not published until 1963, when Duska[5] presented a theoretical basis for the Blondeau–Swartz approach along with some adaptations for applying it more rapidly and conveniently. Hollister[6] has also published a paper on this technique.

The method requires the plotting of time versus distance on log-log paper. If the $1/n$ power relation between velocity and depth holds, the time-distance curve so plotted will be a straight line and n can be determined from it quite simply. Using the slope of the line, one follows a computational procedure that yields the time of a vertically traveling wave through the near-surface formations to a depth which can be assumed to be unquestionably below the weathering. The time thus obtained is used to correct reflection traces to this depth. The method appears to be most applicable where the weathered layer consists mainly of glacial drift or is thick enough to cause differential compaction between top and bottom.

7-3 DETERMINATION OF SEISMIC VELOCITIES

An important step in the interpretation of reflection records is the conversion of reflection times to depths. This step requires the velocity of seismic waves to be known in the material through which the waves travel. In Chap. 2 we considered the factors governing seismic velocities in rocks. Our concern here is the measurement of the actual velocities at the specific locations where the reflection shooting is to be carried out. Such velocities can be measured directly in boreholes, or they can be obtained indirectly by analysis of time-distance relationships on the reflection records themselves. The spreads now used in recording by common-depth-point techniques are generally so long (up to 2 mi) that analytical methods are considerably more accurate than they were when the shorter spreads used for single-coverage shooting were common.

A relatively recent use of analytically determined seismic velocity data is the identification of lithology for discrete formations within the geologic section. The precision obtainable in such determinations makes it possible to obtain useful information of this kind from most modern reflection shooting.

At this point it is desirable to define the different kinds of velocities that enter into seismic interpretation. The following types are referred to most frequently in the geophysical literature:

Average velocity   This is simply the depth z of a reflecting surface below a datum divided by the observed one-way reflection time t from the datum to the surface so that

$$V_{av} = \frac{z}{t} \tag{7-38}$$

If z represents the sum of the thicknesses of layers $z_1, z_2, z_3, \ldots, z_n$, the *average velocity* is defined as

$$V_{av} = \frac{z_1 + z_2 + z_3 + \cdots + z_n}{t_1 + t_2 + t_3 + \cdots + t_n} = \frac{\sum_1^n z_k}{\sum_1^n t_k} \tag{7-39}$$

Interval velocity   If two reflectors at depths z_1 and z_2 give reflections having respective one-way times of t_1 and t_2, the *interval velocity* V_{int} between z_1 and z_2 is defined simply as $(z_2 - z_1)/(t_2 - t_1)$.

Instantaneous velocity   If the velocity varies continuously with depth, its value at a particular depth z is obtained from the formula for interval velocity by contracting the interval $z_2 - z_1$ until it becomes an infinitesimally thin layer having a thickness dz. The interval velocity computed by the formula above becomes the derivative of z with respect to t, and we designate it as the *instantaneous velocity* V_{inst}, defined as

$$V_{inst} = \frac{dz}{dt} \tag{7-40}$$

Root-mean-square velocity   If the section consists of horizontal layers with

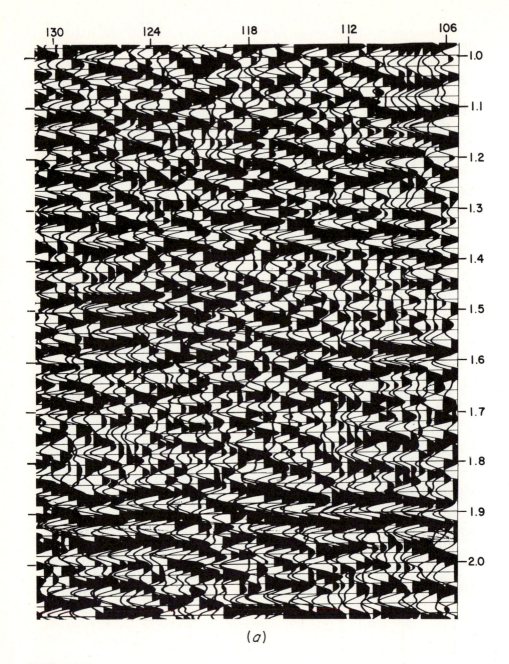

(a)

FIGURE 7-17

Example of improvement possible with automatic static corrections: (a) west Texas record section with conventional static corrections; (b) same section with automatic statics. (*Geocom, Inc.*)

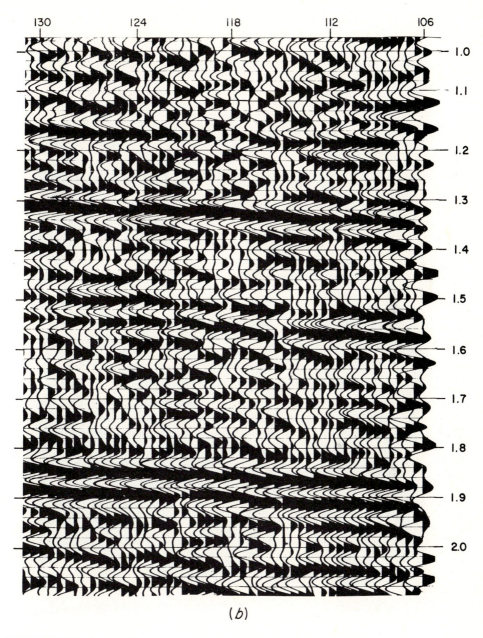

(b)

FIGURE 7-17 (continued)

respective interval velocities of v_1, v_2, v_3, ..., v_n and one-way interval travel times t_1, t_2, t_3, ..., t_n, the *root-mean-square* (rms) *velocity* V_{rms} is obtained from the relation

$$V_{rms}{}^2 = \frac{v_1{}^2 t_1 + v_2{}^2 t_2 + v_3{}^2 t_3 + \cdots + v_n{}^2 t_n}{t_1 + t_2 + t_3 + \cdots + t_n} = \frac{\sum_1^n v_k{}^2 t_k}{\sum_1^n t_k} \qquad (7\text{-}41)$$

This velocity can be obtained by taking the slope of the curve for t^2 versus x^2 at $x = 0$.

Stacking velocity   This velocity V_{st} is based on the relation

$$T^2 = T_0{}^2 + \frac{x^2}{V_{st}{}^2} \qquad (7\text{-}42)$$

where x is the variable separation of shot and receiver for a common-reflecting-point sequence of shots, T is the travel time at x, and T_0 is the vertical time. It is approximately but not exactly the same as the rms velocity and it is the velocity actually used in the stacking process.

Well Shooting

The most direct procedure for velocity measurement is to explode charges of dynamite near the surface alongside a deep borehole and to record the arrival times of waves received by a geophone suspended in the borehole at a number of depths distributed between its top and bottom. Figure 7-18 illustrates the setup and shows interval- and average-velocity curves of the type that are obtained from such data. The interval velocity is obtained by taking the distance between successive detector

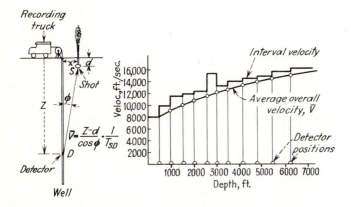

FIGURE 7-18
Well-shooting arrangement with typical interval-velocity and average-velocity curves thus obtained.

positions in the well and dividing it by the difference in arrival times at the two depths after the arrival time has been corrected for angularity of the wave path. The average velocity is either the actual distance from source to receiver divided by the observed time or the vertical component of distance divided by the appropriately corrected time.

Data from velocity surveys must sometimes be applied at great distances from the well actually surveyed. Often values from a number of widely separated wells are averaged to obtain a velocity function which is used over the area in-between. Where enough surveys are available, the average velocities to the horizons which are mapped can be interpolated between wells by contouring. Considerable variation is sometimes found between curves obtained from neighboring wells. These changes often have geological significance which could cast light on the regional geology of the area involved.

Continuous Velocity Logs

Most well velocity surveys carried out since 1955 have made use of continuous velocity logs along with conventional recording in the borehole of near-surface shots, as described in the preceding paragraphs. These logs show the interval velocities of the formations penetrated by the wells as functions of depth either for the entire well below the casing or for selected formations. For seismic applications it is desirable that all of the well below the casing be logged.

Continuous velocity logs have many applications, such as geological correlation, porosity determination, identification of formations, and synthesis of reflections, but in this chapter we shall confine our discussion to their use in converting reflection times to depths. The logs themselves record, as a function of depth, the travel times of seismic waves through the formation between a sound source near one end of the logging tool and each of two receiving units several feet away. The time differential between the two detectors, ordinarily 1 ft apart but sometimes farther, is plotted continuously on a chart in the logging truck or recorded digitally on tape. The reciprocal time, indicated by an appropriate scale on the log, is simply the interval velocity for the 1-ft section straddled by the receivers as the tool is raised in the hole. Figure 7-19 shows a typical velocity log.

Integrating these logs of interval time with respect to depth yields the total travel time between any two points for a seismic wave passing vertically through the section. Integrated times are generally shown for constant intervals on the log itself as pips along a vertical line near its edge. Because no signals can be recorded through surface casing, it is necessary to record a near-surface shot by conventional means with a detector (ordinarily attached to the logging tool) at the bottom of the casing. The integrated time from the log is added to the time measured through the casing, and the sum (after correction to datum elevation) becomes the time from which velocity is calculated to any desired depth in the borehole. Shots of the same type are also recorded at a number of other depths in the borehole to check the integrated times from the logs and to correct for local irregularities in the velocity.

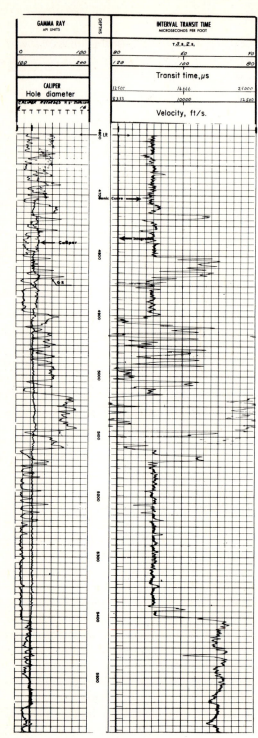

FIGURE 7-19

Velocity–gamma-ray log from well in Michigan. Velocity log is on right side. Transit times between elements 1 ft apart are logged. Velocities (shown on nonlinear scale) are reciprocals of the transit times. (*John W. Mack, Jr.*)

Gretener[7] has compared vertical travel times obtained from well shooting with those determined by integrating velocity logs from 50 wells in the Alberta plains. A statistical analysis of the discrepancies showed an essentially gaussian distribution with a standard deviation of 1.8 ms/1000 ft. An interesting feature of the comparison was a shift of the axis of symmetry of his histogram by 1.7 ms/1000 ft in such a direction that the travel times based on the velocity logs were systematically less than those from the well shots. No plausible explanation could be found by Gretener for this shift, but Strick[8] has endeavored to explain the discrepancies on the basis of inelastic attenuation effects in the formations.

Graphical Determination of Velocity from Reflection Data

Ever since the early days of reflection prospecting, velocities have been determined from data on reflection records using the relationship between reflection times and shot-receiver distances. The earliest procedure, proposed by Green,[9] involves plotting the square of the travel time T to and from a reflector at depth z versus the square of the receiving distance x, using the relation

$$T^2 = \frac{4}{V^2} z^2 + \frac{1}{V^2} x^2 \qquad (7\text{-}43)$$

A plot of this type is shown in Fig. 7-20. It indicates how average velocities are determined, in this case to three successive reflectors. If the velocity V is constant, a plot of T^2 versus x^2 should yield a straight line with a slope $1/V^2$. If $2z/V$ is designated as T_0, the travel time of a reflection received by a geophone at the shot point ($x = 0$), we can write

$$T^2 = T_0^2 + \frac{x^2}{V^2} \qquad (7\text{-}44)$$

A graphical approach that gives interval velocities and rms velocities was developed by Dix.[1] His technique involved recording reflections from a line of subsurface coverage equivalent in length to what would be obtained from a single split spread. Reflections were obtained from the same subsurface line at increasing shot-receiver distances by progressively moving the shot away from the reflecting points in one direction and the receiving points in the other. The techniques used in common-depth-point shooting were anticipated in this procedure except that a common reflecting line was used rather than a common reflecting point.

Dix plotted the squares of the reflection times against the squares of the distances, as shown in Fig. 7-20; however, the slope used for the inverse square of the velocity was not the best average through all the points but the best line that could be drawn tangent to the curve at its origin ($x = 0$). The result of giving greater weight to the data closest to the origin is to obtain a line with an inverse slope equal to the square of the rms velocity rather than of the average velocity.

If there are two reflections having times of T_1 and T_2 with respective rms velocities V_{A1} and V_{A2} (determined as outlined in the last paragraph), Dix showed that the

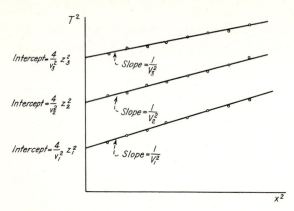

FIGURE 7-20
Determination of average velocities to three horizons by analysis of reflections
on velocity spreads.

interval velocity V_{i12} between the reflections can be obtained from the relation

$$V_{i12}{}^2 = \frac{V_{A2}{}^2 T_2 - V_{A1}{}^2 T_1}{T_2 - T_1} \qquad (7\text{-}45)$$

This formula, known as the *Dix equation*, is widely used to determine interval
velocities between two reflectors from rms velocities.

Automatic Velocity Determination by Digital Computer

With the almost universal digitization of reflection data, it is now customary
to determine velocities directly from such data using computers. A number of
approaches have been programmed for such computations.

Velocity spectra Common-depth-point shooting makes it possible to obtain
velocities from reflection times at various receiving distances much more accurately
than by the earlier techniques described above. Computers can determine velocities
by carrying out calculations based on the time-distance relationship that would be
too tedious to perform manually, and programs are available for automatic plotting
of the velocity-reflection-time relationships thus obtained.

The most commonly used technique for this purpose presents average velocity
as a function of time in a form called a *velocity spectrum*, the technique having been
devised by Taner and Koehler.[10] As with all analytical methods for obtaining
velocities from actual reflection data, the accuracy and usefulness of the results
depend on the quality of the reflection information.

The time T_x of a reflection event at shot-receiver distance x is related to the time
T_0 at the originating point $(x = 0)$ by the equation

$$T_x{}^2 = T_0{}^2 + \frac{x^2}{\bar{V}^2} \qquad (7\text{-}46)$$

Here $\bar{V}^2$ is the stacking velocity, more closely related to the rms velocity than the average velocity, which is always lower. From this equation it is evident that the moveout time, $T_x - T_0$, designated ΔT, can be written

$$\Delta T = \sqrt{T_0^2 + \frac{x^2}{\bar{V}^2}} - T_0 \qquad (7\text{-}47)$$

In common-depth-point stacking, the various signals to be composited into each output trace represent waves reflected from the same subsurface point with different separations between the shot points at which they originated and the geophones by which they were received. Each wave has a different travel time, which differs by an amount ΔT from the time for the path that would be taken by a wave reflected to a receiver located at the shot point. The time-distance relation for this case is shown in Fig. 7-21. If the appropriate value of ΔT is subtracted from the time of each reflection, all the events should have their arrivals at the same time, and if the traces are added after the times are subtracted, the sums should be greater than at any other alignment of the respective events.

In carrying out this manipulation, different values of $\bar{V}$ are assumed, and the computer determines the ΔT's corresponding to each $\bar{V}$ for the respective traces using Eq. (7-47). The ΔT's thus obtained are subtracted from the observed reflection times, and the time-shifted signals for all traces are added. The trial value of $\bar{V}$ which will give the best lineup for each reflection will yield a reflection amplitude after addition that is greater than any other value of $\bar{V}$. All velocities over the entire range that might be plausible for the area are tried at intervals of 100 ft/s.

Figure 7-22 illustrates the principle of the technique. In the top part hyperbola 1, shown by a dashed line, corresponds to an assumed velocity somewhat lower than the true velocity. If amplitude values at times corrected on the basis of this velocity are determined on all the traces and added, the sums will be as shown on the

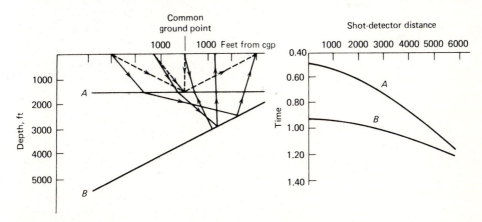

FIGURE 7-21
Schematic time-distance relation for two reflections obtained in common-depth-point shooting arrangement with horizontal and dipping reflectors. (*Modified from Taner and Koehler.*[10])

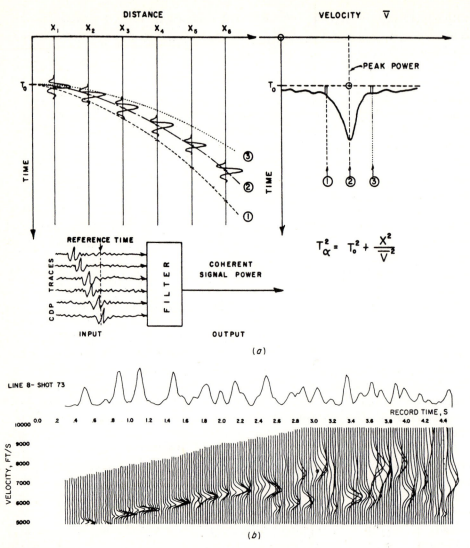

FIGURE 7-22
Velocity spectra: (*a*) alignment of reflections along traces for different shot-receiver distances with shooting configuration for common-depth-point, time-distance relations shown for three assumed velocities; (*b*) typical display of velocity versus time from reflection record. (*From Taner and Koehler.*[10])

plot to the right. Hyperbola 2, which is a full line, shows ΔT versus x for the *correct* velocity. The sum of the amplitudes over all the traces at the times determined by this curve comes out to be greater than for hyperbola 1. Hyperbola 3, a light dotted line, is for a velocity *higher* than the correct value. The sum of trace amplitudes for times determined from this function also turns out, as was the case with hyperbola 1, to be lower than for hyperbola 2. The velocity for which this sum is maximum can thus be considered as the correct velocity for the time chosen for the calculation.

If we choose T_0 values at regular closely spaced intervals over the length of the record, we can obtain a series of such amplitude-versus-velocity curves, one for each T_0 value, indicating the average velocity as a function of reflection time at the shot point. The lower part of Fig. 7-22 illustrates a spectrum of the type obtained when good reflections occur at frequent intervals on the record. The anomalously low velocities which begin to show up at about 2.6 s probably correspond to multiple reflections, as their values are the same as those for times half as great. In determining a continuous velocity function, it is necessary to interpolate velocity values over portions of the time scale where reflections are not recorded.

Velocity scans   An exceptionally simple method for determining average velocity as a function of record time is the velocity scan. An example demonstrating it is shown in Fig. 7-23. A series of trial velocities is assumed with this technique, as with the velocity spectrum, but in a velocity scan the same trial velocity is used for computing moveouts at *all* times on the record. A separate record is made for each velocity. At times for which the trial velocity is too high, a corrected reflection will show upward convexity like half an umbrella. When the velocity is correct, the reflection will have a horizontal or nearly horizontal lineup, and when it is too low, the pattern is like a half-umbrella which points downward. If the records corrected in this way with continuously increasing velocities are placed side by side, as in Fig. 7-23, the increase of velocity with record time is evident to the eye by the shift downward and to the right of the flat segments corresponding to successively deeper reflections. The time and velocity for each such flat event can be plotted to obtain the desired average-velocity–time relationship.

A similar velocity-scanning technique involves the use of a succession of increasing velocities in stacking a set of records shot for common-depth-point processing. The stacking procedure differs from the conventional one in that a single trial velocity is used for stacking all events from the beginning to the end of the record instead of being allowed to vary with time as is conventionally done. Only when the velocity is proper for stacking does a well-defined reflection appear at all. Figure 7-24 illustrates an array of records made for a velocity determination of this type. The increase of average velocity with record time is indicated by the line connecting the points where each reflection seems to show up best.

General aspects of velocity determination by computer   Interval velocities can be determined by any of these techniques from velocity-time relations using Eq. (7-45) (the Dix formula), but it must be realized that the precision of the interval values thus determined is necessarily less than that of the average velocities and that the smaller the time interval between reflections the lower the accuracy.

Both the velocity-spectrum and velocity-scan methods are ordinarily programmed for automatic computation from reflection data recorded in the field. The only

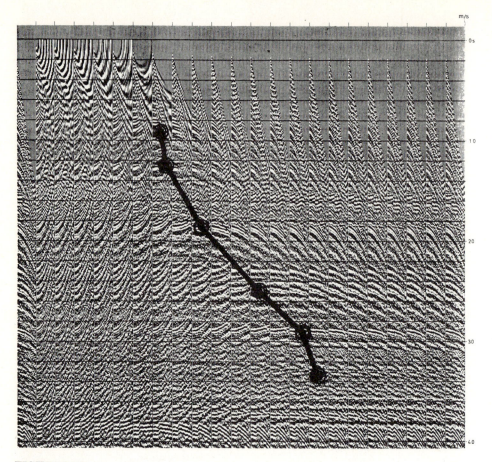

FIGURE 7-23
Velocity scan for typical reflection recording. Each record strip from left to right is corrected for moveout at a successively higher uniform velocity. Average velocity-time function obtained by this method is shown. (*Prakla-Seismos.*)

Time, s	0.90	1.25	1.87	2.50	2.90	3.33
Average velocity, m/s	1870	2010	2170	2700	3370	3550

decisions that must be made are the choice of shot-point position at which the analysis is to be made, the range and separation of the trial velocities to be used, and the interval between the times for which successive determinations are made.

Other computer techniques have also been employed for determining velocity.

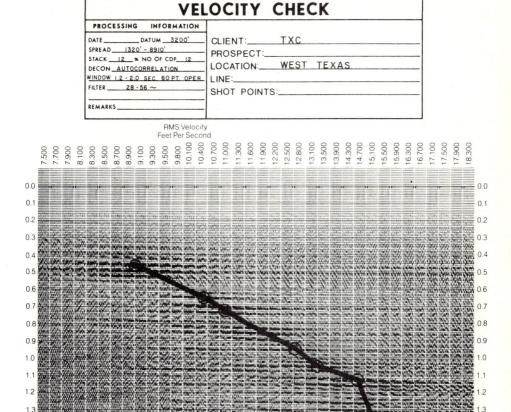

FIGURE 7-24

Determination of average velocity by stacking with series of trial velocities from 7500 to 18,300 ft/s. Trial velocity that gives best stack at each time is the correct one for that time. The velocity function shown by the heavy line was obtained by connecting positions where stack is optimum. (*Teledyne Exploration Co.*)

One method involves the assumption of a velocity function based on whatever information is available and the determination of corrected residual moveouts by cross-correlation of all traces with a reference trace obtained by averaging a number of channels near the shot point. The residual times provide a time-distance relation from which a more accurate velocity function can be computed.

7-4   PRESENTATION OF SEISMIC REFLECTION DATA

For the first 25 years of seismic reflection prospecting, seismic data were presented just as they were recorded in the field, in the form of multitrace oscillographic or wiggly-line seismograms of the type illustrated in Fig. 4-3. The records had no trace corrections of any kind. And as there were no moveout corrections either, all but the deepest reflection events would look like umbrellas on records when split spreads were used in the field and large weathering or elevation changes could be observed as irregularities in the reflection times. The information on each record would represent the output of a single geophone spread.

Record Sections

The introduction of magnetic-tape recording in the 1950s led to a major innovation in presentation that probably had a more direct impact upon seismic interpretation than any other single development up to that time. This was the corrected record section, offering many advantages over the individual records which were the previous end product of the recording process. In the first place, there was no longer any limit to the lateral extent of the coverage that could be presented on a single piece of paper. Such a presentation greatly facilitated examination. Irregularities in the lineups of reflection events caused by normal moveout and near-surface variations such as elevation changes were removed by automatic time correction. Also modes of presentation which were then new, such as variable area, variable density, and combined variable area and wiggly line, aided in the recognition and correlation of reflection events.

Although the variable-area mode is probably the most widely used for plotting record sections, variable-density and superimposed variable-density-oscillographic modes are also employed. The latter is particularly useful where details of the waveform are of interest. Figure 7-25 shows a typical time section made with this mode.

Record sections are conventionally plotted versus time, but computer programs are available for varying the time scale in accordance with the velocity so that vertical distances on the record sections are proportional to depth. As average velocity almost always increases with depth, the time scale is increasingly stretched with time on the record. This has the effect of giving the plotted seismic events wavelengths that appear to get longer with increasing depth. The effect is accentuated by the earth's natural filtering action, which results in lower-frequency signals being observed from greater depths even on time sections. The presentation

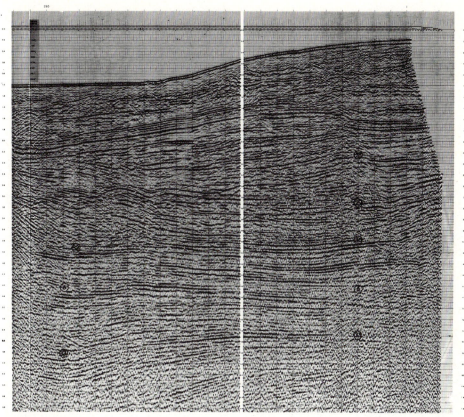

FIGURE 7-25
Record section plotted versus time using superimposed variable-density–oscillographic mode. (*Western Geophysical Co. of America.*)

of seismic signals on a depth scale, although uncommon, can be helpful for interpretation because it demonstrates in geologically meaningful terms the limitation
in the resolution of seismic reflection signals as well as the decrease in the resolution
attainable as depth increases. Figure 7-26 is the section obtained when the data
shown in Fig. 7-25 are plotted versus depth rather than time. The differences in
appearance are quite conspicuous, indicating substantial variations of velocity
within the section.

Before the introduction of corrected record sections, it was customary to plot
reflection times or depths picked from the center traces of individual oscillographic
records as straight-line segments on either a time or a depth scale. Figure 7-27
illustrates a presentation of this kind. Even when the data are taken from corrected
record sections, there are advantages in this type of presentation and it is still
used sometimes in exploration work. Time scales can be expanded to give a good
visual indication of features having such small relief that they would be hard to
detect on the contracted time scale used for most record sections. Also, poorly de-

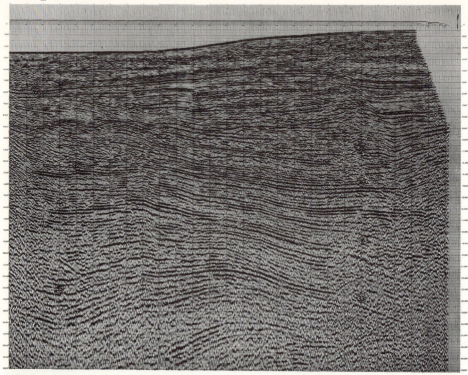

FIGURE 7-26
Record section made from same data as Fig. 7-25 but plotted versus depth
rather than time using velocity information obtained from data. (*Western
Geophysical Co. of America.*)

fined reflection events picked on record sections can be presented in more readable
form on pencil-plotted sections, although the quality and reliability of the actual
reflection data are not evident to the eye as they are on the record sections themselves. Computer programs are available for making plots in depth of the type
shown in Fig. 7-27, the time-depth conversions being based on the velocity functions
introduced as input data.

Positioning of Reflection Points in Space (Migration)

In the initial recording of seismic reflections, there is no way to tell from the information on a single trace where the reflecting point is actually located in space.
On field records and on conventionally plotted record sections, each event will
appear to be directly below the receiving point along a channel representing arrivals
at that receiver. The change in reflection time between adjacent traces (apparent
dip) makes it possible to determine the actual position along the profile of the
reflecting surface. Transformation of apparent reflecting positions to true positions
is referred to as *migration*.

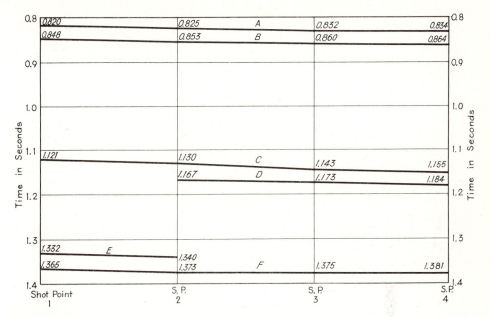

FIGURE 7-27

Plot of reflection time segments corresponding to several typical reflections (*A* to *F*) marked on the records from four adjacent split spreads shot with 100 percent coverage. (*Seismograph Service Corp.*)

Consider a shot and single receiver on the surface of a material having a constant velocity, separated from one another as shown in Fig. 7-28. The locus of the point from which a reflection observed at time *T* originates must be elliptical. This follows from the definition of an ellipse as the locus of a point *C* the sum of whose distances from two other points (*A* and *B*) is a constant (*VT*). The reflecting surface must

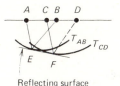

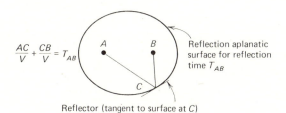

FIGURE 7-28

Locus of reflecting surface for reflection shot at *A* and received at *B* having travel time of *T*. Lower diagram shows how reflecting surface can be specified if two reflections are used.

be tangent to the ellipse. If the shot and receiver are coincident, the ellipse becomes a circle.

A single observation cannot in itself locate the position of C on the ellipse, but a combination of times for two or more paths of waves from the same reflector can do this, as shown in the lower portion of Fig. 7-28. A knowledge of the individual reflection times T_{AB} and T_{CD} makes it possible to draw the respective ellipses corresponding to these times. One approximates the reflecting surface by drawing a line tangent to both ellipses. As times for more reflection paths are recorded, i.e., as the number of traces increase, the accuracy with which the reflecting surface can be detailed by drawing an envelope around the overlapping ellipses becomes progressively greater.

It is evident from this diagram that when the reflector is dipping, the true reflecting points are displaced both laterally and upward from their positions, as indicated by conventional plotting. In Fig. 7-29 we see how reflections from opposite sides of a syncline can appear as two apparently conflicting dip segments when plotted vertically below a reflection spread.

Techniques of migrating reflections   Migration can be carried out graphically with special charts or templates, mechanically with special plotting devices that have moving parts, or by direct plotting of reflections in their migrated positions on the record section. Automatic migration techniques were introduced around 1969 and have come into widespread use for presentation of steep-dip reflection data in the form of migrated record sections. Computer programs designed for such presentation make use of the same principles that were drawn upon for migration of conventionally recorded data before computers became available.

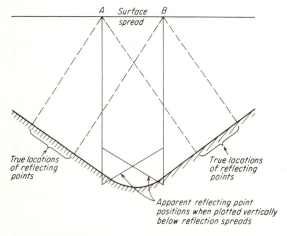

FIGURE 7-29
Distortion of true structure when reflections are not migrated, as illustrated by spread straddling axis of syncline.

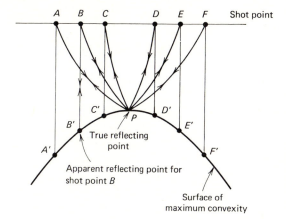

FIGURE 7-30
Surface of maximum convexity for reflecting point at P. Positions along curve are shown for six shot points with coincident geophones.

SURFACES OF MAXIMUM CONVEXITY   Migration using *surfaces of maximum convexity* was introduced by Hagedoorn[11] in 1954, and although it was based on some interesting new concepts, it did not then appear to be a very convenient approach for the routine migration of reflection data. Fifteen years later Hagedoorn's concepts became the basis for most automatic migration by computers.

Figure 7-30 illustrates a surface of maximum convexity as defined by Hagedoorn for the two-dimensional case. Such a curved line represents the time along a slant path that would be required for a reflection or diffraction from a given point in the subsurface to travel to a receiver as its horizontal distances from the point changes. It is sometimes referred to as a *diffraction curve*. Charts of maximum convexity are made by combining individual curves of maximum convexity for successive source depths over the range of interest.

Figure 7-31 illustrates how such charts are used in migrating a single segment plotted vertically on a depth scale. The appropriate curve of maximum convexity is

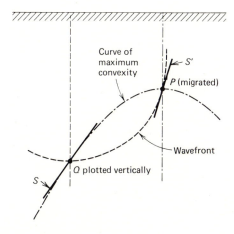

FIGURE 7-31
Use of curve of maximum convexity and wavefront for migrating reflection from apparent position at Q to actual position at P. (*From Hagedoorn.*[11])

the one tangent to segment S at point Q. The true position of Q in space will be at P, which is located on the vertex of the curve. The dip of the migrated reflector is obtained from the curve of equal reflection times (representing the position of a wavefront at the indicated travel time) having a central axis through Q. The migrated segment S' is drawn through P so as to be tangent to the curve of equal reflection time passing through the same point.

In manual migration using Hagedoorn's technique, one needs two charts, a wavefront chart (chart of equal reflection times) and a chart containing curves of maximum convexity for closely spaced vertical reflection times. The two curves are placed in juxtaposition with the zero depth positions coincident. The reflecting point is located along the curve on the wavefront chart corresponding to the time of the reflection along the central (vertical) axis of the chart. The chart with the curves of maximum convexity is slid laterally until a line is found that is tangent to the apparent (unmigrated) dip at the depth of the reflecting point. The migrated position of the point is at the vertex of this line, and the dip of the migrated segment is found by drawing a line tangent to the wavefront circle corresponding to the reflection time where it intersects this vertex.

WAVEFRONT CHARTS   If the velocity is known as a function of depth, it should be possible to migrate reflections with a wavefront chart. Wavefront charts show the successive positions of a wave emanating from a point source at generally uniform time intervals after the instant of wave generation. If the velocity is uniform, the chart will consist of concentric circles, the center being at the shot point. For a velocity increasing linearly with depth, the wavefronts will still be circles, but they will no longer be concentric. The depth of the center below the surface z_c and the radius R of the wavefront circle for a one-way time T can be found from the equations

$$z_c = \frac{V_0}{k} (\cosh kT - 1) \tag{7-48}$$

$$R = \frac{V_0}{k} \sinh kT \tag{7-49}$$

where V_0 and k are the constants for the velocity function

$$v = V_0 + kz \tag{7-50}$$

z being the depth below the surface.

Another family of curves that is usually plotted on the same chart as the wavefronts shows the ray paths for different angles of penetration into the earth from the shot point at the surface. Ray paths are always perpendicular to the corresponding wavefronts, so that all intersections between the respective curves are at right angles. For a constant velocity the ray paths are straight lines that radiate from the center. When there is a linear increase of velocity with depth, the ray paths are circles orthogonal to the circles representing the wavefronts. The positions of the centers of these circles are governed by V_0, k, and the angles of immergence of the respective rays. Figure 7-32 illustrates a typical wavefront chart showing families of wavefronts and associated ray paths. Each wavefront represents the position of the wave at a specified time after its generation at the source. Each ray path represents a different angle of entry of a ray into the earth at its source position on the surface.

Before electronic computers were available, there was no easy way to construct wavefront charts for velocity functions more complex than a linear variation with depth (for which all curves are circles) or a linear variation with time (for which Musgrave[12] devised a mechanical plotter). With proper computer programs and plotters available, it should be feasible to compute and draw charts for almost any velocity function.

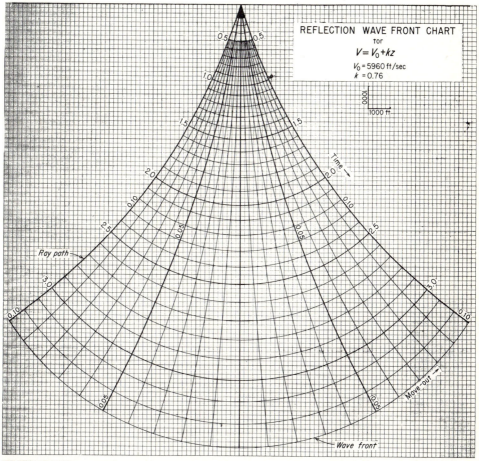

FIGURE 7-32
Typical wavefront chart for linear increase of velocity with depth. (*Reprinted from "Wavefront Charts and Raypath Plotters" by A. W. Musgrave. By permission of The Colorado School of Mines. Copyright 1952 by Colorado School of Mines.*)

A wavefront-and-ray-path chart like that shown for a linear increase of velocity with depth in Fig. 7-32 is quite suitable for migration if the reflection times have been corrected to make the shot point and receiver effectively coincident. The ray paths represent specific emergence angles (at the surface) expressed in terms of moveout time as measured on the surface over a specified horizontal distance. In this chart, the adjacent ray-path curves represent intervals 10 ms/1000 ft apart.

In migrating reflections with the chart one picks the reflection time at zero shot-receiver separation and the moveout time observed over a standard horizontal distance such as 1000 ft. The semitransparent graph paper on which the migrated segments are plotted is superimposed on the chart in such a way that the vertex of the family of ray paths is placed at the shot-point position for the record being migrated, as shown in Fig. 7-33. The wavefront circle corresponding to the reflection time is located directly or by interpolation, and the position along the circle is determined from the ray path corresponding to the moveout time. A straight edge is aligned on a tangent to the wavefronts at the point so specified, and the reflection segment is drawn along it for a distance based on the spread length.

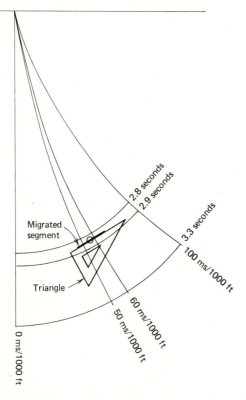

FIGURE 7-33
Migration of reflections at time of 2.835 s and moveout of 57 ms/1000 ft using wavefront chart. Limits of chart (3.3 s and 100 ms/1000 ft) are shown.

If the shot and receiver are both together at the center of the chart, any outgoing ray path must be coincident with that of the returning ray so that the reflecting surface must always be perpendicular to the ray path. As the wavefronts are also perpendicular to the ray path, the reflection surface is tangent to the wavefront.

MIGRATION USING MECHANICAL AIDS ASSUMING CONSTANT VELOCITY ABOVE REFLECTION   If a constant velocity can be justifiably assumed for the reflection path, migration becomes quite simple to carry out with readily accessible drafting aids such as a straightedge and protractor. From Fig. 7-34 it is evident that the dip angle α is related to the difference ΔT in reflection times at shot points a distance x apart by the formula

$$\sin \alpha = \frac{\bar{V} \, \Delta T}{2x} \tag{7-51}$$

where $\bar{V}$ is the average velocity, assumed to be constant over the path.

With α determined, the reflection segment can be located in the proper position on a cross section having a 1:1 horizontal-vertical distance relation. First a line is drawn through the shot point at an angle of α with the vertical. A perpendicular to this line drawn at a distance $\bar{V}T/2$ (where T is the reflection time) from the shot point gives the location of the migrated reflection segment.

A swinging straightedge such as a T square with a protractor at one end (Fig. 7-35) is often employed for migration of this type. The device is pivoted through the center of the protractor at the position representing the shot point on the cross section. For each reflection with a dip α and time T the arm is swung an angle α from the vertical and the segment is drawn in along the top of the triangle at the time T measured along the straightedge.

RESOLVED-TIME METHOD   In areas where the velocity-depth function is not known, a migration technique devised by Rice[13] may provide an adequate approximation to the true structure. Called the *resolved-time* method, this technique involves plotting reflections in time and presenting horizontal distances along the section in equivalent time units obtained by dividing the surface distance by the velocity below the base of the weathered zone. Once this transformation of the coordinate system is made, the migration is accomplished by swinging arcs from the various shot-point positions using the reflection times for the respective shot points as radii. Lines are drawn tangent to the arcs for the successive shot points and show the migrated positions of the reflection segments. The times picked from the lines

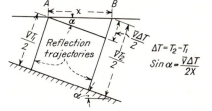

FIGURE 7-34
Measurement of dip angle by taking difference of reflection times at adjacent shot points. V is average velocity and α is dip angle.

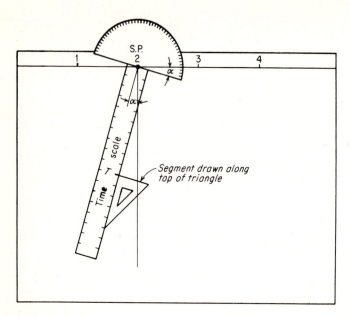

FIGURE 7-35
Swinging arm and protractor for migrating reflection where constant average
velocity is assumed.

directly beneath the shot points can be converted to depths using the best velocity
information available.

MIGRATION OF RECORD SECTIONS BY COMPUTER   Modern third- and fourth-gener-
ation computers make it possible to construct migrated record sections using
programs that are generally based on Hagedoorn's surfaces of maximum convexity.
Automatic computer migration, like all other migration, requires a knowledge of
the velocity-depth relation if it is to be useful. Programs for such migration are
generally based on two expectations: (1) that all dipping events are tangent to
some curve of maximum convexity constructed with this velocity function and (2)
that such events when migrated will appear at the vertex of this curve.

Figure 7-36 illustrates how this principle is applied in designing such a computer
program. In this example, signals on channels 1 through 99 are composited to obtain
data for plotting on a single output trace (trace 50), which is printed out on the
final record section. The sampling time used to determine trace amplitudes on
each of the 99 input channels is that at which this channel is crossed by the curve
of maximum convexity having its vertex at a time of 1.200 s on channel 50. The
samples at these times are composited to obtain an output value which is plotted
on the center channel (50) at 1.200 s. When a reflection or diffraction peak or
trough follows this line or is essentially tangent to it, the sum will be greater than
if the waves along the line of addition are in random orientation. Reflections tan-

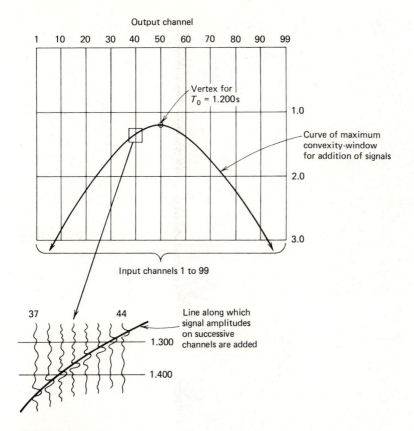

FIGURE 7-36
Principle of computer migration using curve of maximum convexity. Signals
are added at times of each trace intercepted by curve of maximum convexity
associated with time at the vertex, which is located along the output channel.
Any lineup of reflections along this curve will be plotted as a high-amplitude
event at the vertex.

gent to the curve or diffractions falling along it would thus yield a high-amplitude
event at the vertex which represents the true position of the event in space.

The vertex times for which amplitudes are stored are separated by the sampling
interval so that the next summation, in this case for a time of 1.202 s on channel
50, will be shifted so as to allow a different curve of maximum convexity. Each
output channel for a 3-s record involves the summation of points along 1500 curves
of maximum convexity if the sampling interval is 0.002 s. When all output values
are computed for channel 50, the computation is carried out in the same manner
for channel 51, the input window now extending from channels 2 to 100. When
signals covering the time range on all the output channels have been stored in this

(a)

(b)

FIGURE 7-37
Migration by computer: (a) unmigrated time section showing great number of
diffraction patterns; (b) migrated section showing synclinal and anticlinal
features, some of which are almost undiscernible on the unmigrated section.
(*Prakla-Seismos.*)

way, a record section is plotted which should have all reflection and diffraction events located in their true positions in space.

Figure 7-37 compares a record section before migration with one plotted from the same field data after migration by procedures of the type described in the previous paragraphs. Where dips are steep, the migrated sections should be much more useful and reliable for interpretation than the corresponding unmigrated sections. Diffracted events are particularly easy to suppress by such an approach, and this type of processing can be most helpful when the superposition of reflected and diffracted events complicates interpretation. Another example of automatic migration can be seen in Fig. 8-4.

Migration by modeling   The migration procedures thus far considered require that a velocity be assumed down to the reflecting formation that is a single-valued average or else a *continuous* function of depth or record time. In actual geologic sections there may be a series of large discontinuous changes in velocity at steeply dipping boundaries, e.g., between thick shale and limestone layers, so that the velocity-depth relation cannot be represented well enough by a continuous laterally uniform function to allow meaningful migration with any of the techniques we have considered.

The best approach in such cases is to model the configuration of the subsurface, drawing presumed boundaries for each layer to which a discrete velocity can be assigned. Such a procedure is described by Taner et al.[14] The initial geometry proposed for the boundaries is based on reflection patterns on the unmigrated section, and the velocities are the interval values determined automatically from the recorded reflection data using the Dix formula or a similar relationship. Ray trajectories for reflections are determined for closely spaced shot points across the model using Snell's law at each boundary to obtain the angles of refraction, and the times for each reflecting path are predicted on the basis of path lengths and velocities in the various layers.

After comparison of predicted and actual reflection times for the different interfaces, a new configuration of the boundaries is selected to reduce the observed discrepancies, and the ray tracing is repeated with the new model. The process continues by successive approximations until adequate correspondence is observed between model reflection times and those observed on the record section. Final mapping is done from the model that gives the best correspondence. The entire operation is programmed for automatic handling by a computer.

Figure 7-38 illustrates a proposed model, plotted in depth, of a complex structure involving a recumbent diapir, possibly a salt feature. The layering geometry and velocities of each layer constitute the input. A computer program involving ray tracing is used to determine the pattern of reflection times for each interface. The calculated time section (Fig. 7-39) is compared with the actual record section, and if changes are called for in the original pattern, a modified depth section is drawn and another time section is plotted by the computer. This is continued until an

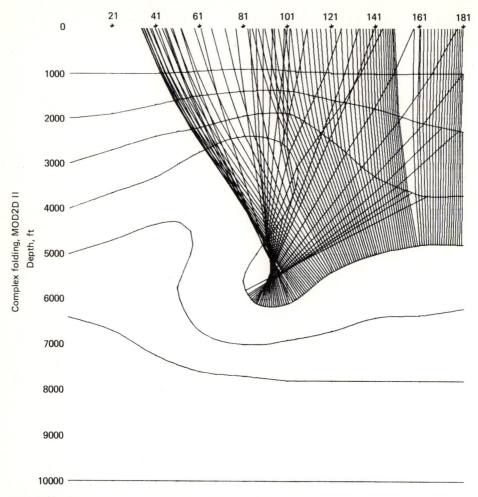

FIGURE 7-38
Model indicating structure of recumbent diapir and deformation of associated layers. Reflection ray paths shown are to the deep side of the interface observed at a depth of 4000 ft along the left edge. (*Seiscom Delta, Inc.*)

adequate fit is achieved, the depth section yielding the best fit being looked upon as the final interpretation.

Presentation of Reflection Data on Maps

The transfer of seismic data from reflection records or cross sections (whether in the form of plotted sections or corrected record sections) to a map involves a transformation from a two- to a three-dimensional representation. Ordinarily the data are presented on the map in the form of contours representing actual reflection

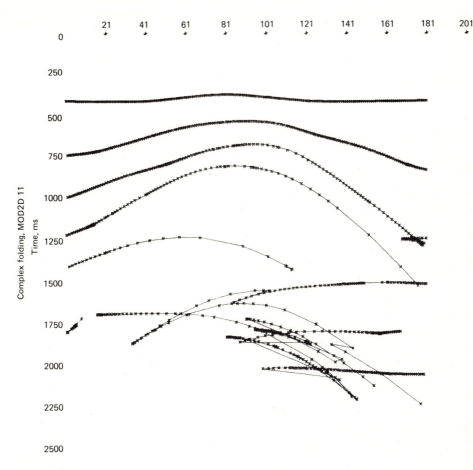

FIGURE 7-39
Computed time section based on ray paths and assumed velocities for structural
model shown in Fig. 7-38. (*Seiscom Delta, Inc.*)

times or depths. Sometimes the reflections are not continuous; where this is the
case, it may be desirable to construct a *phantom* horizon, which is kept parallel to
whatever reflections are nearest it along the section. The phantom horizons must
close around loops like continuous reflections. If they do not, previously unsuspected
faulting could account for the misclosure.

Depths or times are ordinarily entered on maps at convenient intervals along the
shooting lines, often at those shot points (chosen from a common-depth-point
array with much closer spacing) which fall approximately $\frac{1}{4}$ mi apart. The positions
of the reflections will be displaced from the shot points if the reflectors have appreciable dip, and the depth as mapped must be corrected for such lateral and
vertical displacement. In tectonically disturbed areas where strike direction is
not established, it may be desirable to shoot occasional spreads perpendicular to

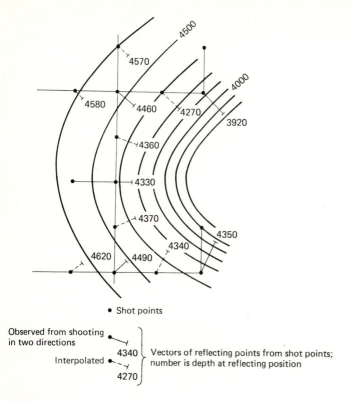

FIGURE 7-40

Mapping of reflection data in areas of steep dip based on shooting along perpendicular lines at alternate shot points. Depths are shown on map in migrated positions obtained from apparent dips and depths along the shooting lines.

the main profiles to determine the magnitude and direction of the true dips. These can be calculated from the respective apparent dips in two directions, as discussed on page 210. The tangents of the apparent dip angles can be plotted as vectors in the directions of the respective lines as shown in Fig. 7-8. If lines are drawn perpendicular to each vector at its end, the intersection of the perpendiculars gives the direction of true dip. Then the line from the center to the point of intersection is proportional in length to the actual dip. Figure 7-40 illustrates how such data are mapped and contoured.

After conversion of times to depths using the best available velocity information, the reflection map should ideally be equivalent to a structure map for the geological horizon that is represented. The reliability of the map as an indicator of true geological structure depends on the quality of the reflections, the extent to which adequate information on seismic velocities is available over the area, and the density of well ties. Faulting or lithology changes will also affect the accuracy of seismic structure maps.

REFERENCES

1. Dix, C. H.: Seismic Velocities from Surface Measurements, *Geophysics*, vol. 20, pp. 68–86, 1955.
2. Slotnick, Morris M.: in Richard A. Geyer (ed.), "Lessons in Seismic Computing," Society of Exploration Geophysicists, Tulsa, Okla., 1959.
3. Hileman, J. A., P. Embree, and J. C. Pflueger: Automated Static Corrections, *Geophys. Prospect.*, vol. 16, pp. 326–358, 1968.
4. Disher, David Alan, and Paul J. Naquin: Statistical Automatic Statics Analysis, *Geophysics*, vol. 35, pp. 574–585, 1970.
5. Duska, L.: A Rapid Curved-Path Method for Weathering and Drift Corrections, *Geophysics*, vol. 28, pp. 925–947, 1963.
6. Hollister, John C.: A Curved Path Refraction Method, pp. 217–230, in Albert W. Musgrave (ed.), "Seismic Refraction Prospecting," Society of Exploration Geophysics, Tulsa, Okla., 1967.
7. Gretener, P. E. F.: An Analysis of the Observed Time Discrepancies between Continuous and Conventional Well Velocity Surveys, *Geophysics*, vol. 26, pp. 1–11, 1961.
8. Strick, E.: An Explanation of Observed Time Discrepancies between Continuous and Conventional Well Velocity Surveys, *Geophysics*, vol. 36, pp. 285–295, 1971.
9. Green, C. H.: Velocity Determinations by Means of Reflection Profiles, *Geophysics*, vol. 3, pp. 295–305, 1938.
10. Taner, M. Turhan, and Fulton Koehler: Velocity Spectra: Digital Computer Derivation and Applications of Velocity Functions, *Geophysics*, vol. 34, pp. 859–881, 1969.
11. Hagedoorn, J. G.: A Process of Seismic Reflection Interpretation, *Geophys. Prospect.*, vol. 21, pp. 85–127, 1954.
12. Musgrave, A. W.: Wavefront Charts and Raypath Plotters, *Q. Colo. Sch. Mines*, vol. 47, no. 4, 1952.
13. Rice, R. B.: New Seismic Computing Method Fast and Efficient, *World Oil*, vol. 137, no. 2, pp. 93–98, 104, 1953; Additional Notes on the Resolved Time Computing Method, *Geophysics*, vol. 20, pp. 104–122, 1955.
14. Taner, M. Turhan, Ernest E. Cook, and Norman S. Neidell: Limitations of the Reflection Seismic Method: Lessons from Computer Simulations, *Geophysics*, vol. 35, pp. 551–573, 1970.

8

GEOLOGICAL INTERPRETATION OF
SEISMIC REFLECTION DATA

In preceding chapters we have considered how seismic reflection data are recorded in the field and enhanced by filtering operations in the processing center. We have also examined techniques for velocity determination, time-depth conversion, and adjustment for distortions caused by ray-path geometry. But there is still another aspect of the geophysicist's work which is perhaps the most important of all, and that is the translation of the seismic information into geological terms. This process, *geological interpretation*, calls for the greatest possible coordination between geology and geophysics if it is to be carried out successfully. It is in this area that the demands for technical competence are so stringent that it is hard for either geologists or geophysicists to meet them unless they work together.

The many technological developments in the seismic art that have virtually revolutionized geophysical exploration during the 1950s and 1960s did not often affect the interpretation of seismic data except peripherally. More recently there has been an upsurge of interest in the application of sophisticated computer techniques to the solution of interpretation problems. The new techniques of direct detection have given a profound stimulus to such activity, as have new uses of reflection data for identifying lithology and studying stratigraphy.

8-1   THE MEANING OF INTERPRETATION

The word interpretation has been given many different meanings by geophysicists who handle seismic reflection records and by geologists who put the information from them to use. To some it is virtually equivalent to data processing and is tied inextricably to computer software. To others it consists of all of the operations we considered in the last chapter such as the mechanical transformation of seismic reflection data into a structural picture by the application of corrections, time-depth conversion, and migration.

To the author of this book, interpretation can be all of these things, subject to the one inviolable condition that it involve some exercise of judgment based on geological criteria. By this conception, interpretation can begin with planning and programming a seismic reflection survey if they are guided by the geology of the area and by the economic or scientific objectives of the survey. It can involve the choice of field parameters, such as the kind of seismic source to be used, the geometry of source and receiver patterns, and the settings on the panels of the recording instruments, as long as such choices are governed by the geological information desired. The selection of processing procedures and parameters is also an important part of the interpretation if it is supported by the same considerations.

Any purely mechanical operations not requiring discretion on the part of the geophysicist would come under the category of reduction, not interpretation. It is possible to make a seismic map, particularly one in time, without carrying out any real interpretation at all if every stage of its preparation is routine or automatic and no decisions have to be made that involve geological considerations.

After a seismic map is constructed, an important part of its interpretation is integrating the seismic data on it with geological information from surface and subsurface sources, e.g., fault traces or geologic contacts. This involves identifying reflections and making ties to wells or surface features. The extent to which this can be done depends on the amount of geologic information available.

The digital computer has made it feasible to use previously unexploited characteristics of seismograms to obtain geological information. Under favorable circumstances, interval velocities can be determined from reflection records with enough precision to permit them to serve as a basis for identifying lithology. Another property of seismic waves that has been employed for studying rock composition is attenuation of seismic-wave amplitudes between successive reflectors, a parameter that can now be measured because of the high dynamic range in modern recording equipment. Thanks to the digital computer, the selection of tools available for seismic interpretation has been greatly extended since the late 1960s.

8-2   CORRELATION OF REFLECTIONS

Before corrected record sections came into universal use, it was customary to place adjacent records edge to edge in the manner shown in Fig. 4-4 and to correlate reflections from the last trace of one record to the first of the next. Record sections, which have almost entirely replaced the individual multitrace records, make it possible to follow events over long distances much more conveniently. The sections are corrected to eliminate such irregularities as the "umbrellas" of Fig. 4-3, caused by variations in the lengths of the ray paths, as well as topographic anomalies.

Even with the new kinds of display, correlation of reflections is not always as simple as it may appear. Often a reflection characterized on the record by a single trough evolves into two troughs over a very few traces, as illustrated in Fig. 8-1. There may then be some doubt which of the two events correlates with the one trough from which they branch. In the case shown, the interpreter may be guided by the pattern of an adjacent reflection of better quality like the one just above. Although such changes in waveform may be associated with the geology of the reflecting formation, the most usual reason for them is noise of one type or another which causes distortion of the signal. Modern data-processing techniques are designed to suppress such noise and thus increase the reliability of correlation. Even so, it often requires considerable experience and good judgment, particularly when the data are marginal, to make correct picks. An error of 1 cycle could mean that the predicted depth to a geological boundary is 100 to 200 ft higher or lower than it should be.

8-3 RESOLUTION AND PRECISION OF SEISMIC REFLECTION MEASUREMENTS

Although current recording and processing techniques have made it possible for the geophysicist to work with high-quality reflection data in a form more suitable for interpretation than was available a decade or so ago, the intrinsic limitations in the reflection process must be recognized. All these limitations are related to the basic physics of seismic reflection in a medium having the characteristics of earth materials. The attenuation of seismic signals at a rate which is proportional to frequency strongly affects the resolution that can be expected from such signals. Seismic waves are generated at their source as pulses of such short wavelength that they can be looked upon as spikes for all practical purposes. If the pulses could actually travel as spikes for large distances through the earth, there would be few problems in resolving reflections. But the continual removal by attenuation of higher-frequency components as the signals propagate through earth materials results in a continual broadening of the basic signal with increasing travel time. This effect has been discussed in Chap. 2 (pp. 46–49).

Composition of reflections The basic seismic signal traveling through the earth is, as we have noted, a pulse with a breadth that increases with distance traveled. After the pulse has passed through 10,000 ft of section, its effective wavelength may be of the order of 200 ft or so. At 15,000 or 20,000 ft, it may be twice this magnitude. If a pair of reflecting surfaces are no closer together than the wavelength of the reflection pulse, they can easily be resolved on a reflection record.

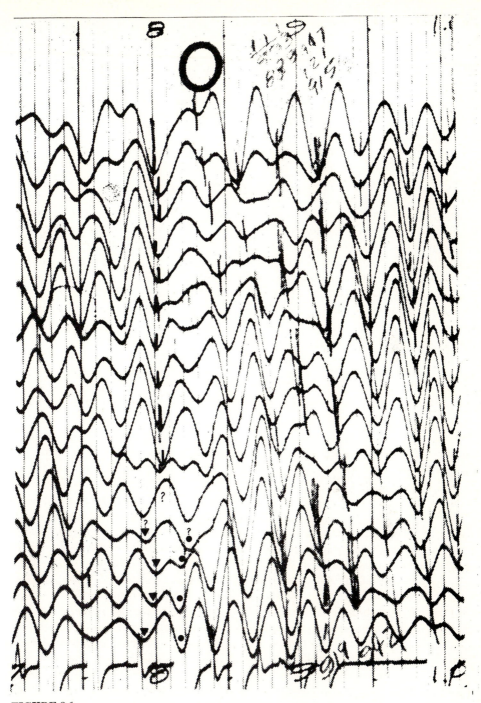

FIGURE 8-1

Break in continuity of reflection with attendant uncertainty in correlation. Event starting at top trace on time line marked 8 can be followed for nine more traces and then seems to terminate. Does continuation of event below this follow series of troughs marked by solid circles or that marked by solid triangles? The trend of other reflections suggests that the former is more likely.

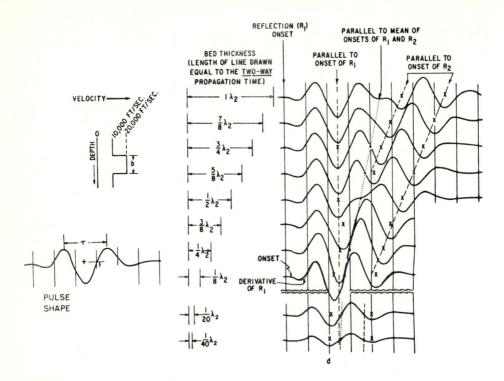

FIGURE 8-2
Effect of varying bed thickness on reflection wave. R_1 is reflection from top,
R_2 from bottom of bed. High-speed layer is sandwiched between beds of much
lower speed, as shown on left. When thickness of high-speed layer reaches
three-eighths of a wavelength, it is no longer possible to resolve reflections.
(*Modified from Widess.*[1])

If, on the other hand, the surfaces are separated by less than a wavelength, the
problem of resolution becomes more complex, the difficulty increasing as the
separations become smaller.

Figure 8-2 shows the reflected signal from a high-speed bed of finite thickness
underlain and overlain by very thick sections of lower-speed material. The waves
on the right are the reflections that would be expected for various ratios of bed
thickness to wavelength. The waveforms have been synthesized by superposition of
reflections from the top and bottom of the pinchout. The signal reflected by the
lower surface is equal in amplitude to the signal reflected by the upper surface
but 180° out of phase with it because there is a decrease in velocity below the deeper
interface and an increase in velocity below the shallower one. When the thickness is
about three-eighths of a wavelength or less, the two signals merge with so little
separation between them that it is no longer possible to resolve the pulse by eye
into two identifiable elements. At an eighth of a wavelength, the amplitude decreases

rapidly because of cancellation as the beginnings of the two out-of-phase reflections approach coincidence. The analysis leading to these conclusions was carried out by Widess.[1]

This model, like most models that represent the earth by discrete layers, is oversimplified. It is quite likely that other reflecting interfaces would be located close enough to the surfaces bounding the high-speed layer for reflections from them to interfere with (and thus complicate) the reflection signals that are shown. Also, despite the success of modern field and processing techniques for suppressing noise, some interference from other sources would almost certainly be superimposed on the reflection signals, making it unlikely that the increasingly weak reflections generated as the high-speed bed gets thinner would stand out enough above the noise level to be observable on real records.

Actually, most sedimentary rocks are so highly stratified that there are many discontinuities in velocity within a vertical distance corresponding to a seismic wavelength. The continuous velocity log introduced early in the 1950s first made this evident to geophysicists. Many velocity logs, like that in Fig. 7-19, show contrasts that would give rise to reflections no more than a few tens of feet apart. Figure 8-3 shows how a simple down-traveling source pulse with a waveform comparable to that of a Ricker wavelet is altered when it is reflected from a sequence of boundaries a short distance apart in comparison with the wavelength of the pulse. Each interface returns a pulse having the same waveform as the source pulse, but the amplitude and phase (whether or not reversed by a lower velocity below the boundary) are governed by the reflection coefficient across it. The sum of all the individual reflections is recorded by the geophones at the surface. The difference between the source pulse and reflected signal is quite pronounced, and it is not possible to isolate the contribution made to the reflection waveform by any of the individual boundaries.

This example shows how a typical seismic reflection should be looked upon as an interference pattern made up of impulses from many sources spread vertically over hundreds of feet rather than as a simple event originating from a single lithological interface. When the separations (in time) or the relative strengths of the individual reflections change with lateral distance, we should expect the character of the resultant reflection event on the record to change also. Model experiments illustrating this process were carried out by Woods[2] to illustrate the composition of reflections for various stratigraphic configurations.

The reflection process in a case like that illustrated in Fig. 8-3 is entirely comparable to the action of a time-domain filter of the type discussed in Chap. 6. The down-going pulse is the input, and the reflectivity-versus-depth plot for the layering as shown on the diagram is equivalent to the impulse response. The individual reflectivities act as weighting functions, and the travel times between the boundaries govern the time shifts between successive pulses returned to the surface. It is thus evident that a layered earth acts as a filter and that the shape of the reflected pulse depends on the reflectivity-versus-depth function. Sengbush et al.[3] have shown that

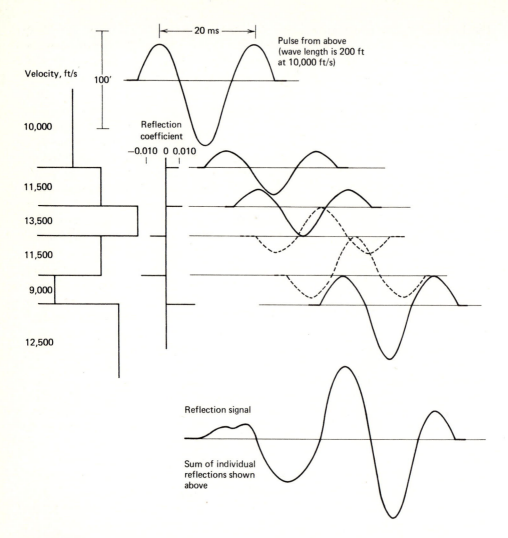

FIGURE 8-3

Composition of reflection from series of five interfaces with separations small in comparison to wavelength. Note change in waveform caused by reflection process. Waveforms indicated by dashed lines represent reflections with phase reversal.

the reflection waveform can be predicted by convolving the source signal with the reflectivity function (of seismic travel time rather than of depth) associated with the earth.

Synthetic seismograms Synthetic seismograms of the type first described by Peterson et al.[4] are artificial reflection records made from velocity logs by conversion of the velocity log in depth to a reflectivity function in time and by convolution of this function with a presumed source pulse. Wuenschel[5] was the first to describe a technique for obtaining synthetic seismograms that include multiple reflections. It is evident that the same computer programs used for time-domain filtering of reflection signals in routine seismic processing can be applied to generate synthetic seismograms for any assumed source signal $f(t)$ if the detailed velocity stratification of the subsurface is known from a velocity log.

Precision of reflection times   Another limitation associated with the seismic reflection process lies in the precision with which reflection times and depths of reflecting surfaces can be determined from events on seismic records.

Times of reflection events are ordinarily recorded for the highest-amplitude troughs (or sometimes peaks) of the oscillatory signals usually associated with them on the records. Such features are easiest to identify and observe, particularly in the presence of noise. Strictly, the times of the *onsets* of the reflections should be recorded rather than any troughs or peaks which follows the onsets. The problem is complicated by the fact that the signals, having been recorded by velocity-sensitive geophones, show peaks or troughs where the particle displacement has its greatest rate of change rather than its greatest amplitude. Also, digital processing operations such as deconvolution may cause phase shifts which make it difficult to identify reflection onsets. Many filter programs are designed when possible for minimum phase shift so as to leave the greatest energy in the reflected wave as close to its onset as possible. Such manipulation can sometimes distort complex waveforms, obscuring identification of events.

As a typical reflection signal will normally be the composite of individual reflections, mostly overlapping, from a series of interfaces extending over a range of a hundred feet or more, it is not likely that any identifiable feature of the signal, such as a peak or trough, can be uniquely related to a specific geological boundary, as was pointed out in our discussion of Fig. 8-3. The filtering action of the earth causes time lags comparable to phase lags in electronic filters. The existence of such lags has been well established by comparing field records with synthetic seismograms made from logs run in wells near the lines where the field records were shot. These lags range from 20 to 80 ms, but there is seldom any way to predict what the lag should be at a location near which there are no well ties.

Thus it is more difficult to determine the *absolute* depth of a reflecting interface in the earth from a reflection signal than it is to measure *relative* depths of such a boundary between two points at which the same reflection has been recorded. Where all individual layers encompassed within the zone contributing to the reflection are conformable, the structural relief can be mapped with an accuracy of 1 or 2 ms if the reflection quality is good. Where velocities do not change laterally, absolute depths can be obtained with comparable accuracy if the lines are tied to at least one well at which the reflection event is related to a particular well top. The differences between the time for an event at the well tie and the times elsewhere on the line can readily be transformed into depth differences reliable to 10 or 20 ft if the reflections are good and the average velocity values are known precisely enough.

8-4 REFLECTION DATA OVER GEOLOGIC STRUCTURES SOUGHT IN OIL EXPLORATION

Over the period of more than four decades during which the reflection method has been used in exploration it has been possible to obtain useful information on many types of structural features which might be responsible for the entrapment of oil. Corrected record sections, particularly those presenting data obtained with modern field and processing techniques, often make the presence of such structures obvious to the eye.

The most common structural targets associated with oil entrapment are anticlines and faults. Anticlines are generally easy to see on record sections, and faults of more than marginal displacement should be discernible, although some types are more readily recognizable than others. Structural deformations caused by salt domes and

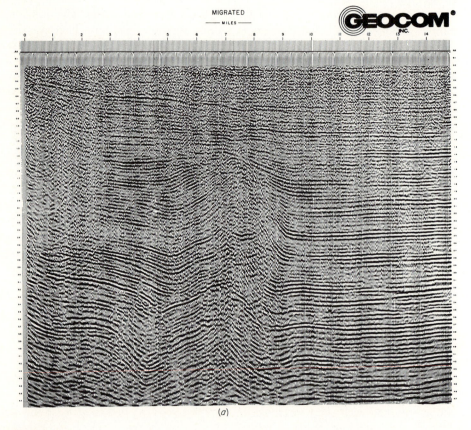

(a)

FIGURE 8-4
Anticline from San Joaquin Valley, Calif.: (a) section as automatically migrated by computer; (b) unmigrated section. The migration has collapsed the many diffraction patterns that concealed the actual structure. (*Geocom, Inc.*)

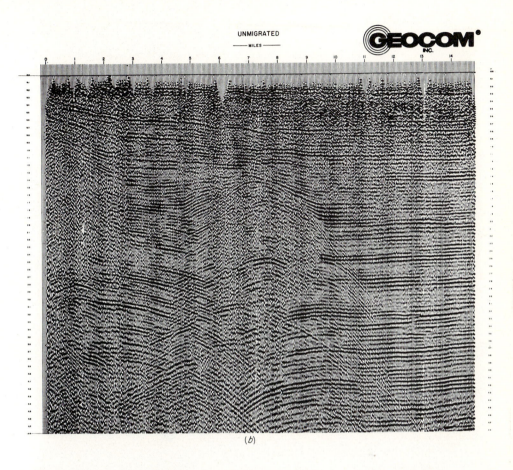

UNMIGRATED

(b)

other intrusives can usually be mapped as well. In the paragraphs to follow we shall consider the interpretation of many such structural features and present examples illustrating their appearance on seismic sections.

Migrated record sections often improve the definition of complicated structures from areas where there has been major tectonic disturbance. They are particularly helpful in resolving complex fault patterns.

Anticlines

Anticlines can easily be mapped by reflection if the data are of good quality and if the closure is greater than spurious irregularities in apparent structure such as are caused by lateral velocity changes. Oil-bearing anticlinal structures may be associated with tectonic forces as well as with deformations due to the upward push of rising salt domes or other diapiric features underneath. Figure 8-4a illustrates an unusual kind of anticline as displayed on a migrated record section. The corresponding unmigrated section, shown in Fig. 8-4b, does not show a very meaningful picture.

Faults

The detection of faulting on seismic sections can be quite easy under favorable circumstances. Often, however, the indications are subtle, and the identification and delineation of such features can be quite challenging. Because of the role faults often play in the entrapment of hydrocarbons, the techniques for finding and mapping them have considerable practical importance.

The principal indications of faulting on reflection sections are the following:

1   Discontinuities in reflections falling along an essentially linear pattern
2   Misclosures in tying reflections around loops
3   Divergences in dip not related to stratigraphy
4   Diffraction patterns, particularly those with vertices which line up in a manner consistent with local faulting
5   Distortion or disappearance of reflections below suspected fault lines

Where discontinuities are well defined, the position of the fault trace may be highly evident on the record sections even to someone entirely inexperienced in seismic interpretation. Figure 8-5, a reflection section 15 mi long shot offshore in the

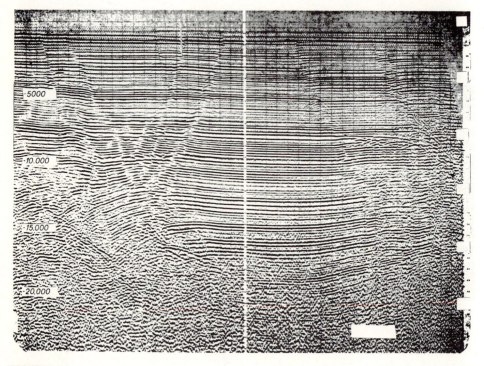

FIGURE 8-5
Pattern of normal faulting in the Gulf of Mexico. The faults appear to be associated with salt structures, indicated but not explicitly visible, at the lower right and lower left. (*Exxon, Inc.*)

Gulf of Mexico, shows two systems of normal faults, one group dipping to the right, the other to the left. Throws, indicated quite clearly by the displacement of the more conspicuous reflections, are as great as 500 ft. The faulting is attributed to uplift of a salt mass (not visible on the section) as well as to subsidence of formations toward the center of the basin. The exceptional detail that can be observed in the faulting pattern reflects the success with which the computer procedures employed in processing the raw data eliminated noise that might otherwise have obscured the more marginal fault indications.

The detection and mapping of thrust faults is generally based on divergences between reflections as well as on the repetition of identifiable reflections above and below the thrust plane. A classic area for faulting of this kind is in the foothills of the Canadian Rockies, where oil and gas are entrapped in allochthonous* limestone sheets at their updip termination against the underlying thrust plane. Another area where such faulting shows up well on seismic recordings is the Wind River Basin of Wyoming, where the section shown in Fig. 8-6 was obtained.

An important aid to identifying and tracing fault surfaces is the family of diffraction patterns that originate from the edges of beds disrupted by faulting. Such an edge can often act as a point source for returning seismic energy by diffraction of the type discussed in Chap. 2 (page 43). The resulting pattern will

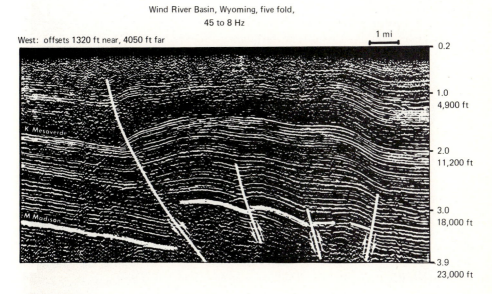

FIGURE 8-6
Overthrusting in the Wind River Basin of Wyoming as indicated on a Vibroseis section with low-frequency sweep range. (*Continental Oil Co.*)

* The term *allochthonous* refers to the formations above an overthrust fault. *Autochthonous* refers to the formations below the fault plane.

have an arcuate shape like the trace of a surface of maximum convexity discussed in the previous chapter. The vertex of the diffraction pattern shows the position of the diffracting edge on the section. A number of such patterns originating from different points along a fault should make it possible to locate the fault edge even in the absence of reflections yielding such information. Figure 8-7 shows diffractions which have vertices that line up in a way that supports the presence of fault trends independently indicated by patterns of displacement in the reflections.

Diffraction patterns need not be complete to be useful in this way. Sometimes fragmental portions of a pattern can be fitted to an appropriate curve of maximum convexity based on the known velocity for the area concerned and the curve can be used to project the position of the vertex even when it is not actually observable on the section.

Unless diffraction patterns are recognized as such, they might be interpreted by the unwary as reflections. A diffraction observed on both sides of its source can look deceptively like a reflection from a symmetrical anticline. The best way to determine whether a suspected feature on a section is a diffraction pattern or a true structure is to calculate the time-versus-horizontal-distance relationship for a diffraction that would originate from the vertex of the feature using the best velocity information available. Then one compares the predicted and observed patterns, preferably by the use of a transparent overlay. The closer the fit the greater the likelihood that the feature is a diffraction.

It is often possible to recognize faulting from divergences of reflections below the fault plane or from disturbances in the quality or character of reflections originating beneath the suspected fault that appear to pass through it. Quarles[6] has shown how faults in the Gulf Coast are identified by divergences in the apparent dip of underlying reflectors, and Laubscher[7] has pointed out the diagnostic value of distortion and deterioration of reflections from below a fault, illustrating this criterion by an example from Venezuela.

Salt Domes and Other Diapirs

Salt domes can show up quite conspicuously on reflection sections. Figure 8-8 shows an example of such a feature. The rim syncline formed around the dome stands out conspicuously on the left side; also the formations show a rise on each side of the figure, indicating other salt features in both directions. It is rare that the salt surface itself yields a clearly defined reflection. More generally the distortion of reflecting formations above the dome and alongside it, as well as the absence of reflections within the salt mass itself, will be sufficiently diagnostic to allow at least an approximate delineation of its surface. Migration of steeply dipping reflections from beds close to the dome that have been tilted upward by it often makes it possible to improve the precision with which one can map the flanks. Faulting of sedimentary formations caused by uplift of the salt can sometimes be observed in the reflections. Structures in sedimentary rocks resulting from the uplift of other

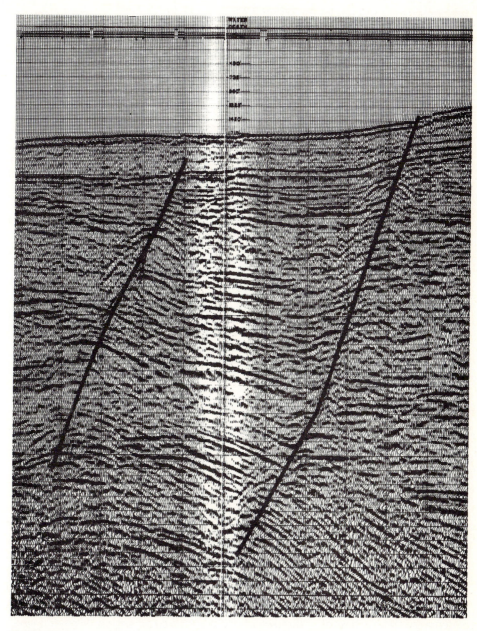

FIGURE 8-7
Tracing fault surfaces by following vertices of diffraction patterns. (*Western Geophysical Co. of America.*)

FIGURE 8-8
Distortion of sedimentary layers due to forces associated with salt-dome
buoyancy. Some of the structures shown, e.g., those below the piercement-type
salt dome and salt pillows (like the deep one on the right) are not real but result
from velocity effects. (*Exxon, Inc.*)

diapiric bodies such as igneous plugs often have an appearance on record sections
that is identical to that caused by the uplift of salt domes. In areas such as previously
unexplored offshore shelves where the geology is little known, it would generally
require gravity or magnetic data to identify the nature of the diapir causing a
pattern of disruption in the reflections.

Basement Structure

Until common-depth-point recording and digital processing became available, it was
seldom possible to identify the surface of the basement from reflection records
because of multiple reflections and other noise dominating the deep portions of the
records. While it is still not always feasible to recognize the basement on record
sections, particularly when it is very deep, the removal of noise and multiples may
make its surface readily observable. Irregularities on the top of the basement often
generate diffraction patterns which give a series of closely spaced arcs on the section

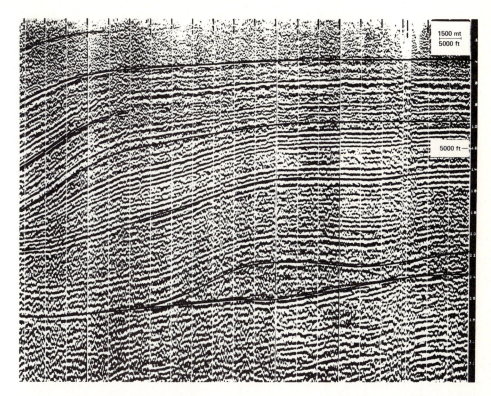

having an envelope that can define the convolutions of the basement quite closely. Figure 8-9 illustrates this effect in a deep-sea area. The absence of reflection events below this envelope makes the identification all the more likely, even though there has been no drilling along the line to verify it.

Pitfalls in Structural Interpretation

The conventional representation of a single channel of seismic reflection information is a plot of signal amplitude versus time on a trace corresponding in its position on the section to that of the receiving geophone group represented by the channel. The record section shows an assemblage of such traces side by side. Such a presentation should not be looked upon as a geological cross section because it can distort the actual geometry of the subsurface in two ways: (1) any variations in velocity, either vertical or lateral, will cause the time section to have a configuration different from the actual geological section plotted in depth; (2) geometric bending of the ray paths away from the vertical will have the same effect as pointed out in our discussion of migration in the preceding section.

Migrated sections and sections plotted in depth rather than time (see Fig. 7-26) are designed to minimize such distortion on conventional record sections. But a large majority of the sections actually used in exploration are of the conventional

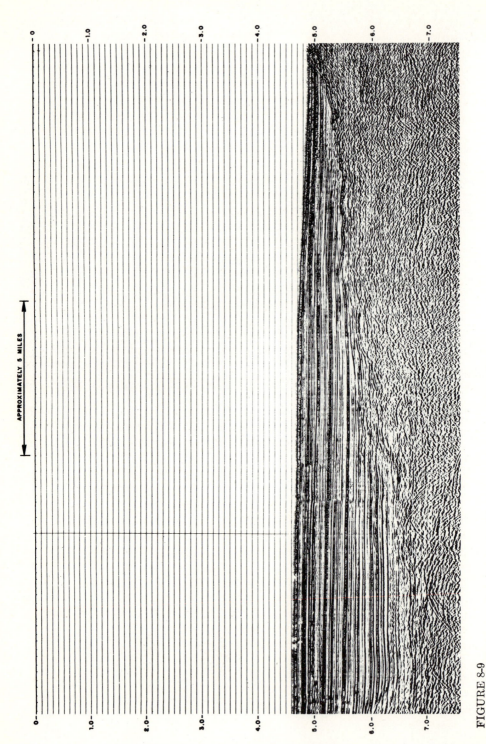

FIGURE 8-9
Identification of basement surface from diffraction patterns on section along
traverse in deep water. (*United Geophysical Corp. proprietary data.*)

type, and it is important that all geophysicists recognize the errors that these distortions can lead to in seismic interpretation. Moreover, many migrated sections display incorrect structural indications because of uncertainties or complexities in the velocity distribution.

Another common pitfall in structural interpretation can result from erroneously chosen processing parameters, e.g., an incorrect stacking velocity.

All these types of hazards have been reported on by Tucker and Yorston.[8] A few of the examples they present will be discussed here to illustrate the consequences that result when some of the pitfalls they point out are not taken into account.

Velocity pitfalls   Let us take another look at Fig. 8-8, which illustrates spurious structures attributable to velocity anomalies in overlying salt bodies. A strong reflection is observed under the large salt dome that is about 150 ms higher at its shallowest point than the correlative events on either side of the dome. Yet there is no real structural uplift below the salt. The fact that the velocity in the salt column is higher than that in the surrounding material can account completely for the apparent structure of the subsalt formation. The observed dropoff in the same reflection near the right-hand edge of the section is explainable by a salt pillow just above it, which happens to have a lower velocity than the surrounding formation, an effect often observed when salt is unusually deep.

Another example from Tucker and Yorston[8] is related to overthrusting. Figure 8-10 shows a section that would lead one to the conclusion that there is an anticlinal feature between 1.0 and 2.0 s. But the authors' interpretation is quite different, as one sees from the marked section in the lower portion of the figure. A low-angle overthrust fault has a high-velocity allochthonous tongue (presumably limestone) on the left side of the section. This tongue causes the flat autochthonous beds below to appear arched upward because of velocity pull-up. Such effects are frequently observed in the Canadian foothills, as in the area around Turner Valley where the Mississippian Rundle limestone is thrust at a low angle over lower-velocity Cretaceous formations.

Geometrical pitfalls   A common type of geometrical pitfall is illustrated by Fig. 8-11. This is what is observed when a section crosses a syncline with such a sharp curvature that the reflection ray paths cross one another on their way to and from the surface. Figure 8-12 shows the ray-path geometry resulting in the "bow tie" that is observed on the section at depths below 2.0 s. The two reflections dipping steeply in opposite directions are reflections from respective sides of the synclinal structure that cross one another on their way to the surface, and the arcuate feature below is a diffraction from a point at the bottom of the syncline, the shortest time being observed over the syncline's deepest point. Properly designed automatic migration would collapse the diffraction arc to a point and would shift the flanks to their true positions, but even in the absence of such migration the bow-tie pattern should enable the knowledgeable geophysicist or geologist to recognize the true nature of the source.

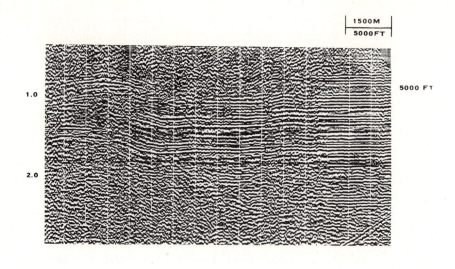

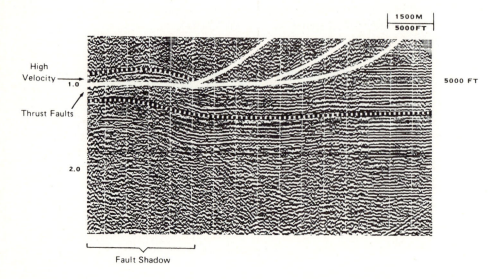

FIGURE 8-10

"Anticline" caused by thrusting of high-velocity material over monoclinal layers. Markings on lower section indicate interpreted structure. (*From Tucker and Yorston.*[8])

Processing pitfalls Figure 8-13 illustrates a processing pitfall. It is an example of how different choices of stacking velocity can lead to greatly different interpretations. On the left the deeper structure appears to be a gently warped monocline. On the right it appears to be a sharp anticline with a strong suggestion of faulting on one side. A careful comparison of the deeper-dipping events on the section at the left

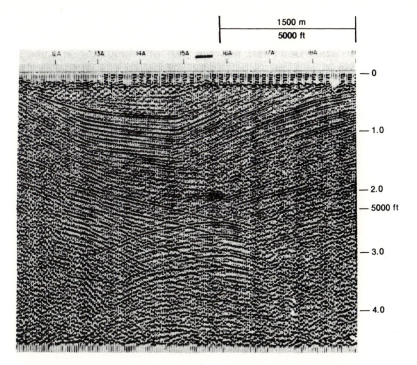

FIGURE 8-11
Bow-tie effect observed over sharp syncline in the Adriatic Sea. Apparent
anticline is actually a diffraction feature. (*Geocom, Inc.*)

with shallower ones at half their time makes the deeper events look like multiples.
This correspondence favors the likelihood that the stacking velocity yielding the
anticlinal structure is the correct one.

Time Versus Depth in Structural Mapping

An important consideration that must be taken into account in evaluating maps
and sections in depth is the reliability of the velocity information on which the
time-depth conversion was made. With computer programs available for determining velocity analytically from regular reflection records, we no longer need closely

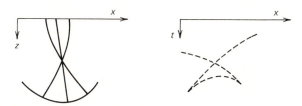

FIGURE 8-12
"Bow-tie" effect. Wave paths and pattern on record section for reflections
from syncline with curvature greater than that of approaching wavefronts.

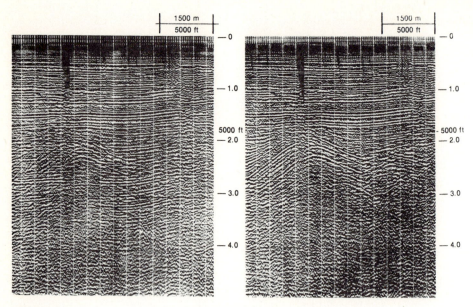

FIGURE 8-13
Which structure is correct? Each represents a different stacking velocity. As gently dipping events on the left have all the characteristics of multiple reflections, the structure on the right is preferred. (*From Tucker and Yorston.*[8])

spaced well-velocity surveys in order to obtain reasonably trustworthy information on reflection depths. Yet the precision of conversion velocities obtained from moveout times alone is subject to certain limitations. Accuracy depends on reflection quality, which is not always good in spite of computer-based data-enhancement techniques. Moreover, velocities determined by computers are based on slant ray paths and uniformly dipping reflectors, which usually differ because of anisotropy and other reasons from the vertical velocities which should be used for time-depth conversions.

For many years there has been a difference of opinion among geophysicists over whether seismic results should be presented in time or in depth. Some have preferred cross sections and maps in time rather than in depth because they are based only on objective data and are not subject to change as new velocity information is acquired. Such a preference is hardly justified now that automatic computer programs like that for the velocity spectrum allow the determination of velocities directly from the reflection data and thus make it much easier to obtain detailed velocity information than was possible before such programs became available. It is even possible, as demonstrated in Fig. 7-26, to plot record sections directly in depth from velocity information obtained with such programs.

The final presentation of seismic structures in terms of reflection times must often be looked upon as an evasion of the geophysicist's responsibility, which is to convert his data into a form that is as geologically meaningful as possible. Well tops and isopachs are expressed in units of distance, not time, and geophysical information to be coordinated properly must be presented in the same way. Maps in time do not incorporate the effect of lateral velocity changes, which in exceptional cases could even account for reversals in the direction of dip with respect to those indicated by time contours. Even admittedly imperfect and incomplete velocity control can prevent grossly erroneous conclusions in geological interpretation that might be made on the basis of time sections and maps alone. The geologist and exploration manager must rely on the geophysicist to provide the best possible interpretation the geophysical art allows. The geophysicist is not taking his professional responsibilities seriously enough if he presents only objective information and leaves it to others, often less qualified than he is, to convert it into geologically meaningful terms.

8-5   REFLECTION AS A TOOL FOR STRATIGRAPHIC STUDIES

Throughout the history of the reflection method, its performance in locating hydrocarbons in stratigraphic traps has been much less favorable than in finding structurally entrapped oil and gas. Only one type of stratigraphic feature associated with hydrocarbon entrapment has been discovered and mapped by reflection with consistent success, and that is the limestone reef. Contrasting lithological characteristics of reefs and the shales they replace have made them favorable targets for seismic location ever since the Leduc Field in Alberta was found by reflection in 1947. The physical limitations of the seismic technique make it less successful in exploring for other kinds of stratigraphic traps, which generally have lithological characteristics giving a less diagnostic response to seismic waves.

Let us consider some examples of the results obtained in actual exploration to illustrate the varying degrees of success for the different types of stratigraphic entrapment. Many of the examples to be reviewed appear in Ref. 9. The general problems encountered in exploring for stratigraphic traps with the seismograph are discussed by Lyons and Dobrin[10] in the same publication.

Stratigraphic oil traps can result from reefs, pinchouts, or other features associated with erosional truncation, facies transitions, and sand lenses associated with buried channels, lakes, or similar sources. The great variety of geological situations giving rise to such entrapment makes it difficult to reduce the problem of finding them seismically to as small a number of basic elements as we would like.

The principal explanation for the poor success of seismic methods in detecting stratigraphic features other than reefs lies in the limited resolution of the seismic pulse. Structural traps generally involve deformations in beds that remain conformable over at least a few hundred feet of section. In most types of stratigraphic traps, however, there is a variation in lithology which is often confined to a distance much shorter than a wavelength, so that resolution becomes a major problem. And it is evident from our discussion of the composition of reflections that any change in stratification could result in the alteration or even the destruction by interference effects of reflection signals associated with the beds on either side of the point where the layering characteristics change.

Use of Reflection Data to Reconstruct Depositional History

The greatest success of the seismic method in stratigraphic studies has not been related directly to the discovery of hydrocarbons but more indirectly in casting light upon the depositional environment and history of deposition in the areas where exploration is being carried out. The patterns shown by reflections often make it possible to understand how the deposition took place in the areas under investigation, and interval-velocity studies often enable the geologist to identify gross lithological features, allowing a more complete reconstruction of the depositional environment.

The various movements of a shoreline, progressive and regressive, are associated with geometrical patterns which are indicative of the types of deposition that took place at various periods of geological history. A knowledge of such history enables the geologist to predict the most favorable areas for oil accumulation, and more detailed exploration, e.g., by stratigraphic drilling, over such areas should lead to the most expeditious location of stratigraphically entrapped hydrocarbons.

Some patterns on seismic sections that should make it possible to reconstruct depositional history are illustrated in Fig. 8-14. Parallel bedding indicates deep-water deposition on a stable surface. A rising basin hinged at the shoreline leaves a wedge that thins in the seaward direction, while a sinking basin so hinged is associating with a wedge that thickens seaward. The arrows on the section indicate these movements during each phase of deposition. Onlap deposition and prograding as in deltaic fans can be deduced from characteristic patterns of cycle termination on seismic records.

Unconformities can also be mapped from the divergent pattern of reflections on a seismic section. The presence of unconformable contacts on a seismic section can often cast important light on the depositional and erosional history of an area and on the environment existing during the time when the movements took place.

Figure 8-15a shows a rather unusual pair of unconformities with large divergences in structure across each discontinuity. The complexity of the geological history is increased even more by folding and possible faulting in the portion of the section lying between the unconformities. In offshore areas where information from exploratory drilling is not available, seismic patterns like this may provide the only basis for reconstructing depositional as well as tectonic history. Figure 8-15b is a closeup of a portion of the section illustrating the depositional pattern in greater detail.

Classification of Stratigraphic Traps

The success of the seismic reflection method in finding stratigraphic traps varies with the type of trap involved. Most such entrapment features fall within four

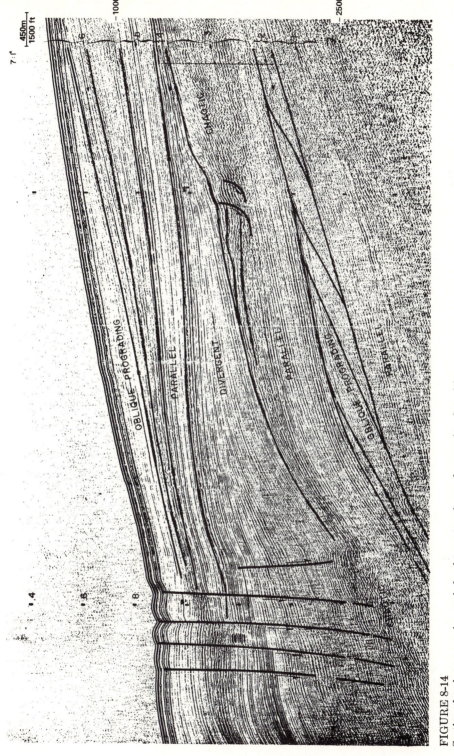

FIGURE 8-14

Section showing successive periods of transgressive and regressive deposition. Divergences and convergences indicate shoreline movements or sinking of basins. Oblique prograding sequences denoted by pattern of updip termination of dipping reflections. (*Exxon, Inc.*)

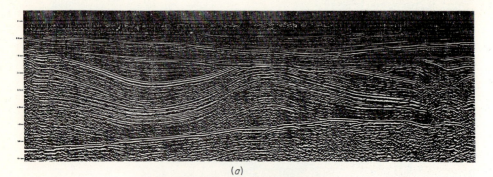

(a)

(b)

FIGURE 8-15
Unconformities at two levels indicated by seismic layering pattern: (a) complete section; (b) detail of central portion. (*Prakla-Seismos.*)

categories: (1) limestone reefs, (2) permeability barriers associated with erosional truncation, (3) sand bodies such as lenses or stream channels surrounded by impermeable material, and (4) facies changes from permeable to impermeable lithology. Seismic exploration for each type of entrapment is discussed in the paragraphs to follow.

Reefs   Limestone reefs consist of the skeletal remains of coral, algae, or similar shallow-water organisms, the buildups generally occurring on shoals or islands surrounded by deeper water. The porous reef material originally deposited is surrounded and overlain by muds, which are subsequently consolidated into shale, resulting in ideal conditions for generation and entrapment of hydrocarbons. Since the discovery by seismic means of oil in the Leduc reef of Alberta, reflection techniques have been successful in finding productive limestone reefs in west Texas, Alberta, Illinois, Libya, and other petroliferous areas.

Two characteristics of reefs facilitate their location by the seismic method. The more widely used one is the contrast between the velocity in reef limestone and that in the shale which often surrounds it at the same stratigraphic level. This contrast causes the reef to act as a lens which can give rise to apparent structures in underlying reflectors. The other characteristic is draping, which frequently occurs in the overlying sediments due to differential compaction over the reef and over the shale. Structures caused by draping effects can, of course, be mapped in the same way as other structures.

The effect of the velocity differential between reef limestone and off-reef shale reefs on reflections from below the reef level has been demonstrated by Skeels.[11] Figure 8-16 illustrates the respective electric logs and the interval-velocity distri-

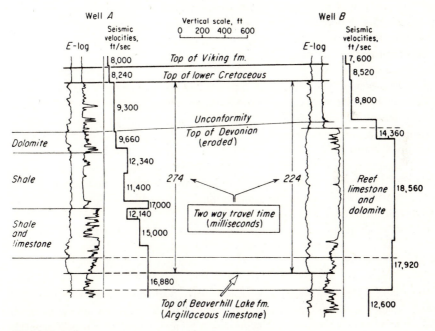

FIGURE 8-16
Electric logs and seismic interval velocities in corresponding portions of nearby off-reef (a) and reef (b) wells in Alberta. (*From Skeels.*[11])

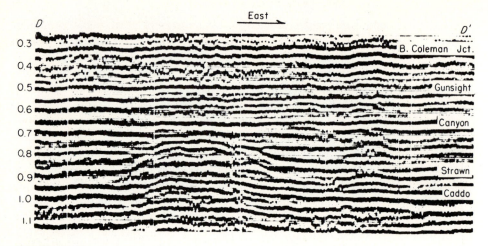

FIGURE 8-17
Seismic section showing reef buildup in North Knox City Field, west Texas.
(*From Harwell and Rector.*[12])

butions in a well (*B*) penetrating a thick Devonian reef as well as in another well (*A*) off the reef about 3 mi away. The seismic velocity through the reef averages about 18,500 ft/s, while the normal velocity through the shale and limestone in the same stratigraphic interval averages about 13,000 ft/s. So large a velocity contrast would lead us to expect a substantial shortening of the time interval between a reflection from below the reef zone and one from above it as the maximum reef buildup is approached from the area outside of the reef.

Reefs are sometimes revealed on seismic records by subtle changes in reflection quality, including the interruption of reflections and possibly by diffraction effects. Occasionally reefs can show up very conspicuously on record sections, as in Fig. 8-17 over the North Knox City Field (see Harwell and Rector.[12]) On this section the reef has a characteristic pattern that reflects the incoherence of the deposition during the time that the reefing took place. Below the reef buildup in the Canyon Formation, a "structure" is shown in the underlying Caddo, which is entirely the result of velocity pull-up in the reef limestone.

A somewhat different approach to reef exploration, proposed by Fitton and Dobrin[13] for areas of poor reflection quality, involves measuring changes in reflection frequency in the zone above the reef where differential compaction might be expected.

In the middle 1960s oil-bearing pinnacle reefs were discovered in the middle Devonian in the Rainbow Lake area in northwestern Alberta. Evans[14] discusses the seismic exploration for reefs on this type which occur in the Keg River Formation. He shows sections displaying conspicuous draping effects in the overlying Slave Point as well as irregularities in the reflections from the Keg River zone.

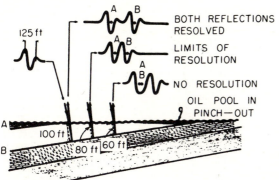

FIGURE 8-18
Following a converging bed toward its termination using seismic reflections from its top and bottom surfaces. (*From Lyons and Dobrin.*[10])

Pinchouts and other erosional truncations It is generally difficult to detail erosional surfaces with the reflection seismograph because of the limitations in resolution inherent in the seismic method. Most reflections observed on records, as previously pointed out, are a composite of returns from individual boundaries which arrive at such a time that they interfere constructively. A single boundary, such as an unconformable surface that is not parallel to other interfaces above or below, would not be expected to generate a reflection with the lateral persistence that is necessary if it is to be useful in reflection mapping. Figure 8-18 illustrates how an erosional unconformity associated with a pinchout would cause the character of a reflection to change continually in the lateral direction because of progressive shifts in the phase relation between constituent signals from different lithological boundaries encompassed by a single seismic wavelength.

Such unconformities have considerable practical significance in oil exploration. This is because oil is often entrapped in the updip wedge that is left when a dipping porous sand is eroded and the erosional surface makes an angle with the base of the sand, eventually truncating it as shown in Fig. 8-18. When this surface is covered by impermeable material in subsequent deposition, we have favorable conditions for oil accumulation. The East Texas Field, one of the world's greatest, had its origin in the updip wedging by erosion of the Woodbine sand and the subsequent deposition of the Eagle Ford shale, which acted as a seal for hydrocarbon entrapment.

To locate such features precisely by seismic reflection would be a most desirable objective, and sometimes this can be done. More generally, however, the limitations in resolution we have discussed make it difficult to do more than narrow the range of uncertainty in where to look for this type of entrapment. Figure 8-19 from a paper by Robinson[15] illustrates the problems that occur when one is looking for such pinchouts by combining geological and seismic information. At 1.1 s there are three peaks on the left side of the section which correlate on the basis of velocity information with a water-bearing sand 290 ft thick at well 1. Following this event in the updip direction, which is to the right, we see that the three peaks soon converge into two peaks near the center of the section, which as they approach the

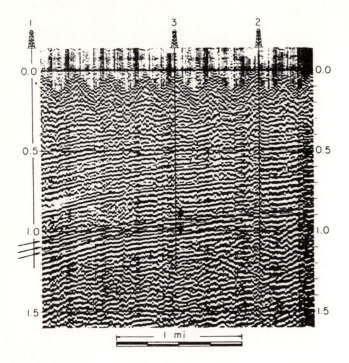

FIGURE 8-19
Pinchout of sand body as seen on seismic section. Sand is indicated by arrows
along left edge of section where well 1 showed it to be 290 ft thick. It is missing
in well 2. In well 3 it was water-bearing, suggesting no oil in pinchout. (*From
Robinson.*[15])

extreme right lose their coherence. This change in waveform should indicate that
the reflecting formation has changed character or disappeared. The second well to
be drilled, well 2, encountered no sand at all, and when well 3 was drilled downdip
from well 2, the sand was found, higher and thinner but water-bearing. No further
drilling has been done on the feature. It should be pointed out, of course, that a sand
pinching out updip can be saturated with water all the way to the wedge's edge just
as an anticline can be water-bearing at its highest point.

If oil should occur farther updip in the sand than well 3, better resolution of the
reflections from this level might have made it possible to predict the position of the
pinchout more closely and locate the third well where it might have brought in oil
rather than water. More up-to-date shooting and processing procedures might
obtain such resolution. In general, however, one must expect more dry holes before
bringing in a stratigraphic oil pool than would usually be required for structural
discoveries. This example illustrates why greater risks are generally involved in
drilling for stratigraphic than for structural oil.

One of the greatest accumulations of oil in the world, Prudhoe Bay on the Alaska
North Slope, is entrapped on one side by a truncational surface between the

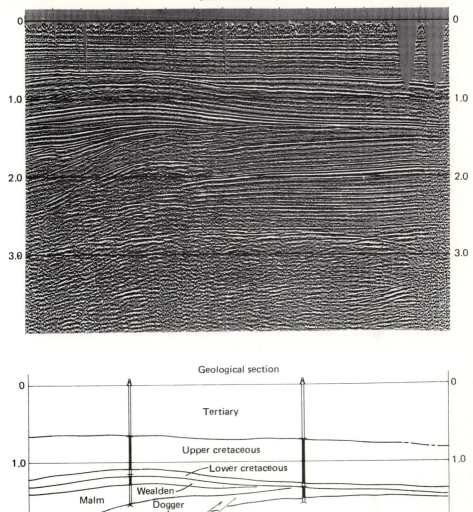

FIGURE 8-20
Record section and corresponding geologic section through the Hohne Field, West Germany. Production is in the Dogger subcrop below a Jurassic-Cretaceous unconformity. (*Prakla-Seismos.*)

impermeable Cretaceous and the productive Jurassic-Mississippian zone. The unconformity responsible for the entrapment shows up very clearly on the seismic section, published by Morgridge and Smith,[16] which crosses the field from southwest to northwest.

Productive truncation traps of a somewhat different type have been found by seismic reflection in the Gifhorn Basin of northwestern Germany. Here the oil-bearing Dogger Beta sand (Jurassic) subcrops below the unconformity at the base of the Albian (Cretaceous). Along the line illustrated in Fig. 8-20, oil has accumulated in the updip wedge-edge of the Jurassic Dogger sand that terminates at the unconformity between the Cretaceous and the Jurassic. A convergence indicative of such wedging was observed on early seismic records, leading to the discovery of Dogger production in the Hohne Field. A case history of this discovery has been published by Hedemann and Lorenz.[17] The seismic section shown in the upper part of Fig. 8-20 is from shooting carried out after the field was developed.

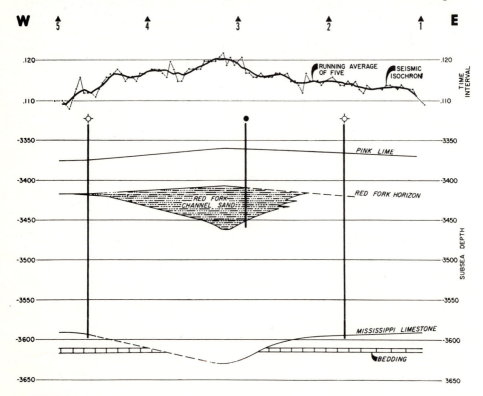

FIGURE 8-21
Cross section made from seismic records defining productive channel-fill (Red Fork sandstone) on basis of anomalies in time interval between two reflections (Pink lime and Mississippi-limestone) straddling sand. (*From Lyons and Dobrin.*[10])

Channel-sand deposits and other sand lenses Where a buried stream channel is covered with an impermeable material it can entrap oil under favorable source and reservoir conditions. Lyons and Dobrin[10] show how an extension to the South Ceres Field in Noble County, Okla., where oil is entrapped in a channel sand of this type, was located on the basis of seismic reflection data. Figure 8-21 is a cross section through the sand which illustrates how the lens was traced by careful measurement of the differential times between a reflection directly above the sand zone and one directly below it. The isochron pattern shows how the position where this time is a maximum coincides with the maximum thickness of the sand body.

A less favorable showing was made by geophysics in the discovery of the Candeias Field in the Reconcavo Basin of Brazil. Vieira[18] has published a case history on this field. Figure 8-22 shows a geological cross section which indicates the positions of the field's four productive sand bodies. The discovery well, 1C-1-BA, was drilled on a seismic high, which was later accounted for by the high velocity associated with the massive sandstone body near the surface at the left edge of the figure. The well happened to penetrate the lowest part of the uppermost producing sand, and this is how the discovery was made. If it had been a few hundred meters farther to the northwest, it would have missed the sand body altogether and the field might never

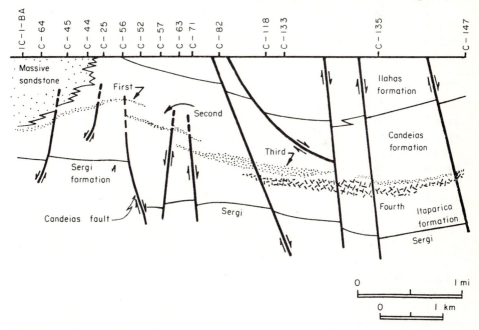

FIGURE 8-22
Structure of the Candeias Field in Brazil. The discovery well, 1C-1-BA at the left edge of the line, was drilled because of a seismic high caused (it was found) by velocity pull-up in the massive sandstone. The productive sands have no seismic expression. (*From Vieira.*[18])

have been discovered. The other sand bodies were located in subsequent development drilling.

In 1969, a number of experimental seismic lines were shot to test whether modern digital recording and processing techniques would show up the known sand bodies. Reflections from the Sergi sandstone (nonproductive) could be followed from the basement, but none of the producing sands showed up on the sections as observable reflection lineups. The negative results must force us to the conclusion that highly sophisticated digital systems and processing programs do not guarantee success in finding productive sand lenses.

Facies changes   In some oil fields the accumulation of oil is governed by lateral changes from permeable to impermeable facies. It is very difficult to observe such a transition directly by the seismograph, but it was possible to do this over the Bramberge Field of West Germany, discussed by Roll.[19] The Bentheim formation changes from sand to shale; the facies change making possible the updip entrapment of the oil in the field. The contours in the original seismic map indicated an updip structural closure against a fault, but the discovery well, drilled to meet an obligation rather than on technical grounds, found oil outside the closure and suggested that the accumulation was actually due to a change in facies from sand to shale. A later experimental seismic survey showed a clear-cut character change between the Bentheim sand surface and the shale surface on the other side of the facies boundary which could have been used to outline the field's boundaries if its significance had been appreciated at the proper time.

The San Emidio Nose Field of California, reported on by Bazeley,[20] has a complicated discovery history. Twenty-four years elapsed between the initiation of seismic work and the drilling of the discovery well. During this period more and more subsurface information was obtained, mainly from dry holes, that made it possible to close in on the productive area. The oil is entrapped in tongues of sand interfingering and terminating updip in shale. The boundaries were not detectable on seismic records, but the existence of a structural nose could be determined, and this information helped in the search for a successful drill site. This discovery may reflect one of the most thorough efforts to integrate geology and geophysics that has ever been described in the literature. The estimated reserves are greater than 50 million barrels of oil, and so it appears that the effort ultimately paid off.

Recent developments in seismic data acquisition and processing have made it possible to trace sand-shale interfingering more successfully in favorable areas than in earlier times. Figure 8-23, from a paper by Sangree and Widmeier,[21] shows how a transition from marine shale (on the right) to fluvial near-shore sandstone (on the left) could be traced on a reflection record section. The section is plotted in depth, facilitating the comparison with the geologic section, which was based on 18 closely spaced wells. The increased density of reflections in the middle of the section indicates the greater number of alternations between sands and shales along this part of the line and shows how the density of reflection events can give information on the location of such transitions.

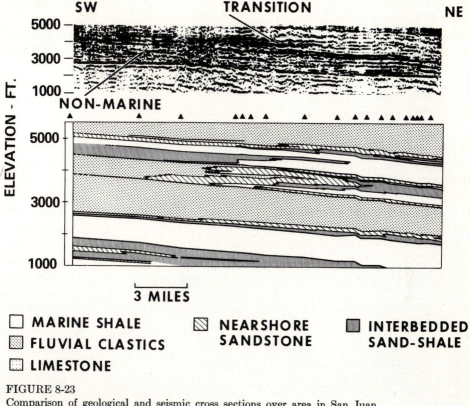

FIGURE 8-23

Comparison of geological and seismic cross sections over area in San Juan
Basin, where sand and shale facies interfinger.   Black triangles show well
locations.   (*From Sangree and Widmeier.*[21])

Study of Lithology from Reflection Data

The identification of lithology in potentially productive strata is an important step
in stratigraphic exploration. In the past there was little that could be learned from
seismic reflection records that would help with such identification, but modern field
techniques and processing procedures now make it possible, under favorable
conditions, to measure such lithological parameters as sand-shale ratios or shale-
limestone ratios. As we have seen in Chap. 2 (pages 49 to 55), interval velocities
between successive reflections have a close relation to the lithology between the
reflecting surfaces. Attenuation of seismic waves between such surfaces can be
similarly diagnostic.

In Chap. 7 we considered techniques of determining interval velocities by
applying such formulas as the Dix equation. When the relationship of velocity and
depth for different rock types in an area is known, it is often possible to identify
lithologic composition from interval-velocity data alone. The longer the spread and

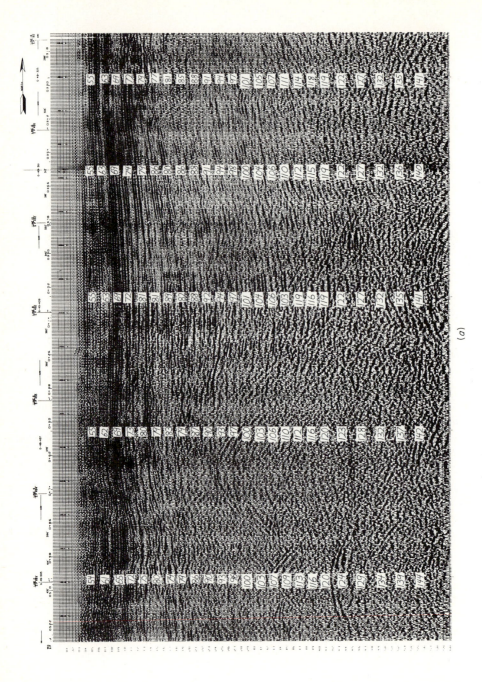

(a)

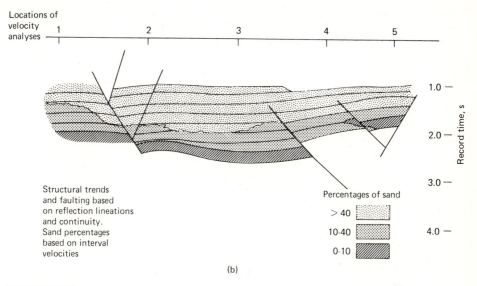

(b)

FIGURE 8-24

Seismic data offshore from Louisiana relating structure and lithology: (a) seismic record section showing interval velocities in hundreds of feet per second; (b) section showing structure and lithology deduced from (a). Sand-shale ratios as estimated from the interval velocities. (*Kerr-McGee Corporation.*)

the better the quality of the data the smaller the time interval between reflections needed for adequate precision in determining the interval velocities.

Figure 8-24a shows a seismic section made in offshore Louisiana, where interval velocities, plotted on the section, were computed at five positions in close proximity. A cross section showing the sand distribution that has been calculated from the velocities is displayed in Fig. 8-24b. Although there is no verification of the predicted lithology, the conformity between the inferred lithologic boundaries and structural trends is impressive.

Another approach, of more recent origin, has been the identification of lithology within a portion of the geologic section by measuring attenuation of seismic waves passing through the section. Using data recorded with binary-gain systems, one can determine changes in relative amplitude between successive reflections that can be interpreted in terms of transitions from sand to shale within the stratigraphic interval that is represented. Figure 8-25 illustrates a case where sand bodies were mapped from attenuation data, the predictions having been verified by well information.

As pointed out by Savit and Mateker,[22] a combination of velocity and attenuation data should yield more reliable lithological information than either kind alone. Modern recording and processing techniques appear to be bringing us closer to the goal of being able to identify subsurface formations as well as to describe their geometry from seismic reflection data. At the present state of the art, a considerable

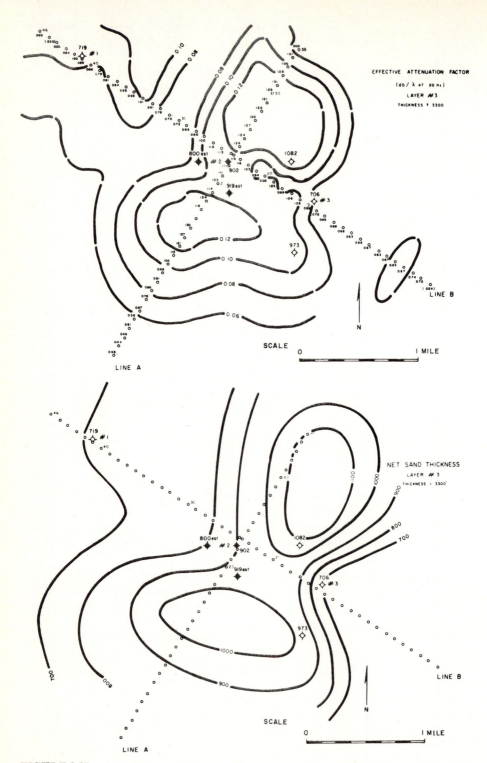

FIGURE 8-25
Comparison of attenuation and net sand thickness in Beer Nose Field, Calif.
(*From Savit and Mateker.*[22])

amount of information is needed from wells before the seismic data can be properly calibrated in terms of lithology.

REFERENCES

1. Widess, M. B.: How Thin Is a Thin Bed?, *Geophysics*, vol. 38, pp. 1176–1180, 1973.
2. Woods, J. P.: The Composition of Reflections, *Geophysics*, vol. 21, pp. 261–276, 1956.
3. Sengbush, R. L., P. L. Lawrence, and F. J. McDonal: Interpretation of Synthetic Seismograms, *Geophysics*, vol. 26, pp. 138–157, 1961.
4. Peterson, R. A., W. R. Fillippone, and F. B. Coker: The Synthesis of Seismograms from Well Log Data, *Geophysics*, vol. 20, pp. 516–538, 1955.
5. Wuenschel, P. C.: Seismogram Synthesis Including Multiples and Transmission Coefficients, *Geophysics*, vol. 25, pp. 106–129, 1960.
6. Quarles, Miller, Jr.: Fault Interpretation in Southwest Texas, *Geophysics*, vol. 15, pp. 462–470, 1950.
7. Laubscher, Hans P.: Structural and Seismic Deformation along Normal Faults in the Eastern Venezuelan Basin, *Geophysics*, vol. 21, pp. 368–387, 1956.
8. Tucker, Paul M., and Howard J. Yorston: Pitfalls in Seismic Interpretation, *Soc. Explor. Geophys. Monogr. Ser. 2*, 1973.
9. King, Robert E. (ed.): "Stratigraphic Oil and Gas Fields: Classification, Exploration and Case Histories," *Mem. 16, Am. Assoc. Petrol. Geol.*, and *Spec. Pub. 10, Soc. Explor. Geophys.* 1972.
10. Lyons, Paul L., and M. B. Dobrin: Seismic Exploration for Stratigraphic Traps, pp. 225–243, in Ref. 9.
11. Skeels, D. C.: Correlation of Geological and Geophysical Data, *Proc. 4th World Petrol. Congr.*, sec. I, pp. 665–673, Geology and Geophysics, Brill, Leiden, Netherlands, 1955.
12. Harwell, J. C., and W. R. Rector: North Knox City Field, Knox County, pp. 453–459, in Ref. 9.
13. Fitton, John C., and Milton B. Dobrin: Optical Processing and Interpretation, *Geophysics*, vol. 32, pp. 801–818, 1967.
14. Evans, Hugh: Zama: A Geophysical Case History, pp. 440–452, in Ref. 9.
15. Robinson, W. B.: Geophysics Is Here to Stay, *Bull. Am. Assoc. Petrol. Geol.*, vol. 55, pp. 2107–2115, 1971.
16. Morgridge, Dean L., and William B. Smith, Jr.: Geology and Discovery of Prudhoe Bay Field, Eastern Arctic Slope, Alaska, pp. 489–501, in Ref. 9.
17. Hedemann, H. A., and H. Lorenz: Truncation Traps on Northwest Border of Gifhorn Trough, East Hannover, Germany, pp. 532–547, in Ref. 9.
18. Vieira, Lauro P.: Candeias Field: Typical Stratigraphic Traps, pp. 354–366, in Ref. 9.
19. Roll, A.: Bramberge Field, Federal Republic of Germany, pp. 286–296, in Ref. 9.
20. Bazeley, William: San Emidio Nose Oil Field, California, pp. 297–312, in Ref. 9.
21. Sangree, J. B., and J. M. Widmeier: Interpretation of Depositional Facies from Seismic Data, *Soc. Explor. Geophys. Conv.*, Dallas, Tex., November 1974.
22. Savit, C. H., and E. J. Mateker, Jr.: From "Where?" to "What?," *Proc. 8th World Petrol. Congr.*, vol. 3, pp. 95–104, Applied Science, London, 1971.

9

SEISMIC REFRACTION PROSPECTING

The first seismic technique to be used in petroleum prospecting was the refraction method. As early as 1923 refraction shooting was introduced for oil exploration in Mexico. Over the following 7 years, it was responsible (along with the torsion balance) for the spectacular success of geophysics in finding a great number of shallow Gulf Coast salt domes, many of them associated with large accumulations of oil.

Before the seismic refraction method was introduced as a tool for oil exploration, its principles had been applied for some time by earthquake seismologists in determining the structure of the earth's interior from records of earthquakes. The times at which the initial signals from earthquakes were recorded at a number of seismological observatories provided a basis for locating the epicenters and times of origin for the earthquakes. Such information made it possible to plot the distribution of seismic-wave velocities as a function of depth and thus obtain clues to the earth's internal constitution. In 1912 Gutenberg had discovered the earth's core and had calculated its depth (about 3000 km) by refraction techniques. This discovery and Mohorovičić's identification, also made with refraction data from earthquakes, of the boundary which has been given his name at the base of the earth's crust are among the most important contributions ever made to our knowledge of the earth's deep interior.

9-1   REFRACTION VERSUS REFLECTION

The principal difference between the geometry of the refraction and that of the reflection methods is in the interaction that takes place between the seismic waves and the lithological boundaries they encounter in the course of their propagation. The waves reflected by the boundaries travel along paths which are quite easy to visualize. Refracted waves of the type used in exploration work follow a somewhat more complicated trajectory that may not be as obvious to one's intuition. Refraction paths cross boundaries between materials having different velocities in such a way that energy travels from source to receiver in the shortest possible time, as required by Fermat's principle. Most refraction prospecting involves the use of waves having trajectories along the tops of layers with speeds that are appreciably greater than those of any overlying formations. The speeds and depths of such layers are determined from the times required for the refracted waves to travel between sources at the surface and receivers which are also on the surface. The distances between the two are almost always several times as great as the depths of the boundaries along which the waves travel.

The wave-path geometry associated with refraction prospecting requires a considerably different source-geophone geometry in the field than is used for reflection surveys. In mapping structures at any particular depth, the shot and geophones must be farther apart for refraction than they would be for reflection. In most refraction work only the initial arrival of seismic energy is recorded, although later arrivals are sometimes used if conditions are favorable. Because of the greater distances traveled, the frequency of refraction signals tends to be lower than that of reflections. The recording requirements are thus different, and the instruments employed are likely to have different characteristics.

Applications   Refraction is often applied as a reconnaissance tool in newly explored areas. It is most useful where there is at least one high-speed bed having geological interest which extends without significant change over a wide area. To be mappable by refraction such a bed must be overlain by formations which have a lower speed. The method has been extensively used for structural mapping in the Alberta foothills, where the productive Rundle limestone, with a velocity around 20,000 ft/s, underlies Cretaceous sands and shales having speeds which are seldom greater than 12,000 ft/s.

The formations within many sedimentary basins have slower speeds than the basement rocks below them. Where this is the case, refraction should be useful for mapping the basement top and thus determining the thickness of the sedimentary section. The introduction of the refraction sonobuoy in the late 1960s has led to a widespread use of the refraction method for measuring basement depths on little-explored offshore shelves.

Since the conclusion of the early fan-shooting campaign in the Gulf Coast in the 1920s, refraction has been used much less frequently than reflection. The unique capabilities of reflection have undoubtedly been responsible for the eclipse. Yet exploration problems often arise for which refraction appears much more suitable than reflection. In many cases, refraction is substantially less expensive, particularly for reconnaissance.

The proper interpretation of refraction data may require a higher degree of skill and ingenuity than is generally needed for interpreting reflection data. The techniques are not only more complex but call for a greater exercise of judgment. Thus refraction interpretation offers a greater potential challenge to the geophysicist and affords a greater degree of satisfaction when a solution is found that fits all the data.

Literature on refraction In 1967, the Society of Exploration Geophysicists published a volume[1] of papers on refraction prospecting that covers many significant aspects of the subject. Among these are field techniques, fundamental principles, interpretation techniques, and case histories. The papers in the volume were all prepared specifically for it and generally do not cover material previously published; thus the book cannot be looked upon as a comprehensive reference work covering the field. But an extensive bibliography is provided to guide readers through the large body of literature on the subject that appeared before 1967.

The geometrical optics of refraction prospecting is covered in considerable detail by Slotnick.[2] Although his book is intended for the beginning student, it is recommended for all who want to learn the fundamentals of refraction from the ground down.

9-2 WAVE PATHS AND TIME-DISTANCE RELATIONS FOR HORIZONTAL LAYERS

Mechanism for Transmission of Refracted Waves

Refraction ray paths are not always as easy to predict as reflection paths. It may not be obvious that in a layered earth rays refracted along the tops of high-speed layers travel down to them from their source along slant paths, approach them at the critical angle, and return to the surface along a critical-angle path rather than along some other path, e.g., at normal incidence. The physical mechanism involved in this type of propagation was first treated mathematically by Muskat.[3] Dix[4] has summarized Muskat's treatment without using any mathematics, and his explanation will be reproduced here.

Let us consider a hypothetical subsurface consisting of two media, each with uniform elastic properties, the upper separated from the lower by a horizontal interface at depth z (Fig. 9-1). The velocity (longitudinal) of seismic waves in the upper layer is V_0 and in the lower V_1, with $V_1 > V_0$. A seismic wave is generated at point S on the surface, and the energy travels out from it in hemispherical wavefronts. A receiving instrument is located at point D a distance x from S. If x is

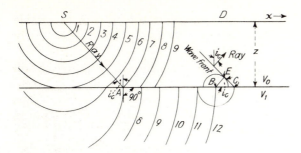

FIGURE 9-1
Mechanism for transmission of refracted waves in two-layered earth.
(*After Dix.*[4])

small, the first wave to arrive at D will be the one that travels horizontally at a speed V_0. At greater distance, the wave that took an indirect path, traveling down to, along, and up from the V_1 layer will arrive first because the time gained in travel through the higher-speed material makes up for the longer path.

When the spherical wavefronts from S strike the interface where the velocity changes, the energy will be refracted into the lower medium according to Snell's law. The process is demonstrated in the diagram for the time corresponding to wavefront 7. At point A on wavefront 7 the tangent to the sphere in the lower medium becomes perpendicular to the boundary. The ray passing through this point now begins to travel along the boundary with the speed of the lower medium. Thus, by definition, the ray SA strikes the interface at the critical angle. To the right of A the wavefronts below the boundary travel faster than those above.

The material on the upper side of the interface is subjected to oscillating stress from below which, as the wave travels, generates new disturbances along the boundary. These disturbances themselves spread out spherically in the upper medium with a speed of V_0. The wave originating at point B in the lower medium will travel a distance BC during the time in which the one spreading out in the upper medium will attain a radius of BE. The resultant wavefront above the interface will follow the line CE, which makes an angle i_c with the boundary. From the diagram, it is seen that

$$\sin i_c = \frac{BE}{BC} = \frac{V_0 t}{V_1 t} = \frac{V_0}{V_1}$$

The angle which the wavefront makes with the horizontal is the same as that which the ray perpendicular to it makes with the vertical, so that the ray will return to the surface at the critical angle $[\sin^{-1}(V_0/V_1)]$ with a line perpendicular to the interface.

As with reflection, the simplest and most useful way to represent refraction data is to plot the first-arrival time T versus the shot-detector distance x. In the case of a subsurface consisting of discrete homogeneous layers, as in Fig. 9-2, this type of plot consists of linear segments.

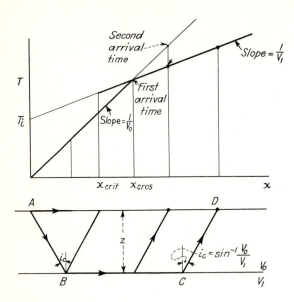

FIGURE 9-2
Ray paths of least time and time-distance curve for a layer separated from its substratum by horizontal interface; x_{crit} is critical distance; x_{cros} is crossover distance.

Two-Media Case

Let us determine the time-distance relations for the case, illustrated in Fig. 9-2, of two media with respective speeds of V_0 and V_1, separated by a horizontal discontinuity at depth z.

Intercept time The direct wave travels from shot to detector near the earth's surface at a speed of V_0, so that $T = x/V_0$. This is represented on the plot of T versus x as a straight line which passes through the origin and has a slope of $1/V_0$. The wave refracted along the interface at depth z, reaching it and leaving it at the critical angle i_c, takes a path consisting of three legs, AB, BC, and CD. To determine the time in terms of horizontal distance traveled we make use of the three relations

$$\sin i_c = \frac{V_0}{V_1} \qquad \cos i_c = \left(1 - \frac{V_0^2}{V_1^2}\right)^{1/2}$$

and
$$\tan i_c = \frac{\sin i_c}{\cos i_c} = \frac{V_0}{\sqrt{V_1^2 - V_0^2}}$$

The total time along the refraction path $ABCD$ is

$$T = T_{AB} + T_{BC} + T_{CD} \qquad (9\text{-}1)$$

One can write Eq. (9-1) in the form

$$T = \frac{z}{V_0 \cos i_c} + \frac{x - 2z \tan i_c}{V_1} + \frac{z}{V_0 \cos i_c} \tag{9-2}$$

$$= \frac{2z}{V_0 \cos i_c} - \frac{2z \sin i_c}{V_1 \cos i_c} + \frac{x}{V_1} \tag{9-3}$$

This can be readily transformed into

$$T = \frac{2z}{V_0 \cos i_c} (1 - \sin^2 i_c) + \frac{x}{V_1} \tag{9-4}$$

$$= \frac{x}{V_1} + \frac{2z \cos i_c}{V_0}$$

$$= \frac{x}{V_1} + \frac{2z \sqrt{1 - (V_0/V_1)^2}}{V_0} \tag{9-5}$$

so that finally

$$T = \frac{x}{V_1} + \frac{2z \sqrt{V_1^2 - V_0^2}}{V_1 V_0} \tag{9-6}$$

On a plot of T versus x, this is the equation of a straight line which has a slope of $1/V_1$ and which intercepts the T axis ($x = 0$) at a time

$$T_i = 2z \frac{\sqrt{V_1^2 - V_0^2}}{V_1 V_0} \tag{9-7}$$

T_i is known as the *intercept time*.

Crossover distance   At a distance x_{cros} (see Fig. 9-2), the two linear segments cross. At distances less than this, the direct wave traveling along the top of the V_0 layer reaches the detector first. At greater distances, the wave refracted by the interface arrives before the direct wave. For this reason, x_{cros} is called the *crossover distance*.

Depth calculation   The depth z to the interface can be calculated from the intercept time by use of Eq. (9-7) or from the crossover distance, a concept introduced on p. 204 to show the relation between refraction and reflection times. This equation can be solved for z to obtain

$$z = \frac{T_i}{2} \frac{V_1 V_0}{\sqrt{V_1^2 - V_0^2}} \tag{9-8}$$

T_i can be determined graphically as shown in Fig. 9-2 or numerically from the relation $T_i = T - x/V_1$.

The depth can also be expressed in terms of x_{cros}, the crossover distance, making use of the fact that the times

$$T_0 = \frac{x}{V_0} \quad \text{and} \quad T_1 = \frac{x}{V_1} + \frac{2z \sqrt{V_1^2 - V_0^2}}{V_1 V_0}$$

are equal at x_{cros}. Then

$$\frac{x_{cros}}{V_0} = \frac{x_{cros}}{V_1} + \frac{2z \sqrt{V_1^2 - V_0^2}}{V_1 V_0} \tag{9-9}$$

and

$$z = \frac{1}{2} \frac{V_0 V_1 x_{cros}}{\sqrt{V_1^2 - V_0^2}} \left(\frac{1}{V_0} - \frac{1}{V_1} \right) \tag{9-10}$$

This simplifies to

$$z = \frac{1}{2} \sqrt{\frac{V_1 - V_0}{V_1 + V_0}} \, x_{cros} \tag{9-11}$$

Three-Media Case

For three formations with velocities V_0, V_1, and V_2 $(V_2 > V_1 > V_0)$, the treatment is similar but somewhat more complicated. Figure 9-3 shows the wave paths. The ray corresponding to the least travel time makes an angle $i_1 = \sin^{-1}(V_0/V_2)$ with the vertical in the uppermost layer and an angle $i_2 = \sin^{-1}(V_1/V_2)$ with the vertical in the second layer, i_2 being the critical angle for the lower interface. The time along each of the two slant paths AB and EF through the uppermost layer is

$$T_{AB} = \frac{AB}{V_0} = \frac{z_0}{V_0 \cos i_1} = \frac{z_0}{V_0 \sqrt{1 - (V_0/V_2)^2}} = T_{EF} \tag{9-12}$$

while that through each of the legs BC and DE crossing the middle layer is

$$T_{BC} = \frac{BC}{V_1} = \frac{z_1}{V_1 \cos i_2} = \frac{z_1}{V_1 \sqrt{1 - (V_1/V_2)^2}} = T_{DE} \tag{9-13}$$

The time for the segment of path CD at the top of the V_2 layer is CD/V_2. The ex-

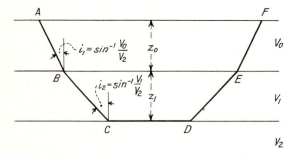

FIGURE 9-3
Ray paths of least time for two layers and a substratum separated by two horizontal interfaces.

pression for the total travel time from A to F is

$$T = T_{AB} + T_{BC} + T_{CD} + T_{DE} + T_{EF}$$

$$= \frac{2z_0}{V_0 \sqrt{1 - (V_0/V_2)^2}} + \frac{2z_1}{V_1 \sqrt{1 - (V_1/V_2)^2}} + \frac{CD}{V_2} \tag{9-14}$$

where $\qquad CD = x - 2z_0 \tan i_1 - 2z_1 \tan i_2$

$$= x - 2z_0 \frac{V_0}{V_2 \sqrt{1 - (V_0/V_2)^2}} - 2z_1 \frac{V_1}{V_2 \sqrt{1 - (V_1/V_2)^2}} \tag{9-15}$$

Rearranging terms, one obtains

$$T = \frac{x}{V_2} + \frac{2z_0 \sqrt{V_2{}^2 - V_0{}^2}}{V_2 V_0} + \frac{2z_1 \sqrt{V_2{}^2 - V_1{}^2}}{V_2 V_1} \tag{9-16}$$

for the overall travel time of the wave refracted along the top of the V_2 zone. The portion of the time-distance curve corresponding to the first arrival of this wave is a straight line with a slope $1/V_2$ and an intercept time

$$T_{i2} = T - \frac{x}{V_2} = \frac{2z_0 \sqrt{V_2{}^2 - V_0{}^2}}{V_2 V_0} + \frac{2z_1 \sqrt{V_2{}^2 - V_1{}^2}}{V_2 V_1} \tag{9-17}$$

Solving for z_1, one obtains

$$z_1 = \frac{1}{2} \left(T_{i2} - 2z_0 \frac{\sqrt{V_2{}^2 - V_0{}^2}}{V_2 V_0} \right) \frac{V_2 V_1}{\sqrt{V_2{}^2 - V_1{}^2}} \tag{9-18}$$

The depth to the lower interface is the sum of z_1 and z_0, where z_0 is computed by the two-media formula [Eq. (9-8)] using the slopes of the first two segments of the time-distance curve and the intercept of the second segment.

Figure 9-4 shows time-distance-depth relations for a refraction profile over two

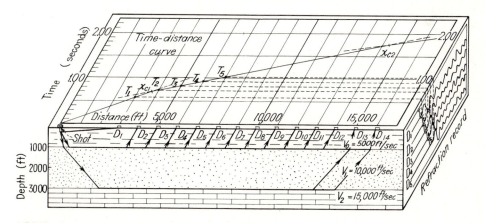

FIGURE 9-4
Wave paths, schematic record, and time-distance curve for subsurface consisting of two horizontal layers overlying a substratum.

subsurface discontinuities (two layers and a substratum). The arrival times for the first five detector positions are measured on the record shown schematically at the right side of the block. The first arrival on each trace appears as a sharp upward displacement from the baseline representing an undisturbed earth. The three segments of time-distance curve represent paths along the tops of the respective media. The two breaks in the slope of the time-distance curve occur at the crossover distances x_{c1} and x_{c2} for the respective interfaces.

Multilayer Case

The time-depth relations just derived for the two- and three-media cases can be readily extrapolated to apply to a larger number of layers as long as the speed in each layer is higher than in the one just above it. This is illustrated by Fig. 9-5, which shows ray paths and time-distance plots for six media, the lowermost designated for generality as the nth. Each segment on the plot represents first arrivals from the top of one of these subsurface layers, the horizontal ray paths corresponding to the respective segments being designated by the letters b to f. The deeper the layer, the greater the shot-detector distance at which arrivals from it become the first to be observed. In other words, the crossover distance for each layer increased with its depth. The slope of each segment is simply the reciprocal of the speed in the

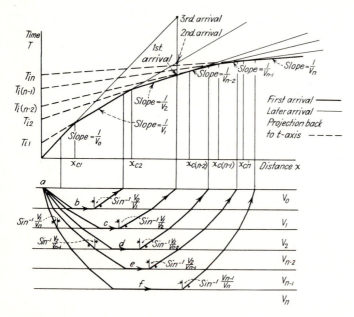

FIGURE 9-5
Ray paths, time-distance curve, and critical distances for multilayered earth. x_{c1}, x_{c2} etc. are crossover distances for successively deeper interfaces.

layer if the wave has traveled along it horizontally. The intercept time of each segment depends on the depth of the interface at the bottom of the corresponding wave path as well as on the depths of all those interfaces that lie above it in the section.

Low-Speed Layer

To calculate depths for additional layers by extending Eq. (9-18) to apply to more than two refracting interfaces would be valid only in cases where successively deeper layers have successively higher speeds. If any bed in the sequence has a lower speed than the one above it, it will not be detectable by refraction shooting at all. This is because the rays entering such a bed from above are always deflected in the downward direction, as shown for the interface between the V_0 and V_1 layers in Fig. 9-6, and thus can never travel horizontally through the layer. Consequently there will be no segment of inverse slope V_1 on the time-distance curve. The presence of such an undetected low-speed layer will result in the computation of erroneous depths to all interfaces below it if only observed segments are used in the calculations.

Blind Zone

A similar kind of error may occur if the thickness of a layer with a speed intermediate between that of the layer overlying it and the one below it is small and/or

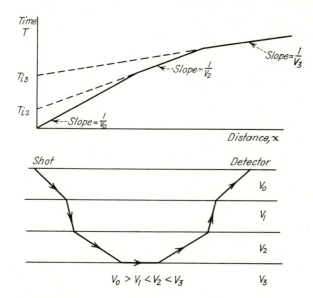

FIGURE 9-6
Ray paths of least time and time-distance curve where low-speed layer V_1 lies below higher-speed layer V_0. $V_3 > V_2 > V_1$.

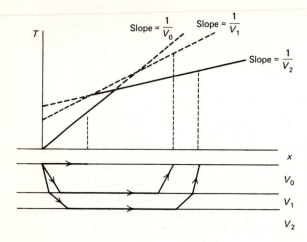

FIGURE 9-7
The blind zone in refraction shooting. It is not possible to observe a
first arrival from the top of the V_2 layer because arrivals from either V_3
or V_1 are earlier at all receiving distances.

the velocity contrast between it and the layer that underlies it is inadequate. If only
first arrivals can be observed on the records, the refracted waves from the intermediate layer will not be discernible because they will always reach the surface at a
later time than from either a shallower or a deeper bed, depending on the shot-
detector distance. Figure 9-7 shows the relation of the time-distance segments from
the respective layers when there is a *blind zone*, as an intermediate layer of this type
is called.

Soske[5] has pointed out the importance of taking such zones into account for
proper accuracy in shallow refraction investigations, particularly for engineering
purposes. Unless second events can be identified on the records it is not possible to
recognize the existence of blind zones from seismic data alone. But if a test borehole
in the area of the survey reveals a layer undetectable by refraction, its effect can be
taken into account in calculating depths to high-speed beds. Soske's technique for
doing this makes use of the wavefront method of Thornburgh,[6] which will be discussed in more detail later in this chapter. Hawkins and Maggs[7] have developed a
nomogram that yields the limiting thickness of a blind zone having any assumed
velocity on the basis of observed arrival times.

A related problem occurs when a high-speed layer on the surface or at shallow
depths masks a deeper marker of equal or only slightly greater velocity which it is
desired to map by refraction. This difficulty is sometimes encountered when surface
volcanic layers, e.g., the basalts in the Parana Basin of Brazil, cover a sedimentary
section which contains deeper high-speed layers such as the basement surface or
sheets of dense limestone that could be mapped by refraction if the volcanic cover
were not present. When the speed of the volcanic material is approximately equal

to or greater than that of the marker to be mapped, rays that could be returned to the surface by refraction should not penetrate below the near-surface material. But if the shallow high-speed layer is thin compared with the wavelength of the seismic impulse, it will be "transparent" to the seismic energy and will not mask arrivals from the deeper horizon. As the distance from source to detector becomes greater, the wavelength naturally increases because of selective absorption of high frequencies. Thus it is possible to overcome masking effects of surface or near-surface high-speed layers if they are thin enough by operating at large shot-geophone distances. Trostle[8] describes an experiment in the Delaware Basin of west Texas where interference from such layers was overcome by recording at distances greater than 100,000 ft from the shot. Dense Devonian limestones at depths of 15,000 ft were mapped through anhydrites of almost the same speed with tops as shallow as 3500 ft.

9-3 DIPPING BEDS WITH DISCRETE VELOCITIES

When the top surface of a refracting marker bed is not horizontal, the angle of dip can be determined from the time-distance data. Consider (as shown in Fig. 9-8) a boundary between two beds having respective speeds of V_0 and V_1 which dips at an angle α. We define z_d as the perpendicular distance from the shot to the interface

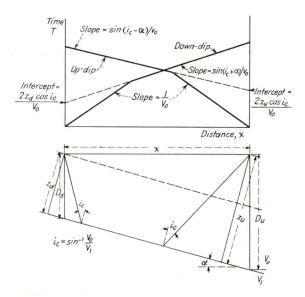

FIGURE 9-8
Refraction along an interface dipping at an angle α. Respective shots are at updip and downdip ends of profile.

at the end of the line at which one shoots downdip, and z_u as this distance at the end from which the ray refracted downdip originates.

As with the horizontal interface, the ray path for the first refracted arrival consists of three linear segments. One of these extends along the sloping interface and corresponds to wave travel at speed V_1. The other two, in the upper medium, make an angle of i_c with the normal to this boundary and the wave travels along them at speed V_0.

Shooting downdip, the total time from shot to detector is

$$T_d = \frac{z_d}{V_0 \cos i_c} + \frac{x \cos \alpha - z_d \tan i_c - (z_d + x \sin \alpha) \tan i_c}{V_1} + \frac{z_d + x \sin \alpha}{V_0 \cos i_c}$$

(9-19)

Making the same transformation used in deriving the formula for a single layer and substratum separated by a horizontal interface, one obtains

$$T_d = \frac{2z_d \cos i_c}{V_0} + \frac{x}{V_0} \sin (i_c + \alpha)$$

(9-20)

Similarly, the time for shooting updip is

$$T_u = \frac{2z_u \cos i_c}{V_0} + \frac{x}{V_0} \sin (i_c - \alpha)$$

(9-21)

where z_u is the perpendicular distance to the interface from the shot from which arrivals are received at the updip end of the line.

Time-distance relations for both trajectories are shown in Fig. 9-8. The respective slopes of the two linear segments for each are used to determine the true speeds, V_0 and V_1, of the two media and the dip α of the sloping interface.

To obtain V_1 and α, one makes use of the fact that the slope m_d of the downdip segment is $[\sin (i_c + \alpha)]/V_0$ while the slope m_u of the updip segment is $[\sin (i_c - \alpha)]/V_0$, so that

$$V_0 m_d = \sin (i_c + \alpha)$$

(9-22)

$$V_0 m_u = \sin (i_c - \alpha)$$

(9-23)

$$i_c + \alpha = \sin^{-1} V_0 m_d$$

(9-24)

$$i_c - \alpha = \sin^{-1} V_0 m_u$$

(9-25)

Solving for i_c, one adds and obtains

$$i_c = \tfrac{1}{2}(\sin^{-1} V_0 m_d + \sin^{-1} V_0 m_u)$$

(9-26)

To solve for α, one subtracts, obtaining

$$\alpha = \tfrac{1}{2}(\sin^{-1} V_0 m_d - \sin^{-1} V_0 m_u)$$

(9-27)

With i_c determined, V_1 is readily obtained from the relation

$$V_1 = \frac{V_0}{\sin i_c} \tag{9-28}$$

The perpendicular distance z_u to the interface comes from the intercept time

$$T_{iu} = \frac{2 z_u \cos i_c}{V_0}$$

and is

$$z_u = \frac{V_0 T_{iu}}{2 \cos i_c} \tag{9-30}$$

A similar expression is obtained for z_d in terms of T_{id}. The depth D_u is $z_u/(\cos \alpha)$, while D_d is $z_d/(\cos \alpha)$, as is evident from Fig. 9-8. When there is more than one dipping layer, the formulas for depth and dip are derived by use of the same principle but the computations are more complex. Mota[9] has published formulas which apply this approach to a large number of layers.

9-4 REFRACTION SHOOTING ACROSS A FAULT

If a high-speed bed (velocity V_1) situated under a low-speed overburden (velocity V_0) is faulted vertically, a refraction profile across the fault should make it possible to locate the faulting and measure its throw. When the source is on the upthrown side of the fault, as shown in Fig. 9-9a, the rays that are refracted along this high side will be received as first arrivals over the portion of the recording profile from B to D. Diffractions from the corner of the fault taking paths like DF will be observed as first arrivals from D to G, at which point they will be overtaken by refracted events from the downthrown side of the fault. The refracted arrivals for each direction of shooting are plotted along two parallel but displaced linear segments on the time-distance curves. These have an inverse slope equal to the speed in the faulted formation. The two segments correspond to rays refracted respectively from the upthrown and downthrown sides of the fault. The amount of throw z_t can be determined from the difference $T_{i2} - T_{i1}$ between the intercept times corresponding to these segments.

The case where the shot is on the downthrown side of the fault is illustrated in Fig. 9-9b. Any waves from the shot received by detectors to the left of point C' cannot emerge at the critical angle because they cannot have traveled along the interface at grazing incidence. But at a large distance from it, such as at F', the path from the base of the fault Q becomes so nearly parallel with the interface that a critical angle with the high-speed surface can be assumed and the intercept time for arrivals on the upthrown side can be determined from the limiting position on the left of the curved segment between C' and F'. Here also the throw can be determined from the difference between the intercept times T_{i2} and T_{i1}.

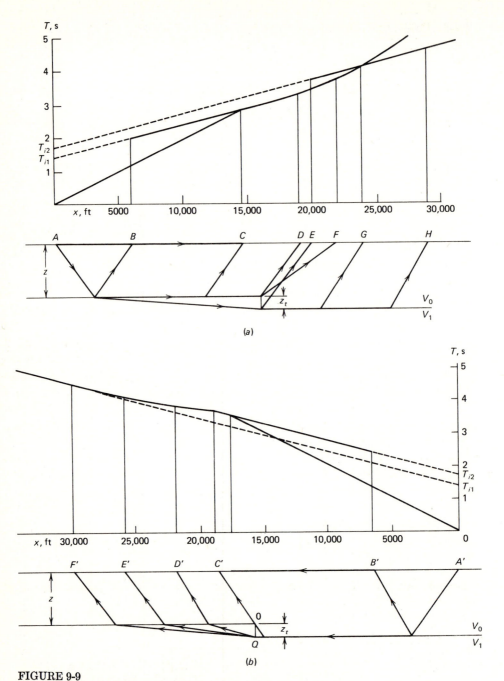

FIGURE 9-9
Refraction across a fault with a throw z_t. Depth to the high-speed surface on its upthrown side is z: (a) shooting from the upthrown side; (b) shooting from the downthrown side. Discontinuities in times and slopes on both time-distance curves are diagnostic of the faulting.

9-5 REFRACTION IN A MEDIUM HAVING CONTINUOUS CHANGE OF SPEED WITH DEPTH

In many areas, e.g., the Gulf Coast of the United States, the lithology tends to change gradually with depth of burial rather than in discrete steps at boundaries of the type we have been considering. In most clastic formations the velocity increases continuously with depth because of differential compaction effects. To visualize the paths of refracted waves through a section with this kind of lithology, we can assume a series of thin layers, each of higher speed than the one above, as in Fig. 9-10, and pass to the limit of an infinite number of infinitesimally thin members. Such a limit corresponds to a section having a continuous increase of velocity with depth. The ray path would then have the form of a smooth curve which is convex downward, and the time-distance curve, also smooth, would be convex upward. In many cases, as Ewing and Leet[10] have pointed out, it is a matter of choice whether a set of observed time-distance points should be connected by linear segments each representing a discrete layer or by a smooth curve representing a continuous increase of speed with depth.

The actual ray trajectory, as well as the form of the time-distance curve, depends on just how the velocity varies with depth. Where the relation between velocity and depth is linear, the ray paths are circles. More complex velocity-depth functions, e.g., those involving a velocity proportional to the square or square root of depth, also give mathematically expressible ray-path trajectories and time-distance relationships. Kaufman[11] has published a useful set of tables relating ray-path geometry and time for a large number of velocity functions.

Time-Distance Relations for Linear Increase of Velocity with Depth

Because of its simplicity and its close correspondence to the actual velocity-depth relationship in most clastic materials, the function

$$v = V_0 + kz \tag{9-31}$$

where v = speed at depth z
V_0 = speed at zero depth
k = constant

is extensively employed to represent the velocity variation in sedimentary basins.

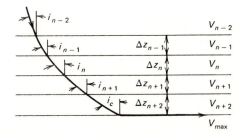

FIGURE 9-10
Ray path on the side toward the shot for series of thin layers with small increments of velocity between them. Ray refracted along bottom interface approaches it at the critical angle i_c which is $\sin^{-1}(V_{n+2}/V_{max})$.

The geometry and time-distance relations for reflection paths in a medium with this type of velocity function were discussed in Chap. 7 (pages 204 to 206).

To demonstrate in an intuitive way how the time-distance and time-depth relationships are derived for this velocity function, we can trace a ray through a sequence of layers each with a thickness Δz and a velocity $v = V_0 + kz_n$, where z_n is the depth to the center of the nth layer. For any particular ray there will be a layer having a speed V_{max} in which the path becomes horizontal. V_{max} is actually a parameter that itself identifies the particular ray under consideration. A ray passing through the nth layer will be bent as shown in Fig. 9-10. The time ΔT_n required for the ray to travel through this layer is

$$\Delta T_n = \frac{\Delta z_n}{V_n \cos i_n} = \frac{\Delta z_n}{V_n \sqrt{1 - \sin^2 i_n}} \tag{9-32}$$

But $i_n = \sin^{-1} (V_n/V_{max})$, where V_{max} is the velocity of the layer in which the ray travels horizontally, so that

$$\Delta T_n = \frac{\Delta z}{V_n \sqrt{1 - (V_n/V_{max})^2}} \tag{9-33}$$

Δx_n, which is the horizontal distance traversed by the ray as it passes through the nth layer, can be expressed as

$$\Delta x_n = \Delta z_n \tan i_n = \Delta z_n \frac{V_n/V_{max}}{\sqrt{1 - V_n^2/V_{max}^2}} \tag{9-34}$$

The total time for a wave to travel through N such layers is

$$t = \sum_{n=1}^{N} \frac{\Delta z_n}{V_n \sqrt{1 - (V_n/V_{max})^2}} \tag{9-35}$$

and the net horizontal distance is

$$x = \sum_{n=1}^{N} \frac{\Delta z_n V_n/V_{max}}{\sqrt{1 - (V_n/V_{max})^2}} \tag{9-36}$$

According to Snell's law,

$$\frac{\sin i_n}{\sin 90°} = \frac{V_n}{V_{max}}$$

so that

$$\frac{1}{V_{max}} = \frac{\sin i_n}{V_n}$$

The ratio $(\sin i_c)/V_{n-1} = \sin i_0/V_0$ is a parameter that is constant for any particular ray with a given initial angle of penetration into the earth. This constant, which we designate as the *ray parameter* p, thus describes each ray in terms of the surface velocity V_0 and the emergence angle i_0.

As the number of beds increases and the thickness Δz_n of each becomes smaller, the limit is approached of a continuous increase of velocity with depth, where the velocity at any depth z can be expressed as a continuous function of z. V_n becomes v, and $1/V_{max}$ becomes p. The time and net horizontal displacement can then be expressed for each ray corresponding to a given p as

$$t = \int_0^{Z_{max}} \frac{dz}{v \sqrt{1 - p^2 v^2}} \tag{9-37}$$

and

$$x = \int_0^{Z_{max}} \frac{pv \, dz}{\sqrt{1 - p^2 v^2}} \tag{9-38}$$

In the case where v is a linear function of depth having the form

$$v = V_0 + kz$$

then

$$t = \int_0^{Z_{max}} \frac{dz}{(V_0 + kz) \sqrt{1 - p^2 (V_0 + kz)^2}} \tag{9-39}$$

and

$$x = \int_0^{Z_{max}} \frac{p(V_0 + kz) \, dz}{\sqrt{1 - p^2 (V_0 + kz)^2}} \tag{9-40}$$

Evaluation of these integrals is straightforward if p is constant. The reader is referred to Slotnick[2] (pp. 206–208) for the details. The resultant expressions are

$$x = \frac{1}{kp} \{ [(1 - p^2 V_0^2)^{1/2}] - [1 - p^2 (V_0 + kz)^2]^{1/2} \} \tag{9-41}$$

$$t = \frac{1}{k} \ln \frac{(V_0 + kz)[1 + (1 - p^2 V_0^2)]^{1/2}}{V_0 \{ 1 + [1 - p^2 (V_0 + kz)^2]^{1/2} \}} \tag{9-42}$$

Equation (9-41) can be rewritten

$$\left[x - \frac{(1 - p^2 V_0^2)^{1/2}}{kp} \right]^2 + \left(z + \frac{V_0}{k} \right)^2 = \frac{1}{k^2 p^2} \tag{9-43}$$

This is the equation of a circle having a radius $1/kp$ and a center along a horizontal

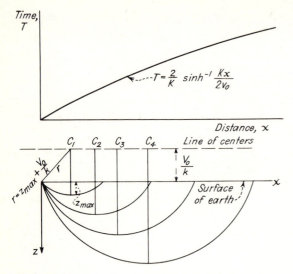

FIGURE 9-11
Ray paths and time-distance curve for linear increase of speed with depth.

line a distance V_0/k above the surface and to the right of the origin a distance $(1 - p^2V_0^2)^{1/2}/kp$. Figure 9-11 illustrates a family of such circles for a number of rays having different angles of immergence into the earth.

The time-distance relation for such a circular ray path between entry into and emergence from the earth can be found by eliminating p and z from Eqs. (9-41) and (9-42). This can be expressed in the form*

$$T = \frac{2}{k} \sinh^{-1} \frac{kx}{2V_0} \tag{9-44}$$

Also the formula for the depth of maximum penetration for a ray with an initial angle of i_0 is

$$z_{max} = \frac{V_0}{k} (\csc i_0 - 1) \tag{9-45}$$

The time-distance curve applying for a linear increase of speed with depth is plotted in the upper part of Fig. 9-11. The inverse slope of the time-distance curve at any point is equal to the velocity at the depth of maximum penetration for the ray reaching the surface at that point.

---

* The notation $\sinh^{-1} u$ means the angle whose hyperbolic sine is u. Hyperbolic functions can be found in most books of mathematical tables, where they are tabulated in the same way as trigonometric functions.

Application of Curved-Path Theory to Refraction Exploration

In most areas where linear relationships exist between speed and depth, k is of the order of 1 ft/(s)(ft). In the Gulf Coast of the United States it is usually around 0.6. Table 9-1 shows travel times and depths of maximum penetration as a function of shot-detector distance x for $V_0 = 6000$ ft/s and $k = 0.6$.

Increase of velocity to indefinite depths For these constants, it appears that the penetration is less than 1000 ft if the detector is within about 9,000 ft of the shot. At a distance of 10,000 ft, the first arrival time is only about 3.5 percent less for a wave along the curved path than it would be for a wave following the earth's surface at a speed of V_0. At a distance of 40,000 ft, the gain in time is 27.5 percent. The penetrations indicated in the table are greater than those actually observed in the Gulf Coast because the true increase in velocity does not continue downward indefinitely but levels off at depths where compaction no longer increases.

Section with linear variation underlain by high-speed marker A geologic section in which the velocity increases continuously with depth for an indefinite distance below the surface would probably be of little practical interest. In oil exploration, refraction surveys would be more likely to be useful in areas where there is some definite discontinuity in velocity associated with discrete changes in lithology. Such discontinuities could be mapped by refraction if the underlying material had a higher velocity than anything above it.

Often a horizontal high-speed marker* bed lies at the base of a section in which there is a linear increase of speed with depth having constants V_0 and k, as shown in

Table 9-1 CURVED-PATH TIME-DEPTH DATA

$k = 0.6$ ft/(s)(ft) $V_0 = 6000$ ft/s

Shot-detector distance x, ft	Arrival time T, s	Depth of maximum penetration, ft
1,000	0.167	12
2,000	0.333	50
5,000	0.827	310
10,000	1.602	1,180
20,000	2.935	4,140
30,000	3.983	8,028
40,000	4.82	12,360

---

* A *marker horizon* is the top surface of a bed with a high enough velocity compared with formations above it to permit waves refracted along it to be observed at the surface.

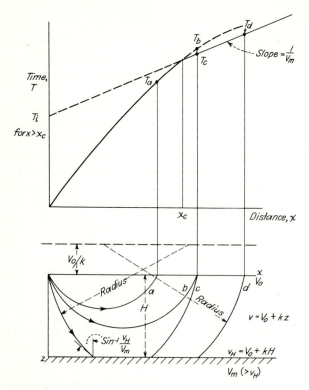

FIGURE 9-12
Time-distance curve and ray paths for high-speed marker below overburden with linear velocity-depth relation.

Fig. 9-12. At a depth H, the velocity changes discontinuously to V_m and remains constant at this value from here on downward. The path of least time to the marker is the arc of a circle, with center V_0/k above the surface, whose tangent at the velocity break makes the angle $i_c = \sin^{-1}(v_H/V_m)$ with the vertical. Here $v_H = V_0 + kH$. The path along the marker follows the horizontal interface, and the return path to the surface follows an arc which is identical to that for the downward trajectory. Barry[12] has presented the derivation of an expression first developed by Gardner[13] for the total travel time of refracted rays for this velocity configuration:

$$T = \frac{x}{V_m} + \frac{2}{k}\left[\cosh^{-1}\frac{V_m}{V_0} - \cosh^{-1}\frac{V_m}{v_H} - \sqrt{1 - \left(\frac{V_0}{V_m}\right)^2} + \sqrt{1 - \left(\frac{v_H}{V_m}\right)^2}\right] \quad (9\text{-}46)$$

The time-distance curve (T versus x) obtained from this equation is shown in the upper part of Fig. 9-12. The intercept time T_i can be obtained by subtracting x/V_m from the right-hand side of the equation.

To put this relationship into useful form it is necessary to express T_i in terms of the depth H. If the velocity v_H is written as $V_0 + kH$, we have

$$T_i = \frac{2}{k}\left[\cosh^{-1}\frac{V_m}{V_0} - \cosh^{-1}\frac{V_m}{V_0 + kH} - \sqrt{1 - \left(\frac{V_0}{V_m}\right)^2} \quad \sqrt{1 - \left(\frac{V_0 + kH}{V_m}\right)^2}\right]$$

(9-47)

This equation shows the functional relationship between T_i and H, but it cannot be solved explicitly for H. To determine H, one can plot Eq. (9-47) using the marker depth H as the independent variable and introduce appropriate values for V_0, V_m, and k. The depth can be determined implicitly from such a plot when the intercept time is known. An equation like this involving a trial-and-error solution is particularly adaptable for handling by electronic computers.

9-6   DELAY TIMES

Refraction interpretation is intrinsically ambiguous because the depths computed from intercept times represent the sum of the respective depths to the marker below the shot and below the detector. If the refracting interface is horizontal, the actual depths are equal at each end and formulas like those in Eq. (9-8) or (9-18) apply. If the interface is an inclined plane, solutions like those developed for this case in a preceding section can be used. Generally, however, the refractor is neither horizontal nor uniformly dipping, as is the case in Fig. 9-13, so that separating the depths at the two ends of the trajectory will require special interpretation techniques.

The concept of *delay time*, introduced by Gardner,[14][15] is convenient for carrying out this separation. The intercept time $T - x/V_1$ may be considered to be made up of two *delay times*: D_1, associated with the shot end of trajectory, and D_2, associated with the detector end. The principle underlying the separation of the intercept time into its component delay times is illustrated in Fig. 9-13.

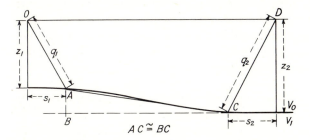

FIGURE 9-13
Separation of intercept times into delay times where depth is different under shot O than that under receiver at D.

The interface shown in the diagram lies at depths of z_1 below the shot and z_2 below the geophone. Immediately below the shot and geophone, the interface is horizontal; between these flat portions the interface slopes so gently that the horizontal distance deviates only negligibly from the actual slant distance.

The intercept time is the difference between the actual travel time of the wave and the time that would be required if the wave traveled horizontally between shot and detector at the highest speed encountered along the refraction path. The portion of the path along the refractor will not enter into the calculation as long as its dip is so small that AC can be considered equal to BC. Virtually all the intercept time is associated with the slant paths q_1 and q_2.

In the case shown, the delay time D_1 for the shot end is simply $q_1/V_0 - s_1/V_1$, while the delay time D_2 for the detector end is $q_2/V_0 - s_2/V_1$. Expressed in terms of depths and velocities,

$$D_1 = z_1 \frac{\sqrt{V_1^2 - V_0^2}}{V_1 V_0} = \frac{z_1 \cos i_c}{V_0} \tag{9-48}$$

and
$$D_2 = z_2 \frac{\sqrt{V_1^2 - V_0^2}}{V_1 V_0} = \frac{z_2 \cos i_c}{V_0} \tag{9-49}$$

Thus the depths at each end can be determined if the intercept time can be separated into component delay times. If the interface is horizontal, each of the two delay times is half the intecept time.

Since delay times are never measurable directly, it is necessary to separate the observed intercept time into its component delay times by indirect means. Some techniques for accomplishing this separation will be discussed in Sec. 9-8.

9-7 REFRACTION OPERATIONS IN THE FIELD

There is little difference between the organization of refraction and reflection parties. Most crews can shift with ease from one type of operation to the other by making some minor changes in their equipment, usually by substituting lower-frequency for higher-frequency geophones. Normally a refraction crew needs more surveyors than a reflection crew, and it may be most efficient for two shooting units to operate at the same time. Radio communication is essential in refraction although it is not always used in reflection shooting.

It is generally desirable to use lower-frequency instruments in refraction recording than in reflection work because refracted signals travel much greater distances than reflections. In early refraction work in Iran, where recording distances were particularly great, the first arrivals had exceptionally low frequencies, often less than 5 Hz, and detectors as large as wastebaskets were employed. In most refraction shooting, however, the distances are such that the first arrivals seldom have frequencies lower than 10 Hz. Many geophones used in reflection recording can handle

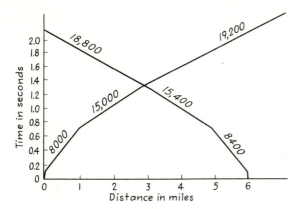

FIGURE 9-14
Typical time-distance curves obtained on refraction profile in west
Texas. Numbers on time-distance segments are inverse slopes or
apparent speeds. Average speed of zone just below uppermost low-speed
layer is 8200 ft/s; of next layer, 15,200 ft/s; and of deepest layer,
19,000 ft/s. Thickness of each zone can be determined from intercept
times. (*After Harris and Peabody.*[16])

refraction signals of 10 Hz or less. Thus there is an overlap between the frequency bands in which reflection and refraction geophones are sensitive that
often makes it possible to use reflection phones for refraction surveys with good
results, especially where recording distances are not much greater than 10 mi.

For reconnaissance surveys in new areas, profiles are usually shot in reverse
directions with progressively increasing shot-detector distances in order to obtain
a complete set of time-distance curves. These curves provide the velocity information needed for calculating depths from refraction times. They also establish the
range of shot-detector distances that should be used over the area when any
particular marker horizon is to be mapped. Figure 9-14 illustrates the type of
information obtained from such profiles. The two shot locations are fixed, and the
geophone spreads are moved progressively to obtain complete two-way coverage
over the line between the shots. Each zone of discrete speed in the earth is represented by a pair of linear segments on the time-distance plot. Suppose, for example,
that the top of the second zone, with an average speed of 15,200 ft/s, is being
mapped. We see from the time-distance plot that the shot-detector distances
should be confined to the range between 1 and 3 mi if this is the only marker of
interest in the area.

9-8   REFRACTION RECORDS; FIRST AND SECOND EVENTS

Figure 9-15 shows two typical refraction records from the foothills of western
Canada which were shot from opposite directions, each being recorded with a

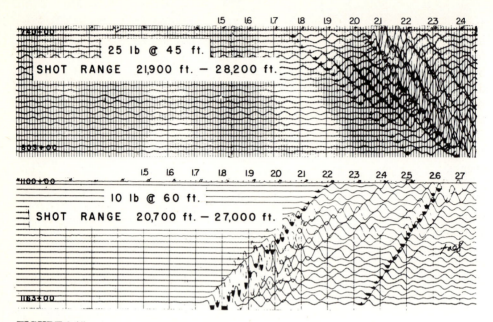

FIGURE 9-15
Two typical refraction records, shot in opposite directions, from foothills of Canadian Rockies. Earliest events on the respective records are from marker with speed of about 14,000 ft/s. Second events are from higher-speed and deeper refractor, in this case the Rundle limestone. (*After Blundun.*[17])

spread of 22 detectors spaced 300 ft apart along a profile in line with the shots. It is evident from the times marked on each record that the shot moments, not shown, occurred about 1.75 s before the first signals from each shot arrived at the nearest detectors. The first arrival at each detector is indicated on the corresponding trace by a pronounced rise in amplitude above the background level, after which the level of ground motion decreases somewhat but still remains much more active than before. On the records shown, each arrival is characterized by an "upkick" followed by a trough and a subsequent peak. The troughs, which are blacked in, are easier to correlate from trace to trace across the records than the actual first breaks are. From the moveouts of about 0.440 s across each of the 6300-ft spreads, it is apparent that the first arrival has an apparent speed of about 14,200 ft/s.

At times 0.2 to 0.3 s later on the upper record and at 0.4 to 0.5 s later on the lower record a similar set of events is observed. These are *second arrivals*. Their apparent velocity is observed to be much higher, as is evident from the shorter times needed for the events from the respective shooting directions to cross the records. The speed turns out to be about 21,000 ft/s, which identifies the second arrivals as coming from the high-speed Rundle limestone.

In the first decade or so of refraction shooting, it was seldom possible to identify any arrival later than the first on the records, and operational and interpretation techniques were based almost entirely on use of first-arrival times. Subsequent instrumental improvements have made detection and resolution of later arrivals feasible in many areas, particularly in marine recording of refractions using sonobuoys. For reasons which are not well understood, second arrivals are not generally distinguishable on seismic traces, even when the background noise level is low at the time on the record when such events would be expected. The foothills area east of the Canadian Rockies, where the records in Fig. 9-15 were shot, appears to be a classic location for good reception of second arrivals, and most of the refraction work carried out there since the middle 1950s has been designed for recording such events.

The earliest description of second-event refraction work was published by Gamburtsev,[18] who discussed techniques used in the U.S.S.R. around 1940. The earliest account of North American refraction work using second arrivals is by Gillin and Alcock.[19]

For second-event shooting the range of shot-detector distances is kept within the crossover distance for the refractor to be mapped, as shown in Fig. 9-16. Because this distance range is shorter than would be necessary for first-event mapping, the

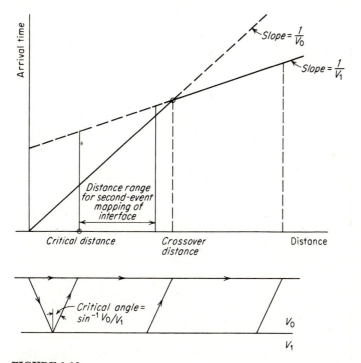

FIGURE 9-16
Range of shot-detector distances for second-event mapping of interface at base of V_0 layer. The range does not extend to the crossover distance because second arrivals become difficult to pick at too short a time after the first arrival.

field operations can be carried out more economically and efficiently, and the inter-pretation of the data is generally not so subject to ambiguity. Less dynamite is necessary for the shorter receiving distances, and radio communication is less subject to interference. Logistics can also be simplified. With the shot-detector distances shorter, the refracting points at the shot and geophone ends of the trajectory may be so close together that the depths corresponding to each will not generally be much different. Thus the problem of converting intercept times into delay times becomes simpler.

Second and even later arrivals are often readily discernible on record sections obtained directly on drum recorders or, after processing, in playback centers from signals received by, and transmitted from, sonobuoys in marine refraction oper-ations. The sections thus obtained are effectively time-distance plots which by preserving a record of the waveforms contain information that can be more useful for interpretation than conventionally plotted time-distance data.

Figure 9-17 is a typical section from a sonobuoy survey covering a distance of about 50,000 ft. The water-wave arrival serves to calibrate the distance scale, as the velocity of sound in water is deduced from the water temperature and salinity. First, second, and later arrivals, as well as some reflections (identified by their characteristic curvature), are readily distinguished by inspection.

The presentation of the data on such record sections makes it easier to recognize refraction arrivals from thin beds, i.e., beds having thicknesses less than a wave-length. Arrivals from such beds lose amplitude more rapidly with distance than those from formations which are at least a wavelength thick, and the loss of energy thus observed on the sections could identify the refractors as thin layers.

9-9 INTERPRETATION FOR COMMON SHOOTING ARRANGEMENTS

In planning refraction surveys in the field, the arrangement of shots and detectors is determined by a number of factors such as the geologic problem involved, the terrain, and the facilities available. Most of the standard arrangements are designed to facilitate the separation of intercept times into delay times. For this reason, different interpretation techniques have been developed for different shooting con-figurations. In this section, a number of widely used field arrangements and the interpretation methods designed for them will be discussed together.

Profile Shooting

In profile shooting, the most widely used of all field techniques for refraction work, the shots and detectors are laid out on long lines. Successive shots are taken at uniform or almost uniform intervals along each line, and successive detector spreads are shifted about the same distance as the corresponding shot points in order to

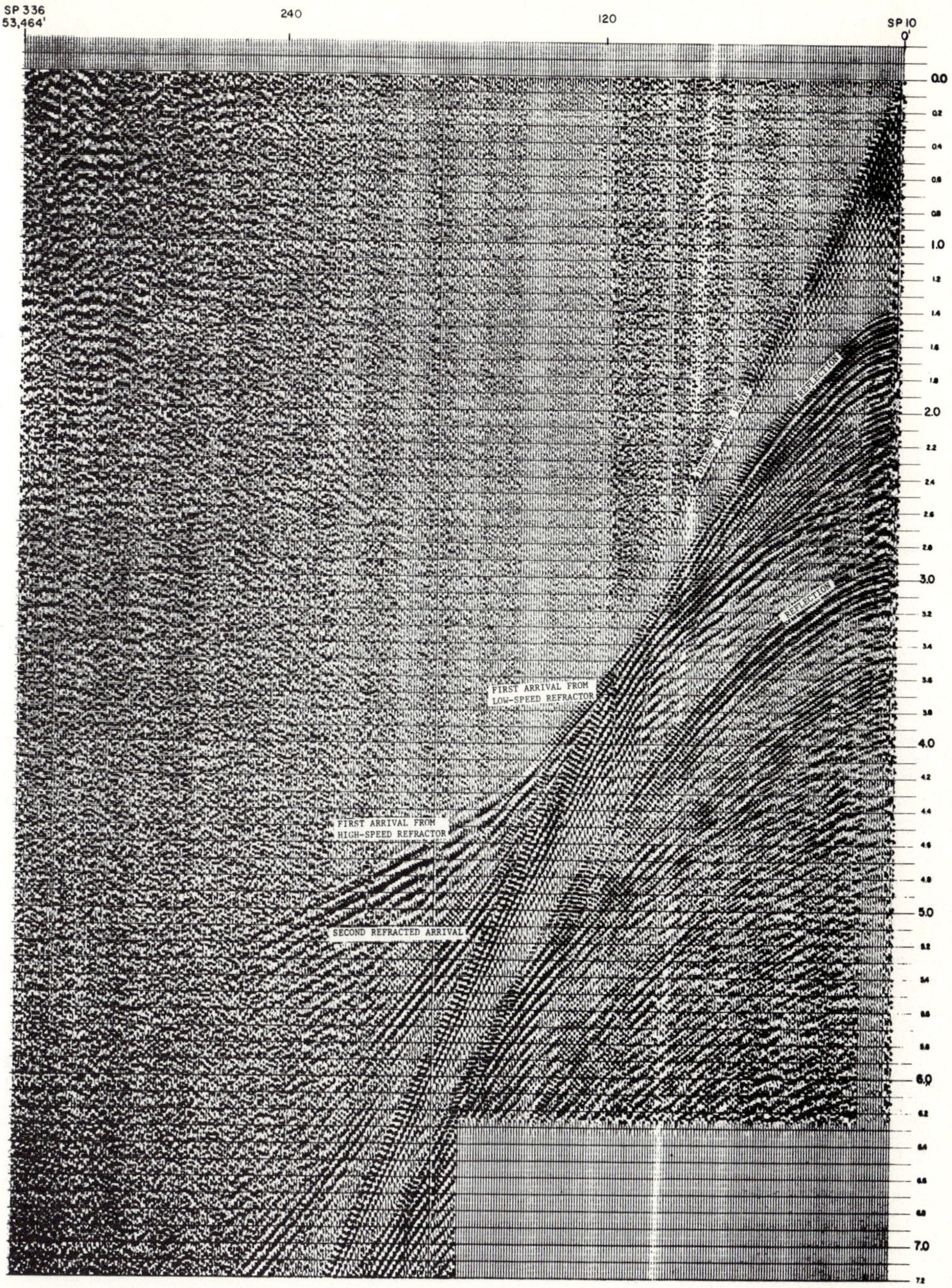

FIGURE 9-17
Typical sonobuoy refraction profile made in processing center from digital tape recording on ship. First arrivals observable to a distance of about 40,000 ft from source. (*Petty-Ray Geophysical Division, Geosource Inc.*)

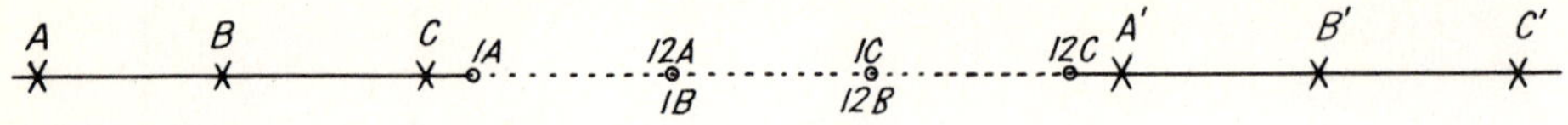

FIGURE 9-18
Typical shooting and receiving arrangement along refraction profile.
Shots A and A', fired successively, are picked up by detectors $1A$ to
$12A$, etc.

keep the range of shot-detector distances approximately the same for all shots.
Shots are generally received from opposite directions on each detector spread.
Figure 9-18 illustrates a typical arrangement.

The distance range is chosen so that the first (or, where desired, the second)
arrivals will be refracted from formations of interest such as the basement or a
high-speed limestone marker. The proper distance is usually determined from a
time-distance plot like that illustrated in Fig. 9-14, which is based on experimental
shooting carried out early in the program.

The principal problem in profile interpretation is to convert the intercept times
into delay times which represent the respective depths to the marker under shot
and receiver. Several techniques for determining the individual delay times from the
intercept times have been reported in the literature and will be reviewed here.

Barthelmes' procedure One of the simplest methods has been described by
Barthelmes.[20] It requires that the depth to the mapped horizon be known at one
point on the profile, as would be the case if the profile were tied to a well. To determine the depths at other points along the profile, it is only necessary to know
the differences between intercept times at the other points and that at the known
point.

The relation between the difference in depth Δz and the differential intercept
time ΔT_i at two detectors a distance x apart is evident from the relation (for the
two-layer case)

$$z = \frac{V_0 V_1}{2 \sqrt{V_1^2 - V_0^2}} T_i \qquad (9\text{-}50)$$

so that

$$\Delta z = \frac{V_0 V_1}{2 \sqrt{V_1^2 - V_0^2}} \Delta T_i \qquad (9\text{-}51)$$

In applying this method, one must bear in mind that the point at which the
depth applies is not directly under the shot or detector but is offset either toward
the shot or toward the detector, as shown in Fig. 9-19, by a distance

$$q = z \tan i_c \qquad (9\text{-}52)$$

in the single-layer case or

$$q = z_0 \tan i_0 + z_1 \tan i_1 + z_2 \tan i_2 \qquad (9\text{-}53)$$

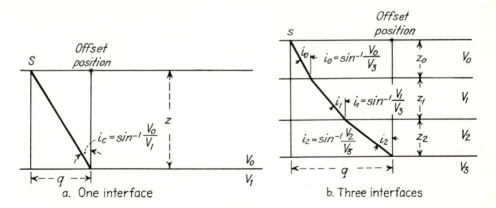

FIGURE 9-19
Horizontal offsets of depth values computed from delay times at receiver S for single and multiple layers.

in the case where the ray path has three slanting segments. The angles are determined by Snell's law from the speeds of the successive layers.

The relative dip curve is actually made up from intercept times computed at all detector positions along the profile. The intercept-time curves for adjacent surface spreads where different shot points have been used will not appear as continuous segments, but the dips, averaged for opposite directions of shooting after the shifts for offset have been made, can be followed to make up a continuous line which is a kind of "phantom" dip horizon. It is this line which is tied to the well or other point where the depth is known.

Wyrobek-Gardner method   Where continuous two-way subsurface coverage is available along the profile over the marker to be mapped and where the marker has dips no greater than $10°$, the separation of the intercept times into delay times is fairly unambiguous. Continuous coverage makes it possible to determine the *differences* in delay time (and hence in depth) between the offset positions along the refractor corresponding to the shots and those corresponding to the various receiving stations. The intercept time observed at each geophone station gives the *sums* of the two delay times corresponding respectively to the shot and geophone positions. As both the sums and differences are thus known, it is a simple matter to compute their values along the profile and to map depths uniquely.

Wyrobek[21] and Gardner[14] have independently developed interpretational techniques for two-way continuous coverage over gently dipping markers. The techniques differ more in details of execution than in principle. The Wyrobek method makes use of special graphical aids designed to facilitate routine computations, and it may be more handy for routine use. Being based on the same principle, both methods should yield about the same information.

As previously noted, the dip of the marker bed must be small (less than $10°$), and the structure of overlying formations must be sufficiently uniform for the section above the marker to be represented as a single homogeneous layer. Continuous control is necessary, preferably two-way, but if the shooting pattern is properly designed, and if the marker velocity has been well established, shooting from a single direction will yield data suitable for interpretation by this method.

Figure 9-20 illustrates the steps employed in the Wyrobek approach, starting with a plot of arrival time T versus detector position for all receiving stations along the profile. These steps are:

1 From the slopes, averaged in both directions, of numerous characteristic segments on the time-distance plot corresponding to the horizon to be mapped, an average marker velocity V_2 is determined for intercept-time calculations.

2 The values for the intercept time, $T - x/V_2$, are computed for all arrivals and are plotted at the respective detector positions along the profile. Values for opposite directions of shooting are plotted at the same receiving positions.

3 It is important to guard against the possibility that arrivals received from the opposite directions have been refracted from different marker horizons. A reciprocity test is therefore applied. The intercept times between any shot and receiver should be the same when the shot point and receiver positions are interchanged. For example, the intercept time for a receiver at shot point K from a source at shot point H should be the same as the intercept time for a receiver at shot point H from a source at shot point K. If it is not, the events picked on the

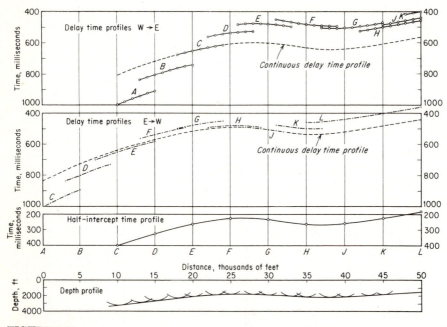

FIGURE 9-20
Steps in conversion of delay-time profiles shot from opposite directions to final depth profile using Wyrobek method. (*After Wyrobek.*[21])

respective records do not correspond to equivalent wave paths in the earth and a new cycle must be picked on one of the records.

4 On the intercept-time plots, continuous lines are drawn having the slopes of the successive intercept-time segments along the profile. These are designated as *relative curves*. The vertical positions of these curves are chosen arbitrarily.

5 Intercept times are now determined at the shot points for the opposite directions of shooting. The respective values at each shot point are averaged, and the averages are divided by 2 and plotted at the shot-point positions as *half-intercept times*. The points thus plotted are used, on a best-fit basis, to provide an absolute reference for the relative delay-time curves.

6 The two relative curves for the respective direction of shooting are transferred to separate pieces of transparent paper or acetate which can be superimposed on the original plot. The lines are then shifted from their original horizontal positions in the directions of the respective shot points for each curve. This shifting takes care of the horizontal offset between refracting point and surface receiver. The proper amount of shift is considered to be that for which the curves come closest to coinciding in shape. If there is a systematic divergence between the curves after the lateral shifting, they are tilted by an equal angle in opposite directions until the best coincidence is observed. Such tilting will be necessary if the velocity chosen for the intercept-time determination is not equal to the actual velocity over the profile. The amount of tilt, incidentally, gives measure of a true average velocity which can be applied in the initial determination of intercept times.

7 After the two relative curves are offset and tilted where necessary, a relative curve averaging the dips of the individual curves is drawn to represent the true dip along the profile. The average curve is fitted to the half-intercept times plotted at the shot points by vertical shifting and by further tilting when required. This represents the delay-time curve for the profile.

8 Values of the delay time D picked at each shot from the final profile thus plotted are multiplied by an appropriate factor F for conversion to depth. For a homogeneous overburden with a velocity V_1, this factor is $V_1 V_2 / \sqrt{V_2^2 - V_1^2}$, where V_2 is the marker speed. Arcs with a radius equal to FD are drawn at regular intervals on a vertical cross section with a 1:1 distance scale. A smooth-line tangent to these arcs gives the depth profile.

The procedure outlined above is applicable only to relatively shallow horizons where there is no appreciable change in dip over a horizontal distance equal to the offset distance. For deep refractors, Wyrobek outlines a more complex technique.

In Gardner's variation upon this technique, delay times are obtained from the averaged relative curve and the observed intercept time at each detector position. The following relation is used to separate the intercept time into a delay time D_1 under the shot and a delay time D_2 under the receiver:

$$D_1 - D_2 = \Delta D$$

along the composite relative curve between the respective ends of the wave trajectory.

$$D_1 + D_2 = T_i = \text{intercept time}$$

Since ΔD and T_i are known, D_1 and D_2 can readily be determined.

Slotnick method   A graphical technique has been developed by Slotnick[22] for interpreting refraction data where successively deeper marker beds have dips that differ from one another. It is quite useful where the beds maintain their identities without change of velocity over appreciable distances along the profile. The method involves the optical tracing of rays from the surface downward through the various layers in the subsurface and back up to the detector. This is done by applying the laws of refraction to velocities observed for opposite directions of shooting along the profile. The depths and dips of successively deeper interfaces are determined by working downward one layer at a time from an arrival time corresponding to wave travel along the top of that layer.

In the Slotnick method the slopes of the various segments of the time-distance curves are used to determine the angle of approach at the surface for the waves refracted along the interfaces corresponding to the respective segments. With the angle of approach to the surface known, a ray can be traced to the marker along which it travels by using Snell's law at each interface. The total travel time is used to calculate the depth of the marker.

The method can be illustrated by a simple example, taken from Slotnick's paper, which is for a two-layer earth with a dipping interface between the layers (Fig. 9-21). From other work in the area it has been established that the true velocities of the two layers represented here are 1700 and 2100 m/s, respectively. The critical angle θ is thus $\sin^{-1} (1700/2100) = 54°$. At point C the emergence angle α is $\sin^{-1} (1700/2130) = 53°$, where 2130 m/s is the apparent speed at C. At point A, the emergence angle β can be obtained from the relation

$$\beta = 2\theta - \alpha = 108° - 53° = 55°$$

We then draw CD so that $\alpha = 53°$ and AD so that $\beta = 55°$. The point E is then located so that DE will equal DC. AE is scaled off and found to be 85 m. The travel

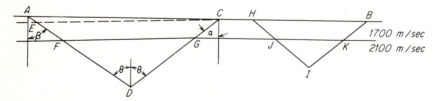

FIGURE 9-21
Construction for applying Slotnick method to two-layer problem.
(*After Slotnick.*[22])

time recorded for the trajectory $AFGC$ is 1.588 s while the time from A to E is

$$t_{AE} = \frac{85 \text{ m}}{1700 \text{ m/s}} = 0.50 \text{ s}$$

The remaining time t along the path EDC is $1.588 - 0.050 = 1.538$ s. This must be separated into two portions, one corresponding to slant-path travel through the upper layer at 1700 m/s and the other corresponding to travel along the interface at 2100 m/s. The distance CE is scaled off and found to be 2930 m.

The slant distance a, which is equal to CG or EF, is determined from the observed time t, the velocities, and the distance CE (here designated as d) from the formula

$$a = \frac{V_1 t - d}{2(V_1/V_0 - V_0/V_1)} \tag{9-54}$$

where V_1 and V_0 are the respective velocities for the upper and lower media.

Applying this to the calculation of CG or EF, we find that

$$a = \frac{2100 \times 1.538 - 2930}{2(2100/1700 - 1700/2100)} = 350 \text{ m}$$

The points G and F are scaled off on CD and AD, respectively, so that $CG = EF = a = 250$ m. The line FG, which is located by connecting F and G, is the required refracting surface.

A similar calculation at B and H for the ray path $BKJH$ on the righthand side of the profile gives the line JK as the refracting surface in the neighborhood of shot point B. These lines may be joined as shown.

The same approach can be applied to map deeper interfaces, each boundary being used progressively in tracing rays to the next lower surface of velocity discontinuity. In the example for which the beginning step is shown above, Slotnick maps four interfaces separating five layers with different velocities.

Wavefront method Although the wavefront technique was first demonstrated as early as 1930 by Thornburgh,[23] it did not take its place as an established method of refraction interpretation until more recently. Rockwell[24] has published a comprehensive survey of this method which presents more complicated applications than will be discussed here.

Figure 9-22 should be helpful in visualizing how the method works. This diagram shows that as successive individual wavefronts reach the first interface, the ray corresponding to the point of contact is bent in accordance with Snell's law as it passes into the second layer. The wavefront perpendicular to it is also bent, as shown. Just below the deepest interface, the wavefront is oriented vertically until it peels off and returns to the surface at the critical angle, as illustrated in Fig. 9-1. Every wavefront is labeled by the time it represents.

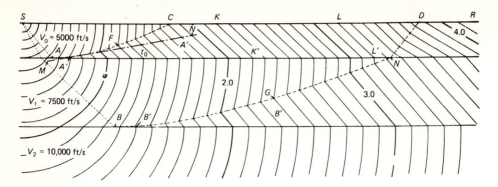

FIGURE 9-22
Wavefront travel in two-interface case. Each continuous line represents
position of wavefront at successive tenths of second after shot. Crossover distances are SC and SD. (*After Thornburgh.*[23])

The times for intersecting wavefronts from sources in opposite directions can be
used to map structures of refracting interfaces by use of the relation demonstrated
in Fig. 9-23. This shows two shot points, A and D, with recording positions laid out
between them along the surface. As the time from A to D will be the same regardless
of which point is the source and which the receiver, we can designate the equivalent
times T_{ABCD} and T_{DCBA} as the reciprocal time T_R.

At a refracting point E from which the respective rays EF and EG leave the
boundary at the critical angle in opposite directions, it is easy to see from the
diagram that

$$T_R = T_{ABEG} + T_{DCEF} - T_{EF} - T_{EG}$$

or
$$T_R = T_{ABE} + T_{DCE} \tag{9-55}$$

if E lies along a straight line connecting B and C.

If the marker does not have excessive relief, E can depart appreciably from such
a line without introducing an unacceptable amount of error in the depth determination. As a rule of thumb, convolutions of the marker surface (as illustrated by the
dashed line $BE'C$) can be accommodated up to the point where the angle $E'CE$ is
as great as 10°.

If wavefronts are constructed from both shot points, any point where the two
wavefronts for respective times which add up to T_R intersect will lie along the
surface of the refractor. The wavefronts for each shot point can be constructed
by starting from the source as was done by Thornburgh (Fig. 9-22). It is more
convenient, however, to work backwards from the receiving point, as illustrated by
Fig. 9-24. We need to know the velocity V_0 of the material above the refracting
surface, and this can be obtained from the time-distance curve.

Consider the segment of the time-distance curve for arrivals from shot A at a
distance of 27,000 ft. As the arrival time at this distance is 3.0 s, the surface trace
of the wavefront for a travel time of 3.0 s must pass through this point. At 23,000

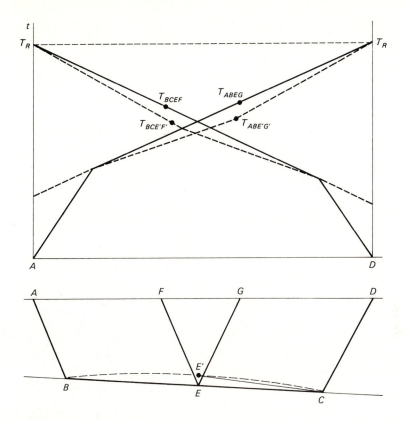

FIGURE 9-23
Use of reciprocal times to locate points along a refracting surface.

ft, the 2.5-s wavefront intersects the surface. Now if the velocity V_0 is 6000 ft/s, the 2.5-s wavefront will also pass through a point below the surface from which the time of travel to the 3.0-s emergence point P is 0.5 s. This subsurface point can be located by striking an arc centered at P having a radius of $0.5V_0$ or 3000 ft. A line through Q tangent to this arc defines the course of the 2.5-s wavefront as it approaches the earth's surface. By the same mechanism, the emerging wavefront corresponding to 2.0 s can be constructed by drawing a line from R that is tangent to an arc 3000 ft in radius corresponding to a travel time from Q of 0.5 s and also tangent to a second arc centered at P with a 6000-ft radius corresponding to a travel time from P of 1.0 s. Applying this procedure, one can draw a family of wavefronts for refracted events from A corresponding to any desired time interval. These are the full lines which are inclined to the right.

The same method can be employed to construct wavefronts for the arrivals from shot point B and a family of such waves at 0.5-s intervals is also shown, being indicated on the diagram by the dotted lines inclined to the left.

According to the time-distance plot in the upper part of Fig. 9-24, the reciprocal

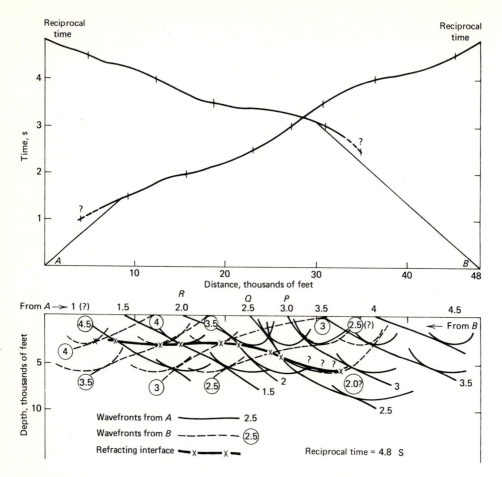

FIGURE 9-24
Example of how wavefront method can be applied to obtain structure
from reversed time-distance curves.

time for shot points A and B is 4.8 s, and the marker surface is located wherever
the *sum* of the times for crossing wavefronts from the respective shot points equals
4.8 s. A series of intersections where the times conform to this condition is indicated
on the cross section of Fig. 9-24, and the smooth curve connecting them is the desired marker surface in the plane of the profile.

The wavefront technique can be used for a multilayered subsurface if the velocity
of each layer is known. Once the structure of the shallowest boundary has been
determined by the method we have illustrated, the next interface is constructed by
extending the wavefronts downward from points along the boundary and repeating
the process, using such points as the source of new arcs, drawn with proper radii,
for tracing the wavefronts into the lower medium. Deeper interfaces can be mapped

in the same way by peeling off successive layers as their bases are mapped. It is only necessary that the velocities of the layers be known, and they can generally be determined from the two-way time-distance segments for the respective layers.

Tarrant method The Tarrant method[25] was designed for cases where the subsurface marker is so irregular that it is difficult or impossible to select segments from the time-distance curves for the respective directions of shooting that correspond to the same part of the refractor.

Going back to the definition of delay time as given on page 314, we see from Fig. 9-13 that

$$D_1 = \frac{q_1}{V_0} - \frac{s_1}{V_1} = \frac{q_1}{V_0} - \frac{q_1 \cos \phi}{V_1} \tag{9-56}$$

where ϕ is the angle made by OA with the horizontal. As $V_0/V_1 = \sin i_c$, it follows that

$$q_1 = \frac{V_0 D_1}{1 - \sin i_c \cos \phi} \tag{9-57}$$

This is the equation of an ellipse with one focus at the origin having a semiminor axis of $V_1 D_1 \tan i_c$ and a semimajor axis of $V_0 D_1/(\cos^2 i_c)$. The center of the ellipse is displaced horizontally from the focus by a distance $V_1 D_1 \tan^2 i_c$.

In applying this method, V_0 and V_1 are determined from time-distance relations over the area or from well-velocity information. Intercept times are separated into delay times by some technique appropriate to the type of data available. An ellipse is then constructed for each detector station using the constants given above. The envelope of the family of ellipses thus obtained is the desired refracting surface.

When the dips are not large, it is often permissible to approximate the ellipses by circular arcs, which are much simpler to draw. The radius R of the circle having the same curvature as the ellipse just below its center is

$$R = V_1 D_1(\tan i_c + \tan^3 i_c) \tag{9-58}$$

Hales' method This graphical technique, devised by Hales[26] for interpreting refraction profiles, is particularly applicable to high-relief structures. With the Hales method, it is only necessary to identify the sections of time-distance curves for opposite directions of shooting that correspond to arrivals from the marker horizon to be mapped. The geometrical relations involved in the method are illustrated in Fig. 9-25.

The technique is designed to locate the position of a refracting point such as F by determining the distance between the two points at which refracted rays originating from shot points A and D, respectively, and traveling upward from F in opposite directions reach the surface. If an arbitrary point C is chosen along the time-distance curve for shot point A, a simple construction is used to locate the

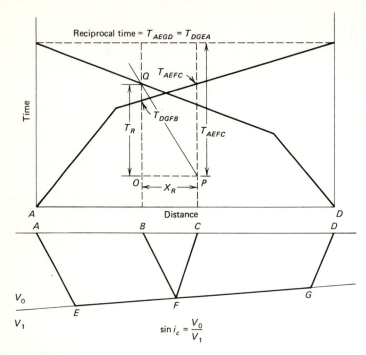

FIGURE 9-25
Finding T_R and X_R from reversed time-distance curves by Hales'
method.

point B, at which the ray originating from shot point D and leaving the marker
surface at F reaches the surface.

The distance X_R is obtained by drawing a line on the time-distance plot from
point P (a distance AC from A and a time T_{AEFC} below the reciprocal time) with
a slope equal to $V_1 \sin i_c$, where i_c the critical angle is $\sin^{-1}(V_0/V_1)$, and noting the
intersection at Q of this diagonal line with the segment of the time-distance plot
from shot point D. The time T_R is found graphically as shown in the figure. The
horizontal distance PO is equal to X_R.

With T_R and X_R known, the refractor position is determined by a construction
of the type shown in Fig. 9-26. B and C are located a distance X_R apart along a
horizontal line. Then a line is drawn from B at an angle i_c with BC and its intersection with the perpendicular bisector of BC becomes the point from which line
DF, having a length R (the formula for which is indicated), is drawn at an angle α
(also defined) with the vertical. The perpendicular to DF at F gives the position of
the refracting surface through point F.

A series of constructions like this are made at intervals along the time-distance
curve, and the refractor is mapped over the entire range where arrivals from it are
received. Derivation of the relations used in the reductions are presented in detail
by Musgrave[27] (p. 268).

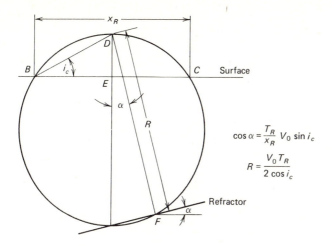

$$\cos \alpha = \frac{T_R}{x_R} \, V_0 \sin i_c$$

$$R = \frac{V_0 T_R}{2 \cos i_c}$$

FIGURE 9-26
Use of Hales' method for locating position of refractor using data from
time-distance curves representing shots from opposite directions.

General Discussion of Profile Interpretation Methods

It is evident that there is a wide variety of techniques available for interpreting data from in-line refraction profiles. Most of them are designed primarily for two-layer problems, although many can be adapted to multilayered configurations. The Slotnick method is most suitable for a multiplicity of layers with the boundaries between them having different dips. Many of the methods, such as the Wyrobek and wavefront techniques, are accurate when the dip of the refractor is gentle (say less than 10°), while others, such as the Hales and Tarrant methods, can be applied where refractor dips are steep. The nature of the subsurface and the type of data available will ordinarily govern the choice of the interpretation method to be employed.

Broadside Shooting

In some areas it is not feasible to use profile shooting techniques to determine the structure across strike. As an alternative, broadside shooting may provide a convenient and economical method of getting the desired information. The broadside lines are often laid out in conjunction with conventional profiles that run along strike. In the Alberta foothills, for example, the steep dips in the high-speed Rundle limestone, which is the refraction marker, make it difficult to get useful information from profile shooting perpendicular to geological and topographic strike.

The field setup is illustrated schematically in Fig. 9-27, and a typical group of correlated records is shown in Fig. 9-28. The shot points and detector spreads are laid out along parallel lines, which are generally across strike. The distance between

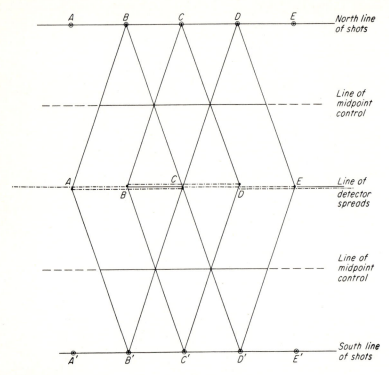

FIGURE 9-27
Shot and geophone pattern for broadside shooting. Adjacent spreads
overlap 50 percent, so that subsurface midpoint control is continuous
along indicated lines on both sides of the receiving line.

each line of shots and the receiving line should be chosen so that it will always be
greater than the double offset distance for the refractor being followed. Ideally the
distance should be only slightly greater so that the primary refracted event will be
received as a second arrival or, as Richards[28] suggests, as a wide-angle reflection
which has taken a path making an angle equal to or just greater than the critical
angle at the marker surface. When this spacing is used, the refracting point associated with the shot will be very close to that associated with the detector and the
delay time for each will be approximately half the intercept time. A single depth
point (based on half the intercept time) is then plotted midway between shot and
receiver. All depth points can thus be plotted along the *control lines* which are located
halfway between the shooting line and the receiving line. For shot-detector distances
much greater than the critical distance, such plotting can lead to some error in
mapping the structure of the marker.

Whenever possible, the broadsides should be tied to an in-line profile where delay
times and depths have been determined at the subsurface tie point by one of

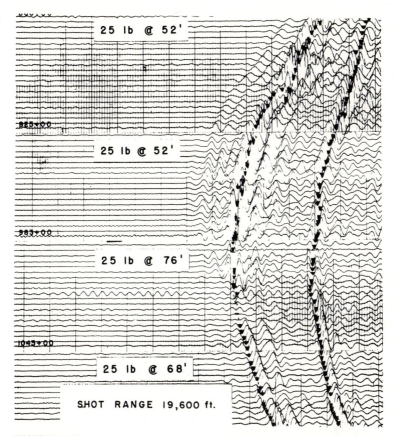

FIGURE 9-28
Four refraction records made in the foothills of Alberta from a single
shot point with adjacent geophone spreads along a broadside. Second
and later events are marked. (*After Blundun.*[17])

the profiling methods. In second-event work, it is particularly desirable to have
such a tie, since it is often difficult to identify the correct peak or trough on the
broadside records. Where the broadside is tied to an in-line, the respective records
at the tie point can be correlated and the apparent speed on the in-line record can
be used to identify the proper pick on the broadside. If the same cycle is not followed
on all broadsides and in-lines, the final map will be in error.

Blundun[17] has described broadside refraction operations in the Alberta foothills,
where this kind of shooting provides the most economical method for obtaining
structural information on a reconnaissance basis. This method makes it possible to
observe faults in the high-speed limestone directly on the records and thus to map
their location without having to wait for completion of the interpretation over the
entire area of the survey.

9-10 CORRECTIONS USED IN REFRACTION ANALYSIS

Refraction times must be corrected for elevation and for changes in the thickness of the weathered layer. The former correction removes differences in travel times due only to variations in the surface elevation of the shots and detector stations. The weathering correction, like the comparable reduction in reflection work, removes differences in travel time through the near-surface zone of unconsolidated, low-speed sediments which may vary in thickness from place to place. Similar corrections are made in reducing reflection times, but the slant paths along which most refraction ray paths leave and approach the surface and the common use of intercept or delay times in refraction interpretation create a need for somewhat more complex correction procedures.

Elevation correction The most usual computational procedure is to put both the shot and the detector on the same imaginary datum plane by subtracting the times that would be required for the wave to travel from the datum to the respective shot or detector locations if they are higher than the datum or by adding the times that would be required if they are lower.

Figure 9-29 demonstrates how this transformation is accomplished. Assume that both the shot and the detector are above the datum plane. We wish to place the shot point at P on the datum plane directly below the shothole and the detector at point Q on the datum plane below E. The hypothetical ray path after the correction is shown by the dotted line. The difference between the time from A to D along the actual path and that from P to D along the hypothetical path is

$$\frac{AB}{V_0} - \frac{CD}{V_1} = \frac{AB}{V_0} - \frac{PB}{V_1} \equiv D_s \qquad (9\text{-}59)$$

which is, by definition, the delay time associated with the layer between the bottom of the shot at elevation $e - h$ and the datum plane at elevation d. This material constitutes a horizontal slab of thickness $e - h - d$. In this case the delay time is

$$D_s = \frac{(e - h - d)\cos i_c}{V_0} = \frac{(e - h - d)\sqrt{V_1^2 - V_0^2}}{V_1 V_0} \qquad (9\text{-}60)$$

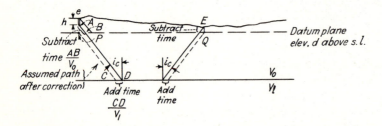

FIGURE 9-29
Elevation correction for two-layer case; e is shot elevation, E the detector elevation above sea level.

Similarly, at the detector end, where the elevation is E, the delay time associated with the path from the surface to the datum is

$$D_d = \frac{(E - d) \sqrt{V_1^2 - V_0^2}}{V_1 V_0} \tag{9-61}$$

The sum of these corrections in delay time should be subtracted from the observed intercept time in order to place both shot and detector effectively in the datum plane. The elevation of the shot is actually the surface elevation at the top of the shothole e minus the depth of the charge in the hole h, so that the final elevation correction to be applied to the intercept time is

$$\text{Elevation corr.} = \frac{(e - h + E - 2d) \sqrt{V_1^2 - V_0^2}}{V_1 V_0} \tag{9-62}$$

Weathering corrections   As with reflection work, uncorrected variations in the thickness of the weathered zone near the surface could easily lead to fictitious structures on horizons mapped from refraction data. There are several methods of correcting for the effect of the weathered layer by determining its speed and thickness and then subtracting the delay time associated with it from the observed intercept time. Because no arrivals from the base of the weathering are ever observed on refraction records, it is necessary to correct for weathering by special shooting with short shot-geophone distances. The thickness t of the low-speed material above the first velocity discontinuity (Fig. 9-30) is calculated from the intercept times using the two-layer depth-intercept-time formula. The delay time associated with this low-speed layer can now be removed from the intercept time for the regular refraction shot recorded at the same position. Elevation corrections are then made from the base of the weathered layer at the detector end of the trajectory and from the bottom of the shothole (which usually penetrates the weathered layer). Thus the total correction (weathering plus elevation), to be subtracted from the intercept time, is the delay time associated with the weathered zone plus the delay time for the layer between weathered zone and datum plane, plus the delay time from the bottom of the shothole to the datum. Adding these

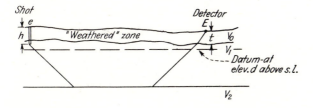

FIGURE 9-30
Correction of refraction times for weathering and elevation.

times together, we obtain a total correction

$$\text{Total corr.} = \left[(e - h - d) + (E - t - d) \right] \frac{\sqrt{V_2{}^2 - V_1{}^2}}{V_1 V_2} + \frac{t \sqrt{V_2{}^2 - V_0{}^2}}{V_2 V_0}$$

$$(9\text{-}63)$$

The term in brackets can of course be simplified to $e + E - h - t - 2d$.

9-11 DETAILING SALT DOMES BY REFRACTION

Although the fan-shooting technique was highly successful during the 1920s in discovering shallow salt domes, it was not suitable for locating domes with tops many thousands of feet deep because of limited penetration in the Gulf Coast sedimentary section. McCollum and LaRue[29] proposed a variation on the technique in which the fans would center around a detector located in a deep well, radiating outward to a group of near-surface shot points placed at intervals around a ring. Any time leads noted from such a three-dimensional fan would suggest the presence of a deep salt dome which might not be detectable using a conventional fan consisting of phones on the surface. They suggested that the time differences at various detector depths in the well could be used to define the shape of such a dome as well as to determine the presence and extent of overhang.

Gardner[30] described a similar technique for detailing the boundaries of a salt dome by using a detector in a deep well which penetrates or flanks the dome. Shots were fired from points near the surface so chosen that the vertical planes determined by the well and the respective shot points will cut the salt dome at well-distributed angles. Since part of the total time from shot to well detector is through the sediments and part through the higher-speed salt, the problem is to determine how much of the trajectory lies in each medium. The locus of the salt boundary must satisfy the condition that the sum of the two times equals the observed time for each shot-detection combination. The envelope of these loci should correspond to the boundary between the salt and the sediments.

Gardner's method has been applied by Musgrave et al.[31] to the delineation of shale bodies as well as salt. In mapping salt domes these authors also made use of careful surface-to-surface profiling to define the top surface of the diapiric body and the upper portion of the flanks.

9-12 PRESENTATION OF REFRACTION DATA

The practice, common since 1960, of recording refraction signals on analog or digital tape makes it a simple matter to present refraction data in variable-density or variable-area modes. Such a presentation can often facilitate interpretation. A series of refraction records made from a common shot point with tying geophone spreads at progressively greater distances can be easily plotted in the form of a variable-area section. This will give all the information obtainable from a time-distance plot plus additional data, such as amplitudes, waveforms, and indications of possible second arrivals that are not observable on conventionally plotted time-distance curves.

The section shown in Fig. 9-17 from a sonobuoy profile illustrates the appearance of a refraction record or a series of records from the same shot point after conversion into variable-area form.

By appropriate time shifting it is possible to present refraction signals on sections as plots of intercept time (absolute or relative) vs. distance, which, as in the case of broadsides, give structural information very similar to that on a geological cross section. As the intercept time is simply $T - x/V$, it is only necessary to shift the arrival time by an interval equal to the distance x over the marker velocity V to obtain a direct plot of this quantity on the final section.

REFERENCES

1. Society of Exploration Geophysicists: "Seismic Refraction Prospecting," Albert W. Musgrave (ed.), Banta, Menasha, Wis., 1967.
2. Slotnick, Morris M.: "Lessons in Seismic Computing," Richard A. Geyer (ed.), Society of Exploration Geophysicists, Tulsa, Okla., 1959.
3. Muskat, Morris: The Theory of Refraction Shooting, *Physics*, vol. 4, pp. 14–38, 1933.
4. Dix, C. H.: Refraction and Reflection of Seismic Waves, II: Discussion of the Physics of Refraction Prospecting, *Geophysics*, vol. 4, pp. 238–241, 1939.
5. Soske, Joshua L.: Discussion on the Blind Zone Problem in Engineering Geophysics, *Geophysics*, vol. 24, pp. 359–416, 1959.
6. Thornburgh, H. R.: Wave-Front Diagrams in Seismic Interpretation, *Bull. Am. Assoc. Petrol. Geol.*, vol. 14, pp. 185–200, 1930.
7. Hawkins, L. V., and D. Maggs: Nomograms for Determining Maximum Errors and Limiting Conditions in Seismic Refraction Survey with a Blind Zone Problem, *Geophys. Prospect.*, vol. 9, pp. 526–533, 1961.
8. Trostle, M. E.: Some Aspects of Refraction Shooting through Screening Layers, pp. 469–481, in Ref. 1.
9. Mota, L.: Determination of Dips and Depths of Geological Layers by the Seismic Refraction Method, *Geophysics*, vol. 19, pp. 242–254, 1954.
10. Ewing, Maurice, and L. D. Leet: Seismic Propagation Paths, *Trans. Am. Inst. Min. Met. Eng.*, Geophysical Prospecting, 1932, pp. 245–262, 1932.
11. Kaufman, H.: Velocity Functions in Seismic Prospecting, *Geophysics*, vol. 18, pp. 289–297, 1953.
12. Barry, K. M.: Delay Time and Its Application to Refraction Profile Interpretation, pp. 348–361, in Ref. 1.
13. Gardner, L. W.: Seismograph Prospecting, U.S. Patent 2,153,920, Apr. 11, 1939.
14. Gardner, L. W.: An Areal Plan of Mapping Subsurface Structure by Refraction Shooting, *Geophysics*, vol. 4, pp. 247–259, 1939.
15. Gardner, L. W.: Refraction Seismograph Profile Interpretation, pp. 338–347, in Ref. 1.
16. Harris, Sidon, and Gwendolyn Peabody: Refraction Exploration in West Texas, *Geophysics*, vol. 11, pp. 52–58, 1946.

17. Blundun, G. J.: The Refraction Seismograph in the Alberta Foothills, *Geophysics*, vol. 21, pp. 828–838, 1956.
18. Gamburtsev, G. A.: Correlation Refraction Shooting (condensed by L. W. Gardner), *Geophysics*, vol. 11, pp. 59–65, 1946.
19. Gillin, J. A., and E. D. Alcock: The Correlation Refraction Method of Seismic Surveying, *Geophysics*, vol. 11, pp. 43–51, 1946.
20. Barthelmes, A. J.: Application of Continuous Profiling to Refraction Shooting, *Geophysics*, vol. 11, pp. 24–42, 1946.
21. Wyrobek, S. M.: Application of Delay and Intercept Times in the Interpretation of Multilayer Time-Distance Curves, *Geophys. Prospect.*, vol. 4, pp. 112–130, 1956.
22. Slotnick, M. M.: A Graphical Method for the Interpretation of Refraction Profile Data, *Geophysics*, vol. 15, pp. 163–180, 1950.
23. Thornburgh, H. R.: Wave-Front Diagrams in Seismic Interpretation, *Bull. Am. Assoc. Petrol. Geologists*, vol. 14, pp. 185–200, 1930.
24. Rockwell, Donald W.: General Wavefront Method, pp. 363–415, in Ref. 1.
25. Tarrant, L. H.: A Rapid Method of Determining the Form of a Seismic Refractor from Line Profile Results, *Geophys. Prospect.*, vol. 4, pp. 131–139, 1956.
26. Hales, F. W.: An Accurate Graphical Method for Interpreting Seismic Refraction Lines, *Geophys. Prospect.*, vol. 6, pp. 285–294, 1958.
27. Woolley, W. C., A. W. Musgrave, and Helen Gray: A Method of In-Line Refraction Profiling, p. 287, in Ref. 1.
28. Richards, T. C.: Wide Angle Reflections and Their Application to Finding Limestone Structures in the Foothills of Western Canada, *Geophysics*, vol. 25, pp. 385–407, 1960.
29. McCollum, Burton, and W. W. LaRue: Utilization of Existing Wells in Seismograph Work, *Bull. Am. Assoc. Petrol. Geol.*, vol. 15, pp. 1409–1417, 1931.
30. Gardner, L. W.: Seismograph Determination of Salt-Dome Boundary Using Well Detector Deep on Dome Flank, *Geophysics*, vol. 14, pp. 29–38, 1949.
31. Musgrave, A. W., W. C. Woolley, and H. Gray: Outlining of Salt Masses by Refraction Methods, *Geophysics*, vol. 25, pp. 141–167, 1960.

DIRECT DETECTION OF HYDROCARBONS BY SEISMIC REFLECTION

For about half a century after geophysical methods were introduced in oil and gas exploration it was almost universally taken for granted that they were not capable of locating hydrocarbons directly. Their usefulness, it was believed, was limited to the mapping of structural or possibly stratigraphic conditions favorable for the accumulation of oil or gas. But only the drill could indicate whether hydrocarbons were present in geophysically located entrapments. Any claims to the contrary (and many were advanced over this period) were dismissed by most responsible geophysicists as naïveté or, more usually, quackery. Since none of the many such claims made over several decades appeared to stand the test of time, this reaction appeared justified.

But late in 1972 it became known that two major oil companies had been successful in predicting the occurrence of offshore gas from seismic reflection data. The predictions were based on anomalies that would be expected in the *amplitudes* of reflections caused by differences between the reflectivity of surfaces bounding sands containing gas and those bounding water or oil-bearing portions of the sands. The reflectivity contrasts were believed to result from differences in the velocities of the sands having pores filled with the respective materials.

Such amplitude variations would be predicted on the basis of the principles governing seismic velocities in heterogeneous media. Yet it was not until the introduction in the late 1960s of wide-dynamic-range recording and processing techniques that it was feasible to present relative amplitudes in usable form on seismic record sections. Actually, a fairly large number of papers reporting such effects and proposing amplitudes as a basis for direct hydrocarbon detection appeared in the Soviet literature from 1965 to 1971, but at the time there was little recognition of these publications by the geophysicists of western countries.

Since the beginning of 1973, direct detection techniques seem to have seized the attention of those involved in the exploration for hydrocarbons as no other technical development in the history of the art. Comprehensive interpretation methods have been developed that make use not only of reflection amplitudes for hydrocarbon detection but other criteria as well. Because so short a time has elapsed between the publicizing of the technology and the writing of this book, little literature can be cited on the subject except for the Soviet papers just mentioned. Too little information has been released on discovery rates based on the new indicators to evaluate their true effectiveness adequately. There is a great need for further publication of such data.

10-1 REFLECTION AMPLITUDE AS AN INDICATOR OF HYDROCARBONS IN POROUS SANDS

The amplitude of a seismic wave reflected from an interface between two materials is governed by the reflection coefficient R, which, as noted on page 41, is expressed for normal incidence by the relation

$$R = \frac{\rho_2 V_2 - \rho_1 V_1}{\rho_2 V_2 + \rho_1 V_1} \tag{10-1}$$

where ρ_1 and ρ_2 are the respective densities on the near (incident) and far sides of the boundary and V_1 and V_2 are the respective velocities for the two sides. The product of ρ and V is known as the *acoustic impedance;* it is evident that the reflection coefficient and hence the reflection amplitude depend on the contrast in acoustic impedance across the reflecting interface. Any lateral change in the velocity or density of one or both of the materials separated by the interface should cause the amplitude of the reflection from this boundary to vary.

Figure 10-1 shows a cross section of velocities across a hypothetical oil and gas entrapment. Reflections would be expected from the five interfaces, A to E, shown in the diagram, across which there are distinctive contrasts in acoustic impedance. Because V_g, V_o, and V_w as well as ρ_g, ρ_o, and ρ_w differ along the base of the entrapping shale, one would expect lateral variations in the amplitudes of the reflection from this boundary.

The dependence of seismic velocity of a porous material upon the fluid contained in the pore spaces (see Gassmann[1] and Geertsma and Smit[2]) has been studied both theoretically and experimentally (see Wyllie et al.[3,4]) for several decades. These

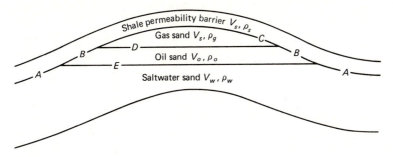

FIGURE 10-1
Anticlinal entrapment of hydrocarbons in porous sand by shale layer.
Interfaces separating different materials designated by letters A to E.
Velocities and densities for each layer indicated by appropriate symbols.

investigations have shown, as might be expected intuitively, that a sand saturated with gas will have a lower velocity than the same sand saturated with oil or water. A paper by Gardner et al.[5] is particularly pertinent to the problem of predicting velocity as a function of porosity for different pore fillers; by combining Gassmann's theory with experimental observation these authors have computed the respective velocity-depth relations for a sand saturated with gas, oil, and brine. Figure 10-2 presents the results of their study. From the curves that are shown it is evident that the contrast between the respective velocities decreases with depth. At depths less than about 5000 ft the difference between the velocity in the gas sand and that in the oil sands is rather large, being more than 20 percent at 4000 ft. The contrast, on the other hand, between the velocity for brine-saturated and for oil-saturated sand at that depth is only 6 percent. At 10,000 ft the velocity contrast between the gas sand and the oil sand is 6 percent, while that between oil and saltwater sands is only about 1 percent. It would appear from the curves that reflections from gas- and liquid-saturated sand surfaces should be diagnostic only at depths less than 5000 ft.

Let us assume the velocities and densities shown in Table 10-1 for the various

Table 10-1

Material	Velocity V, ft/s	Density ρ, g/cm³
Shale	8500	2.50
Gas sand	5700	2.025
Oil sand	7300	2.270
Water sand	7700	2.277

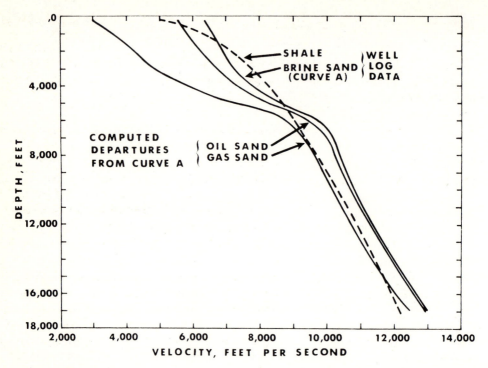

FIGURE 10-2
Variation, predicted from Gassmann's theory, of velocity with depth in shale and in sand saturated with oil, gas, and brine, respectively. (*Gardner et al.*[5])

materials making up the section shown in Fig. 10-1. The reflection coefficients for each interface are indicated in Fig. 10-3. The reflectivity contrast between the water-saturated sand and the gas-saturated sand is considerably greater than that between the water-saturated sand and the oil-saturated sand. One would expect the amplitudes of the reflections to be related closely to these coefficients. The relation will not ordinarily be linear because amplitudes depend on several other factors to be discussed in the next section.

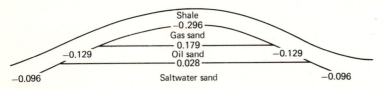

FIGURE 10-3
Reflection coefficients at interfaces marked in Fig. 10-1.

10-2   FACTORS OTHER THAN REFLECTIVITY THAT CONTROL REFLECTION AMPLITUDES

Amplitudes cannot be used as a means of distinguishing sands containing hydrocarbons (particularly gas) from those containing water unless corrections are made for other effects, many of them predictable, that govern the amplitudes of reflections. Among these are spherical spreading, attenuation, interference (constructive or destructive) between reflections from interfaces that are close together, distortions due to processing operation, e.g., deconvolution, and focusing effects. Sheriff[6] has summarized the various factors affecting reflection amplitudes which must be taken into account in the interpretation of amplitude anomalies on seismic records.

Spherical spreading   The amplitude of any elastic wave traveling through a homogeneous, nonabsorptive medium will be inversely proportional to the distance from the source of the wave. This is because the energy of the disturbance is uniformly distributed over the area of the spherical shell which it occupies at any instant of time. This area is proportional to the square of the radius so that the energy density at any point on the spherical surface is inversely proportional to the square of the radius, which is the distance from the source. But as the amplitude is proportional to the square root of the energy, it should bear an inverse first-power relation to the distance.

Absorption   In Chap. 2 we showed that materials in which seismic waves propagate absorb some of the energy of the wave by viscoelastic drag or internal friction. In general this effect causes amplitude A to fall off with distance x from its initial value A_0 according to the relation (for a plane wave)

$$A = A_0 e^{-\alpha x}$$

where α is the attenuation coefficient, which varies with the material in which the wave travels as indicated in Table 2-2. The attenuation coefficient appears to be proportional to the frequency of the wave; i.e., if the frequency doubles, so does the value of α.

The amplitude of a reflection will obviously be affected by the absorptive properties of all the material the reflected wave passes through along its path from the source to the reflector and back to the receiver. If the absorption coefficient varies laterally along any part of the path, one would expect the amplitude of a reflection signal from an interface of uniform reflectivity to show variations that are related to the lithologic changes affecting the absorption. Sand generally has a higher absorption coefficient than shale. Thus a change of facies within a layer from sand to shale could give rise to a corresponding variation in the amplitude of a reflection signal passing through the layer from an interface below. Such variations in amplitude are sometimes put to use for identifying lithology, as pointed out in Chap. 8 (page 289).

Interference effects If two or more reflecting surfaces are close together compared with wavelength, the pulses reflected from them will be superimposed and the waveform of the resulting reflection signal will have an amplitude that depends to some extent on the separation between the respective reflectors. This effect was considered in Chap. 8. The individual reflection coefficients may be positive or negative, depending on whether the acoustic impedance of the lower medium is greater or less than that above. For two interfaces, the maximum amplitude will occur when the superimposed events have a phase difference of one cycle and reinforce one another. A half-cycle difference causes cancellation, resulting in a minimum amplitude. A greater number of interfaces gives rise to a larger potential range of variation in the amplitude. This kind of amplitude effect depends both on geometric and on lithologic factors, the lithology governing the actual amount of time shift between reflectors because of the dependence of velocity upon rock characteristics.

Effects of processing Any filtering of a reflection signal, whether in the earth (as by a water layer) or in digital processing, causes changes in the waveform of the signal and in doing so changes its maximum amplitude. If, for example, an input signal consists of a sharp pulse, a filter will generally produce an output spread out over a longer time interval which consists of an initial single-cycle pulse followed by one or more additional cycles, each of lower amplitude than the one before if the filter is minimum-phase. This spreading of the reflection energy over a number of cycles can result in a distortion of amplitude relations which will depend on the filter characteristics. Deconvolution based on frequency analysis using autocorrelation can give ambiguous output signals depending on whether the program is designed on the assumption of minimum phase shift. If it is, the maximum amplitude is likely to be observed in the first cycle of the output regardless of whether the true reflection waveform has this characteristic. The problems encountered in compensating for the effects of processing in the display of true reflection amplitudes are discussed by Paige.[7]

Geometric effects Curvature of a reflecting interface can have a significant effect upon the amplitudes of the reflections that are recorded from it. If the reflecting surface is convex upward, the rays are spread out because of the defocusing effect of a convex mirror and the energy density at any point above it is decreased. The amplitude should therefore be lower for reflections recorded above the structure than for those recorded from flat areas on either side. If the surface is concave upward, as in a syncline, the rays are concentrated by the focusing effect of a concave mirror and the energy density is increased over the structure so that the reflection amplitudes are anomalously high. The effects of such focusing on reflection amplitudes is discussed by Dix.[8] A simple technique to compensate for such effects is proposed by Hilterman.[9]

10-3   RECORDING RELATIVE AMPLITUDES; BRIGHT SPOTS

The large range of variation in the absolute amplitudes (10^6 or more) encountered at the input of each channel in reflection recording systems over the time interval from the beginning to the end of a typical seismic record has created problems ever since the introduction of the reflection method. The final output used for interpretation has conventionally been an oscillographic presentation on paper or film with a total dynamic range no greater than a factor of 10 (20 dB). To show readable waveforms along the entire length of a record 5 s or more long has required suppression of the early part of the record and expansion of the later part along with automatic gain control to prevent exceptionally strong signals from going off scale. The net effect is to restrict the amplitudes of all reflections as observed on a record or record section to a very limited range. This is good for extracting the best structural information from the data, but it makes the records virtually useless for showing up amplitude variations of the type that would indicate changes in reflectivity. The equalization of amplitudes by expansion, suppression, and automatic gain control makes it all the more unlikely that amplitude variations associated with changes in sand velocity due to gas saturation will be discernible on conventionally processed seismic sections.

Correction for predicted amplitude variations   It is not feasible to record true amplitude variations simply by removing the equalization elements just noted from the recording system. The limited dynamic range available in the ultimate presentation would hardly make it possible to work with more than a very few reflections at a time if this were done. A more practical procedure is to determine, either theoretically or empirically, the normal variation of amplitude with record time due to all causes other than reflectivity changes and to correct the observed amplitudes accordingly. If the normal amplitude thus determined is subtracted from the observed amplitude, the residual variations will be primarily attributable to differences in reflection coefficients.

The simplest model which might be used for the correction is one which takes only spherical spreading into account. As the record is plotted versus time rather than distance, it is necessary to make use of the best available velocity information and convert time to distance in order to make such a spreading correction. In the absence of actual data a reasonable velocity function must be assumed.

A common procedure is to fit an appropriate smooth curve to the observed amplitude-versus-time function. This is done by averaging amplitudes of successive samples over a moving window many cycles long and expressing the smoothed amplitude-versus-time function thus obtained as a low-order polynomial. The value for the resultant polynomial at each sampling time is subtracted from the observed amplitude of the sample. It is the difference between the two that is ultimately plotted on the record section. The polynomial accounts for changes caused by spherical spreading and absorption, but it is not of high enough order to incorporate variations in reflection coefficients at individual boundaries.

Amplitude effects due to interference between events from individual reflecting surfaces in close proximity cannot be taken into account in developing the polynomial. Such effects are studied by constructing models of a type that will be considered in the next section.

10-4 BRIGHT SPOTS AS INDICATORS OF GAS ACCUMULATION

A variable-area or variable-density record section from data processed to accentuate amplitude variations associated with changes in reflectivity should register high-amplitude events from interfaces with a high reflection coefficient as a thicker or "heavier" lineation than that associated with adjacent boundaries where the reflectivity is lower. The stripe or doublet of stripes which indicates a particular reflection event becomes more prominent or "blacker" over a limited horizontal region than it is on either side. This high-amplitude portion of the reflection is referred to as a *bright spot*, a paradoxical designation in view of the fact that it actually appears blacker than surrounding events on the section, but bright spot seems to sound better than black band.

As the reflection coefficient from the top or bottom surface of a sand containing gas is greater than that from the corresponding surface of a water- or oil-bearing sand, the presence of the gas sand should be indicated on the section as a bright spot. Figure 10-4 shows a reflection section processed to preserve relative amplitudes. The strong reflection event at about 1.25 s stands out above the level of the rest of the record. This event comes at a time that would be predicted for a reflection from a known gas-saturated sand, and its lateral boundaries correspond to the known

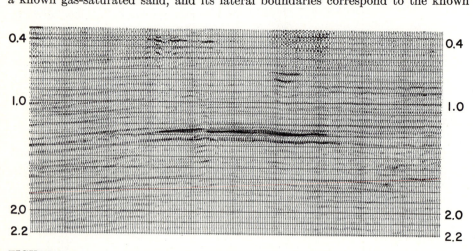

FIGURE 10-4
Record section processed to bring out relative amplitudes. Bright spot at center shows known gas accumulation. (*Western Geophysical Co. of America.*)

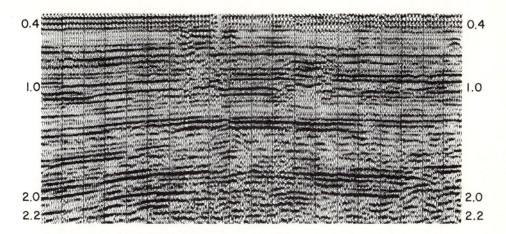

FIGURE 10-5
Same section as in Fig. 10-4 processed by conventional techniques designed to equalize amplitudes. Reflections from the gas zone do not stand out as they do when the special processing is employed. (*Western Geophysical Co. of America.*)

limits of gas accumulation. The same section as conventionally processed is shown in Fig. 10-5.

The principal evidence that bright spots indicate the presence of gaseous hydrocarbons comes from processed sections of reflection lines over known gas accumulations. Occasionally a high-amplitude event is observed at levels where it is known that no hydrocarbons are present. Other causes, such as facies changes, geometric focusing effects, or even the presence of high-velocity contrasts not associated with hydrocarbons (as at a limestone-shale boundary) might account for the amplitude anomaly in such cases.

A more serious limitation lies in the fact that the method has seldom been successful in finding gas from seismic data recorded on land. It is true that amplitude anomalies have been observed over gas fields on land in the Sacramento Valley and in Yugoslavia. Yet most experiments to look for bright spots over such known gas fields have ended in failure. The most probable explanation for this failure lies in the greater homogeneity of seismic transmission through the sea bottom than through the near-surface formations on land.

The velocity of seismic waves in oil is so close to that in saltwater that a higher degree of discrimination than current technology seems to allow would be required for oil accumulations to show up in this manner.

10-5   OTHER INDICATORS OF HYDROCARBONS ON SEISMIC RECORDS

While bright spots are the most conspicuous indicators of gas accumulations on seismic records, other features can also be diagnostic of such accumulations if they are properly interpreted. Let us consider some of them.

Velocity pulldown effects The low velocity of gas-saturated sand compared to water- and oil-saturated sand gives rise to lateral anomalies not only in the reflection coefficients from surfaces bounding the gas sand but also in the time required for seismic waves to pass through the sand. If, as we have assumed, the velocity of the gas sand is 5700 ft/s and that of the water sand is 7700 ft/s, and if the saturated zone is 75 ft thick, it would take 7 ms longer for a reflection from an interface below the sand to reach the surface through the gas zone than through the water sand. This is a small but measurable differential, and at shallow depths, where the velocity contrast is greater, the differential time would be larger and hence even more evident to the eye on record sections.

Figure 10-6 illustrates this effect. Here there is multiple-zone gas production over a depth of 4500 ft, and the effect of the lower seismic velocities is cumulative throughout this interval. The first conspicuous sag in a reflection occurs at about 1.1 s, and the bottom reflection at 1.8 s is depressed almost 0.1 s directly below the drilling rig.

Attenuation effects Because of the poor acoustic coupling between sand grains and gas compared with that between sand grains and water, oil, or other liquids, one would expect the attenuation coefficient of gas-saturated sands to be greater than that of sands whose pore spaces are saturated with liquids. This would cause the amplitude of reflections received from below the level of the sand to be less over gas accumulations than on either side. Such an effect should be much more conspicuous over thick gas columns than over thin ones. Figure 10-6 shows this effect along with the pulldown noted above. Below about 1.1 s the reflections observed near the rig become fainter than on either side, and many of them disappear. The multiple pay zones in this field cause the attenuation effects to be particularly pronounced.

Diffraction patterns Whenever there is a change in acoustic impedance concentrated at a point or along a line, as in the case of faulting in a reflecting interface, observable diffractions are likely to be generated. The side of a gas body which shows a velocity contrast with respect to the formations that enclose it should generate diffraction patterns in the same way. This effect, particularly when the center of the diffraction is located at the end of a high-amplitude stripe along a reflection event, should be particularly diagnostic. Sometimes the diffraction pattern can be observed on both sides of the center, as was demonstrated in model experiments by Hilterman.[11] When this occurs, the two events are 180° out of phase with each other; i.e., a positive cycle, or peak, on the right of the diffracting edge lines up with a negative cycle, or trough, or vice versa. It is rare that this effect is actually observed on record sections where there are indications of gas sands. The author has seen no published examples.

Divergences between horizontal reflections and dipping events just above One of the most diagnostic indications of gas is the presence of a strong horizontal

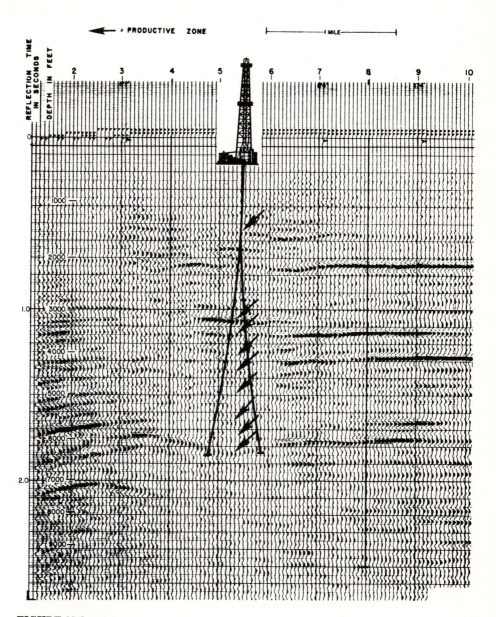

FIGURE 10-6

Depression of deep reflectors under multiple gas zones caused by reduction of velocity in gas-bearing sands. Note attenuation of reflections below well at times greater than 1.1 s. (*Continental Oil Co.*)

reflection just below another reflection which has the dip associated with the lithological boundaries in the portion of the section where it occurs. This type of divergence is observed at a time of 2.0 s on the section in Fig. 10-7. The arching reflection with its crest at about 1.9 s is very probably from the anticlinal shale-gas interface at the top of the gas sand. The horizontal event below it is probably from the gas-water interface within the sand, an interface which must by its nature be horizontal. Both surfaces show a high reflectivity, as would be expected. It is hard

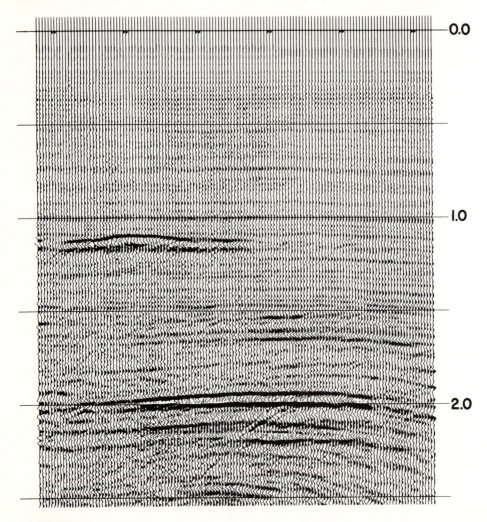

FIGURE 10-7
Horizontal reflections (at 1.2 and 2.0 s) indicative of gas-oil water contacts. Note arched reflections above each flat event, presumably from top of gas sand. (*Teledyne Exploration Co.*)

to think of any more plausible explanation for a horizontal reflection originating from a zone of dipping beds.

Other effects   A number of other indicators have been noted on seismic sections over gas accumulations that can be associated with the presence of the hydrocarbons in the earth. The higher attenuation previously discussed in the gas sand could result in a selective absorption of higher-frequency components of the reflections from interfaces below the productive formation. This filtering effect might show up as a visibly lower frequency for the reflections below the gas than for those from the same interface observed on either side of the gas zone. Also, sudden terminations of reflections may occur at the actual edges of the hydrocarbon accumulations or where the thickness becomes so small that interference effects take over. Polarity reversals might occur for the same reason, and they too would be diagnostic of the lateral termination of gas sands.

10-6   SEISMIC MODELING TO DETERMINE HYDROCARBON CHARACTERISTICS AND TO ESTIMATE RESERVOIR CAPACITY

Thus far we have considered only methods for *detecting* the presence of hydrocarbons from seismic reflections. We have not discussed how to map the geometry and lithology of the reservoirs in which they occur or how to estimate the total reserves in such accumulations. The extraction of such information on the hydrocarbon zones is best initiated by modeling the rock formations associated with them as well as the fluids filling the pore spaces involved.

The basic procedure is to construct a cross section on which interfaces corresponding to observed reflections are drawn and appropriate trial velocities are introduced between the interfaces. Adjustments are made in the geometry of the boundaries initially postulated as well as in the velocities assumed for the spaces between them until a synthetic record is obtained for a model that gives the closest resemblance to the actual processed section. The variables of greatest practical significance are the geometry of the gas sand and the degree of gas saturation in the sand. Such information gives a basis for estimating the total amount of gas in the sand body.

Figure 10-8 shows the portion of a seismic section recorded in offshore Louisiana and processed for preservation of relative amplitude which exhibits a bright spot indicating the presence of gas. Assuming that the gas thus postulated is entrapped in a sand body sandwiched between two shale layers, we would like to determine the size and shape of the body and its gas content. The velocity of seismic waves in the same body depends on its porosity and saturation. If a set of velocities assumed for the respective elements of the model results in a synthetic section giving a good agreement with the actual section (both in the time interval between the reflections from its upper and lower boundaries and in relative amplitudes), it should be

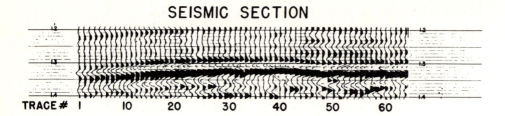

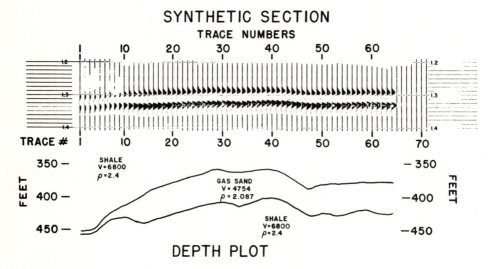

FIGURE 10-8
Model based on actual record (*top*) of a single gas sand with synthetic
section (*center*) based on model parameters shown at bottom. (*Geocom,
Inc.*)

possible to determine the porosity of the sand and/or the extent to which it is
saturated with gas.

The lower portion of Fig. 10-8 shows the structural and lithologic configuration
of the subsurface that gives the optimum agreement with the seismic record section
that has been recorded and processed. The synthetic section resulting from the
particular model parameters indicated on the geological section seems to show a
good similarity to the actual record section at the top of the figure.

The example just considered represents a very simple case of a sand body
saturated with gas that is completely embedded in shale. A more complex reservoir
configuration is modeled in Fig. 10-9. The record to be matched in the modeling was
shot over a known multi-pay gas reservoir. The thicknesses and porosities assumed

for the model (which gives synthetic record section in the middle of the figure) coming closest in appearance to the section recorded in the field were actually about 30 percent lower than those observed in a nearby well. Of the four sands penetrated by the drill, three (the uppermost, second, and lowermost) had a gas content of only 20 percent, but the third showed the pore space to contain 80 percent gas and

SEISMIC DATA

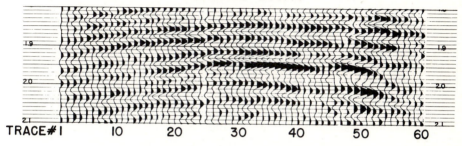

SYNTHETIC RESPONSE

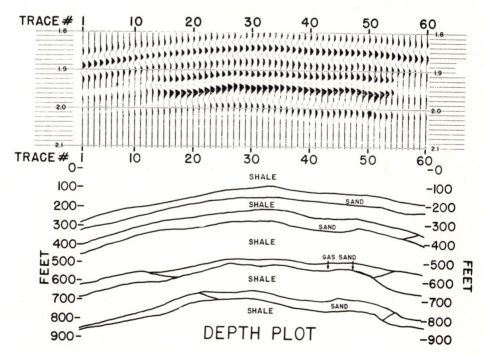

FIGURE 10-9

Modeling of complex reservoir configuration following procedures illustrated in Fig. 10-8. (*Geocom, Inc.*)

20 percent water. The productive sand was associated with the highest-amplitude reflection (at about 1.95 s) indicated on both the synthetic and actual records.

Some modeling can be applied for other purposes, often to examine different hypothetical patterns of hydrocarbon entrapment in actual producing zones such as those modeled in the two examples just considered. An example of this type of model, reproduced in Fig. 10-10, is for a sand 100 ft thick at the left, embedded in shale which wedges out updip on the right side of the model. On its left side the sand contains water, in its central part oil, and in the wedge itself gas. The gas-oil contact is indicated by the fragmental horizontal reflection segment near the center of the figure. Such segments are often observed on real sections (as we saw in Fig. 10-7).

An important application of modeling is to estimate the amount of gas in a reservoir detected on the basis of amplitude anomalies. The waveforms and relative amplitudes are related both to the thickness of the various lithologic members in the reservoir and to their velocities. The velocities of the sands depend on their porosity, the liquid and/or gas filling the pore spaces, and the degree of saturation for each constituent. The amplitudes depend on both the velocity contrasts at the interfaces and on interference effects between individual reflections when the

UPDIP PINCHOUT

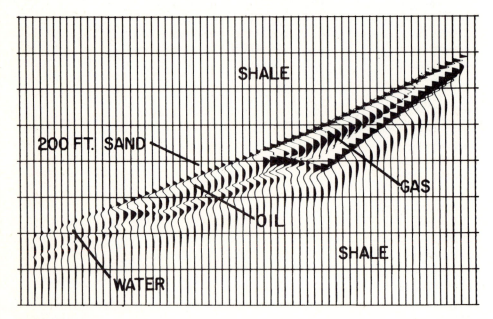

FIGURE 10-10
Modeling of reservoir containing oil and gas entrapped in a pinchout.
(*Teledyne Exploration Co.*)

reflecting interfaces are close together. Modeling makes it possible to vary all these parameters independently and to arrive at combinations which match results obtained by appropriate processing of reflection data recorded in the field. The ambiguities inherent in such procedures and the inability of the technique to model fine changes on a scale smaller than the seismic resolution allows must inevitably limit the precision of reserves estimates made in this way.

10-7   LONG-TERM POTENTIAL OF DIRECT DETECTION TECHNIQUES

It is always hazardous to make predictions on future technological developments, particularly in books designed to be current over a longer time than more ephemeral publications. From this standpoint, it is unfortunate that the direct detection of hydrocarbons by seismic methods is at such an early stage of its development at the time this book goes to press. Evaluation of current techniques is difficult as little statistical information has been released that would enable one to say how successful they actually are. Offshore exploration is so expensive and competitive that we cannot expect pertinent seismic information leading to discoveries on the one hand or dry holes on the other to be released to the public domain in the near future. Contractors involved in data acquisition and processing can sometimes obtain permission to publish seismic data on which good discoveries were based (usually with locations unspecified), but they have understandably little incentive to show promising looking sections which led to dry holes or noncommercial hydrocarbon accumulations. Eventually enough data for evaluating these methods adequately will become available but it is hard to say when.

One recently recognized obstacle to the accurate interpretation lies in the fact that a small amount of gas content in sands filled predominantly with water can lower the velocity in the sand much more than originally expected. Domenico[14] demonstrated this by applying the equation of Biot and Geertsma,[2] which predicts the velocity in sand containing both gas and liquids in terms of the percentage saturation of the gas. His results indicate a substantial lowering of velocity in sands having only a small percent of gas saturation. If this limitation is operative in actual reservoirs, it may not be feasible to predict which sand bodies indicated as bright spots have a high enough saturation to be commercial.

Regardless of these uncertainties, there is little question that the success obtained thus far with direct detection by seismic methods will have a very great effect on the future course of geophysical technology. Investigation in this area will dominate the development of exploration geophysics for a long time to come. The possibility of expanding the present capabilities of direct detection technology to the point that it will be possible to locate oil offshore and both oil and gas onshore offers a tremendous incentive for accelerating research and development in this field. Efforts in this direction, considering the extent of the world's energy shortage, are not likely to diminish until their objectives are achieved or until a long-continued lack of success in meeting them leads to a major reduction in such activity. The problems to be faced are formidable, but so are the technical resources available for seeking the breakthroughs in the art that would have to be made if success is to be achieved.

REFERENCES

1. Gassmann, F.: Über die Elastizität poroser Medien, *Vierteljahrsschr. Naturforsch. Ges. Zur.*, vol. 96, pp. 1–23, 1951.
2. Geertsma, J., and D. C. Smit: Some Aspects of Elastic Wave Propagation in Fluid-saturated Porous Solids, *Geophysics*, vol. 26, pp. 169–181, 1961.
3. Wyllie, M. R. J., A. R. Gregory, and L. W. Gardner: Elastic Wave Velocities in Heterogeneous and Porous Media, *Geophysics*, vol. 21, pp. 41–70, 1956.
4. Wyllie, M. R. J., A. R. Gregory, and G. H. F. Gardner: An Experimental Investigation of Factors Affecting Elastic Wave Velocities in Porous Media, *Geophysics*, vol. 23, pp. 459–493, 1958.
5. Gardner, G. H. F., L. W. Gardner, and A. R. Gregory: Formation Velocity and Density: The Diagnostic Basics for Stratigraphic Traps, *Geophysics*, vol. 39, pp. 770–780, 1974.
6. Sheriff, R. E.: Factors Affecting Amplitudes: A Review of Physical Principles, in "Lithology and Direct Detection of Hydrocarbons using Geophysical Methods," *Geophys. Soc. Houston Symp.*, Oct. 8–9, 1973.
7. Paige, D. S.: The Dark Side of the Bright Spot, in "Lithology and Direct Detection of Hydrocarbons Using Geophysical Methods," *Geophys. Soc. Houston Symp.*, Oct. 8–9, 1973.
8. Dix, C. Hewitt: "Seismic Prospecting for Oil," Harper, New York, 1952.
9. Hilterman, Fred J.: Amplitudes of Seismic Waves—A Quick Look, *Geophysics*, vol. 40, pp. 745–762, 1975.
10. Larner, K. L., Emil J. Mateker, Jr., and C. Wu: Amplitude: Its Information Content, in "Lithology and Direct Detection of Hydrocarbons using Geophysical Methods," *Geophys. Soc. Houston Symp.*, Oct. 8–9, 1973.
11. Hilterman, Fred J.: Three-dimensional Seismic Modeling, *Geophysics*, vol. 35, pp. 1020–1037, 1970.
12. Lindsey, J. P.: Modeling for Lithology, in "Lithology and Direct Detection of Hydrocarbons using Geophysical Methods," *Geophys. Soc. Houston Symp.*, Oct., 8–9, 1973.
13. Barry, Kevin M., and Thomas R. Shugart: Seismic Hydrocarbon Indicators and Models, in "Lithology and Direct Detection of Hydrocarbons using Geophysical Methods," *Geophys. Soc. Houston Symp.*, Oct. 8–9, 1973.
14. Domenico, S. N.: Effect of Water Saturation on Seismic Reflectivity of Sand Reservoirs Encased in Shale, *Geophysics*, vol. 39, pp. 759–769, 1974.

GRAVITY PROSPECTING: PRINCIPLES AND INSTRUMENTS

Seismic prospecting, discussed in Chaps. 2 to 10, requires introducing elastic-wave energy (generally artificial) into the earth and recording the energy returned to the surface from buried interfaces. The gravity method, on the other hand, involves measuring a field of force in the earth that is neither generated by the observer nor influenced by anything he does. The magnetic method and some electrical techniques involve passive observations of the same kind. All such methods of exploration make use of natural potential fields.

The type of information obtainable from gravity or magnetics is quite different from that which seismic prospecting, especially the seismic reflection method, can yield. Whereas seismic reflection records give a picture of the subsurface that shows detailed structure at many levels at once, the field observed in gravitational or magnetic prospecting is a composite of contributions from all depths within the usual range of exploration interest, and such contributions can be individually resolved only in special cases. Hence one cannot expect to obtain the detailed and relatively precise structural picture from gravity or other potential data that is generally obtainable by seismic methods.

In gravity measurements, the quantity actually observed is not the earth's true gravitational attraction but its variation from one point to another, usually at positions along the earth's surface which are close together. Such lateral differences can be measured with a much greater degree of precision than the total gravitational field, and field instruments are designed to measure differences in gravity rather than its actual magnitude.

The variations in gravity observed through such measurements depend only upon lateral changes in the *density* of earth materials in the vicinity of the measuring point. Many types of rocks have characteristic ranges of density which may differ from those of other types that are laterally adjacent. Thus an anomaly in the earth's gravitational attraction can often be related to a buried geological feature, e.g., a salt dome or other diapir, which has limited horizontal extent.

11-1 THE PLACE OF GRAVITY IN OIL AND MINERAL EXPLORATION

The gravity method was initially used in oil exploration for locating salt domes in the Gulf Coast of the United States and Mexico and later for finding anticlinal structures in the midcontinent area. Even now special types of structures in which hydrocarbons are entrapped exhibit such large contrasts in density with respect to surrounding formations that gravity data alone can be used to decide on drilling locations. In northwestern Peru, for example, oil entrapped by block faulting in shallow indurated formations has been found with drilling guided by appropriate anomalies on gravity maps.

Most gravity surveys currently carried out in the search for oil are designed for reconnaissance of large, previously unexplored areas. Where little or no geological information is available in a region, the first question that must be answered is whether a sedimentary basin large enough and thick enough to justify further investigation is present. If the geology is suitable, the gravity method can provide this kind of information rapidly and economically. Most sedimentary rocks have densities lower than basement rocks, and where this condition is met, the density contrast makes it possible to map the boundaries and determine the approximate depth distribution of the sedimentary basins.

Gravity surveys can be particularly useful in the initial exploration of water-covered shelf areas, where no geological information may be available at all. The introduction in the late 1960s of shipborne gravity meters has led to an increased amount of gravity prospecting over continental shelves. Gravity recording can be carried out simultaneously with seismic shooting for only a small incremental cost. The combined gravity and seismic data frequently enable the geophysicist to identify and detail geologic features such as diapiric structures more reliably than would be possible with either method alone.

In analyzing potential fields it is desirable, but not always feasible, to separate effects from different subsurface sources. The greater the difference between the lateral dimensions of the individual sources, the easier it should be to isolate the field associated with each. Techniques have been developed for discriminating between gravity effects from large regional features and those from smaller, generally shallower features having more limited horizontal dimensions.

Gravity appears to have more limited applicability in mineral prospecting. Some ores, such as chromites, have such a high density compared with the material that surrounds them that they can be located directly by detailed gravity surveys. Buried channels which might contain gold or uranium minerals can frequently be located by gravity because the channel fill is less dense than the rock in which the channel has been cut. Regional studies making use of gravity surveys may make it possible to detail major structural features such as faults or lineaments where mineral accumulations are likely to be found.

Gravity surveys are employed in base-metal exploration for a number of purposes. In the southwestern United States they have been used to find shallow bedrock features such as pediments under alluvial cover. In the Canadian Shield they have been carried out to differentiate electromagnetic anomalies caused by massive sulfides from those caused by low-density graphites. Such surveys have also been useful for estimating tonnages in sulfide or iron deposits detected by drilling.

11-2 GRAVITATIONAL FORCE, ACCELERATION, AND POTENTIAL

The expression of gravitational attraction in a form most useful for application to exploration requires an understanding of the basic physical concepts relating force, acceleration, and potential.

Gravitational force   The theory behind gravitational prospecting is based on Newton's law expressing the force of mutual attraction between two particles in terms of their masses and separation. This law states that two very small particles of mass m_1 and m_2, respectively, each with dimensions very small compared with the separation r of their centers of mass, will be attracted to one another with a force

$$F = \gamma \frac{m_1 m_2}{r^2} \tag{11-1}$$

where γ, known as the universal gravitational constant, depends on the system of units employed.

In the centimeter-gram-second (cgs) system, the value of γ is 6.670×10^{-8}. This is the force in dynes that will be exerted between two masses of 1 g each with centers 1 cm apart. Although the law of gravitational attraction was deduced by Newton from astronomical observations, the constant γ cannot be determined astronomically but must be measured in the laboratory. The earliest measurement was by Cavendish in 1797. His apparatus consisted of a horizontal beam with an equal weight at each end suspended at the center by a sensitive torsion fiber. Large external weights were placed alongside the ends of the beam in such a way that their attraction would cause it to rotate. The restoring torque of the fiber, increasing linearly with the rotation, balanced the turning moment of the weights. Cavendish's original value of 6.754×10^{-8} cgs unit is not much different from the best present value, 6.670×10^{-8}, obtained in 1930 by Heyl using a refined version of Cavendish's original apparatus.

Consider the attraction between two billiard balls touching one another so that their centers are about 3 in (7.5 cm) apart. If we assume each ball to have a mass of 225 g (about $\frac{1}{2}$ lb), we can compute the attractive force F in dynes.

$$F = \frac{(6.67 \times 10^{-8})(225)^2}{(7.5)^2} = 6 \times 10^{-5} \tag{11-2}$$

This force is less than 3×10^{-10} times the weight of one of the balls.

Gravitational acceleration   The acceleration of a mass m_2 due to the attraction of a mass m_1 a distance r away can be obtained simply by dividing the attracting force F by the mass m_2 (since force is mass times acceleration), whereupon

$$a = \frac{F}{m_2} = \gamma \frac{m_1}{r^2} \tag{11-3}$$

The acceleration, being the force acting on a unit mass, is the conventional quantity used to measure the gravitational field acting at any point. All masses located at the same position in the field are subject to the same gravitational acceleration. In the cgs system, the dimension of acceleration is centimeters per second per second (cm/s^2). Among geophysicists this unit is referred to as the *gal* (in honor of Galileo, who conducted pioneering research on the earth's gravity). The gravitational acceleration at the earth's surface is about 980 cm/s^2 or 980 gals, but in exploration work we are likely to be measuring differences in acceleration one-ten-millionth or less of the earth's field. For convenience in working with gravity data obtained for geological and geodetic studies, the *milligal* (mgal, $\frac{1}{1000}$ gal) has come to be the common unit for expressing gravitational accelerations.

Gravitational potential   As the intensity of gravitational, magnetic, or electric fields depends only on position, the analysis of such fields can often be simplified by using the concept of potential. The potential at a point in a gravitational field is defined as the work required for gravity to move a unit mass from an arbitrary reference point (usually at an infinite distance) to the point in question.

Let us assume, as illustrated in Fig. 11-1, that two masses, one of unit magnitude and the other of magnitude m_1, are initially an infinite distance apart; the unit mass is moved until it reaches point O a distance R from m_1, which has remained at P. Let r, which is a variable, be the separation of the two masses at any point along the path taken by the unit mass between its initial position at infinity and its final position O.

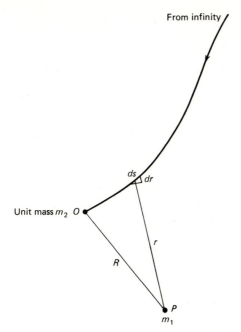

FIGURE 11-1
Gravitational potential is the work
done by the attractive force of m_1 on m_2
as it moves to 0 from infinity.

The force per unit mass, or acceleration, at a distance r from P, is $\gamma m_1/r^2$, and the work necessary to move the unit mass a distance ds having a component dr in the direction of P is $\gamma m_1 \, dr/r^2$. The work v done in moving the mass from infinity to O in the gravitational field of m_1 is

$$v = \gamma m_1 \int_\infty^R \frac{dr}{r^2} = \gamma m_1 \left. \frac{1}{r} \right|_\infty^R = \frac{\gamma m_1}{R} \qquad (11\text{-}4)$$

It can easily be shown that this result will be obtained regardless of the path taken by m_2 in traveling from its starting point an infinite distance from m_1 to its end position at O. The quantity $\gamma m_1/R$ is the gravitational *potential* and depends only on R, the distance from the point source m_1. By differentiating both sides of Eq. (11-4) it can be seen that the gravitational acceleration is the derivative of the potential with respect to r. Any component of the force can be obtained by differentiating the potential with respect to distance in the direction along which the force is desired.

Any surface along which the potential is constant is referred to as an *equipotential surface*. No work need be done by or against gravity in bringing a body from one point on an equipotential surface to another. Sea level, for example, is an equipotential surface, even though the actual force of gravity varies along the sea surface by more than 0.5 percent between the equator and either of the poles.

11-3   APPLICATION OF NEWTON'S LAW TO MASSES WITH LARGE DIMENSIONS

The theory considered thus far is applicable only to the case of an attracting source having infinitesimally small dimensions compared with the distance at which the attraction is measured. When the dimensions of the source are large, it is necessary to extend the theory. The procedure is to divide the mass into many small elements, each of infinitesimal dimensions, and to add the effects of each of the elements. Because force or acceleration is a vector having both magnitude and direction, it is necessary to resolve the force from each element of mass into its three components (most generally its vertical component and its north-south and east-west components in a horizontal plane) before the attraction of the body at any point can be determined.

To illustrate this principle let us consider the attraction of an irregular laminar body (part of a two-dimensional sheet) in the xz plane at an external point P (Fig. 11-2). We must first determine the x (horizontal) and z (vertical) components of acceleration at P associated with this attraction. To do this we divide the plate into N small elements of mass, each of area ΔS as shown. If the density (mass per unit area) σ_n is uniform within the nth element, we can express the x component of acceleration at point P due to the attraction of this element as

$$a_{xn} = \frac{\sigma_n \, \Delta S}{r_n^2} \cos \theta = \frac{\sigma_n \, \Delta S}{r_n^2} \frac{x}{r_n} = \frac{x}{r_n^3} \sigma_n \, \Delta S \tag{11-5}$$

and the z component as

$$a_{zn} = \frac{\sigma_n \, \Delta S}{r_n^2} \sin \theta = \frac{\sigma_n \, \Delta S}{r_n^2} \frac{z}{r_n} = \frac{z}{r_n^3} \sigma_n \, \Delta S \tag{11-6}$$

Adding the accelerations for these elements, we obtain the respective accelerations

$$a_x = \sum_{}^{N} \frac{x \, \sigma_n}{r_n^3} \Delta S \tag{11-7}$$

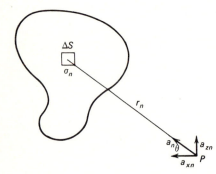

FIGURE 11-2
Determining gravitational acceleration of irregular two-dimensional sheet. ΔS is element of area.

and
$$a_z = \sum^N \frac{z\,\sigma_n}{r_n{}^3}\,\Delta S \tag{11-8}$$

As the elements become smaller and smaller, ΔS approaches the limit dS and we can express the two components of acceleration at P by the respective integrals

$$a_x = \int^S \frac{\sigma x}{r^3}\,dS \tag{11-9}$$

and
$$a_z = \int^S \frac{\sigma z}{r^3}\,dS \tag{11-10}$$

where S is the area of the body and σ is the mass per unit area.

We now apply the same approach to a three-dimensional body with a density (in this case the mass per unit volume) of ρ. It is easy to show by extending the reasoning for a two-dimensional body to one of three dimensions that the components of acceleration in the x, y, and z directions respectively become

$$a_x = \int^V \frac{\rho x}{r^3}\,dV \tag{11-11}$$

$$a_y = \int^V \frac{\rho y}{r^3}\,dV \tag{11-12}$$

and
$$a_z = \int^V \frac{\rho z}{r^3}\,dV \tag{11-13}$$

where V is the volume of the body.

In gravity exploration, only the vertical component of force is measured, so that we are normally concerned only with a_z in determining the attraction at the surface of a buried body. Later in this chapter, it will be shown how Eq. (11-13) can be used to derive the components of the accelerations caused by several bodies having simplified geometrical forms for which the integration can be carried out in a straightforward way.

11-4 THE EARTH'S GRAVITATIONAL FIELD AND ITS RELATION TO GRAVITY EXPLORATION

As all gravity measurements made in exploration work show only *differences* in gravity from one place to another, the attraction of the earth itself is significant only insofar as it varies laterally over the earth's surface. Such variation must be taken into account in evaluating the gravity effect of buried bodies having geological or exploration significance.

If the earth were made up of homogeneous spherical shells and did not rotate, the attraction at the surface of the earth would be the same everywhere and it would not affect the readings of gravity meters, which measure only *differences* in acceleration between one place and another. Actually, the earth rotates (so that centrifugal forces are superimposed on gravitational attractions), is spheroidal (being flattened at the poles), and has lateral irregularities in density which can extend to depths at least as great as the base of the earth's crust. The observed value of gravity can depend on latitude, elevation, topography, and tidal movements as well as on lateral changes in density distribution that are sometimes of economic interest. Variations in gravitational attraction not associated with anomalous geological features have different degrees of predictability. For further information on the relationship between the earth's shape and its gravity field, the reader is referred to a condensed discussion of these topics in Garland[1] or a more detailed one in Heiskanen and Vening Meinesz.[2]

Global Variations in Gravity

To predict the gravitational field of the earth precisely at any point, we must know its shape and density distribution with the greatest possible accuracy. Because of its rotation, the earth is not actually spherical. Its shape can be approximated as an oblate spheroid (a surface generated by revolving an ellipse around its minor axis) with an eccentricity of $\frac{1}{297}$. Both its departure from sphericity and its rotation (because of associated centrifugal forces) cause the earth's gravitational acceleration to have a maximum value at the poles and a minimum at the equator. A good overall fit to the observed relation between gravity and latitude, based on measurements made in all parts of the earth, is obtained by using the International Formula, widely accepted as the standard for theoretical prediction of gravity over a laterally homogeneous spheroidal earth.

This formula, which gives gravity as a function of the angle of latitude ϕ, is

$$g = 978.049(1 + 0.0052884 \sin^2 \phi - 0.0000059 \sin^2 2\phi) \qquad (11\text{-}14)$$

where the number outside the parentheses is the value of gravity in centimeters per second per second, or gals, at the equator. Any variation with longitude is so small that it need not be included in the formula.

The value of gravity obtained from this relation is the one that would be observed at sea level on an earth that has a smoothed spheroidal shape giving the best fit to its actual shape and that has uniform density whenever the depth below the surface is the same. The density can vary with depth in this model, but it cannot vary laterally.

The International Formula is one of several relations proposed for predicting the earth's gravitational force at the surface on a global scale. Others are similar in form but use slightly different constants. The actual difference in gravity between the equator and poles is about 5.3 gals, or 5300 mgal, although a theoretical calculation based only on the earth's rotational characteristics and the difference in distance from the earth's center at the two positions would lead one to expect about twice the contrast that is actually observed between pole and equator. Hammer[3] has shown that the discrepancy can be resolved by considering the attracting effect of the material that exists between a sphere having the polar radius and the actual earth surface which has its maximum radius at the equator. The International Formula, like other formulas used for expressing gravity as a function of latitude, is based on the assumption that the sea-level surface is smooth. Actually this surface has bulges and hollows of the order of hundreds of miles in diameter and up to several hundred feet in elevation (Fig. 11-3). Their presence is related to density irregularities in the earth's interior comparable to the mascons on the moon (although not as large in gravitational effect). Off the tip of Ceylon, for example, there is an area where a ship could sail along a parallel of latitude and climb 250 ft in a day without having to do any work against gravity.

The actual equipotential surface of the earth that corresponds to sea level is called the *geoid*, which is defined under land areas as the surface that would be

FIGURE 11-3
Height of geoid in meters as measured from spheroidal surface giving best fit to all observations.

assumed by the top of the water in a narrow sea-level canal if it were extended inland across a continent to connect one ocean with another. The geoid is by definition everywhere perpendicular to the direction of the plumb line.

The discrepancies between the geoid and the spheroid can be expressed by maps like that in Fig. 11-3. As more precise information on the figure of the earth is obtained from gravity and satellite observations, such maps are periodically revised.

Isostasy

If mountains had to be supported by an entirely rigid earth, the higher ones would collapse because their weight would give rise to pressure variations at great depths that exceed the crushing strength of rock materials. It is much easier to account for the persistence of topographic relief over long periods of geologic time if one assumes that the rigid portion of the earth is an outer shell that is thin compared to the earth's radius, and that it *floats* on an effectively liquid interior. This concept of a thin lithosphere overlying an aesthenosphere forms the basis for plate tectonics, a revolutionary concept that has become widely accepted over the past decade. But its origins go back more than a century when it was proposed to account for otherwise inexplicable variations in the earth's gravitational attraction.

This concept requires that any mass above sea level must be compensated by a deficit of mass below sea level, and any ocean basin, being abnormally light, must be underlain by a mass that is greater than normal. If this were not so, the mountains would sink and the basin bottoms would rise. The total weight per unit area at any depth below the lithosphere must be uniform everywhere if equilibrium is to be sustained. Dutton designated such a state of equilibrium by the term *isostasy*.

Origins of the concept   In the middle of the nineteenth century, a highly precise triangulation survey of India brought to light a discrepancy of about 5 seconds of arc (500 ft) between the separation of two stations, Kalianpur and Kaliana (about 375 mi apart on a north-south line), as measured geodetically and their separation as computed from astronomical readings. Although 500 ft would seem to be an almost trivial error out of a total distance of 375 mi, the precision and internal consistency of the triangulation had been too great for even this small a difference to be disregarded as a geodetic mistie.

Figure 11-4 indicates the positions of the two stations with respect to the Himalayas. Kaliana is in the foothills of the mountains, while Kalianpur is surrounded by hundreds of miles of flat lands. Seeking to explain the discrepancy in the distance between the two stations, J. H. Pratt conceived that the mass of the Himalayas would tend to deflect the plumb line northward at each station but more at Kaliana than at Kalianpur. Such deflection would lead to errors in the astronomic survey, which is based on angles of stars with the vertical. Calculating the discrepancy that would be expected because of the horizontal attraction of the Himalayas, Pratt[4] was surprised to find that the difference in separation should have been about 15 seconds of latitude instead of the 5 seconds actually found. In his paper presenting the results of his calculations he made no attempt to account for this discrepancy.

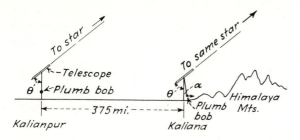

FIGURE 11-4
Pratt's explanation for discrepancy between astronomically and geodetically
determined distances measured in Everest's survey of India. The apparent
angle of the star with the plumb line differs between two stations by amount α,
the angle by which the Himalayas deflect the plumb line at Kaliana (deflection
at Kalianpur assumed zero). This leads to an error in astronomical latitude for
Kaliana.

Airy's hypothesis Less than 2 months after Pratt's paper was delivered,
G. B. Airy[5] submitted his solution to the puzzle. The earth's crust, he said, is a rigid
shell floating on a liquid substratum of greater density. Under mountains the base
of the crust would have penetrated farther into the substratum than under land at
sea level. Figure 11-5 illustrates this principle. He pointed out that the hydrostatic
state of the earth's crust supported by a denser liquid substratum (the mantle)
is like that of a raft made of logs having different diameters floating upon the water.
If the upper surface of one log rises to a higher level than any of the upper surfaces

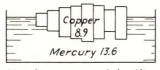

Numbers represent densities

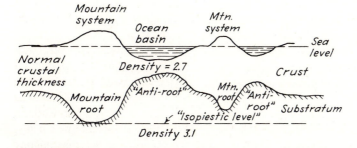

FIGURE 11-5
Airy's theory of compensation. Mountains overlie regions of greater crustal
thickness (roots), while ocean basins overlie sections where crust is thin (anti-
roots). The mechanism is illustrated by copper blocks of various heights and
specific gravity 8.9 floating on mercury with a specific gravity of 13.6. *Isopiestic
level* is the shallowest depth at which pressure is everywhere equal.

of the others, we can be certain that its lower surface lies deeper in the water than the lower surfaces of the others. Similarly, the crust under ocean deeps would be thinner than under land surfaces at or near sea level.

When this concept is applied to the Himalayas, it follows that mountain roots of relatively light crustal material penetrating into a heavier substratum would account for a deficiency of mass below the mountains. The roots of the Himalayas, lighter than the surrounding substratum, would deflect the plumb line in a direction opposite to that of the mountains themselves. If the compensation were complete, the two effects would cancel. In the case we have considered, the compensation seems to have reduced the residual effect to one-third of what it would have been otherwise.

Pratt's hypothesis Four years later Pratt,[6] admitting that the compensation effects are not entirely complete, proposed a somewhat different although equally plausible explanation for his observations. He acknowledged that, as Airy had pointed out, the excess mass of the mountains above sea level had to be compensated by a deficit below sea level, but he hypothesized a crust having a uniform thickness (below sea level) with its base everywhere supporting a uniform weight per unit area. Under the mountains these conditions would call for a deficiency in density of the crustal rocks. The subcrustal density, although variable, should always be such that the total weight of the mountains and crustal material below them is equal to the weight of the crustal rocks alone in an area where the earth's solid surface is flat and at sea level. Under the oceans, the density of the rocks must be greater than average to compensate for the less than normal weight of the ocean water. Figure 11-6 illustrates this hypothesis.

For almost a century the Pratt and Airy hypotheses on the mechanism of compensation have been the subject of controversy. Seismological evidence generally appears to substantiate the Airy theory of roots under mountains and a thinner than average crust under the ocean basins, yet some kinds of seismological data suggest that the crustal thickness is the same under some mountains as it is under the coastal plains. It appears that the most powerful tools available to geophysics have not as yet succeeded in giving a clear-cut answer to the question of who, if either, was right. The problem of resolving this question is complicated because (as in the area south of the Himalayas) perfect compensation is seldom observed. The earth is not everywhere in true isostatic equilibrium, but gravity data give evidence that statistically, at any rate, isostatic compensation does occur even though the actual mechanism cannot be established with certainty.

Anomalies Based on Absolute Gravity Determination

In exploration work, any variation in gravity from that over the larger area immediately surrounding it is referred to as a positive gravity anomaly when higher and a negative gravity anomaly when lower. The determination of what is anomalous and what is normal is not always obvious, and later we shall consider methods of identifying localized anomalies of interest in exploration by separating them from

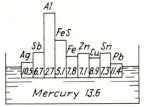

Numbers represent densities

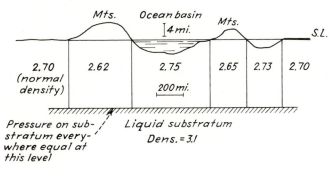

FIGURE 11-6

Pratt's theory of compensation. Elevated blocks such as mountains are underlain by crustal material of less than normal density. Ocean deeps are underlain by greater than normal density. Densities are such that weight per unit area is same at base of each column. Upper diagram illustrates analogy to metals having different densities floating on mercury.

variations which are more likely to be of regional extent. This concept of a gravity anomaly applies where interest is confined to areas of relatively limited dimensions, e.g., those involved in most surveys where oil-bearing structures are sought.

Another concept of a gravity anomaly, oriented toward geodetic applications, made its appearance long before gravity was used as a prospecting tool and has considerable value to geologists studying tectonics on a large scale. It is based on departures of observed gravity corrected to sea level from values determined from the International Formula for the same latitude at which the observation is made. Any difference between the actual reading corrected on the basis of certain predetermined assumptions and the value predicted by the formula is the gravity anomaly for the assumption employed. The assumption which gives the smallest anomaly is considered the most likely to be correct.

The International Formula (or any other such formula which may be used) refers to gravity at sea level, so that it is customary to correct each data point to sea level as a step in computing the anomaly associated with it. The correction actually used takes two effects into account. One is the *free-air correction*, which effectively compensates for the fact that the attraction of gravity above sea level decreases with height because the distance from the earth's center is increasing. For elevation differences small compared with the earth's radius, this is proportional to the height above sea level. The residual between the gravity so corrected and the value predicted from the International Formula is referred to as the *free-air anomaly*.

The second correction designed to place the station effectively on the reference spheroid (considered to be sea level) involves subtracting the attraction of the material (assumed to be a horizontal slab) between the elevation of the station and sea level. This is called the *Bouguer correction*,* after Pierre Bouguer, who made pioneering geodetic observations in the Andes during the eighteenth century. The residual between the predicted gravity and the observed gravity after both the free-air and Bouguer corrections have been applied to it is called the *Bouguer anomaly*.

The International Formula for gravity makes no allowance for lateral density variations in the earth. It is assumed that the density is the same everywhere at the same distance below the surface. Yet where isostatic compensation exists there must be lateral changes in subsurface density, causing this assumption to be violated. To reduce the discrepancy, an *isostatic correction* is made on the basis of surface elevation data, the form of the correction depending on the isostatic model, e.g., Pratt or Airy, assumed. Once the data have been corrected for predicted isostatic effects (along with the free-air and Bouguer corrections) the residual between the observation so corrected and the predicted value at the latitude of the station is referred to as the *isostatic anomaly*. If perfect isostatic compensation exists, this anomaly is zero. If the anomaly is positive over land, there is a deficit of compensation. If it is negative, the area is considered to be overcompensated.

It is seldom, however, that isostatic corrections are made in exploration-type surveys, usually because isostatic variations are of such a nature that they can be included with other regional effects and removed by use of procedures for isolating residual values from regional gravity.

Free-air, Bouguer, and isostatic anomalies may be valuable for deducing the geological history of an area and even for predicting future geological movements. For one thing, the signs and magnitudes of the anomalies and the relationships between them indicate the extent of compensation under the station. Figure 11-7 illustrates the anomalies that would be computed at a station above sea level for the respective cases of no compensation and full compensation. The first has a zero Bouguer anomaly and positive isostatic and free-air anomalies. The second has a zero isostatic anomaly and a negative Bouguer anomaly. Partial compensation would result in anomalies between those indicated for these extremes.

Let us consider a number of examples. The island of Cyprus has a Bouguer anomaly of 120 mgal over its center and almost none near its edge (Fig. 11-8). This evidence has been interpreted as indicating that the island was formed by an intrusion of magma through a feeder column of basic material penetrating lighter crustal rocks beneath the Mediterranean. The magma is believed to have been ultrabasic, spreading out along the sea floor over the area now covered by the island.

---

* Formulas for computing the free-air and Bouguer corrections are presented in Chap. 12.

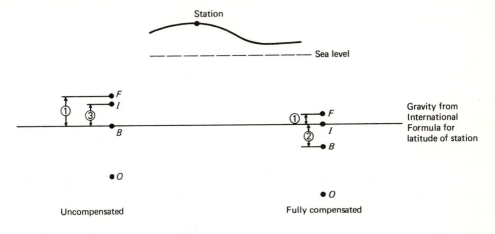

O — Observed gravity
F — Observed gravity after free-air correction
B — Observed gravity after free-air and Bouguer correction
I — Observed gravity after free-air, Bouguer, and isostatic correction

1   Free-air anomaly
2   Bouguer anomaly
3   Isostatic anomaly

FIGURE 11-7
Free-air, Bouguer, and isostatic anomalies in compensated and uncompensated
areas. Distances above or below line representing normal gravity predicted by
International Formula indicate relative magnitudes of anomalies.

Around the eastern Baltic Sea near the portion of it where Norway, Sweden,
and Finland come closest together, there is an extensive negative isostatic anomaly
which appears to result from the recent melting of the Pleistocene glaciers that
covered the area. This would be produced if the compensation developed during
the glaciation is greater than what is actually necessary to balance the land mass

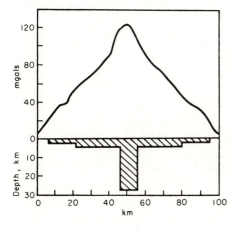

FIGURE 11-8
Bouguer anomaly across island of Cyprus and inferred subsurface distribution of anomalously dense rock. (*After
Harrison, in Garland.*[1])

that extends above sea level after melting of the ice. Careful geodetic observations over a long period of time, as well as the present elevations of ancient structures on the waterfront, indicate that the entire area is rising at a rate of more than 0.5 cm/year in a rebound following its depression during glacial times. In the portion of the region where the anomaly has its greatest negative value it is expected that there will be a further rise of nearly 200 m before equilibrium is reached.

Earth Tides

Gravity-measuring instruments for geophysical prospecting are so sensitive that they respond to the gravitational attraction of the sun and moon and register the periodic variations in the attraction caused by movements of the earth with respect to these bodies. The waters of the earth, having no rigidity, are regularly raised and lowered by such forces in predictable tidal cycles. The earth itself is also acted upon by these tidal forces, and since it is not infinitely rigid, the surface of its solid portion is deformed somewhat in the same way as the free water surface, although not of course to the same extent. The actual tidal movement of a point on the land surface is much smaller than the corresponding fluctuation in water level, being only a matter of a few inches. This displacement in itself causes small but measurable changes in gravity as the distance from the center of the earth is altered. Such changes are, of course, superimposed on changes in gravity caused by the attractive forces of the two celestial bodies. The magnitude of these changes varies with latitude, time of month, and time of year, but the complete tidal cycle is accompanied by a gravity change of only 0.2 to 0.3 mgal.

Heiland[7] has derived the formula for the vertical component of the tidal force Δg caused by the sun (mass M_s, distance from the earth D_s) and the moon (mass M_m, distance from the earth D_m) at any point on a perfectly rigid earth of radius r when the respective celestial bodies make geocentric angles of a_s and a_m with the observation station:

$$\Delta g = \frac{3\gamma r M_m}{2D_m{}^3}\left(\cos 2a_m + \tfrac{1}{3}\right) - \frac{3\gamma r M_s}{2D_s{}^3}\left(\cos 2a_s + \tfrac{1}{3}\right) \tag{11-15}$$

Here γ is the gravitational constant.

Evaluation of the respective coefficients of the terms in parentheses indicates that the moon's attraction is more than twice that of the sun. Wolf[8] has observed that the distortion of the earth due to the same tidal forces increases the effective gravitational pull of the sun and moon 20 percent. Adler[9] has constructed a set of curves based on this formula from which the solar and lunar tidal corrections can be obtained for any specified latitude and longitude if the position of each heavenly body at the time of observation is known.

Wyckoff[10] and Wolf[8] have published a number of representative curves of tidal gravity variation. A comparison of theoretically predicted and observed tidal variations is shown in Fig. 11-9. The divergence between the theoretical tidal variation and the observed one gives an indication of the actual deformation of the

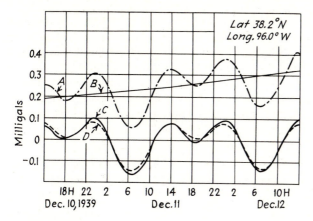

FIGURE 11-9
Comparison of theoretical and observed curves of tidal gravity variation: (*a*)
Gravity-meter readings; (*b*) drift curve; (*c*) observed variation of gravity from
drift curve; (*d*) calculated gravity variation. (*After Wolf.*[8])

earth under tidal influences and makes it possible to determine the earth's true
rigidity.

For many years, tables of worldwide tidal variations were published annually
in *Geophysical Prospecting*, the journal of the European Association of Exploration
Geophysicists. They were based on a theoretical formula for a rigid earth with a
magnification of 1.2 corresponding to an assumed departure from rigidity of 20
percent. As it is now customary to correct for earth-tide effects by incorporating
them into the drift adjustment, such tables are seldom used, and their publication
has been discontinued.

11-5 GRAVITATIONAL EFFECTS OVER SUBSURFACE BODIES
HAVING DISCRETE SHAPES

Long before gravity prospecting came into existence, relationships had been worked
out for the gravitational attraction of bodies having various analytically describable shapes. Exploration applications always involve measurement of differences in the vertical component of gravity fields which are associated with
buried geologic features. The fact that the attraction of the earth is superimposed
upon the attraction of a body buried beneath its surface does not affect these differences as long as any lateral changes in the earth's field (such as those associated
with latitude variation) are taken into account. If all other sources of attraction
can be considered constant over the area of the survey, the effective density to be
used in such calculations is the *contrast* between the density of the body in question
and that of the material surrounding it. Only the *vertical* component of the attrac-

tion need be taken into account, as all gravity meters are designed to respond to vertical forces alone.

One can determine a body's attractive force (or a component of it) at any desired point in space either by direct application of Newton's law or by differentiation of its potential field. The function so computed for geometrical forms having different shapes and depths of burial can be compared with variations of gravitational acceleration actually observed along profiles or on contour maps. Such comparisons with different theoretical models can be used to deduce the shape, depth, and other characteristics of the geological feature responsible for the gravity anomaly.

It is not generally possible to calculate the gravitational fields of bodies having arbitrary or irregular shapes by using simple analytical formulas. Computer programs and graphical devices designed for such a problem will be described in a later chapter. It is relatively easy, however, to calculate fields from generalized geometrical forms such as spheres, sheets having circular or rectangular shapes, and rectangular parallelepipeds. The computation of such fields is ordinarily carried out by integrating expressions for the vertical attraction force from two-dimensional buried sources on a horizontal plane [Eq. (11-10)] or from three-dimensional bodies below the surface [Eq. (11-13)]. The coordinate system and limits employed in the integration depend on the shapes of the bodies.

The sphere It can be proved by potential theory that the attraction at an external point of a homogeneous spherical shell, as well as of a solid sphere, in which the density depends only on the radius, is the same as though the entire mass were concentrated at the center of the sphere. The mass M of a sphere having a radius R and a density (mass per unit volume) ρ is the volume times density or $\frac{4}{3}\pi R^3 \rho$. If the center is at a depth z below the surface, as shown in Fig. 11-10, the total gravitational attraction of the equivalent mass at the center upon a unit mass on the surface a horizontal distance x from the center will be

$$g = \gamma \frac{M}{r^2} = \frac{4\pi\gamma R^3 \rho}{3(z^2 + x^2)} \tag{11-16}$$

Since

$$r = \sqrt{z^2 + x^2}$$

the vertical component g_z, which is simply $g \cos \theta$ or gz/r, will be

$$g_z = \gamma \frac{M}{r^2} \frac{z}{r} = \frac{4}{3} \pi\gamma R^3 \rho \frac{z}{(z^2 + x^2)^{3/2}} \tag{11-17}$$

Nettleton[11] has rearranged this equation in the form

$$g_z \text{ (in milligals)} = \frac{8.53\rho R^3}{z^2} \frac{1}{[1 + (x^2/z^2)]^{3/2}} \tag{11-18}$$

when R, x, and z are measured in kilofeet and ρ in grams per cubic centimeter. This is

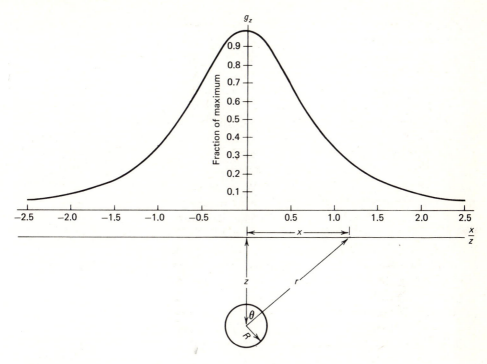

FIGURE 11-10
Gravitational field over sphere. Vertical gravity is plotted versus horizontal
distance from center.

a convenient form for calculation since g can be found at any distance x by multiplying the peak value at $x = 0$, which is $8.53\rho R^3/z^2$, by a factor dependent only on
the ratio x/z. The falloff of gravity along the surface at a distance x from the point
directly over the center is plotted in Fig. 11-10.

For any mass buried in the earth, the effective density ρ, as previously noted, is
the difference between the density of the buried object itself and that of the surrounding material. This quantity is designated as the *density contrast*. If the contrast is negative, the corresponding gravity anomaly will be negative also.

The magnitude of the gravity anomaly that might be expected over a roughly
spherical salt dome can be estimated from Eq. (11-18). If the dome can be represented by a sphere 2000 ft in radius, with a density contrast of -0.25 g/cm³ and
with its center 4000 ft below the surface, the anomaly should have a maximum value
of -1.07 mgal. The effect is thus only about one-millionth the total gravitational
acceleration of the earth itself.

Horizontal cylinder   It is shown in an appendix to this chapter (Sec. 11-9) that
the vertical component of attraction of an infinite wire with mass λ per unit length

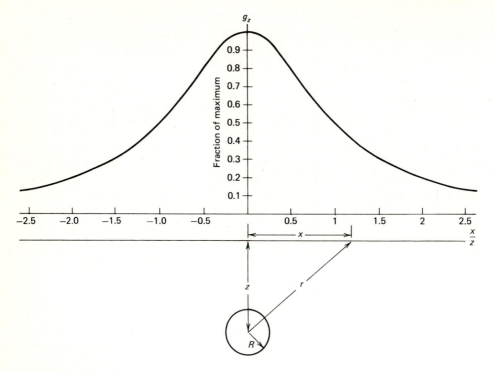

FIGURE 11-11
Gravitational field over horizontal cylinder. Gravity is plotted versus horizontal
distance from the center.

buried at a depth z will be

$$g_z = 2\gamma\lambda \frac{z}{z^2 + x^2} \qquad (11\text{-}19)$$

where x is the perpendicular distance of the measuring point on the surface from
the trace of the vertical plane containing the wire on the earth's surface.

If the wire is expanded to a cylinder of radius R, λ can be expressed as the volume
density ρ times the area, or $\pi R^2 \rho$, so that the formula now becomes

$$g_z = 2\pi\gamma R^2\rho \frac{z}{z^2 + x^2} \qquad (11\text{-}20)$$

If all distances are expressed in thousands of feet and the density in grams per
cubic centimeter, this can be written

$$g_z \text{ (in milligals)} = \frac{12.77\rho R^2}{z} \frac{1}{(1 + x^2/z^2)} \qquad (11\text{-}21)$$

The falloff of the gravity field of a horizontal cylinder with horizontal distance
from the point on the surface directly over its center is plotted in Fig. 11-11. The

anomaly is not so sharp as that from a sphere at the same depth, since the term involving the horizontal distance x, which is in the denominator, is taken to a lower power for the cylinder. It is seen by comparing the respective equations that a horizontal cylinder will have a maximum gravitational effect about $1.5z/R$ times as great as a sphere of the same radius, depth, and density. This should be expected in view of the much greater mass contained in the cylinder.

Buried vertical cylinder   An equally interesting case from a geological viewpoint is that of a buried vertical cylinder. This form is often convenient for computing gravity anomalies from salt domes and volcanic plugs. Consider the cylinder illustrated in Fig. 11-12 having a radius R and height L, with its top surface buried at a depth d. The distance from the point where the axis intersects the earth's surface to the top edge of the cylinder is s_2, and that to the bottom edge is s_1, as shown in the diagram. The formula for the vertical gravitational effect of the cylinder where the axis intersects the measuring surface is shown in Sec. 11.8 to be

$$g_z = 2\pi\gamma\rho(L - s_1 + s_2)$$

or $$\qquad\qquad g_z \text{ (in milligals)} = 12.77\rho(L - s_1 + s_2) \qquad\qquad (11\text{-}22)$$

where ρ is the density contrast in grams per cubic centimeter and all distances are in thousands of feet. If the cylinder is infinitely long, s_1 becomes $L + d$ and the gravity effect along the axis becomes

$$g_z = 12.77\rho(s_2 - d) \qquad\qquad (11\text{-}23)$$

Bowie[12] has prepared some convenient tables for computing the gravity effects of vertical cylinders at the centers of their upper surfaces.

Buried slab   If our vertical cylinder is so extended horizontally that its radius becomes very large compared with its height, s_2 approaches s_1 and Eq. (11-22) becomes

$$g_z = 2\pi\gamma\rho L = 12.77\rho L \qquad\qquad (11\text{-}24)$$

if L is in thousands of feet and ρ in grams per cubic centimeter. This is of course equivalent to an infinite horizontal slab of thickness L. It should be noted that the

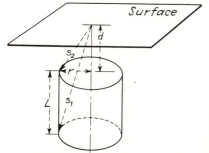

FIGURE 11-12
Definition of quantities used in calculating gravity effects from buried vertical cylinder.

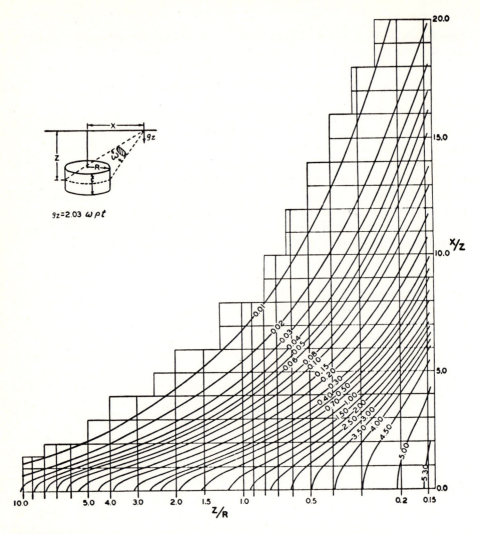

FIGURE 11-13
Chart for calculating gravity at a point on the surface a distance x off the axis
of a vertical cylinder of height t with center buried at depth z. Curves give
solid angles. (*After Nettleton.*[11])

gravity effect depends only on the thickness of the slab, not on its depth of
burial.

Attraction of a buried vertical cylinder at a point off its axis   Returning to
the buried vertical cylinder, let us consider the gravitational field on the surface
at a point off the cylinder's axis. This is a considerably more complex case than
that where the field is computed on the axis, but Bowie[12] has developed a simple
approximation for the case where the top of the cylinder is at the earth's surface.

If the radius of the cylinder having a height L is r and the distance of the measuring point from its axis is x, the vertical gravitational attraction is

$$g_z \approx 2\pi\gamma\rho\left[\sqrt{L^2 + (x-r)^2} - \sqrt{L^2 + (x+r)^2} - 2r\right]\frac{r}{4x} \qquad (11\text{-}25)$$

A case having greater practical interest is that of a cylinder having a top buried a given depth below the surface. An exact formula for this case would be impossible to derive, but when the height of the cylinder is smaller than the depth of its top, it can be approximated by a circular disk and its attraction determined by using the solid angle subtended by the disk at the measuring point. The principle involved will be discussed later in this section. Figure 11-13 shows a chart for ascertaining the subtended solid angle ω in terms of the radius and height of the cylinder, the depth to which its top is buried, and the horizontal distance x of its axis from the measuring point. Using quantities defined in the figure the formula is

$$g_z = 2.03 \ \omega\rho t$$

where t is in thousands of feet.

Attraction of plates and thin slabs   Many geological features can be modeled by a sheet or thin slab of anomalous (although not necessarily constant) density having boundaries that can be specified in various ways. The gravity effects of such forms can be determined analytically without great difficulty. Initially we shall consider bodies with densities contrasting with those of surrounding materials that can be represented by a plane surface (one with thickness small compared to its other dimensions) such as ancient salt-lake beds (now bounded slabs of evaporite), volcanic sills of limited extent, or thin faulted limestone beds.

INFINITELY LONG HORIZONTAL STRIP WITH BOUNDARIES PERPENDICULAR TO PLANE OF PROFILE   Figure 11-14 shows the edge of a sheet extending to infinity in a horizontal plane perpendicular to the page. The boundaries are perpendicular to the x direction. The density contrast of the sheet is σ (mass per unit area). The contribution of a ribbon of width dx to the vertical component of gravity at P is the same as that of an infinite wire perpendicular to the line of measurement. We can write it as

$$dg_z = 2\,\frac{\gamma\sigma \sin\phi \, dx}{r}$$

$\sigma \, dx$, the mass per unit length of the strip, is equivalent to λ in Eq. (11-19) for the wire, and r is the shortest distance from P to the ribbon. Integrating between the limits x_1 and x_2, we obtain

$$g_z = 2\gamma\sigma \int_{x_1}^{x_2} \frac{\sin\phi \, dx}{r} \qquad (11\text{-}26)$$

But $\sin\phi \, dx/r$ is the angle subtended by dx at P, and the summation (or integral)

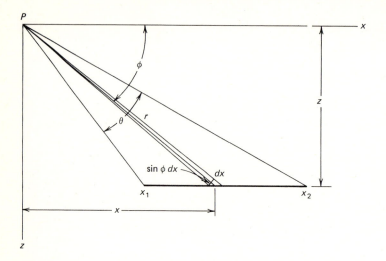

FIGURE 11-14
Attraction at P of a thin sheet of finite width but of infinite length in direction perpendicular to page.

of this quantity over all elements dx from x_1 to x_2 is simply the total angle θ subtended by the strip, so that

$$g_z = 2\gamma\sigma\theta \tag{11-27}$$

If the sheet extends to infinity in the x direction, the line from P to the far edge becomes horizontal and θ can be expressed as $\tan^{-1}(z/x)$ or $\pi/2 - \tan^{-1}(x/z)$.

If the sheet is a thin slab of thickness t which is faulted at $x = 0$ with the right side downthrown an infinite distance, as shown at the bottom of Fig. 11-15, we can write $\sigma = \rho t$ and

$$g_z = 2\gamma\rho t\left(\frac{\pi}{2} - \tan^{-1}\frac{x}{z}\right) \tag{11-28}$$

Far to the left of the fault edge, $x = -\infty$, and here the expression for the attraction of an infinite slab, $2\pi\gamma\rho t$, applies because $\tan^{-1}(-\infty/z) = -\pi/2$. If $x = \infty$, the arc tangent becomes $\pi/2$ and the gravitational effect is zero. Over the fault edge the vertical gravity field is half that of the slab itself. The gravity profile at the top of Fig. 11-15 illustrates these relations.

If the thickness of the faulted slab t is expressed in thousands of feet and the density contrast $\sigma = \rho t$, where ρ is in grams per cubic centimeter, we can write Eq. (11-28) in the form

$$g_z \text{ (in milligals)} = 4.05\rho t\left(\frac{\pi}{2} - \tan^{-1}\frac{x}{z}\right) \tag{11-29}$$

Attraction of thin horizontal plate of finite dimensions   Figure 11-16 shows a bounded thin horizontal plate of irregular shape, for which the attraction at point

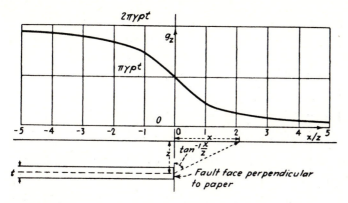

FIGURE 11-15
Gravity profile over thin faulted slab, downthrown side assumed infinitely deep.

P can be determined by summing up the effect of all elements of area on the plate. Consider the element of area dA shown in the figure having a density σ per unit area. The mass is then $\sigma\, dA$. The contribution of this element to the vertical gravity at P is

$$\frac{\gamma\sigma \cos\theta\, dA}{r^2} = \gamma\sigma\, d\Omega \qquad (11\text{-}30)$$

where Ω is the solid angle subtended by dA. The quantity $dA \cos\theta$ is the component of the area perpendicular to R.

Integrating over the entire plate gives

$$g_z = \gamma\sigma\Omega \qquad (11\text{-}31)$$

which indicates that the depth of the sheet is not a unique factor in determining the attraction, the subtended angle Ω being all that counts. Nettleton[11] has published

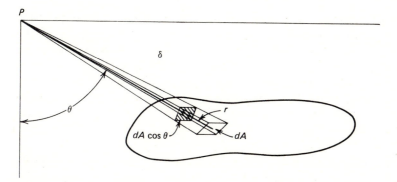

FIGURE 11-16
Determining attraction at P of a thin horizontal plate of irregular shape.

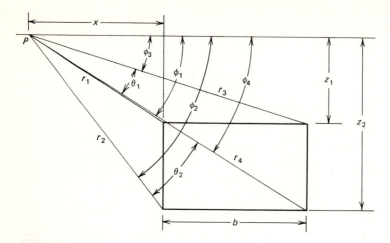

FIGURE 11-17
Determining attraction at P of a "two-dimensional" rectangular parallelepiped
having infinite length in direction perpendicular to paper.

tables for determining the gravity from circular plates as a function of depth, radius,
and horizontal distance from the observing point to the center of the circle.

**Attraction of a two-dimensional body with rectangular cross section in
vertical plane** Many tabular geological features that give rise to gravity
anomalies can be approximated as three-dimensional slabs having elongated plane
sides with upper and lower surfaces that are horizontal. The gravity effect of such a
form can be readily determined by integrating contributions from a series of thin
horizontal laminae, each of the kind shown in Fig. 11-14, which put together make
up the slab. The effect of each lamina is given by Eq. (11-27).

The integration is tedious and will not be carried out here. The attraction of the
body at a point P on the surface along a profile perpendicular to its horizontal axis
can be expressed in terms of the depth of its top and its dimensions as well as by the
positions of its corners with respect to P. Heiland[13] develops the formula for a
slablike body with plane sides that are not necessarily vertical. He shows how the
formula can be applied to a number of geometrical forms that might be used as
models for geologic features.

The relation for the case of a rectangular parallepiped expressed in terms of the
variables defined in Fig. 11-17 is shown by Garland[1] to be

$$g_z = 2\gamma\rho\left(z_2\theta_2 - z_1\theta_1 + x \ln\frac{r_1 r_4}{r_2 r_3} + b \ln\frac{r_4}{r_3}\right) \tag{11-32}$$

This formula is applicable to the case of a rectangular bed bounded by vertical
sides which is much longer than its width b and which has a density that is anomalous with respect to surrounding formations. Certain igneous intrusives as well
as salt stocks can be represented by such a model. If the width b approaches in-

finity, then θ_1 approaches ϕ_1, θ_2 approaches ϕ_2, and r_3 approaches r_4, whereupon Eq. (11-32) becomes

$$g_z = 2\gamma\rho\left(z_2\phi_2 - z_1\phi_1 + x\ln\frac{r_1}{r_2}\right) \tag{11-33}$$

This is the formula for a thick vertically faulted slab with such a large throw that the downthrown side makes no contribution to the gravitational force. As the thickness becomes smaller, we can show that the attraction becomes identical to that expressed by Eq. (11-28) for a thin faulted slab of the type illustrated in Fig. 11-15.

Consider a faulted slab having a finite throw, as shown in Fig. 11-18. At large distances on either side of the fault, the gravity effect is associated only with the thickness of the slab, so that the gravity would be the same at large distances on either side of the fault, as the faulting is assumed to cause no change in the thickness. If we consider the gravity anomalies from the upper and lower slabs separately, we find the only difference, except for the reversal in directions of the high and low sides, is in the gradient across the fault, being steeper for the shallower slab than for the deeper one. Adding the two fields, we obtain a net gravity effect having the form of a doublet with a maximum on the upthrown side and a minimum on the downthrown side. The magnitudes of the positive and negative peaks depend upon

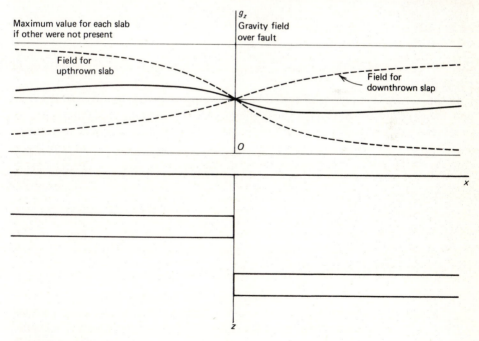

FIGURE 11-18
Attraction of a faulted slab having a finite throw.

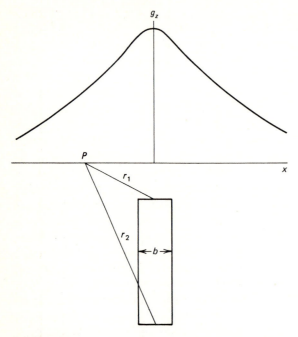

FIGURE 11-19
Attraction at P of a "two-dimensional" rectangular parallelepiped with greater
vertical than horizontal dimensions.

the thickness of the faulted slab, the depth of the upthrown side, and the throw of
the fault.

If the vertical dimension of a slab is large compared with its horizontal extent
and the depth of the bottom is great compared both with its width and with the
depth of its top, θ_1 and θ_2 approach zero. Also ϕ_1 approaches ϕ_3, and ϕ_2 approaches
ϕ_4. We can then write the expression for the vertical component of gravity as

$$g_z = 2\gamma\rho b \ln \frac{r_2}{r_1} \tag{11-34}$$

Figure 11-19 shows the gravity field for this case as a function of distance from the
center of the slab. This formula should give a good approximation in many cases to
the field from a dike with its top and bottom both at finite depths or from thin
vertically elongated salt stocks.

Gradients of gravity The rate of change of gravity with distance, designated as
the *gradient*, has applications to exploration which are likely to assume increasing
importance as the precision of gravity-measuring instruments is improved. Actually,
the earliest use of gravity in exploration (and the first use of geophysics in prospecting for oil) involved direct measurement of gradient (along with a more complex
parameter, the curvature) rather than of gravitational attraction itself. The

Eötvös torsion balance, which was employed with great success in the mid-1920s for locating piercement-type salt domes in the Gulf Coast area of the U.S. and Mexico, was used to map the gradient of gravity. The deepening of the Gulf Coast geosyncline to the south results in a normal gradient that is negative toward the Gulf of Mexico. The anomalous gradient resulting from the density deficit associated with a salt dome turned out to be diagnostic of the shallow domes sought in the area. A further discussion of the principles behind the operation of torsion balances and the interpretation of data obtained with them will be found in earlier editions of this book. Since the introduction of modern gravity meters in the early 1930s, it has been easier to observe gradients by measuring gravity values directly at appropriate intervals along a traverse line. As will be shown in some detail in Chap. 13, the direct interpretation of gravity maps often involves at least a mental scanning of contours for anomalous gradients. A steeper than normal horizontal gradient having a linear trend is often diagnostic of faulting.

In recent years the *vertical gradient* has been looked upon as a promising parameter for direct measurement by gravity meters. The normal vertical gradient is readily obtained by differentiating the expression for the attraction of the earth (considered to be a sphere so that its mass can be treated as if it were concentrated at its center) in terms of its radius r to the measuring point. If we write $g = \gamma M/r^2$, where M is the mass of the earth, then

$$\frac{dg}{dr} = -2\frac{\gamma M}{r^3} = -\frac{2g}{r} \tag{11-35}$$

At sea level the normal vertical gradient of gravity is 0.3086 mgal/m. The maximum deviation from this value at any latitude is -0.0002 mgal/m, a differential that is negligible for all practical purposes. Any deviation of the vertical gradient from its normal value could be associated with anomalous buried masses. A direct measurement of gradient might make it possible to detect such masses with fewer observations than would be needed in conventional gravity measurement. Hammer[14] discusses a number of cases of interest in exploration where it would be advantageous to work with vertical gradients. Such gradients can be measured directly by taking readings at two elevations at the same position and taking the difference between the readings. From a practical standpoint, this can best be done by using two gravity meters, one above the other, on a tower mounted on a truck or by a single meter used for successive readings at the two heights as shown in Fig. 11-20. The practical problems involved are discussed by Thyssen-Bornemisza and Stackler.[15] Thyssen-Bornemisza[16] has proposed a design for a gravity gradiometer that would make it possible to observe the gradients with a single instrument.

11-6 INSTRUMENTS FOR MEASURING GRAVITY ON LAND

Gravity measurements for exploration purposes have been made with three kinds of instruments, each quite different from the others. These are the torsion balance,

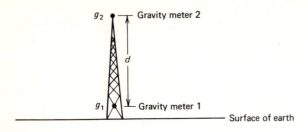

Observed gradient $= \dfrac{g_1 - g_2}{d}$

Theoretical gradient $= .3086$ mgals/m

FIGURE 11-20
Proposed arrangement for measuring vertical gradient of gravity from top and bottom of tower.

the pendulum, and the gravity meter. The first two are obsolete, hardly having been used at all for exploration since the middle 1930s.

Many types of gravity meters have been used in exploration work over the past several decades. Only a few of these are in current use for land operations, but since the middle 1960s several special kinds of meters have been put into service for measuring gravity on board ship or in boreholes. The availability of shipborne gravity meters has led to a resurgence in gravity activity after conventional land work had declined to a small fraction of its activity level in earlier decades.

Gravity Meters for Use on Land

The requirements for precision in gravity exploration are so demanding that few measuring instruments of any kind have specifications which are more difficult to meet. The earth's gravitational field is about 10^6 mgal; yet many anomalies of interest in exploration have a maximum value of 1 mgal or less. To detail such anomalies with the accuracy needed for meaningful interpretation requires readings that are valid to at least 0.05 mgal. Modern land meters have a nominal precision as great as 0.01 mgal when readings are taken with extreme care; i.e., they are responsive to changes of this very small magnitude in the earth's gravitational field. It would be much more difficult and probably impossible to measure absolute gravity that precisely under any conditions, let alone in the field. The precision is attainable because we are measuring *differences* in gravity over a total range of gravity values of only tens to hundreds of milligals.

There are two basic types of gravity meters, the stable and the unstable. A stable meter contains a responsive element (such as a spring) with a change in displacement proportional or approximately proportional to the change in gravity. Since such displacements are always extremely small, they must be greatly magnified by optical, mechanical, or electrical means. Unstable gravimeters are so designed that any change in gravity from its equilibrium value brings other forces into play which increase the displacement caused by the gravity change alone and hence the sensitivity of the instrument. How this is accomplished in practice will be illustrated when specific instruments are discussed. Most current gravity surveying is done with meters of the unstable type.

Stable type   All stable gravimeters have a single element to balance the force of gravity with another force measurable by a displacement (linear, angular, or electrical) which can be magnified and read directly. Any change in gravity is accompanied by a corresponding change in this displacement. For a simple spring this change would be in its length. The usual formula for the restoring force of a weighted spring is

$$F = k(x - x_0) = mg \tag{11-36}$$

where x = length of spring
   x_0 = original length before weight was hung from it
   k = spring constant

Since the mass m is constant, any change in g causes a proportionate change in elongation.

The only type of gravity meter designed on this principle which has had extensive use on land is the Gulf gravimeter,[17] the sensing element of which is a flat spring wound in the form of a helix with the flat surface always parallel to the axis. A weight is suspended at the lower end. Any change in the gravitational pull on the weight causes the spring to rotate as well as elongate. In fact, the rotary motion of the bottom end of the spring is greater and hence easier to measure than the vertical displacement. A mirror attached rigidly to the bottom of the helix makes it possible to measure this rotation by deflecting a light beam. The Gulf instrument has not been used in exploration for several decades.

The Askania Gss3 Seagravimeter, for shipboard use, is also a stable type of meter. It will be discussed on pages 397–398.

Unstable type   In unstable gravimeters, sometimes referred to as *labilized* or *astatized,* the force of gravity is kept in unstable equilibrium with the restoring force. The instability is provided by a third force which intensifies the effect of any change in gravity from its equilibrium value. For small changes of gravity the force called into play by a departure from equilibrium is proportional to the magnitude of change and acts, of course, in the same direction.

It is easy to show that a gravity meter operates like a seismometer except that the displacement is stationary instead of oscillatory. A vertical spring that is elongated by the gravitational pull mg of a mass m will oscillate with a period $T = 2\pi \sqrt{m/k}$, where k is the spring constant. From Eq. (11-36) it is evident that

$$x - x_0 = \frac{mg}{k} = \frac{T^2}{4\pi^2} g \tag{11-37}$$

The elongation is thus proportional to g, and a change in gravity Δg will result in an elongation Δx which is dependent on it according to the formula

$$\Delta x = \frac{T^2}{4\pi^2} \Delta g \tag{11-38}$$

Thus the sensitivity of a meter is proportional to the square of the period of the equivalent seismometer. For a weight hanging from a spring, the ratio of the elongation to the length of the spring would be the same as the ratio of the change in gravity to the earth's gravitational attraction. At the maximum desired sensitivity, this ratio, being of the order of 10^{-7} to 10^{-8}, would require much too long a spring to be practical; this explains why stable meters that have had practical use in the field, like the Gulf gravimeter, do not operate by measuring spring elongation directly.

It follows from Eq. (11-38) that any oscillating system can be adapted by proper damping for use as a gravity meter. A long-period vertical Galitzen-type seismograph developed by LaCoste[18] in 1934 turned out to be the basis for the design of the LaCoste and Romberg gravity meter, which is still widely used, as well as a number of other meters of the same type.

LACOSTE AND ROMBERG GRAVITY METER   The LaCoste and Romberg instrument has a design typical of unstable gravity meters in general. Any change in the force of gravity acting on a weight at the end of a nearly horizontal beam supported by an inclined spring (Fig. 11-21) causes a motion which changes the angle between the beam and the spring (and hence the moment of the spring's pull on the beam), thereby accentuating the moment associated with the gravity increment or decrement, whichever it may be. This reinforcement provides the necessary instability to magnify the effect of the gravity variation. In practice the motion is nulled by an adjustable screw which varies the point at which the mainspring is supported. The angle through which this screw must be turned to restore a light beam, reflected by a mirror on the arm, to its null position is used as a measure of the change in gravity.

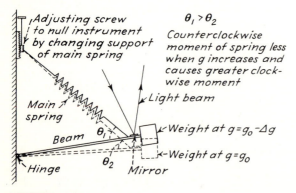

FIGURE 11-21
Operation of LaCoste and Romberg gravimeter (schematic).

An important innovation introduced with the LaCoste and Romberg instrument is the *zero-length spring*. With it the displacement of the spring from equilibrium caused by the weight of the beam in zero position is effectively counteracted by an opposing tension put into the spring when it is wound. It is only with this arrangement that the elongation of the spring caused by any given increment in gravity will be actually proportional to the increment itself. In addition, the deflection will be symmetrical about the equilibrium position; i.e., the positive reading for an increase in gravity over its equilibrium value will be equal to the negative reading for a gravity decrease of the same magnitude.

The LaCoste and Romberg instrument, like the Gulf gravimeter, requires heating elements, a thermostatic regulator, and an insulated case to maintain the temperature sufficiently uniform to avoid significant errors in the gravity readings caused by thermal effects. A change in temperature of 0.002°C will result in a deflection equivalent to a gravity change of 0.02 mgal, which represents the limit of precision for most surveys. To keep the temperature constant, the meter is enclosed in an insulated housing, and a small electric oven powered by a storage battery is regulated by a highly sensitive thermostat.

WORDEN GRAVITY* METER This extensively used instrument, provided with internal temperature compensation, has the advantage of exceptional portability because no storage battery is required for its operation. Instead of a heavy insulated case, the instrument is housed in a thermos flask less than 1 ft high (Fig. 11-22). The flask is sealed and evacuated to eliminate the effect of changes in barometric pressure. The weight of the instrument is less than 6 lb, and the tripod and carrying case add only 8 lb to the total. In the Worden meter a number of torsion fibers are used to balance the torque of gravity upon a mass weighing only 5 mg at the end of the arm attached to a beam around which moments are balanced.

Figure 11-23 illustrates the operation of this meter. The system is in unstable equilibrium around axis *HH*. Any increase in gravitational pull on the mass at the end of the weight arm causes counterclockwise rotation which can be nulled by tightening the pretension spring with an adjusting knob. The amount of rotation of the knob necessary to restore the pointer to its null position provides a measure of gravitational change. The instrument can be read to 0.01 mgal, and the model used for most petroleum exploration work has a dial range of about 60 mgal. A special model is available for geodetic work which has a range of 5500 mgal. Whatever the range may be, the actual values of gravity it covers may have to be set to fit the area of the survey.

Temperature compensation is achieved through the arrangement shown in Fig. 11-24. The two long arms are made of materials having different thermal-expansion coefficients and are connected by a spacing bar at one end. When the temperature changes, the relative lengths of the arms change in such a way that there is an upward or downward movement at the end of the spring which largely compensates

* Trademark, Texas Instruments, Inc.

FIGURE 11-22
Worden gravity meter. (*Texas Instruments, Inc.*)

for any temperature effects in the spring itself. The curved portion of the upper arm represents a short nonlinear spring element which extends the temperature range of the compensation movement.

Calibration of Land Gravity Meters

All readings of gravity meters are in arbitrary scale divisions, and calibration is necessary to express them in milligals. Calibration may be done by tilting the meter or by taking measurements on it at two points where either the absolute or relative values of gravity are known precisely. A pair of locations at which absolute gravity has been established by pendulum observations is often convenient for this purpose. If the absolute gravities differ by an appreciable fraction of the total range of the instrument, it is usually safe to assume linear response and to calibrate the

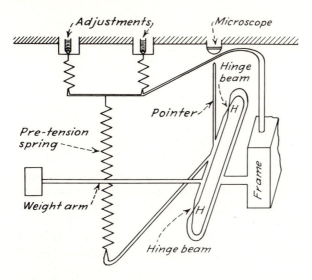

FIGURE 11-23
Principle of operation of Worden gravity meter.

whole scale on the basis of the two readings alone. Greater precision will of course be obtained from a larger number of reference stations. Hammer[19] describes some attempts to calibrate a gravimeter by taking readings at the top and bottom of a tall building and making use of the theoretical difference in gravity between two points at different elevations after correcting for the attraction of the building itself. He concludes that this method is impractical because of variations in vertical gradient and limitations in the precision of the building corrections. A much greater difference in gravity can be obtained between two points at a substantially different latitude. Kuo et al.[20] report some more recent measurements in tall buildings in New York City and its surroundings.

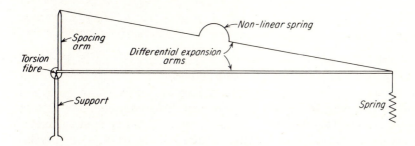

FIGURE 11-24
Temperature-compensation system for Worden gravimeter. (*Adapted by permission from Heiskanen and Vening Meinesz, "The Earth and Its Gravity Field," McGraw-Hill Book Company, 1958.*)

Drift of Gravimeters

If a gravimeter is left undisturbed for an hour or so after a reading and a second observation is taken, the gravity value will apparently have changed by an amount as great as several hundredths of a milligal. If additional readings are taken over a period of hours and the observed gravity is plotted against time, it will be found that the points tend to fall on a smooth curve. This continual variation of the gravity readings with time is known as *drift* and has a number of causes. One is the fact that the gravimeter springs or torsion fibers are not perfectly elastic but are subject to a slow creep over long periods. Another cause is earth tides, and still another is related to uncompensated temperature effects. A sample set of drift curves is reproduced in Fig. 11-25. The drift during the course of a day may be greater than the maximum gravity variation actually observed during the same period. Drift curves are obtained by repeated occupation of a single field station at intervals during the day. Adjustment of readings at other stations is made by recording departures from the drift curves using techniques to be described in the next chapter.

11-7 INSTRUMENTS FOR MEASURING GRAVITY AT SEA

Two types of meters are used for measuring gravity in water-covered areas. One is lowered from a ship to the water bottom in a waterproof housing, whereupon it is leveled and read on board ship. The other measures the gravity on board a ship, being mounted on a platform stabilized by accessory equipment in order to minimize the effect of the ship's motion on the observed acceleration. The first type, for a long time the only one available for marine observations, is still used where an anomaly of small areal extent (such as from a salt dome) must be mapped with the highest possible precision. The shipborne meter is intended for reconnaissance surveys over large areas where the anomalies of interest are so large that high resolution in following lateral variations is not needed.

Bottom Meters

As early as 1941, the Gulf gravimeter was adapted for underwater use by Pepper.[21] The meter was enclosed in a waterproof housing and operated on the water bottom by remote control through cables from a ship. Leveling was accomplished by motor-driven gimbals so designed that the level was controlled independently in two perpendicular directions. The underwater meter now most widely used in exploration work is the LaCoste and Romberg type.[22] Figure 11-26 shows this meter being lowered into the water from a ship by a boom that swings it from the deck.

The LaCoste and Romberg instrument is equipped with a servo system to compensate for vertical motion of the bottom. It also reads gravity automatically and presents it digitally on a shipboard display unit.

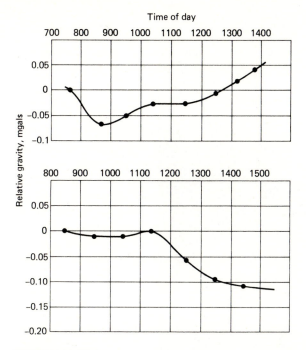

FIGURE 11-25
Typical drift curves for gravity meter. The total excursion in the course of a working day is not much more than 1 mgal. (*Tidelands Geophysical Co.*)

The basic gravity-measuring element operates on the same principle as the LaCoste and Romberg land meter discussed in Sec. 11-6, and the precision is about the same. A light beam from a source attached to the case is directed against the movable arm of the meter in such a way that a photocell is actuated when it moves vertically. The cell controls a servomotor which raises or lowers the meter in a direction opposing the motion of the case. Accuracy of the system under reasonably good sea conditions is 0.1 mgal.

Shipborne Meters

The advantages of measuring gravity on a moving ship are quite evident. With bottom meters, the ship must stop for at least several minutes at each location where an observation is made. Any instrumental system that allows readings without such stops should make it possible to obtain the gravity data much more economically and efficiently.

The principal obstacle to obtaining useful gravity information from an instrument on a ship is the fact that motions of the ship (such as pitch, roll, and heave) are accompanied by accelerations which are themselves much greater than the gravity differences to be measured. For a shipborne gravity meter to work at all,

FIGURE 11-26
A LaCoste and Romberg water-bottom meter being lowered into the water
from a ship. (*LaCoste and Romberg, Inc.*)

it is necessary to neutralize (by compensation or averaging) the accelerations
associated with the motion of the ship itself. These accelerations may be 100,000
mgal or more, depending on the type of ship and the sea state, and they have both
horizontal and vertical components. The horizontal motion is from pitch and roll
and the vertical motion is from heave. All these motions have periods of only a few
seconds.

Effects of pitch and roll are generally eliminated with a gyroscopically stabilized
platform using accelerometers to actuate servomotors which keep the platform
horizontal within a few seconds of arc. Vertical motion of the ship is removed by
averaging many cycles of acceleration over an extended period. This can be done
by introducing a damping that results in a long time constant for the meter; in one
type of meter it is as great as 5 min. This means that about 100 vertical oscillations
would have to be averaged for each observation. Such a long time constant limits
the response of the instrument to gravity anomalies which appear to have a long
period when recorded on a moving chart. For example, a shallow high-density
buried mass 1000 ft wide would be crossed in 1 min by a ship traveling at 10 knots.
If an instument with a 5-min time constant were used, the gravity effect of this
mass would be reduced fivefold by the averaging. The damping thus has the effect

of a low-pass filter, suppressing frequencies associated with the ship's motion but passing frequencies associated with subsurface sources that generate fairly broad gravity anomalies. The type of damping that is employed varies with the instrument.

The LaCoste and Romberg shipborne meter   The shipborne gravity-measuring system most widely used in exploration work is the LaCoste and Romberg meter[22] mounted on a gyroscopically stabilized platform. Like the underwater instrument made by the same firm, the basic gravity-measuring system is that originally used in the LaCoste and Romberg meter designed for use on land. This consists of a weight at the end of a nearly horizontal beam supported by a spring making an acute angle with the beam. Because of the need for filtering out short-period vertical accelerations associated with the ship's motion, the damping must be much greater than with a conventional instrument. The damping mechanism consists of cylinders (one set on the beam and another on the frame) interweaving in such a way as to resist the flow of air during motion.

The heavy damping makes it impractical to read gravity directly from the displacement of the beam, as is done with gravity meters designed for use on land, but the rate of change in the deflection of the beam is responsive to changes in effective gravity (earth's gravity plus other vertical accelerations) regardless of the damping. This follows from the basic equation of motion of the system:

$$m\left(g + \frac{d^2z}{dt^2}\right) = kS - F\frac{dB}{dt} - \frac{d^2B}{dt^2} \tag{11-39}$$

where g = acceleration of gravity
$\quad m$ = mass
$\quad z$ = vertical position of point on meter case
$\quad B$ = displacement of mass with respect to its null position in case
$\quad F$ = damping coefficient
$\quad k$ = spring constant
$\quad S$ = displacement of spring

S can be taken as the length of the spring when it is the zero-length type used in the LaCoste land meter. The term d^2B/dt^2 can be neglected in comparison with the other terms because the time constant is much smaller than the time over which the gravity readings are to be averaged. When this term is removed, the equation can be written

$$g + \frac{d^2z}{dt^2} = \frac{k}{m}S - \frac{F}{m}\frac{dB}{dt} \tag{11-40}$$

The left-hand side is the effective gravity acting on the instrument.

As with the LaCoste and Romberg land meter, the beam is kept in balance by continual automatic adjustment of the spring tension, which is carried out by a servo system that changes the position where the end of the spring pivots. For a given total acceleration, Eq. (11-40) shows that this position varies as the rate of change of the displacement. A beam of light is reflected into a photocell by a mirror on the gravity meter beam. A chopper is used to give an ac output from the cell when the beam deviates from the null position, the magnitude depending on the *rate* of its motion away from null. A rectifier is used to convert this output to the signal, which is recorded as a measure of gravity.

Most of the horizontal acceleration associated with the ship's motion is removed by the action of the stabilized platform, but there will be some residual effects. Translational horizontal accelerations, for example, will not be removed even when the platform holds the instrument in a level position. The most serious effect, however, of residual horizontal accelerations, is *cross-coupling* between the horizontal and the vertical accelerations. When the beam, hinged around a horizontal axis, makes an angle β with the horizontal as in Fig. 11-27, there will be a horizontal torque on the beam of $mgD \sin \beta \, d^2x/dt^2$, where D is the length of the beam. It can be shown (see LaCoste[18]) that the cross-coupling error in the gravity reading caused by this torque will be

$$e = \frac{d^2x}{dt^2} \tan \beta \approx \frac{d^2x}{dt^2} \beta \tag{11-41}$$

This is the horizontal acceleration times the angle of inclination of the beam.

The angle of the beam β varies periodically with the motion of the ship according to the relation $\beta = \beta_0 + \beta_1 \sin (\omega t + \psi)$, where ω is the frequency of the ship's motion and ψ is the phase difference between the horizontal and vertical accelerations. A cross-coupling error e as great as 50 mgal is possible for the largest horizontal accelerations that are likely to be observed during recording. This is known as *inherent cross-coupling*.

In the LaCoste and Romberg instrument the cross-coupling error is evaluated internally by a computer into which data on β and the horizontal acceleration are

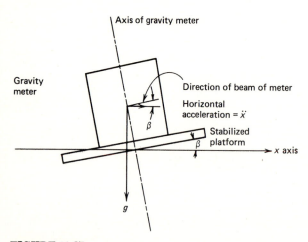

FIGURE 11-27
Cross-coupling in the LaCoste and Romberg shipborne meter.

continually being fed. The error is then removed electrically from the output of the gravity-sensing element.

Askania Seagravimeter   The first shipborne meter to go into service was designed by Graf and manufactured by the Askania Werke of West Berlin.[23] This instrument, the Gss2, which has a horizontal beam as the responsive element and a very strong magnet for damping, has seen extensive use in academic and governmental surveys but has seldom been employed in oil exploration.

In 1971, the Askania Seagravity meter Gss3 was introduced. It differs in principle from the Gss2 in that the sensing element has only vertical motion, eliminating cross-coupling effects. Figure 11-28 illustrates the design. The sensor is a vertically

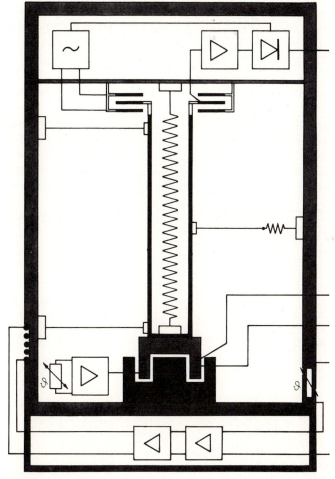

FIGURE 11-28
Principle of operation of Askanian seagravity meter Gss3. (*Bodenseewerk Gerätetechnik GMBH.*)

oriented tubular mass held in place by five fiber ligatures. A vertical spring compensates for the weight of the sensing element. A capacitative transducer is used as a pickup, the signal from it being amplified and rectified. A feedback loop reduces the displacement of the system attributable to ship motion. A filter system is included which is designed to separate the components of voltage due to gravity changes from those caused by accelerations of the ship. The digital output of a voltmeter is fed into a recorder chart and optionally into magnetic tape and punched-tape recorders. An accuracy of 0.5 mgal is claimed for ship accelerations of 0.2 g or less.

Bell ship gravity meter Although this instrument, first employed for geophysical measurements in 1967, is no longer manufactured, it offers a number of features worthy of note. The sensor, instead of being a pivoted beam kept horizontal by a spring, is a disklike mass less than $\frac{3}{16}$ in in diameter surrounded by a coil of wire and attached to the case by three light guide springs (Fig. 11-29). The mass and coil are located between the poles of a pair of conical permanent magnets which are oppositely polarized. The entire system is contained in a cylinder about $1\frac{3}{4}$ in high and $\frac{3}{4}$ in in diameter.

A variable current passes through the coil, and it is adjusted to keep the coil in a null position. If the gravitational force on the coil changes, the mass will move from its equilibrium position and the amount the current must be varied to bring it back to this position gives a measure of the change in acceleration. A pair of capacitative rings attached to the respective pole pieces is used to detect movement of the coil. The change in capacitance brought about by such motion unbalances a bridge, generating a signal that brings the mass back to equilibrium position. The electric

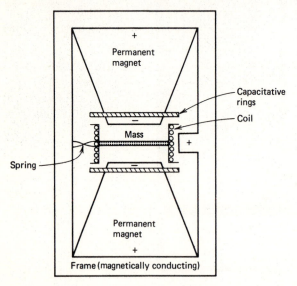

FIGURE 11-29
Principle of operation of Bell shipborne gravity meter.

signal used to obtain this balance, which is proportional to the change in gravity, is digitized and converted to gravity units for final shipboard presentation. Cross-coupling effects are negligible, as the sensing element is very small.

The sensor in the Bell instrument was originally designed for inertial navigation. A number of units have been used in government-sponsored surveys as well as in worldwide geophysical surveys carried out by oil companies.

Wing shipborne acceleration sensor   The Wing meter for measuring gravity on board ship uses a sensor developed for inertial systems. The sensor consists of a pair of weights, one above the other, attached to vertical vibrating strings or wires (of the type used in old-time string galvanometers) at their resonant frequencies. If the tension on the strings varies, the frequency of vibration changes from its resonance value. The string is in a magnetic field, and the frequency of its motion is detected by induction. One of the weights is located above the string to which it is attached, the other below. Thus any increase in gravity raises the vibration period of one string and lowers the period of the other. The difference between the two periods is sensed and converted to gravity units in the output reading. The system is mounted on a stabilized platform. Cross-coupling is effectively eliminated by keeping the masses in precise vertical alignment by the use of ligaments which prevent horizontal movement. This meter, so far as is known, has not been used in oil exploration. Although the sensitivity and precision are adequate, the drift is greater than with instruments actually employed for exploration work.

11-8   BOREHOLE GRAVITY METERS

After more than a decade of development efforts in the United States, the U.S.S.R., and England, the first instruments for precise and reliable measurement of gravity in boreholes became available in the middle 1960s. Two types of meters came into use at this time. One, the Esso borehole gravity meter,[24] makes use of a vibrating filament; the frequency is related to the tension on the filament and changes as gravity varies. The other, developed jointly by the U.S. Geological Survey and LaCoste and Romberg, Inc.,[25] has as its basic sensing element a LaCoste and Romberg gravimeter especially adapted for the space available in a borehole and for the temperatures encountered at the maximum depths where gravity values are to be measured. Both instruments yield gravity measurements with a precision of 0.01 mgal.

The Esso meter has a platinum mass which hangs from a tungsten fiber in a vacuum. It is housed in a cylindrical case with dimensions that allow it to remain vertical in boreholes that deviate from the vertical as much as 4°. The outer diameter of the instrument is 10.2 cm. A thermostat keeps the temperature at 125°C, a value higher than the ambient temperature expected at any depth in the hole.

Gravity variations are measured by determining the precise times required for a specified number of oscillations, actually 10^5, 2×10^5, 3×10^5, etc., cycles. The longer the period of measurement the greater the precision of the gravity value. It takes 20 min to obtain a gravity reading good to 0.01 mgal.

The USGS–LaCoste and Romberg gravity meter is designed according to the same principle as the land and shipborne meters developed by its manufacturer. The sensor, with a zero-length spring and a special adjusting screw to allow a wide range of readings, is mounted on gimbals and can be leveled when the hole deviates as much as 6.5° from the vertical. The outside diameter of the instrument ranges from 15 to 17 cm. Temperature is maintained by a thermostat at 101°C.

Gravity is measured by nulling the beam, as with other LaCoste and Romberg instruments. It requires 5 min to obtain a reading with a precision of 0.01 mgal. Drift is low, and the thermostat keeps the gravity reading from varying more than .04 mgal when the temperature changes from 25 to 95°C. As the increase in temperature with instrument depth is smooth, this effect can be incorporated in the drift and can be eliminated in the same way as other drift effects.

Use of data from borehole gravity meters in determining density will be considered in a subsequent chapter (page 416).

11-9 APPENDIX

Attraction of Horizontal Cylinder at Point on the Surface along a Line Perpendicular to Its Axis

Figure 11-30 shows a buried mass distributed along an infinitely long horizontal line (such as a straight wire) at depth z with a linear density of λ g/cm. Assume that gravity measurements are made on a horizontal surface along the x axis which is perpendicular to the buried line. If a measurement of the *vertical* component of attraction of the line is made at point P, we can define distance R as the perpendicular distance PO from the measuring point P, ϕ as the angle between OP and the x axis, and r as the distance from P to an element of length dl, where l is the distance from O [with coordinates (x, R, z)] to the element of mass.

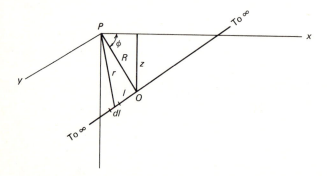

FIGURE 11-30
Gravity at P from infinitely long horizontal wire buried at depth z and parallel to y axis.

The vertical component of attraction from the element dl may be written

$$dg_z = \frac{\gamma \, dm \sin \phi}{r^2} \frac{R}{r} = \gamma\lambda \sin \phi \, R \frac{dl}{(R^2 + l^2)^{3/2}} \tag{11-42}$$

and the total vertical attraction, obtained by integration, is

$$g_z = \gamma\lambda \sin \phi \, R \int_{-\infty}^{\infty} \frac{dl}{(R^2 + l^2)^{3/2}} \tag{11-43}$$

$$= \gamma\lambda \sin \phi R \left[\frac{l}{R^2\sqrt{R^2 + l^2}} \right]_{-\infty}^{\infty} \tag{11-44}$$

$$= 2 \frac{\gamma\lambda \sin \phi}{R} = \frac{2\gamma\lambda z}{R^2} = 2\gamma\lambda \frac{z}{x^2 + z^2} \tag{11-45}$$

If the line is expanded into a cylinder of radius A and density ρ, the mass per unit length becomes $\pi A^2 \rho$ and the attraction becomes

$$g_z = 2\pi\gamma A^2 \rho \frac{z}{x^2 + z^2} \tag{11-46}$$

Attraction over the Center of a Buried Cylinder with a Vertical Axis

Figure 11-31 shows a vertical cylinder with upper and lower surfaces at depths d and f, respectively. Its attraction is the difference between the fields of two infinitely deep cylinders of radius A with top surfaces at depths of d and f. The density of each cylinder is ρ.

Consider the upper cylinder of infinite depth. We see that the slant distance from P (on the axis at the surface) to its top edge is s_2. Its gravity effect can be computed by summing fields from mass elements dm, which are approximate cubes with edges of dimensions dz, $rd\phi$ and dr as defined on the right side of the figure. The attraction at P is obtained from the triple integral

$$g_z = \gamma\rho \int_d^{\infty} \int_0^A \int_0^{2\pi} \frac{d\phi \, rdr \, dz}{(r^2 + z^2)^{3/2}} \tag{11-47}$$

Integration over the cylinder yields

$$g_z = 2\pi\gamma\rho[(d^2 + A^2)^{1/2} - d] = 2\pi\gamma\rho(s_2 - d) \tag{11-48}$$

The lower infinitely deep cylinder with its top edge a distance s_1 from P has vertical attraction $2\pi\gamma\rho \, (s_1 - d)$. We find the field of a cylinder of height L (equal to $f - d$) by subtracting the effect of the deeper cylinder from that of the

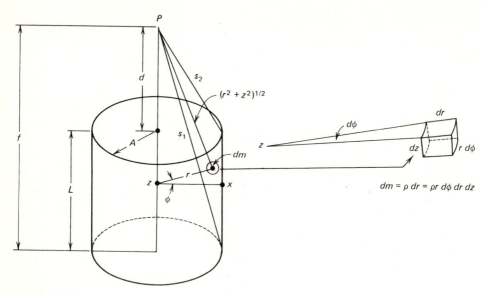

FIGURE 11-31
Quantities used in deriving formula for attraction along axis of buried vertical sphere.

shallower. The difference is

$$g_z = 2\pi\gamma\rho[(s_2 - d) - (s_1 - f)]$$

$$= 2\pi\gamma\rho(s_2 - s_1 - d + f)$$

$$= 2\pi\gamma\rho(s_2 - s_1 + L) \tag{11-49}$$

REFERENCES

1. Garland, G. D.: "The Earth's Shape and Gravity," Pergamon, Oxford, 1965.
2. Heiskanen, W. A., and F. A. Vening Meinesz: "The Earth and Its Gravity Field," McGraw-Hill, New York, 1958.
3. Hammer, Sigmund: Note on the Variation from the Equator to the Pole of the Earth's Gravity, *Geophysics*, vol. 8, pp. 57–60, 1943.
4. Pratt, J. H.: On the Attraction of the Himalaya Mountains and of the Elevated Regions Beyond upon the Plumb-Line in India; On the Computations of the Effect of the Attraction of the Mountain Masses as Disturbing the Apparent Astronomical Latitude of Stations in Geodetic Surveys, *Phil. Trans. R. Soc. Lond.*, vol. 145, pp. 53–55, 1855.
5. Airy, G. B.: *Phil. Trans. R. Soc. Lond.*, vol. 145, pp. 101–104, 1855.
6. Pratt, J. H.: *Phil. Trans. R. Soc. Lond.*, vol. 149, pp. 747–763, 1859.
7. Heiland, C. A.: "Geophysical Exploration," p. 163, Prentice-Hall, Englewood Cliffs, N.J., 1940.

8. Wolf, Alfred: Tidal Force Observations, *Geophysics*, vol. 5, pp. 317–320, 1940.
9. Adler, Joseph L.: Simplification of Tidal Corrections for Gravity Meter Surveys, *Geophysics*, vol. 7, pp. 35–44, 1942.
10. Wyckoff, R. D.: Study of Earth Tides by Gravitational Measurements, *Trans. Am. Geophys. Union, 17th Annu. Meet.*, pp. 46–52, 1936.
11. Nettleton, L. L.: Gravity and Magnetic Calculations, *Geophysics*, vol. 7, pp. 293–310, 1942.
12. Bowie, William: "Isostasy," p. 69, Dutton, New York, 1927.
13. Heiland, C.: "Geophysical Exploration," pp. 151–153, Prentice-Hall, Englewood Cliffs, N.J., 1940.
14. Hammer, Sigmund: The Anomalous Vertical Gradient of Gravity, *Geophysics*, vol. 35, pp. 153–157, 1970.
15. Thyssen-Bornemisza, Stephen, and W. F. Stackler: Observation of the Vertical Gradient of Gravity in the Field, *Geophysics*, vol. 21, pp. 771–779, 1956.
16. Thyssen-Bornemisza, Stephen: Instrumental Arrangement to Measure Gravity with Gradients, *Geophysics*, vol. 35, pp. 713–715, 1970.
17. Wyckoff, R. D.: The Gulf Gravimeter, *Geophysics*, vol. 6, pp. 13–33, 1941.
18. LaCoste, L. J. B., Jr.: A New Type Long Period Vertical Seismograph, *Physics*, vol. 5, pp. 178–180, 1934; also see L. J. B. LaCoste and A. Romberg, U.S. Patent 2,293,437, Issued Aug. 18, 1942.
19. Hammer, Sigmund: Investigation of the Vertical Gradient of Gravity, *Trans. Am. Geophys. Union, 19th Annu. Meet.*, pp. 72–82, 1938.
20. Kuo, John T., Mario Ottaviani, and Shri K. Singh: Variations of Vertical Gravity Gradient in New York City and Alpine, New Jersey, *Geophysics*, vol. 34, pp. 235–248, 1969.
21. Pepper, T. B.: The Gulf Underwater Gravimeter, *Geophysics*, vol. 6, pp. 34–44, 1941.
22. LaCoste, L.: The Measurement of Gravity at Sea and in the Air, *Rev. Geophys.*, vol. 5, pp. 477–526, 1967.
23. Anonymous: First Sea Surface Gravimeter, IGY Bull. 8, *Trans. Am. Geophys. Union*, vol. 39, pp. 175–178, 1958.
24. Howell, Lynn G., K. O. Heintz, and A. Barry: The Development and Use of a High-Precision Downhole Gravity Meter, *Geophysics*, vol. 31, pp. 764–772, 1966.
25. McCulloh, Thane H.: Borehole Gravimetry: New Developments and Applications, *Proc. 7th World Petrol. Congr.*, vol. 2, pp. 739–734, Elsevier, London, 1967.

12

GRAVITY FIELD MEASUREMENTS AND REDUCTIONS

We shall now consider the techniques by which gravity data are obtained in land and marine surveys and converted into a form suitable for interpretation. Chapter 13 will deal with the final stage of gravity exploration, the extraction of geological information from the reduced gravity data.

The development of the equipment and techniques used for making gravity observations in the field has required a rare combination of engineering skill and operational efficiency. The instruments have been designed for ruggedness so they can operate reliably under difficult field conditions, and the procedures used in obtaining data with them have been designed for maximizing the speed with which traverses can be covered.

The operational techniques employed for surveys on land are different in many respects from those used in marine surveys. Since the early 1950s the level of activity in gravity exploration on land (based on the number of crews operating) has dropped substantially in the United States; the fractional decrease since the period of peak activity has been considerably greater than that for seismic work, as is evident from Fig. 1-4. Yet the total number of miles of gravity coverage obtained per year has been much greater since the late 1960s than at any time in the past.

The introduction of shipborne gravity meters has made it possible to record gravity at sea simultaneously with seismic reflection surveys, and a great deal of gravity data has been obtained in the course of such combined surveys.

The cost of making gravity observations and of acquiring seismic reflection data at the same time is only marginally greater than that of carrying out a gravity survey alone because the ship constitutes by far the largest item of expense in any kind of marine geophysical work.

12-1   GRAVITY MEASUREMENTS ON LAND

Determining positions for stations   The placement of stations in a gravity survey depends on two factors: (1) feasibility of access and (2) the spacing pattern necessary to detail the features which the survey was designed to locate and map.

Access depends on the nature of the terrain and the distribution of roads or other thoroughfares (such as streams) through the area being surveyed. Even when portable instruments like the Worden gravimeter are used, the rate of progress is very much impeded if motorized transportation is not available between stations. In most surveys, the objectives can be achieved if gravity stations are confined to existing roads or (in jungle areas) to waterways. This is of course the most economical approach.

The pattern of the stations is generally designed so that they will come as close to forming a rectangular grid as access conditions allow. The spacing within the grid is governed by the depth and lateral extent of the geological features being sought. To obtain the necessary detail on the gravity field of a salt dome 1 mi in diameter and 1 mi deep at its top surface, one might set up a grid $\frac{1}{2}$ mi on the side, although a spacing of 1 mi between stations would probably be adequate for establishing the existence and approximate location of the dome. For small targets such as ore bodies, much closer spacing is in order. A gravity survey in Cuba, where chromite bodies several hundred feet or less in width were sought, is described by Hammer et al.[1] Here the grid spacing, 20 m, was particularly small.

It is often desirable to shift the stations from their position on a uniform grid to avoid topographic features that would distort the gravity readings so greatly that the usual correction procedures for topographic irregularities (discussed on page 419) could not yield sufficiently precise results. A gravity reading, for example, on the summit of a small knoll would be affected much more by the mass of the knoll itself than would readings a short distance to the side, where the uncertainty in the exact shape and density of the knoll would not be likely to contribute nearly as much error. Hubbert[2] has prepared a series of charts showing the magnitude of the correction required at various distances from various kinds of topographic features. These charts were designed for field operators who might need guidance in deciding how far to shift gravity stations from the centers of such features.

Surveying Surveying expense is the largest item in the budgets of most gravity parties. Both elevation and geographical position must be known accurately. Spirit leveling is usually necessary, but in some cases barometric leveling is adequate. The methods of surveying and the cost involved vary greatly, depending on the ease of transportation and on the scale and quality of existing maps and level nets. The degree of necessary precision is established by the sensitivity of observed gravity to elevation and latitude. An error in elevation of 1 ft should change the gravity value at a station by 0.07 mgal, and an error of 100 ft in the north-south coordinate of position should result in an error of 0.03 mgal at middle latitudes. An accuracy of 0.1 ft in elevation and of less than 100 ft in horizontal control is thus required in surveying for gravity stations from which high precision is expected. Such accuracy can be achieved only by close spacing of transit positions and short chain lengths on slopes or by careful adjustment of theodolite data. The cost of such surveying is necessarily higher than that of topographic surveys made for most other kinds of geophysical work.

In regional or academically oriented surveys, where speed and economy are more important than the utmost attainable precision, it is often acceptable to determine elevations barometrically rather than geodetically. Instruments are available which measure differences in barometric pressure with much higher precision than the standard barometers used in geological surveys. Under favorable conditions they can respond to changes in atmospheric pressure equivalent to elevation differences as small as 0.1 ft. The precision of barometric measurements can be increased by comparing readings at field stations with simultaneous readings at bases located on corners of a triangle covering the area of operation. Lateral gradients thus observed can be interpolated for individual stations inside the triangle. Variations in barometric pressure can be removed in the same way that drift is removed from gravity readings.

Surveying and positioning techniques for offshore gravity measurements, whether with shipborne or water-bottom meters, are the same as those used for seismic surveys. The techniques involved were described in Chap. 5. The positioning requirements of shipborne gravity meters are somewhat different than for other offshore geophysical surveys as the need for precision in determining *speed* is greater than that in mapping the location where a reading was taken. This is because of the Eötvös effect, discussed later in this chapter, which requires that the ship's speed be known to a tenth of a knot if this effect is not to limit the overall accuracy obtainable from the gravity readings.

Transportation and operation of the instrument The heavier types of gravimeter, which generally use storage batteries for temperature regulation, are set up inside a truck or passenger automobile on a tripod which is lowered to the ground through holes in the floor. The heaters for thermostatic control required by such instruments are connected by cable to the vehicle's battery and operate continuously. There are many localities, however, where other means of transportation are required on account of special terrain conditions. In the North Slope of Alaska, gravimeters are flown from station to station in light airplanes on skis or pontoons. In the desert areas of Australia, transportation is by airplane and helicopter. On a gravity survey in the foothills of Alberta, described by Hastings,[3] a Gulf party transported its instruments by pack horses. For the marsh country of southern Louisiana, where wheeled vehicles cannot travel at all, the problem of transporting gravimeters and other geophysical equipment has been solved for more than three decades by marsh buggies like the one shown in Fig. 12-1. In some areas, finally, there seems to be no effective substitute for carrying gravimeters on foot. The close station spacing required in the Cuban chromite survey just described made this the only practical means of locomotion. A light instrument like the Worden gravimeter, which does not use batteries for temperature control, is especially suitable for such transport.

While helicopters have many advantages for transporting gravity meters in areas of difficult access, there can be serious problems in determining station positions with the requisite degree of accuracy, particularly as regards vertical control. Unless barometric altimeters are employed (and they will generally not be accurate

FIGURE 12-1
Marsh buggy used for gravity operations in swampy areas. (*Ardco.*)

enough to satisfy the requirements in any but regional surveys), it is possible to obtain elevations only by continuous surface traverses between stations. Electronic systems, e.g., tellurometers, will provide satisfactory horizontal control, but they are not designed to give the more critical elevation data except within the line of sight.

The surveyors normally keep several days ahead of the instrument operators, who find their station locations marked by flags which the surveying crew has left. At each station, the instrument must be carefully leveled before it can be read. It is standard procedure to take three readings in rapid succession at each station. When all stations are alongside roads, it is often possible to cover as many as a dozen of them in an hour.

Adjustment for drift   The instrumental readings as made in the field require a correction for the drift of the gravimeter (including earth-tide effects). All gravity

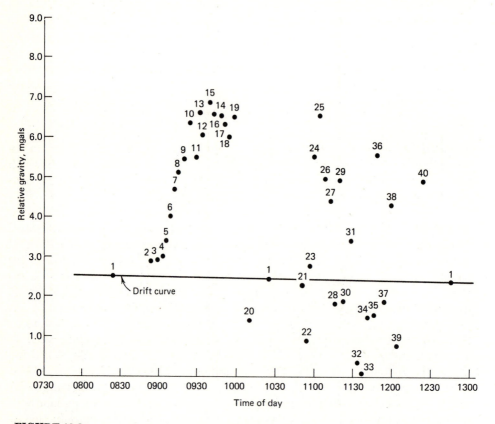

FIGURE 12-2
Typical plot for drift adjustment on day when 40 stations are occupied. Station 1, as base station, is reoccupied twice during the day. Drift curve based on readings at station 1. (*Petty-Ray Geophysical Division, Geosource Inc.*)

values observed during the course of a given day's operations, including those obtained in reoccupations of stations, are plotted against time. In surveys where high precision is desired, each station is reoccupied several times during the course of a day, and thus the instrument must be moved along the profile in a zigzag trajectory. Figure 12-2 shows a drift curve for a typical survey made over the portion of a day in which stations were occupied. The curve, a straight line, was drawn to pass through the points plotted for the original occupation and two reoccupations of station 1. If the gravity difference between stations 1 and any other station is wanted, one simply reads the vertical displacement of the plotted point from the drift curve at the time when the station of interest was occupied. The drift of the instrument used for the observations plotted in Fig. 12-2 was exceptionally small. For most instruments the drift tends to be somewhat larger, having magnitudes of the order indicated in Fig. 11-25.

For surveys where drift is large or irregular with respect to the precision desired, more complicated field techniques are necessary to eliminate its influence. In such cases an oscillating sequence of stations is established. A method proposed by Roman[4] requires four readings at each station. In the intervals between them, readings are taken at other nearby stations.

If stations lie along a profile in the order *a*, *b*, *c*, etc., they would be observed in the order *ab*, *abc*, *abcd*, *abcde*, *edef*, *defg*, etc. Numerical adjustment by standard methods such as least squares is then used to reduce the error resulting from drift to a minimum.

Regional gravity surveys with gravimeters Ever since the 1870s, when pendulums were used for gravity measurements, regional gravity surveys, in which station spacings are measured in tens or even hundreds of miles, have been conducted in many parts of the world for studying the figure of the earth as well as its large-scale geologic features. From such studies valuable information has been obtained which has improved our understanding of plate tectonics and crustal spreading. The early surveys with pendulums were cumbersome and only allowed a precision of several milligals. In recent years, the gravimeter has been used for regional surveys giving readings that are much more precise and convenient to obtain. Often specially designed geodetic gravimeters with a more extended range than those used in detailed exploration are employed for regional studies.

Gravity measurements well distributed around the world provide information that can be used to determine the vertical separation of the geoid and the spheroid at any desired point and hence the true shape of the earth. Stokes developed the theory for this type of analysis, which is summarized by Garland.[5] The denser and more widely distributed the gravity data the better the precision of the geoidal shape thus obtained. Regional gravity studies have been conducted on land and sea by various geodetic institutes with this objective in view. The introduction of guided missiles has given military motivation to the acquisition of precise information on the shape of the earth.

Academically oriented surveys are also carried out to gain a better understanding of regional geology than is obtainable from surface observations alone. Such surveys have been conducted in all parts of the world. The methods are similar to those used in oil exploration, but the spacing of stations is ordinarily greater and the elevation control is usually obtained either from existing bench marks or from barometric measurements. Several typical surveys carried out for purposes of this kind are described later in this chapter (pages 432 to 433).

Such gravity surveys, whether run for geodetic or geological purposes, are most useful when tied to reference stations where absolute gravity is established. The determination of absolute gravity by standard pendulum observations is cumbersome in the field, and for a long time there was a serious shortage of reference points established by this means. In the late 1940s, however, Woollard,[6] at the University of Wisconsin, initiated a project to set up a worldwide network of stations by flying long loops with specially designed Worden gravimeters having a total range of 5400 mgal. These provided ties to established pendulum stations where absolute gravity is known to a high degree of precision. Drift effects were eliminated by returning to the reference station after closing loops sometimes as much as 20,000 mi long. Woollard and Rose[7] have catalogued the stations for which gravity values were established during the course of this project.

12-2 MEASUREMENT OF GRAVITY AT SEA

The earliest measurements of gravity in water-covered areas were in the shallow protected bays and bayous of Louisiana. Tripods and platforms were initially constructed so that readings could be taken above the water level using land meters. For a while gravity meters and their operators were lowered to the water bottom (as much as 125 ft deep) in diving bells.

Since 1941, remote-controlled bottom meters have been employed for work both in protected and open waters. Shipborne meters, first introduced in the late 1950s, have been employed for most marine gravity measurements since the later 1960s. The precision obtainable from shipborne meters is not generally as great as that from bottom meters, but for most regional or reconnaissance surveys the accuracy of 1 mgal or better that can be obtained from shipborne surveys is quite adequate.

Observations with bottom meters   The LaCoste and Romberg bottom meter (LaCoste[8]) is by far the most widely used instrument of this type. As shown in Fig. 11-26, it is lowered into position at the water bottom from the deck of the ship by a boom which holds it far enough out from the ship's side to protect it from banging when it is lowered to or raised from the bottom. When the meter reaches the bottom, it is leveled by a built-in servo system, and the reading is displayed digitally on a panel which is generally located in the ship's wheelhouse.

The principal limitation in the accuracy of a bottom meter is from motion of the sea floor, usually from water waves. The wave motion gives rise to a pressure change at the water bottom which at long periods (such as those of tides) approaches the difference in the static pressure between highest and lowest water elevations. Within the usual range of periods for waves and swell, however, the pressure is much attenuated, the effect decreasing with increasing depth and decreasing wave period. Thus wave effects should not be of concern except in shallow water. A soft bottom will transmit more of the wave motion to the meter than a hard one; thus a lower accuracy can be expected when the meter is standing in mud. Automatic elevators will compensate for wave motion under moderate sea conditions at water depths greater than 25 ft.

It is important to have accurate measurements of water depth at the point where the bottom meter is situated so that corrections will be as precise as possible. A pressure gauge built into the gravimeter case is designed to give digital readings of water depth on the same indicator panel on which gravity is read.

The speed with which observations can be made and the time between successive station occupations depends on the water depth and the grid spacing. If good handling equipment is available, it should be possible to make as many observations in a day as can be obtained on land under ordinary operating conditions. Accuracy of gravity readings is generally no better than 0.1 mgal for a bottom meter, compared with a few hundredths of a milligal in a land meter.

Observations with shipborne meters There are a number of important differences, many of them obvious, between marine gravity surveys with bottom meters and those with meters on board ship. In the first type of survey, discrete readings are made at the individual stations at which the ship stops, just as in land operations. In shipborne surveys, continuous gravity readings are obtained while the ship is under way. It is important that the ship maintain as constant a speed and as straight a course as possible because changes in speed or direction give rise to spurious accelerations which affect the precision of the gravity measurements. The most important consideration, however, in operating a shipborne survey is accuracy in measuring speed and heading. These must be known with the greatest possible precision because of the Eötvös effect, which results from the fact that the east-west component of the ship's speed causes an increase or decrease in the earth's centrifugal acceleration, depending on whether the motion is with or against the direction in which the earth rotates. The Eötvös effect on gravity Δg_E depends on the velocity V of the ship, its latitude ϕ, and its heading α with respect to the north-south direction in accordance with the relation

$$\Delta g_E = 7.487V \cos \phi \sin \alpha \qquad (12\text{-}1)$$

the constant applying when V is in knots and Δg_E in milligals.

This formula is used to compute the Eötvös correction, which compensates for the effect. The correction is positive when the ship is moving to the east (because when it moves with the earth, centrifugal acceleration is increased and the downward pull is decreased) and negative when its motion is westward. An error of 1 knot in measuring the ship's speed when it is moving in the east-west direction at the equator results in an error of 7.5 mgal in gravity, which would be entirely unacceptable for most exploration purposes. As current speed-measuring techniques, such as those using continuous-wave electronic systems and Doppler sonar, have an accuracy that is seldom better than 0.1 knot under ideal conditions, it is evident that the precision obtainable in measuring the ship's speed can be the principal limitation in the accuracy of shipborne gravity work.

The precision that can be expected in shipborne gravity measurements also depends upon the state of the sea. In very calm seas, rms agreement at crossing points where the traverse lines are laid out in rectangular grids has been as good as 0.5 mgal. In heavier seas the agreement is in the range 0.7 to 1.0 mgal, but when the sea state is 4 or greater, it may be necessary to shut down gravity-measuring operations altogether if high accuracy is desired. The question of whether to continue gravity observations in rough seas can become complicated in combined seismic and gravity surveys. Seismic recording can be carried out successfully under sea conditions that will not allow the gravity instruments to operate properly. When this happens during such combined surveys, the decision must be made whether to do without gravity data over the traverse involved and continue the seismic recording or whether to shut down altogether. The importance of the gravity data in the overall exploration picture will usually determine which course is followed.

The output readings of the gravity system may be registered both on a pen-writing chart recorder and on digital tape. The visible chart record allows monitoring of results and gives an instantaneous check on how well the various components of the system are performing. Figure 12-3 shows a typical chart from a recorder used to monitor shipborne gravity measurements. It registers beam deflection, spring tension, total correction, and corrected gravity. Some shipborne charts record additional parameters, e.g., accelerometer outputs, cross-coupling, the ship's speed, and the water depth. Most systems record the data simultaneously on digital tape with formats designed for subsequent input into an electronic computer that will make all necessary corrections and present the corrected gravity values in a form suitable for mapping.

12-3 DETERMINATION OF DENSITIES

The terrain and Bouguer corrections made in the reduction of gravity data require a knowledge of the densities of the rocks near the surface. In many areas near-surface densities are sufficiently homogeneous for an average density value to be obtainable from a few well-spaced determinations. In others, there are such sharp local variations in lithology that use of an average density value can introduce considerable error. White[9] has described a gravity survey in Great Britain where three different densities had to be used for surface corrections within a small area.

When one wishes to determine density directly, representative samples of rock from surface outcrops, mines, or well cores and cuttings may be collected for measurement with a pyknometer or a Schwarz or Jolly balance.

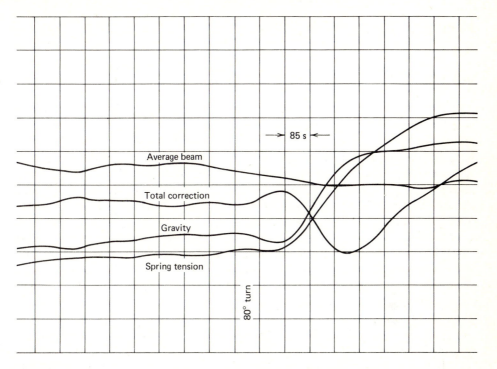

FIGURE 12-3
Typical chart made by pen recorder on board ship to monitor operation of
LaCoste and Romberg gravity meter. The *average-beam* trace is used to get the
slope correction, which is simply the derivative of the beam deflection. The
total correction is the cross-coupling correction plus the slope correction. The
gravity is the *spring tension* (bottom trace) minus the total correction. Filtering
of the correction curves results in time lags that must be taken into account
when the corrections are applied. (*Tidelands Geophysical Co.*)

Nettleton[10] has proposed an indirect means of density determination which may
be more satisfactory for gravity reductions than direct measurements on small
samples. It is illustrated in Fig. 12-4. A closely spaced gravity traverse is run over
a topographic feature, such as a small hill or valley, with dimensions that have
been measured accurately. When the profile of measured gravity is plotted, the
predicted gravity field of the surface feature alone is calculated at each observation
point along the profile and removed from the value observed at that point. The
calculation is repeated a number of times, different densities being assumed for
each computation. The density value at which the hill has the least observable
effect on the gravity profile is considered to be most nearly correct. This method
has the advantage of averaging the effect of density variations more accurately
than can be done from surface samples. Even so, it gives information on densities
only at relatively shallow depths and can be used only when the near-surface

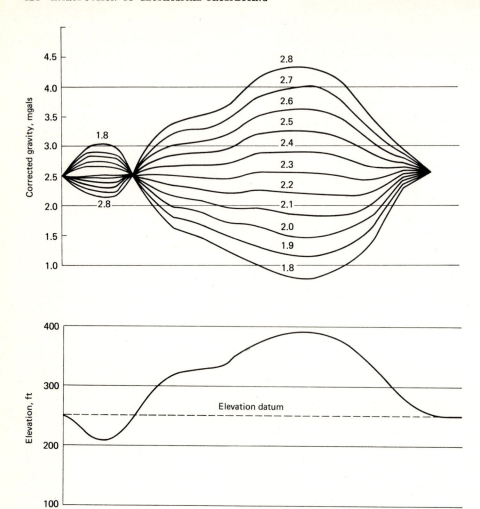

FIGURE 12-4
Nettleton method for determining density of near-surface formations by use of
different densities in making Bouguer corrections over topographic feature.
Corrections are referred to elevation of 250 ft. Correct density appears to be
about 2.3 g cm⁻³.

lithology is homogeneous. Since in many areas the topographic features owe their
existence to outcrops with anomalous lithology, the density value so determined
may not be correct.

The density logger For some years, tools have been available for logging
formation densities in boreholes (see Baker[11]). Logs thus obtained indicate back-
scattered gamma radiation which is a simple function of formation density. The
density logger consists of a radiation source, usually cobalt 60, at one end of the

tool and a detector, generally a geiger counter, about 18 in away at the other end, as shown in Fig. 12-5. The outer wall of the tool is lined with lead shielding, which has two slits so positioned that the only radiation from the source which reaches the detector is that deflected back from the formation by Compton scattering. The principles of Compton scattering are reviewed by Faul and Tittle.[12] The amplitude of the scattered radiation depends on the electron concentration in the formation, and this in turn is roughly proportional to the density of the formation material. The energy is proportional to $1 - \cos \theta$, where θ is the angle between the incident and the scattered ray. This indicates that the maximum energy is at $\theta = 90°$, corresponding to emergent and returning rays which each make an angle of 45° with the borehole wall.

The maximum penetration of the gamma radiation into the borehole wall is 6 in, and the effective sampling volume for each reading is about 1 ft³. Close contact is maintained between the tool and one side of the hole by a spring on the other side. This may not be effective in zones where shale is washed out. Actually, comparisons of logged densities and densities measured from cores at the same level in the borehole generally show an agreement within several hundredths of a gram per cubic centimeter in all formations except shales, which are particularly subject to

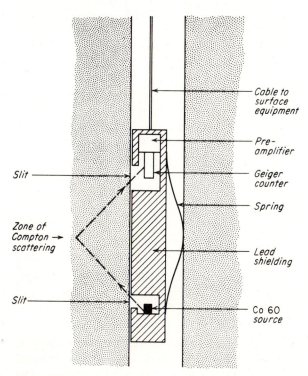

FIGURE 12-5
Schematic diagram of gamma-ray density logger.

caving from washouts. The irregularity of the log itself is usually so great that it is necessary to smooth values by averaging over depth ranges of 100 ft or more to get results that are useful for gravity interpretation. In soft formations, frequent washing out of the borehole walls affects the reliability of the density readings. Thus, where there is any substantial amount of shale in the section the integrated density would be of questionable accuracy. Moreover, the small volume of formation sampled in a single reading may not be representative of the much larger volume of rock contributing to the observed gravity. This hazard also holds for all density determinations made in the laboratory whether from core samples or cuttings out of wells.

Borehole gravity readings The principal application of the borehole gravity meters described in the preceding chapter (see pages 399 to 400) is to determine density versus depth over a greater volume of the formation than is possible with the limited lateral penetration which a gamma-ray logger allows. There are a number of applications for the density information thus obtained. For production and reservoir studies, it is easy to determine porosity from the density if the fluid content and saturation are known or assumed. McCulloh[13] has shown how borehole density data can be used to determine zones of gas saturation. Most important for our purposes, the data can be used as an aid in the interpretation of gravity readings obtained at the earth's surface.

The density is computed from the differences between gravity readings at two depths (generally with separations of the order of 100 ft) in the borehole. After corrections are made for free-air, topographic, and other known effects which might account for some portion of the observed differential, the remaining gravity difference can be associated with the attraction of a fictitious slab (considered to be infinite) of earth material respectively bounded at its top and bottom by horizontal planes in which the two readings are taken. The formula for the density ρ of the slab bounded by horizontal surfaces between the levels of boundaries Δh apart, as shown in Fig. 12-6, is expressed in terms of the difference Δg in gravity at those levels:

$$\rho = \frac{0.094\Delta h - \Delta g}{0.0254\Delta h} \tag{12-2}$$

where ρ is in cgs units when Δh is in feet and Δg is in milligals. The large depth intervals usually involved here and the great lateral extension of the slab make densities determined in this way more applicable to gravity interpretation than those observed by other means such as density logging.

12-4 REDUCTIONS OF GRAVITY DATA

In order to be most useful in prospecting, gravity data as obtained in the field must be corrected for elevation, the influence of nearby topography, and latitude.

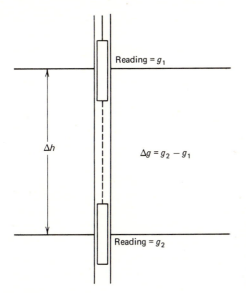

Reading = g_1

Δh

$\Delta g = g_2 - g_1$

Reading = g_2

FIGURE 12-6
Determination of density of an "infinite" slab penetrated by a borehole by taking readings with a borehole gravity meter at its top and bottom surfaces.

The corrections are in principle quite similar to those used for reducing absolute-gravity values to the geoid (see pages 368 to 370). In general all gravity values are reduced to a datum plane, but this plane need not necessarily be the sea-level surface. Absolute values of gravity are seldom presented on gravity maps made for exploration purposes; in general it is only necessary that all values shown on a map be referred to the same datum.

Elevation correction Let us assume that gravity observations are made at two stations S_1 and S_2, each located on terrain that is entirely flat except for the cliff midway between them, as shown in Fig. 12-7a. The elevation of S_1 above sea level is h_1, and that of S_2 is h_2. The stations are so far apart that each is effectively on a horizontal surface of infinite extent. Because the elevations are different, however, there would be a difference in the two gravity readings which must be removed to avoid an indication of a subsurface structure that is not really there. This source of error can be removed very simply. A datum plane is introduced with an elevation above sea level of d, all material above the datum plane being effectively "removed" so that the readings of both instruments are adjusted to be the same as if they were on the datum surface. The adjustment actually consists of two parts: (1) the *free-air correction*, which accounts for the fact that each station is a different distance from the earth's center than the datum plane, and (2) the *Bouguer correction*, which removes the effect of a presumed infinite slab of material between the horizontal plane of each station and the datum.

The free-air correction is based on the fact that the attraction of the earth as a whole can be considered to be the same as if its mass were concentrated at its center. If the elevation of a gravity meter is changed, its distance from the center

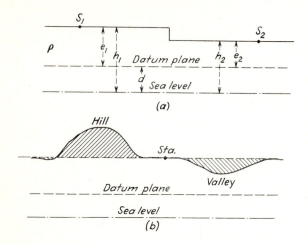

FIGURE 12-7

Elevation and terrain corrections: (*a*) Bouguer correction (to datum plane) at S_1 is $2\pi\gamma\rho e_1$, or $0.025e_1$ if ρ is 2 and e_1 is in feet. At S_2 it is $2\pi\gamma\rho e_2$. Free-air correction is $0.094e_1$. (*b*) Terrain correction removes effect of hill by adding its upward attraction at station and compensates for valley by adding attraction it would exert at station if filled in. Result is to flatten all topography to level of station, so that Bouguer correction will effectively place station on datum.

of the earth changes by the same amount. The inverse-square law enables us to predict how much the acceleration of gravity will change as a result.

If the radius of the earth (at sea level) is R, the height above sea level is h, and the value of gravity at sea level is g_0, then g, the gravity at h, can be expressed as

$$g = g_0 \frac{R^2}{(R + h)^2} = g_0 \frac{1}{(1 + h/R)^2} = g_0 \frac{1}{1 + 2h/R + h^2/R^2} \tag{12-3}$$

$$g \approx g_0 \frac{1}{1 + 2h/R} \qquad \text{as} \quad \frac{h^2}{R^2} \ll \frac{2h}{R} \tag{12-4}$$

Using the first term of the binomial expansion

$$g \approx g_0 \left(1 - \frac{2h}{R}\right) \tag{12-5}$$

one obtains

$$g - g_0 \approx \Delta g \approx - \frac{2hg_0}{R} \tag{12-6}$$

If R, the radius of the earth, which is about 4000 mi, is expressed in feet and g_0, about 980 cm/s², in milligals, it is easy to show that $\Delta g/h$ is 0.09406 mgal/ft.

This effect will be independent of whether or not there is any rock material between the sea-level datum and the station at elevation h. It is referred to for this reason as the *free-air* effect, and the portion of the elevation correction which compensates for it is referred to as the *free-air correction*. If h is positive, i.e., the elevation is above sea level, the effect is to lower the earth's attraction. To compensate for this lowering the free-air correction must be *added*.

If the station is located on an extended horizontal surface having an elevation h above sea level, one must also correct for the attraction of the slab of material having a bottom surface at sea level and a top surface at the elevation of the station. If the material in this slab has a density of ρ, this attraction will be [according to the slab formula, Eq. (11-24)] $2\pi\gamma\rho h$ or $0.01277 \rho h$ (with h in feet); for a density of 2.0 cgs units, the correction will be 0.02552 mgal/ft. This is referred to as the *Bouguer correction*, and it is subtracted from the observed reading. It is evident that this correction is very much dependent upon the density assumed for the material in the slab between the observing station and the datum.

The elevation correction is the sum of the free-air and Bouguer corrections, and for a near-surface material of density 2.0 cgs units it will be 0.0940 − 0.0255, or 0.0685 mgal/ft. For stations above the datum plane, whether at sea level or at some other elevation, the correction is added. For stations below the datum plane, it is subtracted.

Terrain correction In most gravity surveys, the terrain in the vicinity of a station is sufficiently flat for the elevation correction to provide adequate compensation for topographic effects. Nearby hills rising above the level of the station, however, give an upward component of gravitational attraction that counteracts a part of the downward pull exerted by the rest of the earth. Also, nearby valleys below the level of the station correspond to holes in the slab between station and datum level which are responsible for a smaller downward pull at the station than is accounted for by the Bouguer correction. It is necessary to compensate for such effects if the topographic features causing them are sufficiently close to cause distortions in the observed gravity that are large enough to affect the interpretation of anomalies from buried features of interest.

The usual procedure (Fig. 12-7b) in making corrections for such distortions is to calculate the attraction of all the mass that would have to be added to the valleys below and all that would have to be removed from the hills above to give perfectly flat topography having the same elevation as the station. The terrain so reconstituted would then correspond to the slab assumed in making the Bouguer correction. Either correction, whether from a hill or a valley, is *added* to the observed gravity. Calculation of the attraction of irregular topographic elements is greatly facilitated by the use of special templates and tables designed for the purpose. The first terrain-correction chart and set of tables published by Hammer[14] is still in widespread use. His chart, printed on transparent sheeting, is superimposed on a topographic map of the area around the gravity station and has the same scale as the chart. It consists of a series of concentric circles with radial lines dividing the zones between the circles into compartments, as shown in Fig. 12-8. The center of the circles is placed over the gravity station on the map. Hammer's original tables

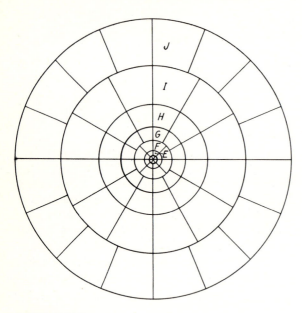

FIGURE 12-8
Terrain correction-zone chart designed by Hammer,[14] used in conjunction with
Table 12-1 for zones through *J*. Scale is $\frac{1}{175\,000}$. (*Gulf Research & Develop-
ment Co.*)

have been extended by Douglas and Prahl[15] to allow corrections over a greater
range of elevations in each zone.

Using the chart, the computer estimates from contours on a topographic map the
average elevation of the land surface within each compartment and, taking the
difference between this and the station elevation, determines the gravity effect of
the prism of land within that compartment from Table 12-1, compiled by Hammer
for the purpose. The correction is always added, regardless of the sign of the difference, except for the rare case of features so distant that earth curvature enters in.
The outermost circle of the chart corresponds to a distance on the map of about 14
mi, whereas the innermost circle (bounding zone *E*) represents a distance of only
558 ft. Topography closer to the station than this can be corrected for when necessary by using another chart constructed on the same principle but with a different
scale.

Often the terrain in the neighborhood of a gravity station has a systematic slope
which can be approximated by an inclined plane. Where this is the case, the topographic correction will be more accurate if the blocks corresponding to the zones
in Hammer's tables are assumed to have sloping instead of horizontal tops. Sandberg[16] has prepared a set of tables to be used with Hammer's chart which is computed for all integral slope angles from 1 to 30°, inclusive.

Latitude correction   As shown on page 364, the International Formula for the
variation with latitude ϕ of normal gravity g, in gals, on the International Ellipsoid

is

$$g = 978.049(1 + 0.0052884 \sin^2 \phi - 0.0000059 \sin^2 2\phi) \qquad (12\text{-}7)$$

Let us differentiate this equation with respect to ϕ, converting the angular differential to one in terms of distance. In this way we obtain the following expression for w, the rate of change of gravity with north-south displacement as a function of latitude:

$$w = 1.307 \sin 2\phi \qquad \text{mgal/mi} \qquad (12\text{-}8)$$

At 45° latitude, the variation is about 0.1 mgal for each 400 ft of north-south displacement. Even for such a short distance, this large a gravity change could be significant in exploration surveys if no correction for it were made. And it indicates that positions must be known to better than 100 ft in the north-south direction where an ultimate precision of 0.02 mgal is desired. In practice, an arbitrary reference latitude is chosen in the vicinity of the survey area, and all readings are corrected to this latitude. If the actual latitude is within a degree or so of this reference, a uniform gradient can be assumed and all stations can be corrected simply by multiplication of the gradient computed for the reference latitude by the north-south distance of the station from the reference line. For surveys involving a larger range of latitudes, variation tables like those of Lambert and Darling[17] can be employed. These give the values of theoretical gravity on the International Ellipsoid for every 10 minutes of latitude from the equator to the pole. Interpolation formulas make it possible to obtain a correction that is accurate enough for commercial gravity work.

Earth-tide correction   As pointed out on pages 372 to 373, the normal value of gravity at any point will vary cyclically during the course of the day by as much as 0.3 mgal because of the tidal attraction of the sun and the moon. In a high-precision survey, this much variation might well be a significant source of error in the relative gravity between two points at which measurements are made at different times. There are two methods of correcting for the tidal effect. One is to construct daily charts of the tidal variation in gravity with time from readings on a stationary instrument and to correct all readings in the field by means of such charts. The more usual method is for the observer to return to the base station so often that earth-tide effects will be fully incorporated into the instrumental drift curve.

Use of reference data in mapping gravity   All gravity-measuring instruments give readings convertible to *relative* values of gravity. In an exploration-type survey covering an isolated area of limited size it may only be necessary to assign an arbitrary value (such as zero) to the gravity at some randomly chosen station and refer all other readings to one at this station. In regional surveys it is common to refer all values to that predicted for the latitude of the station by the International Formula and to map the differences (after appropriate corrections have been applied) as anomalies of various types. But for tying together surveys over adjacent areas it is very useful to have gravity base stations which are themselves tied into a large-scale grid in the same way as geodetic base stations. They may have to be

Table 12-1 TERRAIN-CORRECTION TABLES TO BE USED WITH FIG. 12-8*

Zone B, 4 compartments, radius 6.56–54.6 ft		Zone C, 6 compartments, radius 54.6–175 ft		Zone D, 6 compartments, radius 175–558 ft		Zone E, 8 compartments, radius 558–1280 ft		Zone F, 8 compartments, radius 1280–2936 ft		Zone G, 12 compartments, radius 2936–5018 ft	
± h, ft	T	± h, ft	T	± h, ft	T	± h, ft	T	± h, ft	T	± h, ft	T
0–1.1	0	0–4.3	0	0–7.7	0	0–18	0	0–27	0	0–58	0
1.1–1.9	0.1	4.3–7.5	0.1	7.7–13.4	0.1	18–30	0.1	27–46	0.1	58–100	0.1
1.9–2.5	0.2	7.5–9.7	0.2	13.4–17.3	0.2	30–39	0.2	46–60	0.2	100–129	0.2
2.5–2.9	0.3	9.7–11.5	0.3	17.3–20.5	0.3	39–47	0.3	60–71	0.3	129–153	0.3
2.9–3.4	0.4	11.5–13.1	0.4	20.5–23.2	0.4	47–53	0.4	71–80	0.4	153–173	0.4
3.4–3.7	0.5	13.1–14.5	0.5	23.2–25.7	0.5	53–58	0.5	80–88	0.5	173–191	0.5
3.7–7	1	14.5–24	1	25.7–43	1	58–97	1	88–146	1	191–317	1
7–9	2	24–32	2	43–56	2	97–126	2	146–189	2	317–410	2
9–12	3	32–39	3	56–66	3	126–148	3	189–224	3	410–486	3
12–14	4	39–45	4	66–76	4	148–170	4	224–255	4	486–552	4
14–16	5	45–51	5	76–84	5	170–189	5	255–282	5	552–611	5
16–19	6	51–57	6	84–92	6	189–206	6	282–308	6	611–666	6
19–21	7	57–63	7	92–100	7	206–222	7	308–331	7	666–716	7
21–24	8	63–68	8	100–107	8	222–238	8	331–353	8	716–764	8
24–27	9	68–74	9	107–114	9	238–252	9	353–374	9	764–809	9
27–30	10	74–80	10	114–120	10	252–266	10	374–394	10	809–852	10
		80–86	11	120–127	11	266–280	11	394–413	11	852–894	11
		86–91	12	127–133	12	280–293	12	413–431	12	894–933	12
		91–97	13	133–140	13	293–306	13	431–449	13	933–972	13
		97–104	14	140–146	14	306–318	14	449–466	14	972–1009	14
		104–110	15	146–152	15	318–331	15	466–483	15	1009–1046	15

Zone H, 12 compartments, radius 5018–8578 ft		Zone I, 12 compartments, radius 8578–14,662 ft		Zone J, 16 compartments, radius 14,662–21,826 ft		Zone K, 16 compartments, radius 21,826–32,490 ft		Zone L, 16 compartments, radius 32,490–48,365 ft		Zone M, 16 compartments, radius 48,365–71,996 ft	
± h, ft	T	± h, ft	T	± h, ft	T	± h, ft	T	± h, ft	T	± h, ft	T
0–75	0	0–99	0	0–167	0	0–204	0	0–249	0	0–304	0
75–131	0.1	99–171	0.1	167–290	0.1	204–354	0.1	249–431	0.1	304–526	0.1
131–169	0.2	171–220	0.2	290–374	0.2	354–457	0.2	431–557	0.2	526–680	0.2
169–200	0.3	220–261	0.3	374–443	0.3	457–540	0.3	557–659	0.3	680–804	0.3
200–226	0.4	261–296	0.4	443–502	0.4	540–613	0.4	659–747	0.4	804–912	0.4
226–250	0.5	296–327	0.5	502–555	0.5	613–677	0.5	747–826	0.5	912–1008	0.5
250–414	1	327–540	1	555–918	1	677–1119	1	826–1365	1	1008–1665	1
414–535	2	540–698	2	918–1185	2	1119–1445	2	1365–1763	2	1665–2150	2
535–633	3	698–827	3	1185–1403	3	1445–1711	3	1763–2086	3	2150–2545	3
633–719	4	827–938	4	1403–1592	4	1711–1941	4	2086–2366	4	2545–2886	4
719–796	5	938–1038	5	1592–1762	5	1941–2146	5	2366–2617	5	2886–3191	5
796–866	6	1038–1129	6	1762–1917	6	2146–2335	6	2617–2846	6	3191–3470	6
866–931	7	1129–1213	7	1917–2060	7	2335–2509	7	2846–3058	7	3470–3728	7
931–992	8	1213–1292	8	2060–2195	8	2509–2672	8	3058–3257	8	3728–3970	8
992–1050	9	1292–1367	9	2195–2322	9	2672–2826	9	3257–3444	9	3970–4198	9
1050–1105	10	1367–1438	10	2322–2443	10	2826–2973	10	3444–3622	10	4198–4414	10
1105–1158	11	1438–1506	11	2443–2558	11				
1158–1209	12	1506–1571	12	2558–2669	12				
1209–1257	13	1571–1634	13	2669–2776	13				
1257–1305	14	1634–1694	14	2776–2879	14				
1305–1350	15	1694–1753	15	2879–2978	15				

SOURCE: S. Hammer, *Geophysics*, vol. 4. pp. 190–191, 1939. Reproduced by permission of the Society of Exploration Geophysicists.

* Each zone is a circular ring of given radii (in feet) divided into 4, 6, 8, 12, or 16 compartments of arbitrary azimuth. h is the mean topographic elevation in feet (without regard to sign) in each compartment with respect to the elevation of the station. The tables give the correction T for each compartment due to undulations of the terrain in units of $\frac{1}{100}$ mgal for density $= 2.0$. This correction, when applied to Bouguer anomaly values which have been calculated with the simple Bouguer correction, is always positive.

established for the purpose, making use of ties to stations where absolute gravity has been determined directly or indirectly. Or it may be possible to make use of the network of absolute-gravity values that Woollard and Rose[7] have established. When such ties are used, it may be most feasible to map absolute gravity or to use values based on an arbitrary reference so that smaller numbers will appear on the map.

Where the data are tied to more than one base station, there are likely to be misclosures, which must be adjusted in the same way as misclosures in other kinds of surveying.

Conclusions The object of all corrections is to obtain a picture of the variations in the earth's gravitational field which depends only on lateral departures from constancy in the densities of the subsurface rocks below the datum plane. The most likely source of error in the corrections usually lies in the choice of the near-surface density to employ in the Bouguer and terrain corrections, particularly where the lithology of the near-surface formations is not well known. Sometimes the terrain corrections are inadequate because of limited topographic data near the gravity stations in areas of high relief.

12-5 TYPICAL GRAVITY ANOMALIES FOR VARIOUS GEOLOGICAL FEATURES

After appropriate corrections are made, gravity data are generally presented on maps or cross sections. As the corrections generally do not involve any substantial exercise of judgment, presentations of these kinds which incorporate only free-air, Bouguer, and latitude corrections are looked upon as objective. Interpretation enters in when regional effects are removed to point up residual anomalies associated with the subsurface sources of primary interest. Methods for doing this will be discussed in the next chapter. Bouguer maps are based on gravity data to which both the free-air and Bouguer correction have been applied. Such maps are commonly considered to represent "raw gravity." The information on them is used as the basis for separating anomalies by various analytical techniques which we shall consider later. All such techniques require a certain degree of interpretation, regardless of whether they involve graphical or computer-based methods.

In the presentation of gravity results it is desirable where possible to display the initial Bouguer map along with any residual maps derived from them so that the reader will be able to see the corrected data before the residual gravity was extracted from them.

In this section we shall examine gravity patterns over a number of typical geological features, both regional and local. These examples include maps and cross sections for which Bouguer, free-air, and latitude corrections have been made but for which regional effects have generally not been removed.

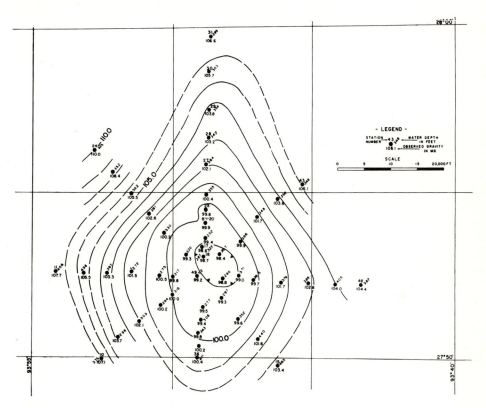

FIGURE 12-9

Gravity map obtained over inferred salt dome causing anomaly in water-bottom topography in Gulf of Mexico. Contours 1 mgal apart. (*After Nettleton.*[18])

Salt domes　Figure 12-9 shows the gravity picture mapped by Nettleton[18] over a topographic mound in the Gulf of Mexico about 150 mi off Galveston and a short distance inside the edge of the continental shelf. The water depth on strike with the center of the mound is 400 ft, but pinnacles project from the mound to depths as little as 60 ft. The origin of the topographic feature was not established until the gravity survey indicated a large closed minimum coincident with the contours of the elevated mound that could be accounted for only by assuming a piercement-type salt dome. The survey was not extensive enough to define the gravity closure on all sides, but judicious extrapolation indicates the maximum negative anomaly to be about 9 mgal. Figure 12-10 shows Nettleton's interpretation of the salt structure giving rise to the anomaly. This is not strictly a "known" source for the anomaly, but the near-surface geology of the coast and continental shelf of the Gulf of Mexico is so well established from extensive drilling that no identification for such a gravity feature other than a salt dome would be geologically plausible. Because the top of the dome is so shallow, it is probable that the uppermost part of

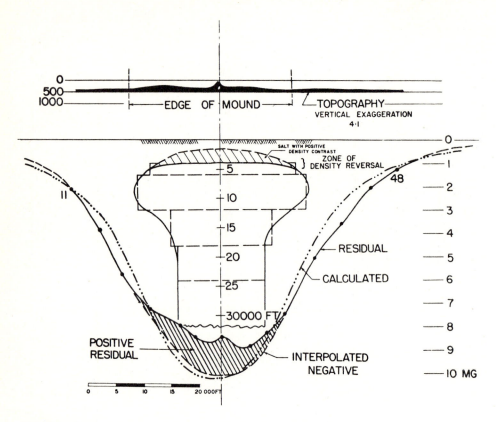

FIGURE 12-10
Assumed structure of salt dome believed to be causing offshore gravity anomaly
of Fig. 12-9. Agreement between observed and calculated gravity profiles supports choice of model for structure. (*After Nettleton.*[18])

the salt (crosshatched in the figure) has a higher density than that of the surrounding sediments at the same level and hence gives rise to a positive anomaly above
the center of the dome. This feature is referred to as the Way Dome.

The Grand Saline salt dome in east Texas, with a top surface averaging only
250 ft in depth, has been mined extensively. Figure 12-11 shows observed gravity
contours over the feature. The cap rock over Grand Saline is only 28 ft thick, so
thin that no effects from it are observable on the gravity map.

Anticlines When the geologic section contains successive formations having
appreciable contrasts in density, any major folding should be reflected in the gravity
picture. If formations having greater than average density are brought nearer the
surface at the crest of an anticline, its crest line should be the axis of a gravity
maximum. If beds of less than normal density are uplifted, there should be a
gravity minimum along the axis.

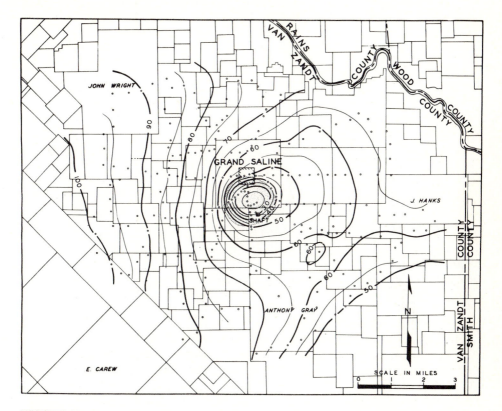

FIGURE 12-11
Gravity contours over Grand Saline salt dome. Free-air, Bouguer, and latitude
corrections have been made, but there is no adjustment for regional trend. Each
gravity unit is 0.1 mgal, so that contour interval is 1 mgal. Stippled area represents position of salt mass. (*After J. W. Peters and A. Dugan, "Geophysical Case
Histories," vol. 1, Society of Exploration Geophysicists, Tulsa, Okla., 1949.*)

The Kettleman Hills–Lost Hills area of California (Fig. 12-12) shows both types
of gravity feature over two parallel anticlines associated with the same structural
trend. The Kettleman Hills trend has a prominent gravitational high along its entire
length, which extends southward from the Fresno-Kings county line at the northwest of the map across the Kings-Kern county line, terminating about a township
beyond. The Lost Hills structure, which begins only a few miles farther south,
marks the axis of a pronounced gravity low, which is quite conspicuous in the
southernmost two townships of the map. Boyd[19] explains the Kettleman Hills high
as resulting from anomalously dense shales and sands of the 600- to 800-ft-thick
Reef Ridge formation. The minimum over the Lost Hills is attributed to a considerable thickness of very light diatomaceous shale, which is found in the same
formation at this point. Figure 12-12 illustrates the association between the structure and its gravity anomaly.

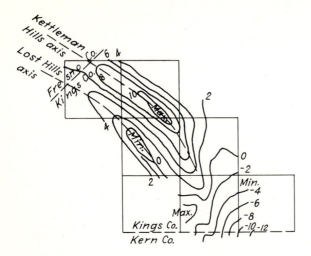

FIGURE 12-12
Contours from gravity survey, Kettleman Hills–Lost Hills area, Calif. Contour
interval, 2 mgal. (*After L. H. Boyd, "Geophysical Case Histories," vol. 1, Society
of Exploration Geophysicists, Tulsa, Okla., 1949.*)

Often the gravity contours over an anticline do not show the closure characteristic of the structure because of regional gravity trends. Coffin[20] shows the gravity
map obtained over the Altus Oil Field in Oklahoma. Here the gravity contours
(Fig. 12-13) exhibit nosing instead of closure over the highest portion of the anticline in which oil is trapped. Means for removing such regional gravity variations,
which can easily mask significant structures, will be discussed in the next chapter.

Limestone reefs When oil was first discovered in ancient limestone reefs, it
became desirable to determine whether such features could be located by gravity
surveys. This would be possible only if there is a density contrast between the reef
limestone and the sedimentary formations that surround it. Unfortunately, the
density and porosity of reef limestone and the material it replaces can be so variable
that it is not possible to specify what the contrast will generally be. Nevertheless,
significant gravity patterns may be associated with reef masses in specific geological
provinces. Brown[21] conducted an experimental gravity survey over the productive
Jameson reef in Coke County, Tex. The gravity effects he attributed to the reef
limestone were so small as to be observable only with most stringent standards of
accuracy in the surveying and instrumental operations. His residual contours indicated a high over the reef of about 0.3 to 0.5 mgal surrounded by an even smaller
annular low.

Pohly[22] has reported on a gravity survey conducted in Ontario which resulted in
the discovery of a productive reef. Here the reefs are exceptionally shallow (less
than 2000 ft deep) and are surrounded by salt, which has a density much lower
than that of reef limestone. The residual high over the reef was several tenths of a

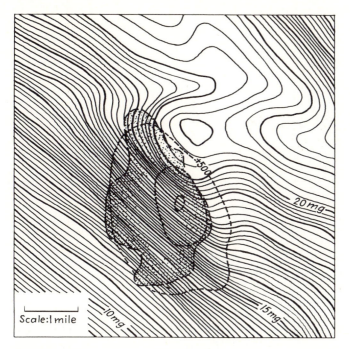

FIGURE 12-13
Observed gravity over Altus pool in Oklahoma. Note strong regional trend.
Depth contours (*dashed*) show structure on top of Canyon Formation at 100-ft
intervals. Stippling bounds area of production. (*After Coffin.*[20])

milligal. The most conspicuous gravity effect over such a reef is the low attributable
to the salt in association with the reef limestone rather than to a high from the reef
material itself. Ferris[23] and Nettleton[24] show examples of diagnostic gravity effects
from reefs in many areas of the United States and Canada.

Ore bodies   During World War II the Gulf Oil Corporation carried out a gravity
survey to locate chromite ores in the Camaguey district of Cuba. The high density
of the mineral, averaging 3.99 g/cm³, made it a particularly favorable prospect for
gravity exploration. Preliminary calculations, which showed that anomalies having
a relief as small as 0.05 mgal might be commercially significant, indicated that extremely stringent standards of accuracy would be necessary in the field work. A
station spacing of 20 m was chosen, and particular pains were taken in both the
gravity readings and surveying to secure the necessary resolution. The probable
error of a single observation was 0.016 mgal. Figure 12-14 shows the gravity anomaly
obtained during this survey over a known chromite deposit.

   In prospecting for many minerals, gravity surveys are carried out in conjunction
with other geophysical surveys such as magnetic or electrical. In a case history on
the discovery of the Pyramid ore bodies in the Northwest Territories of Canada,

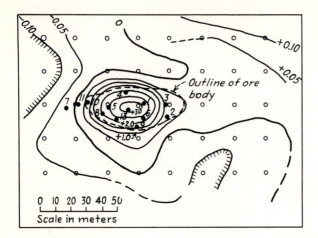

FIGURE 12-14
Gravity anomaly over known chromite deposit. Contour interval 0.05 mgal.
Open circles show locations of gravity stations; solid circles, of drill holes.
(*After Hammer et al.*[1])

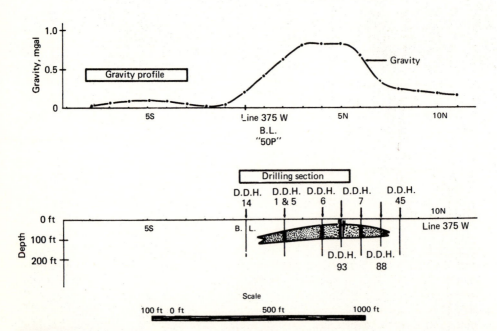

FIGURE 12-15
Cross section showing gravity anomaly over lead-zinc ore body in Pine Point
area, Northwest Territories, Canada. (*After Seigel et al.*[25])

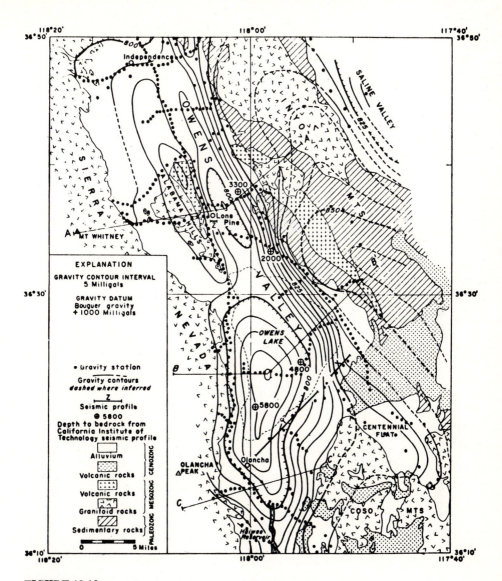

FIGURE 12-16
Bouguer gravity over southern Owens Valley of California and its relation to
the geology of the area. Gravity low over alluvium-covered area interpreted as
coming from thick sedimentary trough. (*After Kane and Pakiser.*[26])

Seigel et al.[25] show how gravity was used to obtain detailed information on the
structure of a lead-zinc ore body which had initially been discovered by induced-
polarization techniques. The density of the lead-zinc ore is 3.95 g/cm³, while
that of the surrounding limestones and dolomites is 2.65. The contrast resulted
in a gravity picture quite diagnostic of the ore body's boundaries, as is indicated

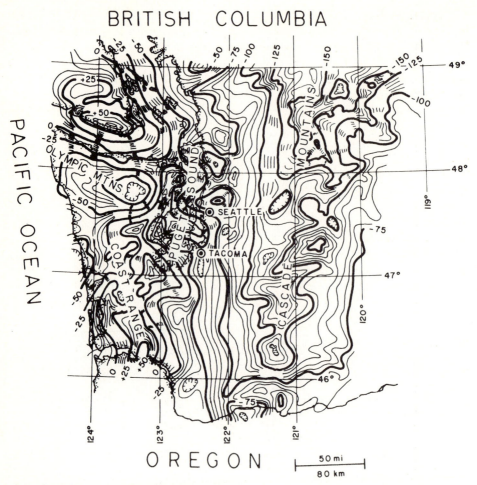

FIGURE 12-17
Regional Bouguer gravity over western Washington. Contour interval, 5 mgal.
(*After Daneš.*[27])

in Fig. 12-15. The depth to the top of the body is about 30 ft, and its maximum thickness is about 80 ft.

Regional surveys Many gravity surveys are undertaken to study regional geology rather than to detect or delineate individual features such as salt domes, anticlines, or reefs. Such surveys may be academically oriented or they may have been designed to obtain reconnaissance information about little-explored areas of economic interest. Information of this kind can serve as the basis for planning subsequent geophysical or geological surveys that might yield more detailed information.

A typical regional study for which gravity provided useful data was carried out by Kane and Pakiser[26] in the southern part of the Owens Valley in California. Here again it was desired to determine the structure of the bedrock surface below great thicknesses of clastic cover in an intermontane valley. Figure 12-16 is a gravity map over the valley. The contours show an unsymmetrical closed low having its axis along the center of the valley. The interpretation of the data indicated that the valley bedrock is faulted and is covered with lower-density clastics which reach a maximum thickness of almost 10,000 ft.

Regional studies of gravity sometimes show major structural trends over large areas and make it possible to demonstrate relationships between major geological features where there are no exposures on the surface. Figure 12-17 shows a map of Bouguer gravity in western Washington. The contours indicate a low trend coincident with the Cascades, a high trend along the Coast Range, and lows along Puget Sound and the Olympic Mountains. The gravity information makes it possible to distinguish the various geologic provinces associated with these features and to determine the relationships of faults and intrusives in the area. Combining such information with other data makes it possible to deduce the geologic history of the region with greater reliability.

REFERENCES

1. Hammer, Sigmund, L. L. Nettleton, and W. K. Hastings: Gravimeter Prospecting for Chromite in Cuba, *Geophysics*, vol. 10, pp. 34–39, 1945.
2. Hubbert, M. King: Gravitational Terrain Effects of Two-dimensional Topographic Features, *Geophysics*, vol. 13, pp. 226–254, 1948.
3. Hastings, W. K.: Gravimeter Observations in the Foothills Belt of Alberta, Canada, *Geophysics*, vol. 10, pp. 526–534, 1945.
4. Roman, Irwin: An Observational Method to Overcome Zero Drift Error in Field Instruments, *Geophysics*, vol. 11, pp. 466–490, 1946.
5. Garland, G. D.: "The Earth's Shape and Gravity," Pergamon, Oxford, 1965.
6. Woollard, G. P.: Recent Regional Gravity Surveys, *Trans. Am. Geophys. Union*, vol. 29, pp. 727–738, 1948.
7. Woollard, G. P., and J. C. Rose: "International Gravity Measurements," *Soc. Explor. Geophys. Spec. Pub.*, Tulsa, Okla., 1963.
8. LaCoste, L. J. B., Jr.: A New Type Long Period Vertical Seismograph, *Physics*, vol. 5, pp. 178–180, 1934; also see L. J. B. LaCoste and A. Romberg, U.S. Patent 2,293,437, 1900.
9. White, Peter H. N.: Gravity Data Obtained in Great Britain by the Anglo-American Oil Co., Ltd., *Q. J. Geol. Soc. Lond.*, vol. 104, pp. 339–364, 1949.
10. Nettleton, L. L.: Determination of Density for Reduction of Gravimeter Observations, *Geophysics*, vol. 4, pp. 176–183, 1939.
11. Baker, P. E.: Density Logging with Gamma Rays, *Am. Inst. Min. Met. Eng. Tech. Publ. 4654*, reprinted in *J. Petrol Technol*, vol. 9, no. 10, October 1957.
12. Faul, Henry, and C. W. Tittle: Logging of Drill Holes by the Neutron, Gamma Method, and Gamma Ray Scattering, *Geophysics*, vol. 16, pp. 260–276, 1951.

13. McCulloh, Thane H.: Borehole Gravimetry: New Developments and Applications, *Proc., 7th World Petrol. Congr.*, vol. 9, pp. 735–744, Elsevier, London, 1967.

14. Hammer, Sigmund: Terrain Corrections for Gravimeter Stations, *Geophysics*, vol. 4, pp. 184–194, 1939.

15. Douglas, Jesse K., and Sidney R. Prahl: Extended Terrain Correction Tables for Gravity Reductions, *Geophysics*, vol. 37, pp. 377–379, 1972.

16. Sandberg, C. H.: Terrain Corrections for an Inclined Plane in Gravity Computations, *Geophysics*, vol. 23, pp. 701–711, 1958.

17. Lambert, W. D., and F. W. Darling: Tables for Theoretical Gravity According to the International Formula, *Bull. Geod.*, vol. 32, pp. 327–340, 1931; also included in L. L. Nettleton, "Geophysical Prospecting for Oil," pp. 137–143, McGraw-Hill, New York, 1940.

18. Nettleton, L. L.: Gravity Survey over a Gulf Coast Continental Shelf Mound, *Geophysics*, vol. 22, pp. 630–642, 1957.

19. Boyd, Lewis H.: Gravity-Meter Survey of the Kettleman Hills: Lost Hills Trend, California, *Geophysics*, vol. 11, pp. 121–127, 1946.

20. Coffin, R. Clare: Recent Trends in Geological-Geophysical Exploration, *Bull. Am. Assoc. Petrol. Geol.*, vol. 30, pp. 2013–2032, 1946.

21. Brown, Hart: A Precision Detail Gravity Survey, Jameson Area, Coke County, Texas, *Geophysics*, vol. 14, pp. 535–542, 1949.

22. Pohly, Richard A.: Gravity Case History: Dawn No. 156 Pool, Ontario, *Geophysics*, vol. 19, pp. 95–103, 1954.

23. Ferris, Craig: Use of Gravity Meter in Search for Stratigraphic Traps, pp. 252–267, in Robert E. King (ed.), "Stratigraphic Oil and Gas Fields", *Am. Assoc. Petrol. Geol. Mem. 16* and *Soc. Explor. Geophys. Spec. Publ. 10*, 1972.

24. Nettleton, L. L.: Use of Gravity, Magnetic, and Electrical Methods in Stratigraphic-Trap Exploration, pp. 244–251, in Robert E. King, (ed.), "Stratigraphic Oil and Gas Fields", *Am. Assoc. Petrol. Geol. Mem. 16* and *Soc. Explor. Geophys. Spec. Pub. 10*, 1972.

25. Seigel, H. O., H. L. Hill, and J. G. Baird: Discovery Case History of the Pyramid Ore Bodies, Pine Point, Northwest Territories, Canada, *Geophysics*, vol. 33, pp. 645–656, 1968.

26. Kane, M. F., and L. C. Pakiser: Geophysical Study of Subsurface Structure in Southern Owens Valley, California, *Geophysics*, vol. 26, pp. 12–26, 1961.

27. Daneš, Zdenko F.: Gravity Results in Western Washington, *EOS Trans.*, vol. 50, pp. 548–550, 1969.

THE INTERPRETATION OF GRAVITY DATA

To be carried out to best advantage, the acquisition, reduction, and mapping of gravity data require highly precise instrumentation, skill in planning and executing the field measurements, and expertise in proper correction of the data observed in the field. We shall now consider the final step, the most challenging of all, which is the deduction of geological structure from the corrected gravity observations.

The geological interpretation of gravity data is more difficult and involves more uncertainties than the interpretation of seismic records. Gravity maps have enough resemblance to structural maps that one can easily fall into the mental trap of identifying gravity contours as being indicative of structure. In evaluating gravity maps it is important that one keep in mind the true nature of the contours; specifically, one must not forget that they depict a potential field rather than a subsurface structure.

Two characteristics of potential fields in general and gravitational fields in particular make their interpretation more of an art than a science: (1) The field observed at any point represents the summation of the gravitational attractions of all subsurface sources detectable by the instrument employed. Yet our object in interpreting such a field is to obtain information on the individual sources contributing to it. Except in very simple cases, the separation of the observed field into its component parts is quite difficult and sometimes not possible at all. (2) The lack of uniqueness in the gravity field from a subsurface source means that an infinite number of different configurations can result in identical gravity data at the surface. To resolve such ambiguity, other information than that from gravity is needed. For this reason the value of gravity data usually depends on the amount of independent geological information available.

Thus interpretation of gravity data is not a clear-cut process but one which requires a great deal of intuition, both physical and geological, and which is most useful where other kinds of subsurface information, e.g., that obtained from wells or seismic shooting, are available.

13-1 SEPARATION OF ANOMALIES

The extraction from an observed field of anomalies associated with individual subsurface sources involves filtering operations similar in principle to those we have considered in the chapter on seismic data processing. Yet the three-dimensional nature of a potential field such as gravity requires some special adaptation of the filtering theory on which seismic signal processing is based. Also the sampling of the field, being along a two-dimensional surface, is somewhat different from the simple time-series transformation employed in digitizing seismic data.

Two types of problems are often encountered in gravity interpretation where anomalies must be separated from one another. The most usual one is where the lateral extent of one anomaly is much greater than the other. This generally occurs when the source with the larger dimensions is a regional geologic feature, such as a basin or geosyncline, and the smaller one is a local feature, such as an anticline or salt dome. In such a situation, the first anomaly can be considered to have a low spatial frequency (equivalent to a large lateral extent or a long wavelength) and the second a high spatial frequency (corresponding to a short lateral distance or wavelength). The most common objective in such cases is to isolate the anomaly from the smaller source, a process which involves high-pass spatial filtering. Techniques for carrying out such a separation were first developed some time before it was generally recognized that they were really filtering operations. The component of the gravity anomaly having the longer effective wavelength is usually referred to as the *regional*, while the narrower, or shorter-wavelength, component having a more localized source is referred to as the *residual*. Extraction of residual from regional is done both by graphical and computational methods.

In regional studies, it may be desirable to remove anomalies from features of small lateral extent so as to bring out larger-scale structures more clearly. Techniques equivalent to low-pass filtering of the observed gravity field should achieve this objective.

A more complex type of filtering may be required when two sources of approximately the same size and buried at about the same depth are so close together that the field appears to come from a single source rather than from two separate ones. Resolution of such individual sources is not always possible, but when it is, rather sophisticated filtering techniques may sometimes be applicable with useful results.

The widespread use of electronic computers for filtering geophysical data of various kinds has led to the introduction of numerous programs for separating anomalies observed in measuring potential fields such as those from gravity. An even more significant application of computers to gravity interpretation has been in geological modeling. Various assumed subsurface structures are automatically tested and the one giving the best fit to the observed data is presented as the output of the computation. It is in this aspect of gravity prospecting that the greatest technological advances have been made since the late 1960s.

Isolation of Residual Gravity Effects

There are numerous areas where large-scale deep-seated structural features cause significant regional variations in the gravitational field. The Amarillo-Wichita uplift, a 200-mi-long basement trend crossing western Oklahoma and the Texas Panhandle, has a regional gradient along its southwestern flank of about 10 mgal/mi. In the Rio Grande Valley, there is a decrease in regional gravity of about 1 mgal/mi toward the Gulf of Mexico as one moves eastward from a point 70 mi inland. A comparable effect is generally observed as one approaches the Gulf anywhere along its periphery.

There are several sources of regional gravity variation. In addition to large-scale geologic structures, there are density effects caused by intrabasement lithologic changes as well as isostatic variations with sources that may often be indeterminate.

Regional gradients like these often distort or obscure the effects of structures, such as salt domes or buried ridges, that are sought in oil exploration. For this reason it is customary to subtract out the regional effects from the observed gravity in order to isolate more clearly the smaller structural features in which we are most interested. Both graphical and mathematical methods are employed for separating residual fields associated with the smaller geological features.

Regardless of the origin of the regional anomaly, and regardless of the method of calculating residuals, it is important to recognize that the basic criterion for separating regional and residuals is the area covered by each type of feature.

Graphical methods In the graphical approach, the regional effect must be estimated from plotted profiles or contour maps of observed gravity; regional contours are interpolated more or less arbitrarily, being superimposed over the original gravity field. A considerable amount of judgment and in many cases some geological information may be necessary to extract the regional properly from the observed data. The process is so subjective and empirical that one geologist has been quoted by Nettleton[1] as remarking that "the regional is what you take out in order to make what's left look like the structure."

Where the contours at a distance from a local anomaly are quite regular, it is possible to take out the regional trend by drawing lines which connect the undis-

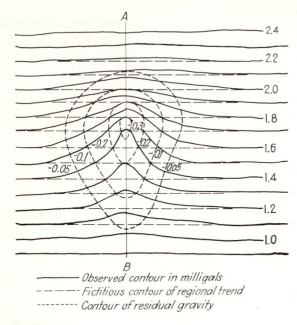

————— Observed contour in milligals

– – – – – Fictitious contour of regional trend

------- Contour of residual gravity

FIGURE 13-1
Determination of residual gravity by subtracting fictitious contours representing regional trend from observed contours.

turbed contours outside the area within which the anomaly is confined, as illustrated in Fig. 13-1. Where the smoothed contours cross contours of observed gravity, the differences between the two, which have discrete values at each intersection, are marked and themselves contoured. The resulting map gives residual gravity. In the example shown in Fig. 13-1, the removal of the regional converts an elongated gravity minimum into a closed low. The best results can usually be obtained if the smoothing is done on typical profiles rather than on contour maps. If a gravity cross section is drawn through the center of the anomaly on a map, the regional trend can be represented by the best straight line connecting the ends of the profiles on either side of the anomalous feature. Figure 13-2 illustrates this. The residual profile plotted below the observed cross section was obtained simply by subtracting the estimated regional value from the observed gravity at all points along the profile.

Often a series of profiles is drawn on the contour map as a network of intersecting lines. These lines may be profiles along which observations have been made, or they may be drawn arbitrarily across a gravity map with gravity values picked off at intersections with contour lines. The regional is estimated for each profile as shown in Fig. 13-2, but there is additional control from the fact that the values on each pair of intersecting profiles must be adjusted to be the same at the points of intersection. Such additional control, although in a sense arbitrary, is particularly helpful when residuals have low relief.

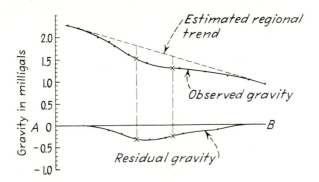

FIGURE 13-2
Gravity profile showing removal of regional trend across the anomaly shown in
Fig. 13-1.

The graphical methods have the advantage that all available geological information from an area can be put to use in drawing the regional. This is particularly
important where the survey covers an area that is smaller than that of a major
structural feature governing the regional trends. But experienced and highly trained
personnel are required for good results. And even when such personnel are available,
subjective processes involve the risk of erroneous geologic conceptions or mistakes
in judgment.

Analytical methods   With analytical methods of determining residual gravity,
numerical operations on the observed data make it possible to isolate anomalies
without such a great reliance upon the exercise of judgment in carrying out the
separations. Such techniques generally require that gravity values be spaced in a
regular array, and templates are designed so that personnel with little training or
experience can interpolate values from maps on a uniform grid. Programs are sometimes used that operate directly from data mapped from irregularly spaced gravity
readings.

Four analytical approaches are in common use: (1) the direct calculation of
residuals by techniques such as the center-point-and-ring method, described in the
following paragraph; (2) the determination of second derivatives, for which several
standard computational formulas are available; (3) polynomial fitting; and (4)
downward continuation, which transforms the gravity field as measured at the
surface to the field that would be observed on a horizontal plane buried at some
specified depth.

DIRECT COMPUTATION OF RESIDUALS   The simplest analytical approach is to calculate the residual directly from a regular grid of gravity observations. Griffin[2] has
described a technique which involves the averaging of gravity values along the
periphery of a circle or regular polyhedron with its center at the point for which
the residual is computed. Figure 13-3 illustrates the method. The average value
around the circle is simply the arithmetic mean of a finite number of equally spaced

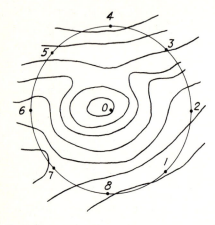

FIGURE 13-3
Direct computation of residual from map. Residual gravity at $O = g_0 - (g_1 + g_2 + g_3 + g_4 + g_5 + g_6 + g_7 + g_8)/8$. (*After Griffin.*[2])

points about its circumference. The residual value is the observed value at the center minus this average. The principal problem is in the choice of a radius. The spacing of sampling points around the circumference is arbitrary also; it should be large enough to ensure that the circle will lie entirely outside the anomaly but not large enough to include irregularities from other sources. It is not always possible to meet these conditions.

The significance of the residual obtained in this way can be understood most readily by considering a plot of gravity along a line crossing the center of an anomaly, as shown in Fig. 13-4. The average of the gravity values A and B at the respective points where the section crosses the circle is C, as shown at the center point. The difference between D, the observed gravity at the center point, and C is the residual.

If the circle has a diameter smaller than that of the anomaly and crosses the profile where the respective values are E and F, the average, also at the center point, will be G and the residual will now be $D - G$, which is obviously not an accurate measure of the true residual. If the circle is larger than the diameter of the anomaly, the residual anomaly will not be affected. It is rare, however, for the

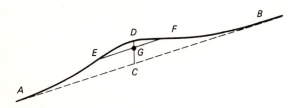

FIGURE 13-4
Residuals observed with circles of different diameter over a gravity anomaly. With circle such as EF having a diameter smaller than the breadth of the anomaly, the apparent anomaly DG is smaller than the true residual DC. A circle with a diameter larger than the breadth of the anomaly gives a more correct indication of its magnitude.

regional field to be a plane, as represented in the diagram. Moreover, there would normally be other anomalies or irregularities superimposed upon the inclined regional baseline that would cause distortion of the residual obtained by averaging values around the larger circle. It is important for this reason to select the optimum radius if the results are to be reliable.

When the data points are located on a square grid, as in Fig. 13-5, a residual can be computed by averaging the values at the corners of the respective squares for which distances from the center point are r, the grid spacing, and $r\sqrt{2}$. This is done by use of the formula

$$\text{Residual gravity} = g_0 - \frac{g_1 + g_2 + g_3 + g_4 + g_5 + g_6 + g_7 + g_8}{8} \qquad (13\text{-}1)$$

Such an approximation gives a result that is very little different from that which would be obtained by averaging the same number of gravity values around a circle, such as the dotted one shown in Fig. 13-5, midway between the two circles circumscribing each square. Use of the grid is much better adapted, however, for calculation by an electronic computer.

SECOND DERIVATIVE COMPUTATION AND MAPPING   The second derivative of a gravity field can be shown to be a measure of the curvature of the field. Consider a plot of gravity versus distance along a profile. Where the curvature of the line thus plotted is greatest (radius least) the second derivative has its higher value. Where there is no curvature (radius infinite) the second derivative is zero. If a shallow geological feature of limited lateral extent (like a salt dome) has a gravity anomaly with greater curvature than the regional field (which originates from deeper sources) on which it is superimposed (as in Fig. 13-6), the second derivative will be greater over the localized feature than over the part of the area where the gravity variation follows the regional trend. Plotting the second derivative will have the effect of making the gravitational anomaly from the local feature stand out more conspicuously.

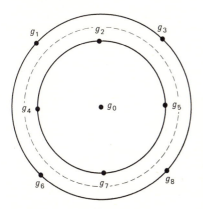

FIGURE 13-5
Approximate determination of residual from values sampled on a square grid. Residual around center point is obtained by averaging eight gravity readings g_1 to g_8 and subtracting g_0 from the average. Average approximates gravity along dotted circle.

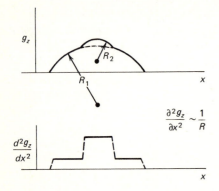

FIGURE 13-6
Relation between curvature of gravity field and second derivatives on a hypothetical profile where the curvature suddenly changes. R_1 and R_2, radii of the respective regions of the gravity-distance curve, are reciprocals of the corresponding curvatures.

To understand the significance of the second derivative of a two-dimensional gravity field, one should consider the grid of Fig. 13-7. With gravity values b_1, b_2, b_3, b_4, and c at the grid points shown, one can express this derivative in terms of differences between gravity differentials at adjacent corners of the grid as follows:

$$\frac{d^2g}{dx^2} \approx \frac{\Delta g_1/r - \Delta g_2/r}{r} = \frac{1}{r}\left(\frac{b_1 - c}{r} - \frac{c - b_3}{r}\right)$$

$$\approx \frac{b_1 + b_3 - 2c}{r^2} \tag{13-2}$$

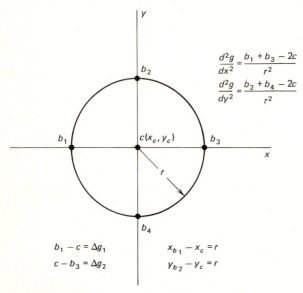

FIGURE 13-7
Grid showing relation between gravity differentials in x and y directions and second derivatives.

Similarly

$$\frac{d^2g}{dy^2} \approx \frac{1}{r}\left(\frac{b_2 - c}{r} - \frac{c - b_4}{r}\right) = \frac{b_2 + b_4 - 2c}{r^2} \tag{13-3}$$

Laplace's equation, which holds for any potential field, states that

$$\frac{d^2g}{dx^2} + \frac{d^2g}{dy^2} + \frac{d^2g}{dz^2} = 0 \tag{13-4}$$

or

$$\frac{d^2g}{dz^2} = -\left(\frac{d^2g}{dx^2} + \frac{d^2g}{dy^2}\right)$$

so that

$$\frac{d^2g}{dz^2} \approx \frac{4}{r^2}\left(c - \frac{b_1 + b_2 + b_3 + b_4}{4}\right) \tag{13-5}$$

If we rotated our x and y axes by a small angle $\pi/2n$, we would obtain a new grid of gravity values where the new axes cross the circle. We might designate these respectively as b_5, b_6, b_7, b_8. Then

$$\frac{d^2g}{dz^2} \approx \frac{4}{r^2}\left(c - \frac{b_5 + b_6 + b_7 + b_8}{4}\right) \tag{13-6}$$

If we repeated this process n times and added the results, we would get

$$\frac{d^2g}{dz^2} = \lim_{n\to\infty} \frac{4}{r^2}\left[c - \frac{(b_1 + b_2 + b_3 + b_4) + \cdots + (b_{4n-3} + b_{4n-2} + b_{4n-1} + b_{4n})}{4n}\right] \tag{13-7}$$

The larger the value of n the closer the fraction in Eq. (13-7) approaches the true average value of gravity around the circle; in the limit we can write

$$\frac{d^2g}{dz^2} = \lim_{n\to\infty} \frac{4}{r^2}\ (c - \text{average gravity along circle of radius } r) \tag{13-8}$$

This derivation was first published by Swartz.[3]

Elkins[4] has shown that if the gravity vertical g_z can be considered to have radial symmetry about a point, it can be expressed as a polynomial in ascending even powers of r as follows:

$$gz(r) = a_0 + a_2 r^2 + a_4 r^4 + \cdots$$

Then the second vertical derivative can be represented by the formula

$$\frac{d^2g_z(r)}{dz^2} = -4a_2 \tag{13-9}$$

where a_2 is the slope where r^2 approaches zero of the line obtained by plotting $g_z(r)$

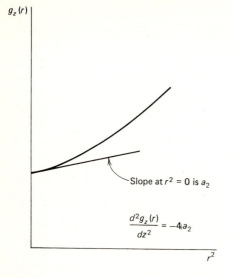

FIGURE 13-8
Relation between slope a_2 of g_z-versus-r^2 curve and second vertical derivative.

versus r^2. The actual gravity field seldom shows radial symmetry around a point, but the formula can give meaningful results if each value of $g_z(r)$ to be plotted is obtained by averaging gravity values around a circle of radius r. If the averages for a series of such circles with different radii are plotted against r^2, one can obtain the second derivative from the plot of $g_z(r)$ versus r^2 by measuring the slope a of the curve at $r^2 = 0$ as in Fig. 13-8 and putting this value into Eq. (13-9).

Determining Second Derivatives from Grids of Gravity Values For Eq. (13-9) to yield entirely accurate values of the second derivative at a point, even where there is radial symmetry around it, there should be a very large number of rings with their centers at the point and continuous gravity data around each ring. In practice, of course, gravity readings are obtained at discrete points,

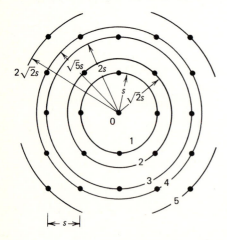

FIGURE 13-9
Five rings made up by passing circles from common center through appropriate points on square grid. Radii of rings are given in terms of grid spacing s.

usually well separated, and the averages must be computed from this restricted sampling of the continuous field. The radii of the rings are determined by the spacing of the gravity observation points.

After the corrected gravity values are put on a map and contoured, a square grid is constructed with corners a distance s apart. The interpolated value of gravity is recorded at each grid point. Derivatives are computed at each point from rings centered around it. Averages are taken around rings having respective radii of s, $s\sqrt{2}$, $2s$, $s\sqrt{5}$, $2s\sqrt{2}$, etc., as shown in Fig. 13-9. This procedure is handy both for manual computation and for programming electronic computers. As there is seldom good symmetry or regularity of gravity fields measured on a real earth, it is not likely that the average gravity value $g_z(r)$ for each ring of radius r will fall exactly on a smooth curve having a slope a_2 at $r^2 = 0$ which can be looked upon as minus one-fourth the *true* second derivative at the center of the rings, as indicated in Fig. 13-10. It is much more probable that the points so computed will show a random scatter about the best curve through them, as shown in Fig. 13-11. The question then is how to draw the best line through the scattered points. One way to construct this line is by least squares, weighting all points equally. This may give an undue weight to the center point at $r^2 = 0$, where any local or accidental irregularity cannot be adequately reduced by averaging with other points. Also points at large values of r^2 should not be given too much weight in determining the slope at $r^2 = 0$ if it is likely that sources of gravitational attraction at a distance from the center point would have a greater effect on gravity values observed along the outer rings than they have at the point where the derivative is actually to be computed.

Weighting Factors   The problem is thus how to weight the various rings. There is no clear-cut answer. The choice of weighting factors depends on the preference of the interpreter and the nature of the data. For example, an area where a large number of individual anomalies are close to one another requires a smaller weighting

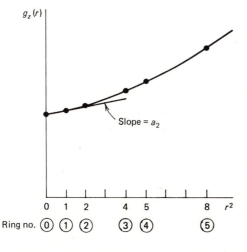

FIGURE 13-10
Average gravity values around rings of different radii falling along smooth line when plotted versus r^2.

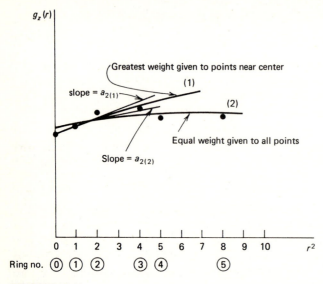

FIGURE 13-11
Uncertainty in determining best value of a_2 when averages of gravity around different ring do not fall on smooth curve. Two different weightings of the points result in different least-squares curves yielding separate a_2 values.

factor for the rings of larger radius than one where the sources of local anomalies are well separated.

Consider the grid of Fig. 13-9, where gravity is observed at the center point and is averaged around rings with radii s, $s\sqrt{2}$, and $s\sqrt{5}$. If a_2 is uniform and $s = kr$, where $1/k$ is the map scale, we should be able to determine a_2 graphically by the equations

$$a_2 s^2 = g_z(s) - g_z(0) \tag{13-10}$$

$$2a_2 s^2 = g_z(s\sqrt{2}) - g_z(0) \tag{13-11}$$

$$5a_2 s^2 = g_z(s\sqrt{5}) - g_z(0) \tag{13-12}$$

A least-squares solution of these three equations gives

$$\frac{d^2 g_z}{dz^2} = -4a_2 = \frac{1}{60 k^2 s^2} \left[64 g_z(0) - 2 g_z(s) - 4 g_z(s\sqrt{2}) - 5 g_z(s\sqrt{5}) \right] \tag{13-13}$$

Nettleton[1] has tabulated the weighting constants used by various authors who have worked out their own formulas for second-derivative computation. A general formula for the derivative, which can be designated D, is

$$D = \frac{C}{s^2} \left(W_0 H_0 + W_1 H_1 + W_2 H_2 + \cdots \right) \tag{13-14}$$

Here H_0 is the value of gravity at the center point, and H_1, H_2, etc., are the averages of the gravity values around the respective circles surrounding the center point. W_0, W_1, and W_2 are the weighting factors for the center point and the respective rings. As used here, s is the distance corresponding to a unit grid spacing, and C is a numerical constant.

The C/s^2 factor indicates the difference between the units used for the second-derivative values and those of observed or residual gravity. The contour maps may look quite similar in shape, but one map shows gravity values divided by a distance squared and the other shows actual values of relative gravity.

Because any grid system operates on selected discrete points, the derivative map at best represents an approximation to its true mathematical value. The grid spacing, the number of rings, and the weighting factors, all empirically chosen, have an important bearing on the configuration of the final map. The effect of using different grid spacings and weighting factors is comparable to using different electric filters which will pass some frequencies and reject others.

Relation between Second Derivatives and Residuals   The relation between the derivative and the residual gravity anomaly when both are determined by circumscribing grid points with rings can also be illustrated by a cross section of the type shown in Fig. 13-12. Here two rings of radius s_1 and s_2, respectively, are drawn around a center point O at the highest point of a gravity field that can be represented three-dimensionally by a sphere of radius R. The residual gravity for circle s_1 (based on the principle illustrated in Fig. 13-4) is h_1, and that for s_2 is h_2. It is easy to show that for values of h which are small (compared with R)

$$h = \frac{s^2}{2R} \tag{13-15}$$

The residual divided by s^2 is thus shown to be inversely proportional to the radius of curvature of the field or directly proportional to its curvature and hence its second derivative.

Henderson and Zietz[5] have demonstrated that a rather simple and widely used formula for weighting gravity values within a center-point-and-two-ring sampling configuration is applicable both to residuals and second derivatives. The only difference, as we have noted, is that one has the dimensions of gravity and the other the dimensions of gravity divided by a distance squared.

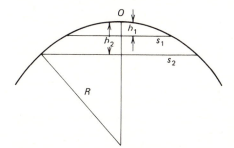

FIGURE 13-12
Relation between residual gravity and radius of ring for gravity variation resembling sphere when mapped. s_1 and s_2 are radii of respective rings. (*After Nettleton.*[1])

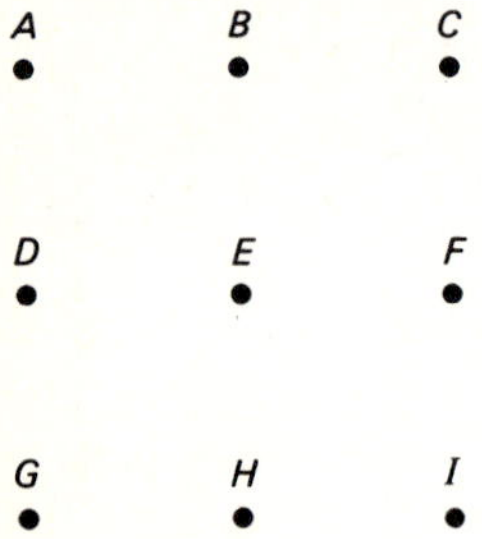

A   B   C
•   •   •

D   E   F
•   •   •

G   H   I
•   •   •

FIGURE 13-13
Computation of second derivative from nine-point grid with simple weighting functions. For unit spacing

$$\frac{d^2 g_z}{dz^2} = \tfrac{1}{2}(3Q - 4R - S)$$

where $Q = 4E$
   $R = D + B + F + H$
   $S = A + C + G + I$
Letters represent gravity values at stations with the same designation.

For rings with respective radii s and $s\sqrt{2}$ the residual can be written as

$$R = \tfrac{1}{9}[(8g_0 - 4g(s) - 4g(s\sqrt{2})] \tag{13-16}$$

It is computed in the same way as the residual shown in Fig. 13-5, except that the center point is averaged along with the samples around the two rings. This does not affect the value of the average significantly.

The formula for the residual is to be compared with the second-derivative value when the inner ring is given the same weight as the outer ring:

$$\frac{d^2 g_z}{dz^2} = \frac{1}{3s^2}[(8g_0 - 4g(s) - 4g(s\sqrt{2})] \tag{13-17}$$

Note that the expression within the brackets is the same as in Eq. (13-16).

Thus the *shape* of the contours on a residual map computed in this way would be the same as that for the second derivative having the particular weighting specified. The numerical values, of the contours, however, would be entirely different because of the grid-spacing factor as well as the factor-of-3 difference in the numerical constant.

Consider the case illustrated in Fig. 13-13, where the expression

$$\frac{d^2 g_z}{dz^2} = \frac{2}{s^2}[3g_0 - 4g(s) - g(s\sqrt{2})] \tag{13-18}$$

is obtained when the inner ring is given a weight 4 times as great as the outer ring. The computation can be carried out manually using a template which shows nine appropriate grid values through windows. It is particularly easy to program it for an electronic computer if the gravity input consists of values on a square grid.

Even more important than choosing optimum weighting coefficients in such analysis is the determination of a grid spacing that will result in the best separation

(a)   (b)   (c)

FIGURE 13-14
Comparison of observed gravity, residual gravity, and second vertical deriva-
tives over closely spaced salt domes in Mykawa-Friendswood-Hastings area,
Tex. (*After Elkins.*[4])

of residual gravity from the observed field. The size of the anomalies that will be
brought out is critically dependent on this spacing. Small ring diameters, for example (as well as coefficients predicating heavy weighting of the gravity averages
around the smallest rings), will emphasize the sharp details which are generally
associated with shallow sources.

The usefulness of second-derivative methods is illustrated by the group of maps
shown in Fig. 13-14, which compares observed gravity, residual gravity, and the
second vertical derivative of gravity in the area of the Mykawa-Friendswood-
Hastings-Manvel salt-dome fields on the Texas Gulf Coast. The maps of both observed and residual gravity show a large minimum at the center, which is located
between the fields. The second-derivative map, on the other hand, resolves the
individual domes and removes the central feature. It should be noted that several
deep dry holes were actually drilled on the basis of the misleading minimum at the
center of the gravity map. This is an exceptional example, however, in that the
difference between the residual and derivative maps is much greater than is observed with most gravity anomalies mapped in exploration work.

POLYNOMIAL FITTING   One of the most flexible of the analytical techniques for
determining regional gravity is polynomial fitting. Here the observed data are used
to compute, usually by least squares, the mathematically describable surface giving
the closest fit to the gravity field that can be obtained within a specified degree of
detail. This surface is considered to be the regional gravity, and the residual is the
difference between the gravity field as actually mapped and the regional field thus
determined.

In practice the surface is expressed mathematically as a two-dimensional polynomial of an order that depends on the complexity of the regional geology. If the
regional field were a simple inclined plane, it would be a first-order surface of the
form

$$z = Ax + By + C \qquad (13\text{-}19)$$

The next stage of complexity would involve representation by a second-order polynomial

$$z = Ax^2 + By^2 + Cxy + Dx + Ey + F \qquad (13\text{-}20)$$

Polynomials of much higher order would probably be necessary for a large area over which the regional has numerous convolutions. The approach is quite similar to that used in trend-surface analysis, a technique that has seen increasing use in reconstructing geologic structures as they should have appeared before regional deformation took place.

We shall illustrate the use of this method, following Agocs,[6] for the simplest possible regional surface, an inclined plane. Here the mathematically specified regional surface z is the linear function expressed in Eq. (13-19). The residual function R for the observed gravity field G is

$$R = G - z = G - (Ax + By + C) \qquad (13\text{-}21)$$

We need to determine the constants A, B, and C that will make R a minimum. To do this we set up the three least-squares equations:

$$\sum \frac{\partial}{\partial A}(R^2) = 0 \quad \text{or} \quad \sum R \frac{\partial R}{\partial A} = 0 \qquad (13\text{-}22)$$

$$\sum \frac{\partial}{\partial B}(R^2) = 0 \quad \text{or} \quad \sum R \frac{\partial R}{\partial B} = 0 \qquad (13\text{-}23)$$

and

$$\sum \frac{\partial}{\partial C}(R^2) = 0 \quad \text{or} \quad \sum R \frac{\partial R}{\partial C} = 0 \qquad (13\text{-}24)$$

Consider the first equation (13-22), $\partial R/\partial A = -x$, so that

$$R \frac{\partial R}{\partial A} = -[G - (Ax + By + C)]x$$

Summing up the resulting polynomial terms, we obtain

$$\sum Ax^2 + \sum Bxy + \sum Cx - \sum Gx = 0 \qquad (13\text{-}25)$$

and by similar operations

$$\sum Axy + \sum By^2 + \sum Cy - \sum Gy = 0 \qquad (13\text{-}26)$$

and

$$\sum Ax + \sum By + \sum C - \sum G = 0 \qquad (13\text{-}27)$$

Knowing G as a function of x and y, one can solve these equations for the unknowns A, B, and C. Once the three constants are determined, residuals can be readily mapped.

It is evident that this procedure would become quite tedious as one goes to higher-order polynomials if the computations had to be carried out manually. High-capacity electronic computers make it feasible to apply this approach to polynomials of much higher order than could be realistically put to use without such aids.

Coons et al.[7] have used this technique to study a large midcontinent linear gravity high which extends for several hundred miles through Iowa and Minnesota. They computed residuals for a number of polynomial fits up to the thirteenth order. Figure 13-15 shows the Bouguer anomaly map for a portion of this area and the seventeenth- and thirteenth-order polynomial surfaces. The resemblance of the polynomial surface and the Bouguer map increases with the order of the polynomial. Residuals computed for the thirteenth-order polynomial show a number of closed features having apparent geological significance which were hardly discernible on maps of lower-order residuals such as those based on the tenth-order surface.

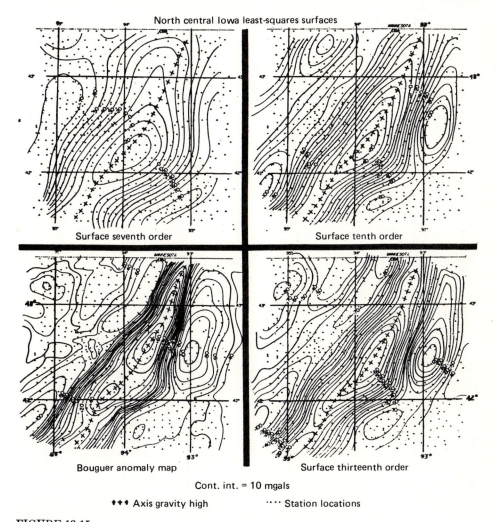

North central Iowa least-squares surfaces

Surface seventh order   Surface tenth order

Bouguer anomaly map   Surface thirteenth order

Cont. int. = 10 mgals

♦♦♦ Axis gravity high   ···· Station locations

FIGURE 13-15
Bouguer anomaly and least-squares surfaces for three high-order polynomials over portion of north central Iowa. (*After Coons et al.*[7])

FREQUENCY FILTERING   All the methods of regional-residual separation discussed thus far involve spatial filtering of one kind or another. But the techniques differ somewhat from the direct filtering of seismic signals by Fourier-transform theory (multiplication of the spectrum of the signal by a filtering function in the frequency domain or convolution of the signal by a filter operator in the time or space domain), as discussed in Chap. 6. With the development of high-speed computer programs for Fourier transformation it has become feasible to filter mapped gravity data by extension to two dimensions of the principles employed in seismic filtering.

When a gravity map is transformed from the space to the spatial-frequency domain, the field can be represented in a two-dimensional coordinate system with the frequency axes (in cycles per kilometer) extending both in the north-south direction and in the east-west directions.

Seismic filtering can be accomplished by multiplying the signal after transformation to the frequency domain by the frequency characteristic of an appropriate filter, using the methods considered in Chap. 6. Transformation of the product back to the time domain gives the filtered signal. It is easy to handle gravity data in an analogous way with modern computer techniques. The data on the gravity map are transformed into the spatial-frequency domain by Fourier-transform methods, and the spatial frequencies (usually very low ones) corresponding to regional trends are removed by appropriate two-dimensional filtering, leaving residual gravity. The fact that components of the regional field often cover a wide range of frequencies may make it difficult to accomplish the separation satisfactorily in this way. In some cases the higher-frequency components of the regional cannot be removed without also removing significant portions of the residual. Moreover, shallow features to be enhanced will have low-frequency components of considerable significance in their spectra which produce an overlap with regional spectra that limits the usefulness of the technique. Thus, as Grant[8] points out, it is not as simple as it might seem to design a spatial filter that will be effective in separating residual from regional fields.

Fuller[9] has shown that the various formulas for taking the second derivative by weighting and adding average values of gravity around a series of rings can be expressed in the form of two-dimensional filter characteristics. Such filtering action can be presented as a map of attenuation versus spatial frequency in the directions of the respective coordinates. The map illustrates the point previously made that separation of residuals by second derivatives is a form of two-dimensional spatial filtering. Differences between the spectra thus presented for different sets of weighting coefficients, as illustrated by Fuller and by Darby and Davies,[10] indicate the real distinctions between various published formulas for obtaining second derivatives from gridded data.

DOWNWARD CONTINUATION OF GRAVITY   In a classic paper published in 1949, Peters[11] showed how a potential field (in his paper, the magnetic field) measured at the earth's surface can be analytically projected upward or downward, i.e., mathematically transformed into what it would be if it could be respectively measured either above the earth's surface or along a horizontal plane inside the earth. The latter procedure, known as *downward continuation*, can extract anomalies originating with shallow sources from the regional background if the continuation is to a depth just above that of the source. A deeply buried body would not be expected to yield a gravity effect that stands out prominently above background because the field at the surface is spread out too much. If the same body were at a shallow depth, however, its gravity field would show much more relief. The function of downward continuation is to convert the broad gravity field of the deep source to the sharper field that would be observed if the body were shallower. Such a conversion helps to separate the anomaly associated with a feature at the depth of continuation from anomalies caused by sources at greater depths.

The computation procedure is similar to that used in second-derivative calculations, a series of circles with appropriately spaced windows being superimposed over a grid of observed gravity values. Appropriate weighting coefficients are applied, and the choice of ring spacing and coefficients is governed by the same considerations that apply for the derivative methods. The separation of data points limits the detail that can be projected downward. The continuation technique is sometimes used to define the shape of a buried feature yielding a gravity anomaly which is very thin compared with its lateral dimensions. This kind of direct interpretation will be discussed in more detail later in this chapter.

Gridding versus smoothing   The relative merits of graphical and analytical techniques for separating anomalies have been argued at some length in the geophysical literature. A convincing case for the graphical approach was made by Vajk,[12] in a paper published before most computer techniques had been introduced. He claimed that the regional effect by its very nature cannot be determined uniquely and hence cannot be properly eliminated by analytical procedures alone. While none of the techniques of anomaly separation carried out with electronic computers have the sophistication to make this objection entirely invalid, it is now possible to select computer programs for this purpose that incorporate geological constraints which should make the residual anomalies more acceptable to the geologist.

A good compromise between the two approaches is computer modeling. This involves setting up various models of the geology, all compatible with established information. Residual maps are rapidly generated for each model by the computer. The best interpretation is presumably the one which gives the most convincing fit with the residual as computed by standard methods. Unless the geology is quite well known, however, limitations in the uniqueness of gravity make it difficult to remove background effects in this manner with any real confidence.

Among those concerned with analytical procedures, controversies have sometimes developed over the "best" coefficients for derivative and continuation calculations. Actually, there is no best choice of coefficients, the selection depending on the nature of the problem at hand.

Finally, it is important in choosing techniques for determining residual gravity to take into account the precision of the original data. Errors in the instrumental readings or local irregularities in gravity could result in misleading residual and derivative maps, particularly if the most appropriate ring diameters and weighting coefficients are not used. Grant[13] has developed a statistical technique analogous to least squares which smooths out local disturbances and observational errors in the course of extracting residual gravity.

All analytical methods involving interpolation of grid values from contour maps have one disadvantage in common: the mode of contouring the Bouguer gravity itself influences the final maps (such as the second-derivative maps) very strongly. As Romberg[14] has remarked, the second-derivative map may be more an interpretation of contouring than of the gravity when the separation of data points is greater than the grid spacing.

The removal of regional effects is one of the two most important problems in gravity interpretation. The other is obtaining information from the residual anomaly on the structural configuration and density distribution of its source.

13-2 DETERMINATION OF DENSITY FOR GRAVITY INTERPRETATION

In the interpretation of gravity anomalies, it is necessary to estimate the densities of the subsurface rocks before one can postulate their structure. For this reason we shall give some attention to the densities of representative rocks in regions where gravity surveys are ordinarily made. For applications to interpretation, it is not the absolute densities but the density contrasts that are significant. In the sedimentary section, these contrasts are almost always small, seldom exceeding 0.25 g/cm³. Moreover, the variation in the possible densities of almost any type of rock sometimes makes it difficult to choose a suitable value to use in actual computations. Tables 13-1 to 13-3 illustrate the range of density values in a number of typical rock materials as well as the differences that can be expected between the different types.

It is evident from these tables that each type of sedimentary rock covers a wide range of densities and that the ranges are quite similar for sandstones, shales, and limestones. Rocks containing massive sulfides often have densities around 4 g/cm³. Igneous rocks are generally denser than sedimentary formations. Even here, however, there is substantial overlap between the densities of sedimentary and igneous types as well as between those of the various kinds of igneous rocks themselves.

Despite such variations, it is often necessary to assume average densities in gravity calculations for areas where the kind of rock is known but no densities have actually been measured. Figure 13-16 indicates the average densities obtained from a large number of laboratory measurements on core and surface samples tabulated by J. W. Peters of Mobil Oil Company. The range of densities and the number of samples for each type rock are tabulated at the base of the diagram.

Table 13-1   BULK DENSITY OF SEDIMENTARY ROCKS

Formation	Age	Locality	Depth of sample, ft	Saturated bulk density (average), g/cm³
Sandstone				
Mount Simon	Cambrian	West Virginia, Wood County	13,005–13,065	2.70
Southern Potsdam	Cambrian	Wisconsin		2.41
Northern Potsdam	Cambrian	Wisconsin		2.32
St. Peter	Ordovician	Arkansas, Ozark Plateau		2.50
Bradford	Devonian	Pennsylvania	~600–~2300	2.40
				*2.65
Chemung Formation	Devonian	Pennsylvania	~1700–~2300	2.51
Berea	Mississippian	Ohio, West Virginia	0–2160	2.39
Atoka Formation (and others)	Pennsylvanian	Arkansas:		
		Ozark Plateau		2.44
		Arkansas Valley		2.51
		Ouachita Mts.		2.56
Bartlesville sand	Pennsylvanian	Oklahoma	1570–2680	2.40
Bunter	Triassic	Great Britain		2.29
Keuper	Triassic	Great Britain		2.25
Woodbine sand	Cretaceous	Texas	2436–3701	2.25
Sandstones and siltstones	Cretaceous	Montana, eastern		2.17
Sandstones	Cretaceous	Wyoming	0–3187	2.32
Sandstones	Miocene	Switzerland		2.37
Limestone, Dolomite, Chalk, and Marble				
Ellenburger Group (limestone and dolomite)	Ordovician	Texas, Llano County		2.75
Beekmantown Group (dolomite)	Ordovician	West Virginia, Wood County		2.80
Black River limestone	Ordovician	Ontario		2.72
Niagara dolomite	Silurian	Wisconsin		2.77
Limestone	Carboniferous	Great Britain, Midlands		2.58
Marl	Carboniferous	Russia		2.63

Table 13-1—Continued

Formation	Age	Locality	Depth of sample, ft	Saturated bulk density (average), g/cm³
Limestone, Dolomite, Chalk, and Marble—Continued				
Oolites	Jurassic	Great Britain		2.44
Limestones	Jurassic	Switzerland		2.66
Glen Rose limestone	Cretaceous	Texas	20.5–30.5	2.37
Chalk	Cretaceous	Great Britain		2.23
Limestone	Cretaceous	Switzerland		2.65
Green River Formation (marlstone)	Eocene	Colorado		2.26
Shale, Claystone, and Slate				
Shale	Pennsylvanian	Oklahoma	1000	2.42
			3000	2.59
			5000	2.66
Shales	Cretaceous	Wyoming, Montana		2.17
Shale, nearly horizontal and undisturbed	Oligocene and Miocene	Venezuela	∼600	2.06
			∼2500	2.25
			∼3500	2.35
			∼6100	2.52
			∼7850	
Sand, Clay, Gravel, Alluvium, and Soils				
Cape May Formation (sand)	Pleistocene	New Jersey		1.93
Loess soil	Quaternary	Idaho		1.61
Fine sand	Quaternary	California		1.93
Very fine sand				1.92
Sand-silt-clay				1.44
Mud	Quaternary	Hudson River		1.20[†]
Silt	Quaternary	Hudson River	50 ft below river	1.79[†]
Newly deposited material	Quaternary	Mississippi River delta		1.26[†]
Soft mud	Quaternary	Clyde Sea	0–2.5 cm in mud	1.31[†]
			22.5–25 cm in mud	1.43[†]

Table 13-1—Continued

Formation	Age	Locality	Depth of sample, ft	Saturated bulk density (average), g/cm³
		Miscellaneous		
Marble	?	U.S.A., Great Britain		2.75

SOURCE: G. Edward Manger, in "Handbook of Physical Constants", rev. ed., *Geol. Soc. Am. Mem.* 97, 1966.

* Assumed grain density, g/cm³.   † Density computed from measured porosity.

In many areas there is a regular increase with depth in the density of the sedimentary section on account of compaction. Hedberg[15] has studied the density variation observed in a well penetrating more than 6000 ft of a shale formation in Venezuela (see Fig. 13-17). Nettleton[16] has made a study of the average variation with depth of the sedimentary rocks in the Texas Gulf Coast area by combining gravity data, density measurements, and Hedberg's compaction theory. Figure 13-18 shows the relationship of density to depth as determined from well samples

Table 13-2   AVERAGE DENSITIES OF HOLOCRYSTALLINE IGNEOUS ROCKS

Rock	Number of samples	Mean density, g/cm³	Range of density, g/cm³
Granite	155	2.667	2.516–2.809
Granodiorite	11	2.716	2.668–2.785
Syenite	24	2.757	2.630–2.899
Quartz diorite	21	2.806	2.680–2.960
Diorite	13	2.839	2.721–2.960
Norite	11	2.984	2.720–3.020
Gabbro, including olivine gabbro	27	2.976	2.850–3.120
Diabase, fresh	40	2.965	2.804–3.110
Peridotite, fresh	3	3.234	3.152–3.276
Dunite*	15	3.277	3.204–3.314
Pyroxenite	8	3.231	3.10 –3.318
Anorthosite	12	2.734	2.640–2.920

SOURCE: R. A. Daly, in "Handbook of Physical Constants", rev. ed., *Geol. Soc. Am. Mem.* 97, 1966.

* From Birch, *J. Geophys. Res.*, vol. 65, p. 1083, 1960.

Table 13-3 AVERAGE DENSITIES OF METAMORPHIC ROCKS

Rock	Number of samples	Mean density, g/cm³	Range of density, g/cm³
Gneiss, Chester, Vermont	7	2.69	2.66 –2.73
Granite gneiss, Hohe Tauern, Austria	19	2.61	2.59 –2.63
Gneiss, Grenville, Adirondack Mts., New York	25	2.84	2.70 –3.06
Oligoclase gneiss, Middle Haddam area, Connecticut	28	2.67	
Quartz-mica schists, Littleton Formation, New Hampshire (high-grade metamorphism)	76	2.82	2.70 –2.96
Muscovite-biotite schist, Middle Haddam area, Connecticut	32	2.76	
Staurolite-garnet and biotite-muscovite schists, Middle Haddam area, Connecticut	22	2.76	
Chlorite-sericite schists, Vermont	50	2.82	2.73 –3.03
Slate, Taconic sequence, Vermont	17	2.81	2.72 –2.84
Amphibolite, New Hampshire and Vermont	13	2.99	2.79 –3.14
Granulite, Lapland:			
Hypersthene-bearing	7	2.93	2.67 –3.10
Hypersthene-free	5	2.73	2.63 –2.85
Eclogite	10	3.392	3.338–3.452

SOURCE: Sydney P. Clark, Jr., in "Handbook of Physical Constants", rev. ed., *Geol. Soc. Am. Mem.* 97, 1966.

in the Gulf Coast and in east Texas. Since the density of rock salt is about 2.15 g/cm³, the salt will generally be lighter than the surrounding sediments at depths greater than about 2500 ft in the Gulf Coast and at depths greater than about 2000 ft in east Texas.

13-3 DETERMINING SUBSURFACE STRUCTURE FROM GRAVITY DATA

Ambiguity of Gravity Information

In Chap. 11 it was shown how one can predict the gravitational effect at any point on the surface from a given subsurface mass distribution by application of potential theory. In interpreting gravity data obtained in the field, one would like to reverse

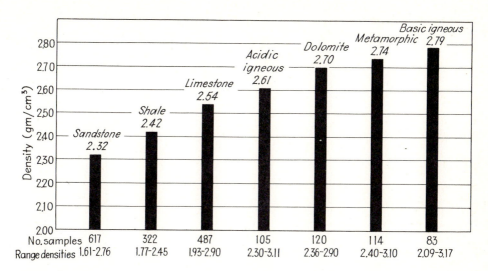

FIGURE 13-16
Average densities of surface samples and cores based on laboratory measurements. (*Mobil Oil Co.*)

the procedure. Unfortunately, the process is not entirely reversible. Whereas a buried mass with a specified position, shape, and density distribution will give a predictable gravity field on the surface, any observed gravity profile could be produced by an infinite number of possible mass distributions. To consider a simple example, any buried sphere will give the same gravity profile as a point source

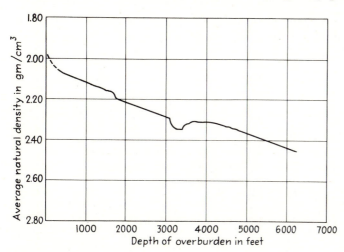

FIGURE 13-17
Natural density versus depth in a Venezuelan oil well penetrating into Tertiary shales. (*After Hedberg.*[15])

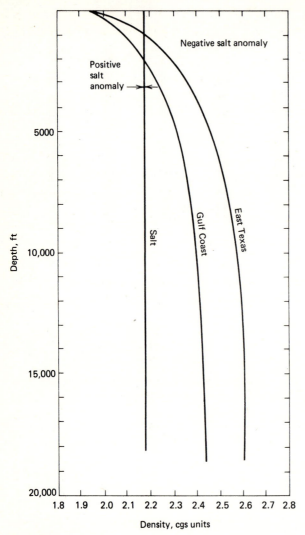

FIGURE 13-18
Density versus depth for sediments in Gulf Coast and east Texas. Note depths
at which sediment densities are equal to density of salt. (*Exploration Techniques,
Inc.*)

having the same mass located at its center. Thus, if the mass and the position of
the center of the sphere are fixed, its density and radius can have any combination
that will yield this mass and there is no way of determining its true dimensions
using the gravity anomaly alone. If we knew the density, we could then determine
the radius, but even then we could not be certain that the body is a sphere just
because the gravity profile has the form predicted for a sphere. Entirely different
geometrical distributions of the subsurface mass can yield the same gravity field
at the surface.

Examples Showing Lack of Uniqueness in Gravity Data

The limitations of the gravity method in defining subsurface structure follow from standard potential theory; they have been well illustrated in a much-quoted paper by Skeels[17] entitled Ambiguity in Gravity Interpretation. He shows how widely different subsurface mass distributions can give identical gravity pictures on the surface. In Fig. 13-19 we see a number of subsurface structures, each of which could account equally well for the indicated gravity anomaly if it were assumed that the anomaly results from undulations in the surface of a basement which has a fixed density contrast with the sediments above. The structure producing it could then have any of the shapes shown in the middle part of the figure. The lower part of the diagram shows an entirely different geological situation, a schist penetrated by an intrusive and overlain by sedimentary rock, and yet this could yield the same anomaly.

Need for Independent Geologic Information

It is seldom that a geological interpretation would be based on gravity data alone. Any independent control, like that obtained from borehole logs or seismic data, reduces the ambiguity in the interpretation. With such information the range of

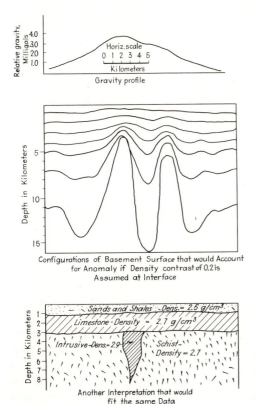

FIGURE 13-19
Alternative interpretations of a hypothetical gravity anomaly. (*After Skeels.*[17])

uncertainty is narrowed, and the number of variables can be reduced to the point where the gravity makes it possible to choose between alternative geological solutions (both lithological and structural) that fit all other kinds of observations. In the example shown in the upper part of Fig. 13-19, if a single well could be drilled to the basement somewhere along the profile, data would then be available on the densities of the sedimentary section and basement rocks as well as on the depth to the interface at one point. If the densities remain laterally uniform beneath the entire area, it should be possible to calculate the shape of the basement surface from the gravity data with considerable reliability.

It is often possible to use information obtained from surface geology to limit the possibilities and guide the assumptions upon which a gravity interpretation is based. Suppose, for example, that surface observations reveal an anticline coincident with a gravity high, as in Fig. 13-20. From regional geology the surface feature is believed to result from a buried basement ridge. The shape of the ridge at each of several possible depths can be estimated from the surface dips. The only unknown is the depth of the basement, and this can be estimated from the gravity data by placing the ridge successively at a number of depths and computing each of the resulting gravity profiles graphically or analytically. It is seen that one of these profiles gives a closer fit to the observed gravity profile than any of the others but does not coincide with it exactly. Although the geological control makes the final picture considerably less ambiguous than it would otherwise be, the structure so deduced is still not unique. Density inhomogeneities or structural features not reflected in the surface geology could make the interpretation quite unreliable.

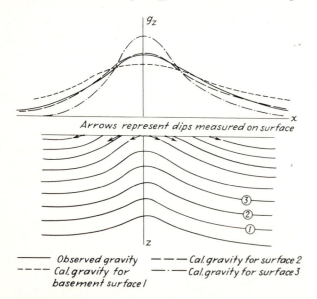

FIGURE 13-20
Estimating depth of buried igneous ridge by extrapolating surface structure downward and comparing calculated gravity for various assumed depths with that on the observed profile. Surface 2 appears to give the best fit.

Barton[18] has cautioned the interpreter on this point in the following words:

Calculations in regard to the mass causing an observed anomaly are of great value in the interpretation of gravity data, but there are no panaceas for the uncertainties of interpretation. The geophysicist should keep the limitations and uncertainties constantly in mind and should see that the users of the results of the calculations are conscious of those limitations and uncertainties.

He then points out that the chief value of such calculations is that they

... definitely throw out possibilities that previously had looked plausible and bring to mind unthought-of new ones that are much more plausible.

This is why drilling locations are seldom selected on the basis of gravity indications alone. Gravity surveys are excellent for reconnaissance of previously unexplored areas, but anomalies thus obtained should be tested by applying other geophysical or geological information before the interpretations based on them are given much weight in decision making.

Direct and Indirect Interpretation

There are two basic approaches to gravity interpretation. One is to determine a plausible mass distribution directly from the gravity data. The other is to assume various models conforming to all known constraints and to match gravity effects predicted for each model with the gravity field that has actually been observed. The model that gives the best fit is then considered to be the most probable one even though it cannot provide a definitive subsurface picture.

Direct interpretation   Upon initial consideration, the first approach would appear to be impractical because of the inherent limitations in direct gravity interpretation associated with the ambiguity we have just discussed. It is true that it would be impossible to derive a valid subsurface picture in this way if the source were three-dimensional in form. There is a special case, however, for which a unique solution is possible. This is when the buried source material is confined to a horizontal plane or, more realistically, is thin compared with its lateral dimensions. A buried evaporite bed left by an ancient salt lake or a thin volcanic sill of finite dimensions would be examples of such a model. In both cases, of course, there would have to be a density contrast between the material in the sheet and that of the surrounding formations.

To show the principle involved here, let us consider the formula [Eq. (11-30)] for the gravity effect at a point P on the surface of a buried mass having the form of a bounded horizontal sheet:

$$g_z = \gamma \sigma(x, y) \Omega \qquad (13\text{-}28)$$

where γ is the gravitational constant, $\sigma(x, y)$ is the surface density contrast (in mass per unit area), which is a function of position on the plane, and Ω is the solid angle subtended at P by the sheet. Now if the measuring point P is brought closer and closer to the surface, as in Fig. 13-21, an *infinitesimal* element of area dA, small enough to have a uniform density, will subtend an angle that approaches 2π as P

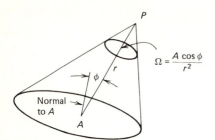

$$\Omega = \frac{A \cos \phi}{r^2}$$

FIGURE 13-21
Definition of solid angle Ω subtended by
area A at point P.

approaches the source plane itself. In the limit then

$$g_z(x, y) = 2\pi\sigma(x, y) \tag{13-29}$$

Thus the gravity as measured on the plane itself is proportional to the density contrast at the point of measurement, and there will be a unique correspondence between the two.

It is not feasible, of course, to measure gravity on the actual surface of a buried sheet, but it is possible to achieve the same result mathematically by downward continuation of data obtained on the surface of the earth above the sheet (see page 452). This is practical only to the extent that the continuation, based as it is on a sampling of gravity at the surface, gives an accurate description of the field below. This approach should give a unique determination of the density distribution at the depth to which the field is transformed if there is no anomalous mass between the surface and the depth of continuation. Downward continuation, as previously noted, is also used for regional-residual separation and for determining whether multiple sources are actually present for an anomaly which appears to originate from a single source.

Thus it is possible to apply the direct approach to determine the mass distribution in horizontal tabular bodies with a thickness small compared with their lateral dimensions. Determining the depth at which the anomalous sheet lies is not a problem, as one usually calculates the field over a range of depths from the surface to the actual depth of the source. The sharpness of the anomaly increases as the depth to which the continuation is carried out gets greater until the true depth of the source is reached. Below this depth, the anomaly becomes unstable, oscillating with increasing amplitude, and it is no longer useful.

Two methods of downward continuation have been applied to this problem, and they will be outlined briefly here. One involves Fourier analysis, the other the computation of vertical derivatives using grid and ring methods similar to those applied for determining second derivatives.

FOURIER ANALYSIS   Continuation by Fourier analysis requires that the field on the surface be expressed as the sum of Fourier components of the form $g_0 \cos px$ and $g_0 \cos qy$, where p is the spatial frequency in the x direction and q is its value in the y direction. If the variation is only in the x direction, the magnitude of a component with spatial frequency p in this direction will be increased by a factor

e^{ph} when continued downward to depth h. The higher the frequency the greater the amplitude of the component of the anomaly at that frequency after continuation. It is evident that the entire process, being frequency-dependent, is a high-pass filtering operation.

Where the observed field is expressed as a function both of x and y, downward continuation to a depth h involves multiplication of the spectrum of the field at the surface by a factor of the form $\exp \sqrt{p^2 + q^2}\, h$.

The usefulness of this approach depends on the range of spatial frequencies represented on the map of observed gravity. This is related most directly to the spacing of the observation points and also to the precision of the gravity readings and corrections. Subsurface bodies, particularly shallow ones, having gravity fields with significant Fourier components at a higher frequency than that of the measuring grid cannot be well defined by continuation methods.

As the Fourier transform represents the limit of the Fourier series for a transient nonrepetitive function, it is sometimes most convenient to carry out the continuation by Fourier-transform methods. A transform is obtained for the gravity field as continued downward by division of the transform of the observed field by a geometrical frequency function associated with the inverse-square attraction law. The transform thus computed is then converted by Fourier methods to a two-dimensional function of position which is directly dependent on the areal distribution of density on the buried sheet. Special computer programs have been designed for this procedure.

TAYLOR'S-SERIES METHOD OF CONTINUATION   The gravity field at the earth's surface $g_z(x, y, 0)$ can be projected downward to any desired depth by use of a Taylor expansion. With this approach, the field at depth h, which is $g_z(x, y, h)$, can be expressed by the series

$$g_z(x, y, h) = g_z(x, y, 0) + \frac{\partial g_z}{\partial z}(x, y, 0)h + \frac{\partial^2 g_z}{\partial z^2}(x, y, 0)\frac{h^2}{2} + \cdots \tag{13-30}$$

The highest-order derivative that must be used depends on the spacing of the data points and the accuracy desired. Only the first and second derivatives are generally needed; sometimes the first alone will suffice.

Techniques for obtaining the second derivative of gravity at any point where the field is mapped have been described earlier in this chapter. A similar procedure has been proposed by Evjen[19] for computing the first derivative, taking mean values along circles covering a range of radii. If the circles have radii r_1, r_2, r_3, etc., and the average gravity values around the circles are $\bar{g}_{z1}$, $\bar{g}_{z2}$, $\bar{g}_{z3}$, etc., with g_{z0} the value at the center, the first derivative at the center, with coordinates $(0, 0, 0)$ approaches the value

$$\frac{\partial g_z}{\partial z}(0, 0, 0) = \frac{\bar{g}_{z0}}{2r_1} + \frac{\bar{g}_{z1}}{2r_2} + \frac{\bar{g}_{z2}}{2r_3} + \cdots \tag{13-31}$$

It is evident that the weighting of the gravity contributions decreases as the circles around which they are averaged become larger. For data on a square grid having points a distance s apart the circles are generally so inscribed as to have radii of s, $s\sqrt{2}$, $s\sqrt{5}$, etc., as with the second derivative method illustrated by Fig. 13-9.

The downward continuation of gravity fields on the surface can be readily programmed for electronic computation. The operation is generally carried out for a series of increasing depths. When the depth is reached for which the field begins to oscillate or otherwise show instability, the process is discontinued. Even where the anomalous mass cannot be assumed to be tabular, the depth of continuation at which an anomaly shows the sharpest definition can generally be looked upon as the depth to the top of the feature from which it originates. Such information can in itself be useful for interpretation.

Indirect methods Because of the limitations that prevent the use of direct methods except for special cases, the indirect approach is all that can be tried for most interpretation problems. This consists of assuming a model for the buried source, predicting the gravity effect from it at the surface, comparing the observed and computed effects, and modifying the model, often by iterative procedures, to minimize the differences that are initially observed.

The choice of the model might be based on some geological hypothesis. A gradient trending along a line, for example, might be assumed to originate from a vertical fault scarp if available geological information from the area makes such a model plausible. A number of depths, throws, and density contrasts could be assumed for computation, and the combination yielding the best agreement would ordinarily be considered to be the most plausible source configuration. It must be emphasized once more that this approach is subject to all the inherent ambiguity of potential methods discussed earlier in this chapter (page 458). It must also be pointed out that indirect methods are seldom very useful unless there is independent geological or geophysical information that will keep this ambiguity within manageable limits.

ANALYTICAL METHODS OF INDIRECT INTERPRETATION There are two methods of computing gravity at the surface from source models in the subsurface. One is analytical, the other graphical. The analytical technique, which we shall consider first, calls for approximation of the geological feature considered to be the source by assigning it a simple geometrical form for which the gravity field can be computed mathematically. A salt dome, for example, might be approximated by a sphere or a vertical cylinder of specified dimensions. In Chap. 11, we derived formulas for the gravity effects of generalized forms as spheres, cylinders, laminae of various kinds, and faults with both inclined and vertical faces. Various values can be assigned to the parameters describing the geometry of these bodies, and the fields computed for such values are compared with measured anomalous fields. The parameters giving the best fit may be looked upon as plausible if they violate no constraints. Figure 13-22 tabulates the formulas for the gravity fields from representative geometrical forms that often have geological significance.

1. Sphere

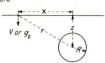

$$g_z = 8.52 \frac{\rho R^3}{z^2} \left[1 + \left(\frac{x}{z}\right)^2 \right]^{-\frac{3}{2}}$$

$$V = 8.38 \times 10^5 \frac{R^3 I}{z^3} \frac{1 - \frac{x^2}{2z^2}}{\left(1 + \frac{x^2}{z^2}\right)^{\frac{5}{2}}}$$

2. Horizontal cylinder

(*Same diagram as for sphere but center of circle now represents axis of infinitely long cylinder perpendicular to paper*)

$$g_z = 12.77 \frac{\rho R^2}{z} \left[1 + \left(\frac{x}{z}\right)^2 \right]^{-1}$$

$$V = 6.28 \times 10^5 \frac{R^2 I}{z^2} \frac{1 - \frac{x^2}{z^2}}{\left(1 + \frac{x^2}{z^2}\right)^2}$$

3. Vertical fault (*third dimension infinite*) approximated by horiz. sheet

$$g_z = 12.77 \rho t \left[\frac{1}{2} + \frac{1}{\pi} \tan^{-1} \frac{x}{z} \right]$$

$$V = 2 \times 10^5 \frac{It}{z} \frac{x}{z} \frac{1}{1 + \frac{x^2}{z^2}}$$

4. Vertical sheet (*third dimension infinite*)

$$g_z = 4.68 \rho t \left[\log \left(1 + \frac{z_2^2}{x^2} \right) \Big/ \left(1 + \frac{z_1^2}{x^2} \right) \right]$$

$$V = 2 \times 10^5 It \left[\frac{1}{z_1 \left(1 + \frac{x^2}{z_1^2}\right)} - \frac{1}{z_2 \left(1 + \frac{x^2}{z_2^2}\right)} \right]$$

5. Vertical circular cylinder

$$g_z = 2.03 \rho t w$$

$$V = I \times 10^5 (w_1 - w_2)$$

ρ = Density g_z = Vertical component of gravity in milligals
I = Intensity of magnetization (assumed vertical) in cgs units
V = Vertical component of magnetic intensity in gammas
w, w_1, w_2 = Solid angles as shown in 5
Linear dimensions arbitrary for magnetic formulas, in kilofeet for gravitational formulas

FIGURE 13-22
Summary of formulas for vertical components of gravitational and magnetic fields from buried bodies having simple geometric forms. The magnetic formulas will be discussed in Chap. 16. (*L. L. Nettleton, Geophysics, vol. 7, 1942.*)

DEPTH ESTIMATION   Such formulas may frequently be used to estimate the depth to the bodies to which they relate. Consider, for example, the gravity field over a long horizontal cylinder of radius R and density along a line on the surface perpendicular to its axis. The formula for the gravity at a horizontal distance x from an axis buried at a depth z (Fig. 13-23) is

$$g_z = 12.77\rho \frac{R^2}{z} \left[1 + \left(\frac{x}{z}\right)^2 \right]^{-1} \tag{13-32}$$

This formula was derived on page 376, where the terms in it are defined, the numerical constant being applicable when the distances involved are in thousands of feet. The gravity field g_z will have its maximum value over the axis, where $x = 0$. We shall define $x_{1/2}$ as the distance at which g_z has fallen off to half its maximum

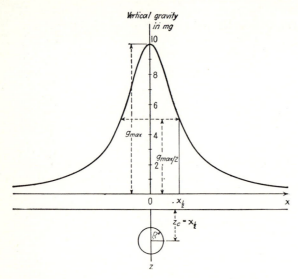

FIGURE 13-23
Determining depth of horizontal cylinder from half width $x_{1/2}$ on gravity profile.

value. This distance is designated as the *half width* of the anomaly. Taking 2 as the ratio of gravity at $x = 0$ to gravity at $x = x_{1/2}$, we can write

$$2 = 1 + \left(\frac{x_{1/2}}{z}\right)^2 \tag{13-33}$$

whereupon
$$x_{1/2} = z \tag{13-34}$$

The depth is equal to the half width of the anomaly when measured along a line perpendicular to the axis. The radius of the cylinder can be determined from the maximum gravity value g_{max} and z using the formula

$$R = \sqrt{\frac{g_{max}^2 z}{12.77\rho}} \tag{13-35}$$

It can be shown by a similar procedure that the half width for a spherical source is

$$x_{1/2} = \frac{z}{1.305} \tag{13-36}$$

$$z = 1.305 x_{1/2} \tag{13-37}$$

The formula is applied in the same way as the corresponding relationship for a cylinder. It may be applicable to salt domes or knolls on the basement surface with heights and diameters that are approximately equal.

In the case of a thin faulted layer, the depth to the center of the faulted slab can be estimated from the gradient of the gravity anomaly at the point where it is steepest.

The formula for the gravity of a thin faulted horizontal slab of thickness t and depth (to its center plane) z was shown on page 380 to be

$$g_z = 2\gamma\rho t \left(\frac{\pi}{2} - \tan^{-1}\frac{x}{z}\right) = 12.77\rho\left(\frac{t}{2} + \frac{t}{\pi}\tan^{-1}\frac{x}{z}\right) \qquad (13\text{-}38)$$

when all distances are in thousands of feet. The gradient is the derivative of g_z with respect to x

$$\frac{dg_z}{dx} = 12.77\rho\,\frac{t}{\pi}\frac{1}{z}\frac{1}{1 + (x/z)^2} \qquad (13\text{-}39)$$

and at $x = 0$, this becomes

$$\frac{12.77\ \rho t}{\pi z} \qquad (13\text{-}40)$$

The required slope can be obtained by plotting g_z versus x on a profile perpendicular to the contours defining the linear fault trend and taking the tangent to the gravity curve at the point where the gradient is maximum.

COMPOSITE FORMS   It is often convenient to fit together a number of generalized geometric forms with different dimensions and (in some cases) different densities to simulate a presumed structure and to adjust the dimensions of the different component parts until the theoretical and observed gravity curves fit. Nettleton[20] accounted for the anomaly over the Minden salt dome in Louisiana by a salt and cap-rock configuration which he reduced to a number of equivalent cylinders, as shown in Fig. 13-24. The interpretation was facilitated by previous knowledge (from drilling) of the depth to the salt layer at the base of the dome. Here again the same gravity picture might have been matched just as well by an entirely different subsurface mass, but the well data made this picture the most plausible. A similar approach to detailing the structure of a salt dome is illustrated in Fig. 12-10.

INDIRECT INTERPRETATION FOR SOURCES WITH FORMS NOT ANALYTICALLY DESCRIBABLE   Few actual subsurface bodies have shapes that can be described so closely by analytical geometry that the observed gravity anomalies even come close to fitting those predicted by the use of simple equations like those we have just considered. In the general case, it is most advisable to assume a model of irregular shape and to predict the gravity field by integrating the elements of mass of which the model is composed. Each element must be small enough that we can consider its gravity effect to be concentrated at a single point near its center without any loss of desired accuracy.

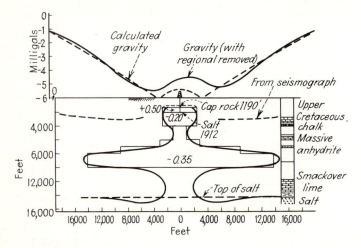

FIGURE 13-24
Comparison of observed gravity anomaly over Minden dome, Louisiana, with
that predicted from salt dome having shape and density contrasts approximated
by series of cylinders as shown. (*After Nettleton.*[20])

Two approaches might be employed for carrying out this integration. One, used
almost entirely for two-dimensional models, is to construct a cross section (in a
vertical plane perpendicular to its axis) of the presumed source and to superimpose
on it a specially designed transparent template called a *graticule*, which can be
used for graphical integration. The other is to describe the geometry of the assumed
source in a form suitable for input into a computer which is programmed to yield
the gravity field as a function of observing position on the surface.

Graticules These are devices for numerical integration, used in practice over
cross sections of two-dimensional bodies such as the infinitely long horizontal cylinders considered in Chap. 9. They consist of cells or compartments with varying
sizes and shapes, each of which covers an area on the cross section corresponding
to a known (usually uniform) contribution to vertical gravity at the measuring
point on the surface. A graticule is generally a transparent template consisting of
a fan-shaped pattern of lines which form a series of compartments with areas that
increase with distance from the vertex. It is superimposed on a cross section at the
same scale which outlines the body for which the gravity effect is to be computed,
the vertex being located at the position of the gravity-measuring station at the
surface.

One of the simplest types of graticule consists of trapezoidal compartments
formed by a system of horizontal lines, equally spaced, and a system of radial lines
emanating from the origin in such a way that each line makes an equal angle with
the next one.

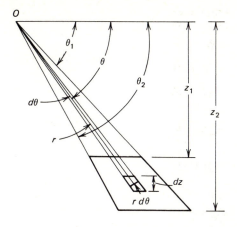

FIGURE 13-25
Calculation of attraction of element in
graticule made from sets of horizontal
lines and lines radiating from point O,
where attraction is to be computed.

The gravity effect of each cell is expressed by the integral (illustrated in Fig. 13-25):

$$g_z = 2\gamma\rho \int_{\theta_1}^{\theta_2} \int_{z_1}^{z_2} \frac{r\,d\theta\,dz}{r} = \int_{\theta_1}^{\theta_2} \int_{z_1}^{z_2} d\theta\,dz \qquad (13\text{-}41)$$

$$= 2\gamma\rho\,(\theta_2 - \theta_1)\,(z_2 - z_1) \qquad (13\text{-}42)$$

Thus if the angular intervals between successive radial lines and the vertical separations between the horizontal lines were respectively equal, the gravity effect of each compartment would be the same. If the two intervals are known, it is a simple matter to determine the calibration constant for the individual cells. For example, if $\theta_2 - \theta_1$ is 5.7° (0.1 rad) and $z_2 - z_1$ is 500 ft, it is easy to show that the gravity effect of each cell would be 0.02202 ρ mgal.

Figure 13-26 illustrates the use of the chart for computing the vertical gravitational effect at the observing point O from the irregular body shown at the center of the diagram. Compartments inside the body are counted, and fractional parts

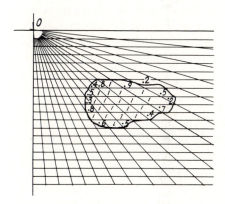

FIGURE 13-26
Use of graticule for which attraction of
each cell is computed as shown in Fig.
13-25 for determining gravity effect at
point O. Decimal parts of compartments cut by boundary of body are
indicated.

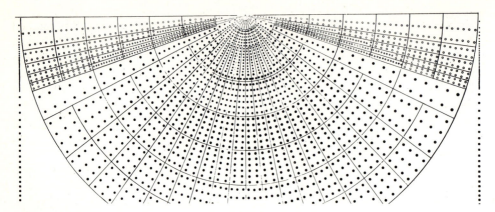

FIGURE 13-27
Dot chart for calculating gravity effect at any point on surface from a two-dimensional buried body. The chart, printed on a transparent sheet, is superimposed on a cross section, the vertex being placed over position on surface where gravity effect is desired. (*Gulf Research & Development Co.*)

of those cut by its boundary are estimated and added. If the buried mass becomes very shallow, it may be necessary to introduce another graticule with larger cells corresponding to greater increments of gravity as the trapezoids on a graticule used for deeper sources could become too small near the earth's surface for accurate counting.

Figure 13-27 shows another type of graticule, developed for gravity calculation by the Gulf Research & Development Co. It consists of cells bounded by radial lines radiating from the vertex and circular arcs with their centers at the vertex. If each compartment corresponded to an equal contribution to vertical gravity, the radial lines would be so drawn that the cosines of their angles with the horizontal would be equally spaced for successive lines and the circular arcs so drawn that increments of radius for successive curves would be equal. In this case it was more convenient to use cells of unequal gravity effect, the contribution for each compartment being indicated by the number of dots inside it.

If the scale of the profile is $1/k$ and the density contrast is ρ, the gravity effect at the center (in milligals) corresponding to a single full dot is $10^{-5}k\rho$. Assume, for example, that the section is drawn to a scale of $\frac{1}{10\,000}$ and the density contrast of the anomalous buried mass is 0.25; each compartment containing 12 full dots will contribute 0.3 mgal to the gravity at the center of the fan. If the boundaries of the feature cut the interior of a compartment, one interpolates the gravity contribution by counting only the dots inside the boundaries of the mass in question.

Use of Computers for Predicting Gravity Effect of Three-dimensional Bodies of Irregular Shape If a three-dimensional body is represented by contours of depth to its bounding surface, a technique developed by Talwani and

Ewing[21] can be used to determine the field at any point on the surface. The method is particularly suitable for electronic computation. The contours can be looked upon as representing the edges of thin laminae bounded by horizontal planes above and below having elevations that correspond to the levels of the contours. A typical lamina is illustrated in Fig. 13-28. The outline is taken as that of the contour line defining its base.

The computation of the gravity effect is facilitated by a theorem of Green for transforming a surface integral over the top of the lamina into a line integral around the contour that bounds it.

Numerical computation of the line integral is simplified for purposes of computer programming if the contour generating the lamina is represented by a polygon instead of a curved loop. A sampling of points along the contour, expressed in terms of their xy coordinates, provides inputs for the corners of each polygon. The computer carries out the integration for each lamina and adds the gravity contributions for the laminae corresponding to each contour so as to yield the total effect from the buried body at any specified point on the surface.

As with all other methods of indirect interpretation, it is possible to modify the presumed shape or position of the subsurface source to improve the fit to the observed gravity on the surface. Even if a perfect fit were obtained, however, it would not be at all certain that the form hypothesized for the source is correct without other evidence.

Elongated bodies of irregular but uniform cross section can also be modeled for computer integration. One technique divides the cross-sectional area into a series

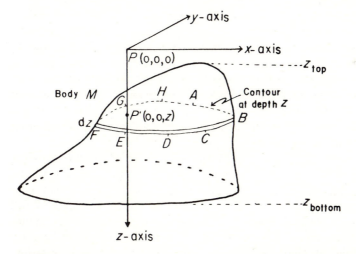

FIGURE 13-28

Principle of Talwani and Ewing method of determining gravity at P from a three-dimensional body of irregular shape that can be represented as a series of horizontal laminae bounded by contour lines on map. Computer carries out integration of elements of height dz as shown. Curved surface of lamina is approximated as polygon $ABCDEFGH$. (*Talwani and Ewing.*[21])

of rectangles or other polygons and programs the computer for formulas based on slant distances from the measuring point on the surface to various corners of the rectilinear cells. Talwani et al.[22] have devised a computer program based on this approach which makes use of the formula for the attraction of a horizontal line of infinite length that was derived on pages 400 to 401.

REFERENCES

1. Nettleton, L. L.: Regionals, Residuals, and Structures, *Geophysics*, vol. 19, pp. 1–22, 1954.
2. Griffin, W. R.: Residual Gravity in Theory and Practice, *Geophysics*, vol. 14, pp. 39–56, 1949.
3. Swartz, Charles A.: Some Geometrical Properties of Residual Maps, *Geophysics*, vol. 19, pp. 46–70, 1954.
4. Elkins, T. A.: The Second Derivative Method of Gravity Interpretation, *Geophysics*, vol. 16, pp. 29–50, 1951.
5. Henderson, Roland G., and Isidore Zietz: The Computation of Second Vertical Derivatives of Geomagnetic Fields, *Geophysics*, vol. 14, pp. 508–516, 1949.
6. Agocs, W. B.: Least Squares Residual Anomaly Determination, *Geophysics*, vol. 16, pp. 686–696, 1951.
7. Coons, Richard L., George P. Woollard, and Garland Hershey: Structural Significance and Analysis of Mid-Continent Gravity High, *Bull. Am. Assoc. Petrol. Geol.*, vol. 51, pp. 2381–2399, 1967.
8. Grant, F. S.: Review of Data Processing and Interpretation Methods in Gravity and Magnetics, 1964–71, *Geophysics*, vol. 37, pp. 647–661, 1972.
9. Fuller, Brent D.: Two Dimensional Frequency Analysis and Design of Grid Operators, Mining Geophys., *Soc. of Expl. Geophys., Tulsa, Okla.*, vol. 2, pp. 658–708, 1967.
10. Darby, E. K., and E. B. Davies: The Analysis and Design of Two-dimensional Fitters for Two-dimensional Data, *Geophys. Prospect.*, vol. 15, pp. 383–406, 1967.
11. Peters, L. J.: The Direct Approach to Magnetic Interpretation and Its Practical Application, *Geophysics*, vol. 14, pp. 290–320, 1949.
12. Vajk, Raoul: Regional Correction of Gravity Data, *Geofis. Pura Appl.*, vol. 19, pp. 219–143, 1951.
13. Grant, Fraser S.: A Theory for the Regional Correction of Potential Field Data, *Geophysics*, vol. 19, pp. 23–45, 1954.
14. Romberg, F. E.: Key Variables of Gravity, *Geophysics*, vol. 23, pp. 684–700, 1958.
15. Hedberg, H. D.: Gravitational Compaction of Clays and Shales, *Am. J. Sci.*, vol. 31, pp. 241–287, 1936.
16. Nettleton, L. L.: Fluid Mechanics of Salt Domes, *Bull. Am. Assoc. Petrol. Geol.*, vol. 18, pp. 1175–1204, 1934.
17. Skeels, D. C.: Ambiguity in Gravity Interpretation, *Geophysics*, vol. 12, pp. 43–56, 1947.
18. Barton, Donald C.: Case Histories and Quantitative Calculations in Gravimetric Prospecting, *Trans. Am. Inst. Min. Met. Eng.*, vol. 164, Geophysics, 1945, pp. 17–65, 1945.
19. Evjen, H. M.: The Place of the Vertical Gradient in Gravitational Interpretations, *Geophysics*, vol. 1, pp. 127–136, 1936.

20. Nettleton, L. L.: Recent Experimental and Geophysical Evidence of Mechanics of Salt-Dome Formation, *Bull. Am. Assoc. Petrol. Geol.*, vol. 27, pp. 51–63, 1943.
21. Talwani, Manik, and Maurice Ewing: Rapid Computation of Gravitational Attraction of Three-dimensional Bodies of Arbitrary Shape, *Geophysics*, vol. 25, pp. 203–225, 1960.
22. Talwani, M., J. L. Worzel, and M. Landisman: Rapid Gravity Computations for Two-dimensional Bodies with Application to the Mendocino Submarine Fracture Zone, *J. Geophys. Res.*, vol. 64, pp. 49–59, 1959.

14

MAGNETIC PROSPECTING: FUNDAMENTAL PRINCIPLES AND INSTRUMENTS

Magnetic prospecting, the oldest method of geophysical exploration, is used to explore for both oil and minerals. In prospecting for oil, it gives information from which one can determine the depth to basement rocks and thus locate and define the extent of sedimentary basins. Such information is of particular value in previously unexplored areas such as continental shelves newly opened for prospecting. It is sometimes employed, although not always successfully, to map topographic features on the basement surface that might influence the structure of overlying sediments. Sedimentary rocks exert such a small magnetic effect compared with igneous rocks that virtually all variations in magnetic intensity measurable at the surface result from topographic or lithologic changes associated with the basement or from igneous intrusives. In mining exploration, the magnetometer is most often used to prospect for magnetic minerals directly, but it is also effective in the search for useful minerals that are not magnetic themselves but are associated with other minerals having magnetic effects detectable at the surface.

Until the middle 1940s all magnetic exploration was carried out on the ground using field methods similar to those in gravity surveys. Today, virtually all magnetic prospecting for oil is done from the air or from ships, as are most reconnaisance surveys for minerals. The speed, economy, and convenience of airborne and marine techniques are responsible for this trend.

The magnetic method of prospecting has a great deal in common with the gravitational method. Both make use of potential fields. Both seek anomalies caused by changes in physical properties of subsurface rocks. Both gravity and magnetics have similar applications in oil exploration. While they are extensively used as reconnaissance tools, there has been an increasing recognition of their value for evaluating prospective areas by virtue of the unique information they provide. Seismic data makes it possible to map an area structurally and, within certain limits, to determine seismic velocities. Gravity data make it possible to assign densities, magnetic data, and magnetic susceptibilities to seismically defined features. The lithologic identifications that can be made by combining these different kind of information have great value in evaluating possibilities for hydrocarbons over a prospective feature. The most common objective in using gravity and magnetics at the present time is this type of evaluation.

Gravity and magnetics employ fundamentally similar interpretation techniques. The magnetic method, for a number of reasons, is more complicated both in principle and in practice. The characteristic of a rock which determines its magnetic effects, the intensity of magnetization (dependent on its susceptibility), has both magnitude and direction, while the corresponding characteristic governing its gravitational pull, the mass (dependent on density), has magnitude only. Magnetic force involves both attraction and repulsion, while all gravitational force is attractive. Also, magnetic effects from rocks may be greatly influenced by small traces of certain minerals, while gravitational effects originate mainly from the rocks' primary constituents.

14-1 BASIC CONCEPTS AND DEFINITIONS

Understanding the magnetic effects associated with earth materials requires a knowledge of the basic principles of magnetism. In this section we shall review the elementary physical concepts that are fundamental to magnetic prospecting.

Magnetic poles If one sprinkles iron filings at random on a sheet of paper which rests on a simple bar magnet, they will tend to line up as shown in Fig. 14-1. The lines along which the filings orient themselves are called *lines of force*. Each of these follows a curved path from a point near one end of the magnet designated as a *pole* to a corresponding pole near the other end. The filings have this orientation because

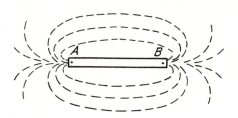

FIGURE 14-1
Lines of force around bar magnet; *A*
and *B* are poles.

each of them is itself a small magnet with an alignment determined by the force field coming from the bar magnet. The latter, if properly balanced, can be oriented by the magnetic lines of force of the earth, which itself acts as a great magnet. Thus if a simple magnet is pivoted at its center so that it can rotate freely in all directions, it will assume a direction governed by that of the magnetic force associated with the earth. One end will always point in the general direction of the North Pole. Near this end one will find the north-seeking, or positive, pole of the magnet. Near the other end is the south-seeking, or negative, pole.

Poles always exist in pairs, but in a very long magnet the lines of force around the positive pole will not be perceptibly affected by the presence of the negative one, and each can be considered to be isolated.

Magnetic force   If two poles of strength P_0 and P, respectively, are separated by a distance r, the force F between them will be

$$F = \frac{1}{\mu} \frac{P_0 P}{r^2} \tag{14-1}$$

The constant μ, known as the *permeability*, depends upon the magnetic properties of the medium in which the poles are situated. The units of pole strength are determined by the specification that F is 1 dyne when two unit poles 1 cm apart are situated in a nonmagnetic medium such as air or vacuum (for which $\mu = 1$). Note the similarity between Eqs. (14-1) and (11-1), the latter expressing the gravitational attraction between two particles. If the poles are of like type, the force is repulsive; if they are unlike, it is attractive.

Magnetic field   The *magnetic field strength* at a point is defined as the force per unit of pole strength which would be exerted upon a small pole of strength P_0 if placed at that point. Thus, the field strength H due to a pole of strength P a distance r away is

$$H = \frac{F}{P_0} = \frac{P}{\mu r^2} \tag{14-2}$$

The magnetic field strength is often expressed in terms of the density of the lines of force or flux representing the field. The unit of H is then expressed as one line of force per square centimeter. It may also be designated in the cgs system as one dyne per unit pole or as one oersted. In the mksg system, the unit of flux density is the tesla, which is 10^4 Oe.

Magnetic moment   Since isolated poles do not exist, actual magnets are generally considered to be magnetic dipoles. A dipole consists of two poles of equal strength P and of opposite sign separated by a short distance L. We define the product PL of the pole strength by the separation as M, the magnetic moment of the dipole. The direction of the moment is along the line between the poles and by convention is toward the north-seeking pole.

Intensity of magnetization (or polarization) Any magnetic material placed in an external field will have magnetic poles induced upon its surface (Fig. 14-2). In the moderately magnetic materials and weak fields we are generally concerned with in geophysical work, this induced magnetization, sometimes called polarization, is in the direction of the applied field, and its strength is proportional to the strength of that field. The intensity of magnetization I may be considered to be the induced pole strength per unit area along a surface normal to the inducing field. It is also equivalent to the magnetic moment per unit volume. This type of magnetization may be looked upon as a lining up of elementary magnets or dipoles, which originally had random orientation, in the direction of the field. It is assumed that the number of magnets thus aligned depends on the strength of the magnetizing field.

Susceptibility In the case of a homogeneous external field H which makes an angle θ with the normal to the surface of a material capable of being magnetized the induced pole strength per unit area is

$$I = kH \cos \theta \tag{14-3}$$

or for a field normal to the surface,

$$I = kH$$

where k, the proportionality constant, is called the *susceptibility*. For a vacuum and for entirely nonmagnetic substances, k is zero. Magnetic materials having positive susceptibilities are known as *paramagnetic*. Grains of such materials tend to line up with their long dimension in the direction of the external field. A paramagnetic material of very high susceptibility may be referred to as *ferromagnetic*. A few substances, such as rock salt and anhydrite, have negative susceptibilities and are designated as *diamagnetic*. Grains of such materials tend to line up with their long dimensions across the external field. Paramagnetic and diamagnetic effects can only be observed in the presence of an external field.

Magnetic induction The magnetic poles induced in a material by an external field H will produce a field of their own, H', which is related to the intensity of magnetization I by the formula

$$H' = 4\pi I \tag{14-4}$$

The total magnetic flux inside the material, as measured in a narrow cavity having

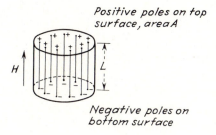

Positive poles on top surface, area A

Negative poles on bottom surface

FIGURE 14-2
Polarization induced in cylindrical pillbox by field perpendicular to ends. Lines of length L represent separation of poles and are used in calculating effective magnetic moments for uniform field and homogeneous material within box.

an axis perpendicular to the field, is called the *magnetic induction B*. This is the sum of the external and the internal fields and is proportional to the external field strength in moderately magnetic materials, as shown by the relation

$$B = H + H' = H + 4\pi I = H + 4\pi k H$$
$$= (1 + 4\pi k)H = \mu H \tag{14-5}$$

Permeability The proportionality constant $1 + 4\pi k$ is equivalent to the permeability μ introduced in Eq. (14-1). Equation (14-5) can thus be written

$$\mu = \frac{B}{H} = 1 + 4\pi k \tag{14-6}$$

The permeability is a measure of the modification by induction of the force of attraction or repulsion between two magnetic poles. Its magnitude depends on the magnetic properties of the medium in which the poles are immersed.

Residual magnetism The direct proportionality between B and H indicated by Eq. (14-5) is at best an approximation, and it breaks down entirely in highly magnetic materials. The behavior of a ferromagnetic substance undergoing cyclic magnetization and demagnetization is illustrated in Fig. 14-3. An unmagnetized sample of magnetic material is placed between the poles of an electromagnet; it produces an external field H that can be controlled by increasing, decreasing, or reversing the current. Magnetization is begun by introducing a current in the coils of the magnet that will increase H from zero (step 1). The induction, which is measured by a ballistic galvanometer connected to a coil wound around the specimen, increases more or less linearly, following the relation $B = \mu H$, until the sample is magnetized to saturation, at which point the curve approaches a horizontal line. The external field is then brought back to zero (step 2), but B does not return to zero; instead it retains the value R, which we call the *remanent magnetization*. If the current and hence H are now reversed (step 3), B will decrease until it is also reversed, eventually approaching saturation in the negative direction. A decrease in the reversing field to zero (step 4) will bring B to $-R'$, while the ap-

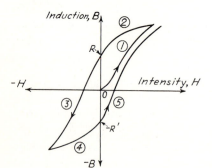

FIGURE 14-3
Hysteresis loop for ferromagnetic material. R and R' are residual inductions.

plication of a positive magnetizing field at this stage (step 5) will reverse the direction of B again and result in a second phase of positive saturation. This entire pattern of magnetization is called a *hysteresis loop*. The curve shows how a body with magnetic susceptibility can remain magnetized after the disappearance of the original magnetizing force.

Magnetization of rocks   Magnetic rocks have almost always acquired their polarization from the earth's field, the exceptions being the rare cases where the magnetization has resulted from lightning. Often the polarization is of the induced type, and its magnitude and direction are determined entirely by the magnitude and direction of the earth's field as it is today. When the earth's field changes, this kind of magnetization changes accordingly. Other magnetic rocks display a remanent magnetization that is not related to the earth's present field but is governed instead by the field that existed when the rock was formed. If such a rock is igneous, its direction of magnetization will be that of the earth's field at the time it cooled from its initial molten state to a temperature below the Curie point. This is called *thermoremanent magnetization*. If the rock is sedimentary, any orientation of its magnetic grains during deposition, generally in quiet water, would have been in alignment with the field that existed when the deposition occurred. This is *depositional remanent magnetization*.

The existence of such remanent magnetization in rocks gives earth scientists a powerful tool for studying the history of the earth's field and by inference the geologic history of the earth itself. This type of study is referred to as *paleomagnetism*. Many of our modern concepts in geology, such as crustal spreading and plate tectonics, were developed, at least in part, with the aid of paleomagnetic evidence. For further information on this application of magnetic data, the reader is referred to a nonmathematical discussion by Strangway.[1]

Units of magnetic intensity   In magnetic prospecting, one usually measures variations in the intensity, or some component of the intensity, of the earth's magnetic field. The conventional unit of field intensity is the oersted, although much of the geophysical literature uses the numerically equivalent gauss. The oersted is too large a unit for practical use in prospecting, since the variations in which we are interested are often less than a thousandth of this amount. The *gamma*, defined as 10^{-5} Oe, is more convenient and has become the most common unit of field intensity for geophysical work. The total magnetic field of the earth is normally about $\frac{1}{2}$ Oe or 50,000 gammas. The gamma is equivalent to the *nanotesla* of the mksg system.

The total field and its components   Any magnetic field associated with a buried source is of course superimposed on that of the earth. The resultant field observed on the surface is a vector which has both magnitude and direction. Early instruments for measuring the field on land like the magnetic balance were designed to read its horizontal or vertical components. Modern airborne and ship-towed instruments, on the other hand, measure the actual field, which is referred to as the *total field*. Interpretation of total fields, which can vary both in magnitude and direction, is necessarily more complex than that involving individual components, such as the vertical, which was the one generally recorded in most early measurements on land.

Electromagnetism   Every electric current generates a magnetic field in a plane perpendicular to it, as shown by the orientation of the compass needles around the wire in Fig. 14-4. The strength of the field is proportional to the current and in the case of a long, straight wire is inversely proportional to the distance from the wire. This principle is important in magnetic prospecting only insofar as it forms the basis for certain types of geomagnetic instruments and enables us to account for the earth's field on the basis of electric currents in its interior.

The magnetic potential and Poisson's equation   Like the gravitational field, the magnetic field has a potential, which is simply the work necessary to bring a unit magnetic pole from infinity to a point a distance r from another source of magnetic polarity of strength P. The magnetic potential U can be expressed by the relation

$$U = \frac{1}{\mu}\frac{P}{r} \tag{14-7}$$

The magnetic field is the spatial derivative of the potential. The direction of the field is determined by the direction associated with the derivative. The x component of the field is simple $\partial U/\partial x$, where x is in the direction of the resultant magnetic field.

The magnetic potential and hence the magnetic field strength associated with a magnetized body can be found at any point in terms of the gravitational potential

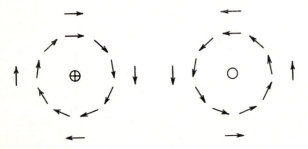

Current into paper—
compasses take
positions shown

Current out of paper—
directions of compass
needles reversed

FIGURE 14-4
Orientation of compass needles around straight wire (perpendicular to paper) carrying current. This experiment shows that the current creates a magnetic field with circular lines of force having their center at the wire.

by use of *Poisson's relation*. This is particularly valuable for predicting the magnetic effect of buried bodies. Magnetic fields from such bodies, even those with simple geometrical forms, are more difficult to derive directly than the corresponding gravitational fields. According to Poisson, the magnetic potential U can be expressed in the form

$$U = -\frac{I}{\gamma\rho}\frac{dV}{di} \tag{14-8}$$

where V = gravitational potential
   i = direction of magnetic polarization
   I = magnetization or polarization
   ρ = density
   γ = universal gravitational constant

The corresponding magnetic field component in any direction s is

$$H_s = -\frac{\partial U}{\partial s} = \frac{I}{\gamma\rho}\frac{\partial}{\partial s}\left(\frac{\partial V}{\partial i}\right) \tag{14-9}$$

If the body is polarized in the z (vertical) direction, and if the horizontal component H_x of the magnetic field is desired, it can be obtained from the equation

$$H_x = -\frac{\partial U}{\partial x} = \frac{I}{\gamma\rho}\frac{\partial}{\partial x}\left(\frac{\partial V}{\partial z}\right) \tag{14-10}$$

while the vertical component H_z will be

$$H_z = -\frac{\partial U}{\partial z} = \frac{I}{\gamma\rho}\frac{\partial^2 V}{\partial z^2} \tag{14-11}$$

These relations will be put to use later in this chapter to develop equations for the magnetic fields of polarized bodies having simple generalized forms for which gravitational fields have been computed in earlier chapters.

14-2 MAGNETISM OF THE EARTH

As the first step in studying magnetic anomalies from shallow features having economic interest, we must subtract larger-scale variations in the geomagnetic field from the magnetic intensities actually observed. To determine what to subtract, we must be familiar with the earth's magnetism on a global scale. Fortunately, a large body of data of this kind has accumulated over the past century from magnetic observatories and field measurements.

In this section we shall summarize those aspects of the earth's magnetism which have the greatest bearing on magnetic prospecting. For more detailed information the reader is referred to Runcorn.[2]

The Magnetic Elements and Their Characteristics

At every point along the earth's surface, a magnetic needle free to orient itself in any direction around a pivot at its center will assume a position in space determined by the direction of the earth's magnetic field F at that point. Normally, this direction will be at an angle with the vertical, and its horizontal projection will make an angle with the north-south direction. Since measuring instruments conventionally installed in magnetic observatories respond only to the horizontal or vertical components of the actual field, it is customary to resolve the total field F into its horizontal component H (separated into X and Y projections) and its vertical component Z (see Fig. 14-5). The angle which F makes with its horizontal component H is the inclination I, and the angle between H and X (which by convention points north) is the declination D.

The quantities X, Y, Z, D, I, H, and F, known as *magnetic elements*, are related as follows:

$$H = F \cos I \qquad Z = F \sin I = H \tan I$$

$$X = H \cos D \qquad Y = H \sin D$$

$$X^2 + Y^2 = H^2 \qquad X^2 + Y^2 + Z^2 = H^2 + Z^2 = F^2 \tag{14-12}$$

All these relations are derivable from the diagram. The vertical plane through F and H is called the *local magnetic meridian*.

If we take observations with our magnetic needle at various points over the earth, we shall find that in the magnetic northern hemisphere the north-seeking end of the needle will dip downward, while in the magnetic southern hemisphere the south-seeking end will be lowermost. In between, there will be a location along every meridian where the needle is horizontal; i.e., the inclination is zero. The curve around the earth connecting all such points is called the *magnetic equator*. It is rather irregular in shape and runs roughly (but not exactly) along the geographic equator. As one follows the compass needle north or south from the equator, the angle of inclination becomes increasingly larger until the point is reached, in Arctic or Antarctic regions, where the needle becomes vertical. These respective points are the earth's north and south magnetic poles. Both poles are displaced

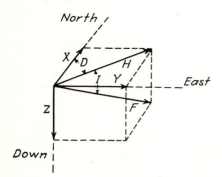

FIGURE 14-5
The magnetic elements.

from the geographic poles by about 18° of latitude. They are not diametrically opposite one another, as the line that joins them passes about 750 mi from the earth's center.

As a first approximation the field of the earth might be looked upon as that of a homogeneous magnetized sphere, as illustrated in Fig. 14-6. Such a picture is much too idealized for practical applications, either in navigation or prospecting, because the earth is *not* a homogeneously magnetized sphere. The irregularities of the earth's field are quite evident on standard isomagnetic charts of the type used in navigation. Plotted on them are lines along which various magnetic elements are equal. They may be lines of equal declination, equal inclination, or equal horizontal or vertical intensity. World maps showing the variation of total intensity F and inclination I are reproduced in Figs. 14-7 and 14-8, respectively.

Variations with Time in the Earth's Magnetic Field

In the early days of navigation with the compass, it was recognized that the earth's magnetic intensity changes its direction slowly and irregularly. Later measurements at magnetic observatories showed many changes in the field that have shorter periods than those originally observed. The variations may be resolved into secular changes, solar diurnal changes, lunar diurnal changes, and changes resulting from magnetic storms.

Secular variation   Slow changes in the earth's field which take place progressively over decades or centuries are known as *secular variations*. Such changes are noted in all the magnetic elements at magnetic observatories everywhere in the world. The rates of change vary with time. In 1925, an annual change in declination of 13 minutes of arc was observed at an observatory in Great Britain. Figure 14-9 shows the rate at which the total field changed around the world during 1965. Note the area of maximum change in the Southern Ocean. Figure 14-10 illustrates how declination and inclination have varied cyclically in London over the past four centuries.

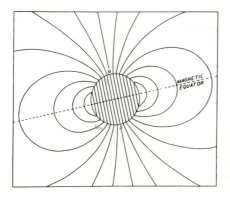

FIGURE 14-6
Magnetic field of an earth having characteristics of a homogeneous sphere. (*After S. Chapman and J. Bartels, Geomagnetism, Oxford, 1940.*)

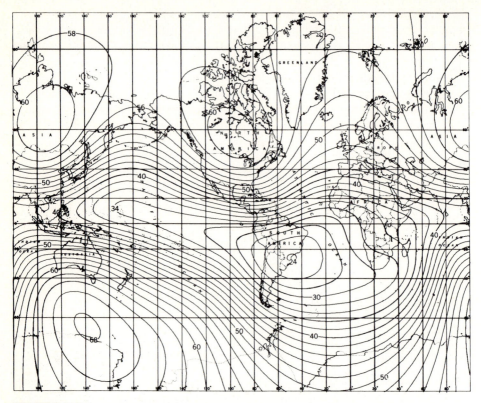

FIGURE 14-7
Total magnetic intensity over earth in 1965. Contour labels in thousands of gammas. [*After E. B. Fabiano and N. W. Peddie, National Oceanographic and Atmospheric Administration National Ocean Survey (formerly C&GS) Technical Report C&GS 38, April, 1969.*]

The projection of current rates of secular variation is not a reliable means of determining past fields or predicting future ones. Paleomagnetic observations show that there have been repeated reversals of the earth's field over geological time, the most recent one having been about 700,000 years ago. Each reversal appears to have taken place in a very short time. The mechanism for the reversals is not well understood, although most theories associate it with the flow of electric currents within the earth which are induced by conducting material in the earth's core set into motion by convection.

Diurnal variation Of more direct significance in magnetic prospecting are the smaller but more rapid oscillations in the earth's field which have a periodicity of about a day and an amplitude averaging about 25 gammas. *Diurnal variations,* as these are called, are regularly recorded at magnetic observatories. The records generally show two types of variations, the *quiet day* and the *disturbed day.* The

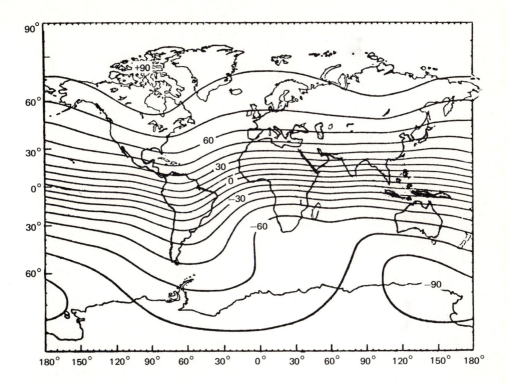

FIGURE 14-8
Magnetic inclination over earth. Contours labeled in degrees. (*Data from U.S.
Naval Oceanographic Office Chart 1700, 8th Ed., 1966.*)

quiet-day variation is smooth, regular, and low in amplitude; it can be separated
into predictable components having both solar and lunar periodicities. The dis-
turbed-day variation is less regular and is associated with magnetic storms. Figure
14-11 shows a magnetogram made on a quiet day at a magnetic observatory in
Tucson, Ariz. Horizontal and vertical intensity and declination are shown on this
record.

Analysis of variometer records on magnetically quiet days shows a definite 24-h
periodicity that depends, to a close approximation, only on local time and geographic latitude. Because of its correlation with the period of the earth's rotation
as referred to the sun, this portion of the variation is referred to as the *solar diurnal
variation*. The average range of this variation in magnetic intensity is of the order
of 30 gammas, the amplitude being intensified in each hemisphere during the local
summer. Most of the elements appear to vary simultaneously but in opposite phase
in the Northern and Southern Hemispheres. The nature of the variation at different
latitudes is illustrated in Fig. 14-12.

Another component in the periodic variation of the earth's magnetic elements
has about one-fifteenth the amplitude of the solar diurnal variation and a periodicity

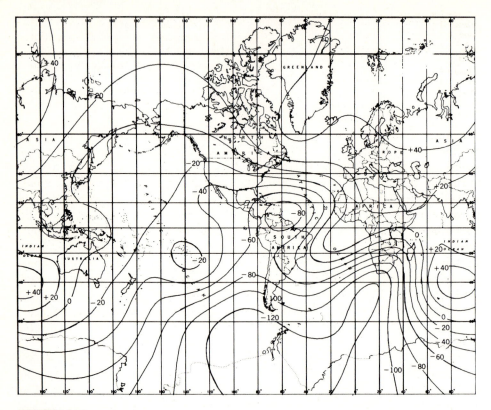

FIGURE 14-9
Annual change of total magnetic field in gammas/yr for 1965. [*After E. B. Fabiano and N. W. Peddie, National Oceanographic and Atmospheric Administration National Ocean Survey (formerly C&GS) Technical Report C&GS 38, April, 1969.*]

of approximately 25 h corresponding to the length of the lunar day. This component of the variation has been related to the earth's rotation with respect to the moon and is referred to as *lunar diurnal variation*. Its low amplitude makes it less important as a source of disturbance in magnetic prospecting than the solar component.

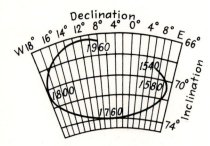

FIGURE 14-10
Changes in declination and inclination at London since 1580.

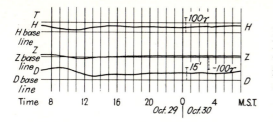

FIGURE 14-11
Magnetogram of typical quiet-day variation in horizontal and vertical intensities H and Z and in declination D at Tucson, Ariz., for Oct. 29, 1947. [*NOAA National Ocean Survey (formerly C&GS).*]

Magnetic storms   In addition to the predictable short-term variations in the earth's field, there are transient disturbances which by analogy with their meteorological counterparts are called *magnetic storms*. Such storms cause considerable disruption in magnetic prospecting operations. The oscillations that take place while they are going on are so rapid and unpredictable that it usually is not feasible to correct for them as with diurnal variations. Magnetic surveys must generally be discontinued during storms of any severity. From the equator to latitudes of 60°, the oscillations during such storms may have amplitudes as great as 1000 gammas. In polar regions, particularly during auroral displays, the storms may be accompanied by much greater amplitudes of magnetic variation. Magnetic storms are not predictable, but they tend to come at intervals of about 27 days. Their frequency correlates with the extent of sunspot activity. The more intense storms begin suddenly, rage simultaneously all around the world, and usually last for several days.

Figure 14-13 shows magnetograms of the horizontal intensity at five well-separated magnetic observatories during the first day of a magnetic storm on Nov. 9, 1947. All records show considerable similarity except for the one from Sitka. The apparent time differences between onsets can be explained by the fact that the times indicated on the records are all local.

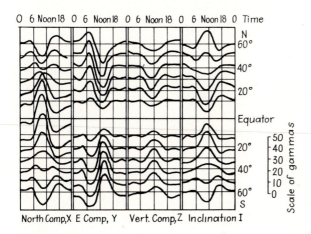

FIGURE 14-12
Solar diurnal variation of four magnetic elements of latitudes 10° apart from 60°N to 60°S at the equinox. (*After S. Chapman and J. Bartels, Geomagnetism, Oxford, 1940.*)

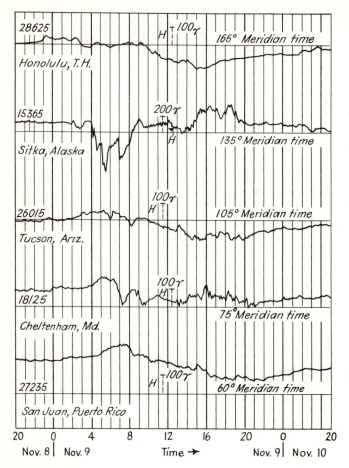

FIGURE 14-13

Magnetograms of horizontal magnetic field variations at five stations during magnetic storm starting on Nov. 9, 1947. Times are all for respective local time zones. [*Data from NOAA National Ocean Survey (formerly C&GS).*]

14-3 MAGNETIC SUSCEPTIBILITY OF ROCKS

The most significant magnetic property of rocks is their susceptibility. This can be measured with the rocks in place by an induction balance of the type developed by Mooney.[3] Pulverized samples of rocks can be placed near a field magnetometer and the deflection of the magnetometer needle caused by the sample can be used to calculate the susceptibility. In situ measurements are preferable wherever it is feasible to make them.

In the laboratory, a number of instruments are employed to measure the susceptibility of rock samples. In one, primary and secondary coils are wound around the space where the sample is placed. When a known current is sent through the primary coil, the voltage induced in the secondary is measured and related to the susceptibility by appropriate calculations. Barret[4] describes an apparatus for measuring susceptibility which is based on this principle.

When an external magnetic field is used for measuring susceptibility, it is customary to specify the strength of this field in tabulating the results. The polarization that governs the response of the sample to the measuring field consists of two parts, the susceptibility polarization kH, dependent on the external field H and the susceptibility k, and the remanent polarization I_p, which governs the residual magnetism in the absence of an external field.

Table 14-1 gives some representative values of susceptibility for several different kinds of rocks and minerals. Note that many of the measurements were made with field of 1 Oe, which is higher than that of the earth by a factor of around 2, while one (that on serpentine) was measured at a field of 30.15 Oe, about 50 times that of the earth. There is some question how applicable such values are in predicting magnetic effects observable in the field. Susceptibilities determined by a method proposed by Slichter[5] are probably more reliable for such a purpose than those measured in fields greater than that of the earth, as were most of the values listed in Table 14-1. Assuming that the magnetism of most rocks is attributable to their magnetite content, he obtained the susceptibility of a rock type by multiplying the volume

Table 14-1 MEASURED SUSCEPTIBILITIES OF ROCK MATERIALS

Material	$k \times 10^6$, cgs units	At H, Oe
Magnetite	300,000–800,000	0.6
Pyrrhotite	125,000	0.5
Ilmenite	135,000	1
Franklinite	36,000	
Dolomite	14	0.5
Sandstone	16.8	1
Serpentine	14,000	30.5
Granite	28–2700	1
Diorite	46.8	1
Gabbro	68.1–2370	1
Porphyry	47	1
Diabase	78–1050	1
Basalt	680	1
Olivine-diabase	2000	0.5
Peridotite	12,500	0.5–1.0

SOURCE: C. A. Heiland, "Geophysical Exploration," Prentice-Hall, Inc., 1940, and L. B. Slichter, "Handbook of Physical Constants," *Geol. Soc. Am., Spec. Paper* 36, 1942.

Table 14-2 CALCULATED SUSCEPTIBILITIES OF ROCK MATERIALS

Material	Magnetite Content and Susceptibility, cgs units						Ilmenite, average	
	Minimum		Maximum		Average			
	%	$k \times 10^6$	%	$k \times 10^6$	%	$k \times 10^6$	%	$k \times 10^6$
Quartz porphyries	0.0	0	1.4	4,200	0.82	2,500	0.3	410
Rhyolites	0.2	600	1.9	5,700	1.00	3,000	0.45	610
Granites	0.2	600	1.9	5,700	0.90	2,700	0.7	1000
Trachyte-syenites	0.0	0	4.6	14,000	2.04	6,100	0.7	1000
Eruptive nephelites	0.0	0	4.9	15,000	1.51	4,530	1.24	1700
Abyssal nephelites	0.0	0	6.6	20,000	2.71	8,100	0.85	1100
Pyroxenites	0.9	3000	8.4	25,000	3,51	10,500	0.40	5400
Gabbros	0.9	3000	3.9	12,000	2.40	7,200	1.76	2400
Monzonite-latites	1.4	4200	5.6	17,000	3.58	10,700	1.60	2200
Leucite rocks	0.0	0	7.4	22,000	3.27	9,800	1.94	2600
Dacite-quartz-diorite	1.6	4800	8.0	24,000	3.48	10,400	1.94	2600
Andesites	2.6	7800	5.8	17,000	4.50	13,500	1.16	1600
Diorites	1.2	3600	7.4	22,000	3.45	10,400	2.44	4200
Peridotites	1.6	4800	7.2	22,000	4.60	13,800	1.31	1800
Basalts	2.3	6900	8.6	26,000	4.76	14,300	1.91	2600
Diabases	2.3	6900	6.3	19,000	4.35	1?,100	2.70	3600

SOURCE: L. B. Slichter and H. H. Stearn, "Geophysical Prospecting," *Am. Inst. Mining Met. Engrs., Trans.*, 1929.

percentage of the magnetite in it by the susceptibility of magnetite (taken as 0.3 cgs unit). He found good agreement between values calculated in this manner and those measured directly in fields having the same strength as the earth's. Stearn[6] has tabulated the magnetite and ilmenite content of a large number of igneous rocks. Susceptibilities calculated on the basis of Stearn's data by Slichter's method are presented in Table 14-2, which shows the range of variation in susceptibility for any given type of rock.

Peters has prepared a bar graph (Fig. 14-14) showing rock susceptibilities from laboratory measurements on a large number of rock samples, igneous, metamorphic, and sedimentary. Igneous and some metamorphic rocks generally have higher susceptibilities than sedimentary rocks, but there is such a large range of variation that it is not possible to identify even the type of rock from magnetic information alone. Under exceptional circumstances sedimentary formations have a high enough magnetite content to be mappable by properly designed magnetic surveys, particularly when high-sensitivity instruments are used. In general, however, the mag-

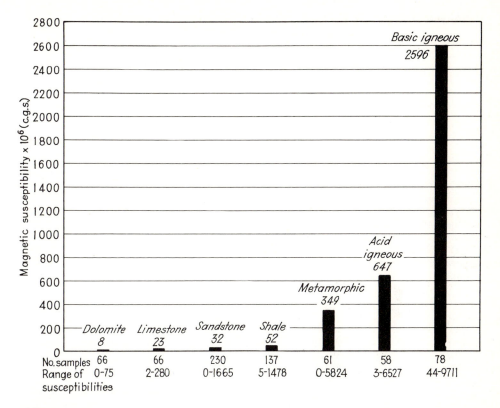

FIGURE 14-14
Average magnetic susceptibilities of surface samples and cores as measured in
the laboratory. (*Compiled by J. W. Peters, Mobil Oil Corp.*)

netization of sedimentary rocks is so small that structural features confined to the
sedimentary section will seldom be reflected in magnetic profiles. A more recent
statistical summary of data on rock susceptibility has been published by Lindsley
et al.[7] It also shows substantial overlap between different rock types. Table 14-3
illustrates the distribution of susceptibility among samples of each type.

14-4   MAGNETIC EFFECTS FROM BURIED MAGNETIC BODIES

Analytical Methods of Computation

The magnetic effects at the earth's surface from buried magnetized bodies of known
shape and susceptibility can be predicted from potential theory by methods similar
to those used for determining the gravitational effects from density contrasts associated with the same bodies. The computations in the magnetic case are considerably

**Table 14-3   RANGE OF MAGNETIC SUSCEPTIBILITY IN MAJOR
ROCK TYPES IN CGS UNITS**

Rock type	Number of samples	Percentage of samples with susceptibility $k \times 10^6$			
		<100	100–1000	1000–4000	>4000
Mafic effusive rocks	97	5	29	47	19
Mafic plutonic rocks	53	24	27	28	21
Granites and allied rocks	74	60	23	16	1
Gneisses, schists, slates	45	71	22	7	0
Sedimentary rocks	48	73	19	4	4

SOURCE: Data from Lindsley et al., "Handbook of Physical Constants," *Geol. Soc. Am. Mem.* 97, 1966.

more difficult, however, since the magnetic dipoles distributed through the body have both attraction and repulsion. The gravitational case is simpler, as might be expected, because all elements of mass attract. The direction of magnetization introduces another complication. Fortunately many of the formulas for the magnetic fields from generalized geometric forms such as spheres can be derived from the corresponding gravitational formulas, making use of Poisson's relationship between magnetic and gravitational potentials, presented on page 482. Because of the complexity involved in predicting total fields mathematically, we shall first consider the analysis of vertical fields from subsurface sources that can be described by simple analytical formulas.

Vertical Intensities from Vertically Polarized Bodies

We shall first consider the magnetic profile that would be obtained from measurements with a vertically sensitive magnetometer on a horizontal plane over an isolated negative pole at depth z, as shown in Fig. 14-15. From Eq. (14-2), remembering that the vertical anomaly $H_z = (z/r)H$, where H, equal to P/r^2, is the

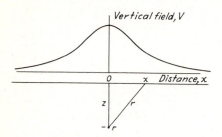

FIGURE 14-15
Vertical magnetic field of buried isolated negative pole.

total field from the pole, we have

$$H_z = \frac{Pz}{(x^2 + z^2)^{3/2}} \tag{14-13}$$

Vertical bar magnet   The field of a thin vertical bar magnet of length L buried below the surface, as shown in Fig. 14-16, can be calculated by assuming the respective poles to be at opposite ends of a dipole and by adding the fields for each pole as determined by Eq. (14-13). In this case

$$L = z' - z$$

If the dipole has been magnetized by the earth's field in the Northern Hemisphere, the north-seeking, or positive, pole will be deeper than the negative pole. Since a buried negative pole reinforces the earth's field, its field is defined as positive. For a dipole having the dimensions shown in Fig. 14-16, the net field will be positive for a distance of about 3500 ft on either side of the buried magnet and negative at greater distances.

In calculating magnetic effects of extended bodies, the pole strength must be expressed in terms of I, the intensity of magnetization, as there is no way of measuring the strengths of equivalent poles directly. For a thick vertical homogeneous magnet of length L, cross-sectional area A, and volume Q, we can assume that the poles are effectively distributed across the faces at the opposite ends of the magnet, the negative poles being on the top and the positive ones on the bottom. Designating the total strength of all the poles of a given sign as P and remembering that I is the magnetic moment (PL) per unit volume, we have

$$I = \frac{PL}{Q} = \frac{PL}{LA} = \frac{P}{A} \tag{14-14}$$

so that $P = IA = kH_zA$, where H_z is the earth's vertical field.

Sphere   A homogeneous sphere vertically polarized in the earth's field with a center buried at a depth large compared with its diameter has an equivalent distribution of dipoles which can be approximated by a single vertical magnet with its

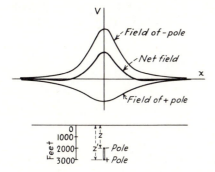

FIGURE 14-16
Vertical intensity over buried vertical dipole or bar magnet. This is also the approximate vertical field of a vertically polarized buried sphere.

negative pole near the center of the upper half of the sphere and its positive pole near the center of its lower half. If these equivalent poles, each of strength P, are a distance L apart and the radius of the sphere with susceptibility k is R, then

$$I = \frac{M}{Q} = \frac{PL}{\frac{4}{3}\pi R^3} \tag{14-15}$$

and

$$P = \frac{4}{3}\pi R^3 \frac{I}{L} = \frac{4}{3}\pi R^3 \frac{k}{L} H_z \tag{14-16}$$

If these approximations are valid, the field of the sphere should be the same as that of a vertical dipole of the type shown in Fig. 14-16, the center of the sphere coinciding in position with the center of the dipole.

This representation of the field is not particularly useful for predicting the magnetic effect of a buried spherical body having a known susceptibility because there is no simple way to estimate L, the effective separation of the poles, accurately. For quantitative analysis it is much more satisfactory to use a formula based on Poisson's relation [Eq. (14-11)] between the vertical magnetic field H_z and the gravitational potential V

$$H_z = -\frac{I}{\gamma\rho}\frac{\partial^2 V}{\partial z^2}$$

For a sphere having a radius R, density ρ, and mass m, the gravitational potential at a distance r from the center is

$$V = \frac{\gamma m}{r} = \frac{\frac{4}{3}\pi R^3 \rho \gamma}{r} \tag{14-17}$$

As $r = (x^2 + z^2)^{1/2}$ in the plane of a profile on the surface of the earth passing through the vertical axis of the sphere,

$$\frac{\partial V}{\partial z} = -\frac{4}{3}\pi R^3 \rho \gamma \frac{z}{(x^2 + z^2)^{3/2}} \tag{14-18}$$

and

$$\frac{\partial^2 V}{\partial z^2} = \frac{4}{3}\pi R^3 \rho \gamma \frac{2z^2 - x^2}{(x^2 + z^2)^{5/2}} \tag{14-19}$$

so that

$$H_z = \frac{\frac{4}{3}\pi R^3 I}{z^3} \frac{2 - \left(\frac{x}{z}\right)^2}{\left[1 + \left(\frac{x}{z}\right)^2\right]^{5/2}} \tag{14-20}$$

Figure 14-17 illustrates the relationship between H_z and x/z for a sphere having a center at a depth z. The field shifts from positive to negative at $x = \sqrt{2}z$.

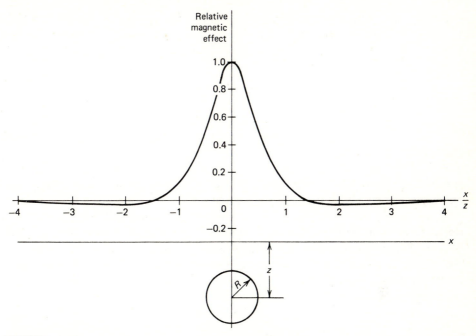

FIGURE 14-17
Vertical magnetic anomaly from a vertically polarized sphere with center buried at depth z. Magnetic effect is plotted versus horizontal distance divided by depth of center.

Horizontal cylinder   Poisson's relation can also be used to derive the vertical magnetic field from a buried horizontal cylinder. Consider a straight wire, infinitely long, with a mass per unit length λ parallel to the y axis, buried at a depth z, and displaced horizontally a distance x from the observing position. By integrating the contributions $(\gamma\lambda\, dl/r)$ to the gravitational potential V at P from each element of length dl one obtains

$$V = 2\gamma\lambda \ln \frac{1}{r} = 2\gamma\lambda \ln \frac{1}{(x^2 + z^2)^{1/2}} \tag{14-21}$$

Then

$$\frac{\partial V}{\partial z} = -2\gamma \frac{\lambda x}{x^2 + z^2} \tag{14-22}$$

$$\frac{\partial^2 V}{\partial z^2} = 2\gamma\lambda \frac{z^2 - x^2}{(x^2 + z^2)^2} \tag{14-23}$$

and from Poisson's relation

$$H_z = \frac{2\lambda I}{\rho} \frac{z^2 - x^2}{(x^2 + z^2)^2} \tag{14-24}$$

Now if the wire is expanded to a cylinder of radius R, a distance constrained to be small compared with z, the mass per unit length becomes

$$\lambda = \pi R^2 \rho$$

so that

$$H_z = 2\pi R^2 I \, \frac{z^2 - x^2}{(x^2 + z^2)^2} = \frac{2\pi R^2 I}{z^2} \, \frac{1 - (x/z)^2}{[1 + (x/z)^2]^2} \tag{14-25}$$

Here the magnetic field changes sign when x is equal to $\pm z$. The field is plotted versus x/z in Fig. 14-18.

Vertical cylinder For a buried vertical cylinder magnetized vertically (Fig. 14-19), the magnetic dipoles within the body can be represented as uniformly distributed vertical magnets parallel to the axis of the cylinder with poles effectively distributed along sheets of uniform pole density coinciding with the respective end surfaces. This simplified representation makes it possible to take advantage of the approach developed on page 381 for determining the vertical gravity at a specified point above a horizontal sheetlike body of arbitrary shape. It was shown there that the vertical component of gravity is proportional to the solid angle subtended by the

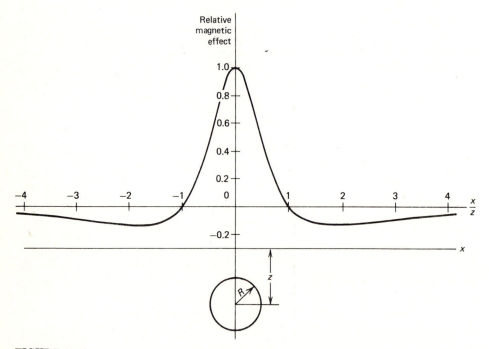

FIGURE 14-18
Vertical magnetic anomaly from a vertically polarized horizontal cylinder along a line on the surface perpendicular to its axis. Magnetic effect is plotted versus horizontal distance x divided by depth z of center.

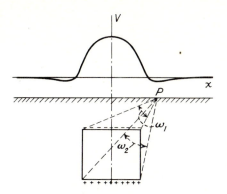

FIGURE 14-19
Approximate vertical field on surface
from buried vertical cylinder. ω_1 and ω_2
are solid angles subtended respectively
by top and bottom surfaces. $H_z = I \times 10^5 (\omega_1 - \omega_2)$ gammas.

body multiplied by its surface density. The vertical magnetic effect is also proportional to the solid angle subtended at the observing point, but the angle is multiplied by the intensity of magnetization of the surface. The magnetic effects of the top and bottom surfaces of the cylinder are treated separately, each sheet having magnetization of a different sign and subtending a different angle at the measuring point. Directly above the body the greater angle will be subtended by the top sheet, consisting of negative poles, and the net vertical effect will be positive; at a large horizontal distance from the axis the bottom sheet, consisting of positive poles, will subtend a larger angle and the effect will be negative. Where the two contributions are equal, the net field will of course be zero. The vertical-anomaly field H_z at P can be predicted by the formula $H_z = I(\omega_1 - \omega_2)$, where the ω's are the respective solid angles. If the length of the cylinder becomes so great that the angle ω subtended by the bottom surface approaches zero, the vertical component of the anomaly, which is now from the top surface, can be expressed by the relation

$$H_z = I\omega = \frac{\pi R^2 z I}{(x^2 + z^2)^{3/2}} \qquad (14\text{-}26)$$

where R is the radius of the cylinder and z the depth to its top. This formula was obtained by determining the component of the area of the top surface that is perpendicular to a line from the measuring point to the center of the upper surface. The solid angle is this component divided by the square of the length of this line.

Faulted horizontal slab In a similar way, the vertical magnetic field on the surface caused by a thin, extended horizontal slab of magnetic material can be computed near its edge by determining the angles subtended at points on the surface by the respective magnetized sheets at the top and bottom. This representation is useful in computing the effect that would be expected over a fault.

For a thin faulted horizontal slab of thickness t buried at a depth z, the field on the surface at a distance x from a line directly above and parallel to the faulted edge of the slab is

$$H_z = It\frac{x}{x^2 + z^2} \qquad (14\text{-}27)$$

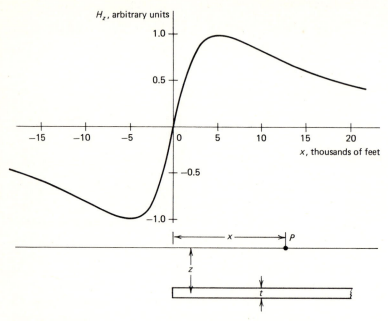

FIGURE 14-20
Vertical magnetic field from thin faulted slab of thickness t, vertically polarized, as a function of horizontal distance x from the fault edge.

as shown in Fig. 14-20. The down-thrown part of the slab is assumed to be too deep to have an observable magnetic effect.

Vertical sheet It is quite simple to compute the vertical magnetic effect of a body having the form of a vertical magnetized sheet with a horizontal top buried at depth z and a bottom considered to be infinitely deep. This is because the magnetic effect of the sheet, as with the very long vertical cylinder, can be represented as coming from negative poles distributed along the top edge of the body. The contribution from the positive poles on the bottom does not need to be considered in this model as they are too far away to affect the field at the surface.

Thus the shape of the magnetic profile along a line on the surface perpendicular to the plane of the sheet is the same as that of a gravity profile from a buried source having the form of a horizontal line except that the strength of the field is now proportional to the intensity of the magnetization rather than to the density.

For a two-dimensional vertical slab of thickness t with top and bottom having respective depths of z_1 and z_2 (Fig. 14-21), the field at a distance x from the center of the slab measured perpendicular to the surface trace of its center plane is

$$H_z = 2It \left(\frac{z_1}{z_1{}^2 + x^2} - \frac{z_2}{z_2{}^2 + x^2} \right)$$

(14-28)

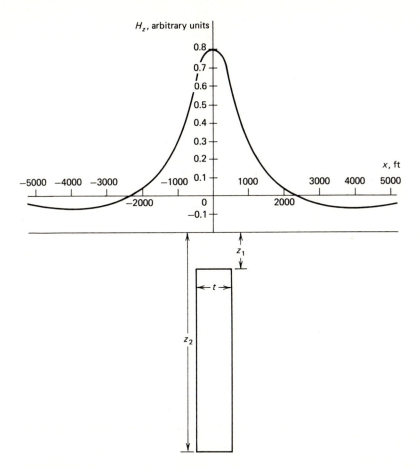

FIGURE 14-21
Vertical magnetic field from vertically polarized vertical magnetic slab (long
dimension perpendicular to page) with thickness t, height of top surface z_1, and
height of bottom surface z_2.

Summary and examples The generalized forms for which vertical magnetic
fields have been derived can be used to approximate a number of geologic features.
A knoll on the basement surface might be represented by a sphere, a basement ridge
by a horizontal cylinder, and intrusives of different types by vertical cylinders or
sheets. A fault affecting the basement surface could correspond to the edge of a
rectangular slab.

To demonstrate how these formulas can be used to predict the maximum fields
to be expected from a number of typical basement structures, ore bodies, or intrusive features, several examples will be considered.

EXAMPLE 1 Assume a two-dimensional vertical dike containing 12 percent
magnetite ($k = 0.5$). It is 50 ft wide and extends in depth from 50 to 300 ft. The

angle from the upper surface is ω_1 and from the lower surface ω_2. What would be the maximum anomaly over the dike? Let $k = 0.5 \times 12\% = 0.06$, and H_z, the vertical component of the earth's field, be 0.6 Oe, so that

$$I = kH_z = 0.036 \text{ cgs unit}$$

Then from Eq. (14-28), taking x as zero,

$$H_{z,\max} = 2 \times 10^5 It\left(\frac{1}{z_1} - \frac{1}{z_2}\right) = (2 \times 10^5) \times (36 \times 10^{-3}) \times (50)\left(\tfrac{1}{50} - \tfrac{1}{300}\right)$$

$$= 6000 \text{ gammas}$$

EXAMPLE 2   If a magnetic ore body (30 percent magnetite) is in the shape of a sphere of 200 ft diameter with its center buried 200 ft below the surface, then

$$k = 0.5 \times 0.30 = 0.15 \text{ cgs unit}$$

$$I = kH_z = 0.15 \times 0.6 = 0.09 \text{ cgs unit}$$

$$H_{z,\max} = 8.38 \times 10^5 \frac{R^3 I}{z^3} = (8.38 \times 10^5)\left(\tfrac{100}{200}\right)^3 (0.09)$$

$$= 0.0945 \times 10^5 = 9450 \text{ gammas}$$

The formula used here is obtained from Eq. (14-20), with $x = 0$ and the numerical constant applied that was computed for distances in thousands of feet.

EXAMPLE 3   If the same sphere were of serpentine, the magnetization would be 14,000/150,000 (about one-tenth) as great as in the case above and the vertical field would then be about 900 gammas. If the center of the serpentine mass were buried 400 ft instead of 200 ft below the surface, the field would be $(200/400)^3$ or one-eighth as great, i.e., about 110 gammas.

EXAMPLE 4   For an example more applicable to petroleum prospecting, consider a buried basement ridge of basalt ($k = 0.005$, $I = 0.003$ cgs unit) with a shape that could be approximated by an infinitely long horizontal cylinder 3000 ft in radius with a center 6000 ft deep. Then, letting $x = 0$ in Eq. (14-25) and using a numerical factor based on expressing R and z in thousands of feet, we have

$$H_{z,\max} = 6.28 \times 10^5 \frac{R^2 I}{z^2} = 6.28 \times 10^5 \frac{(3000)^2}{(6000)^2} (0.0003)$$

$$= 46.5 \text{ gammas}$$

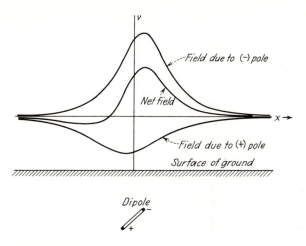

FIGURE 14-22
Vertical magnetic field of inclined dipole.

If the radius of the cylinder were 1500 ft, the anomaly would be one-fourth as great.

We thus see that in prospecting for minerals, anomalies as high as thousands of gammas may be encountered, but in petroleum exploration topography on a deep-seated basement is not likely to yield anomalies greater than tens, or at the most hundreds, of gammas. Changes in susceptibility associated with lithology contrasts in the basement can account for much larger magnetic anomalies than are normally observed from topographic features.

Inclined polarization   It is only at the magnetic poles that the earth's field is actually vertical. At other places, it makes an angle with the vertical that increases with distance from the poles until it becomes 90° at the magnetic equator. For induced magnetization, the polarization of a buried magnetic body will have the same direction as the earth's field. The calculation of vertical fields from buried masses which have inclined polarization is generally more complex than for vertically polarized bodies, although it is simple (as shown in Fig. 14-22) to determine the field from an inclined dipole, for which the observed vertical field is the resultant of the fields from the respective poles at either end.

It is often difficult to distinguish between inclined polarization and actual dip in the interpretation of asymmetric fields such as that in Fig. 14-22. An inclined dike should result in an unsymmetrical magnetic anomaly along a line across its strike. Figure 14-23 illustrates the relations of vertical intensity to distance as predicted by Cook[8] for profiles crossing dikes having different strikes and dips.

Prediction of total fields   Thus far we have considered only the vertical components of the magnetic fields at the surface which originate from various types of buried bodies. Actually, only a small proportion of the magnetic surveys currently

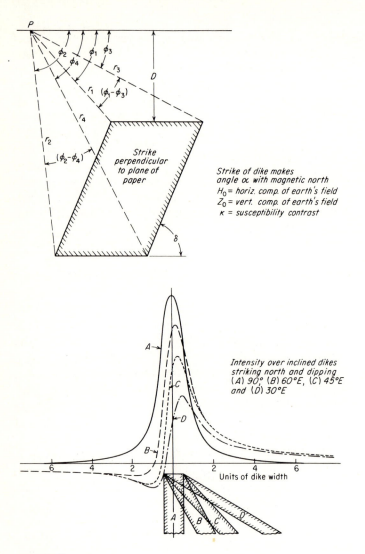

FIGURE 14-23

Formulas and typical profiles for vertical and inclined dikes. (*Adapted from Cook.*[8])

Vertical components of anomaly at P:

1. General case

$$V = 2k \sin \delta \left[(H_0 \sin \alpha \sin \delta + Z_0 \cos \delta) \ln \frac{r_2 r_3}{r_1 r_4} \right.$$

$$\left. - (H_0 \sin \alpha \cos \delta - Z_0 \sin \delta)(\phi_1 - \phi_2 - \phi_3 + \phi_4) \right]$$

2. Vertical dike infinitely deep, striking north

$$V = 2kZ_0(\phi_1 - \phi_3)$$

carried out for exploration utilize instruments measuring the vertical component of gravity. To the extent that they are used at all, vertical magnetometers are now employed almost entirely for mining exploration in surveys where specific ore bodies are sought. Total-field instruments are also used for such surveys on land. By far the greatest amount of current activity is with airborne or shipborne instruments that measure total fields only.

It usually requires more sophistication to interpret the total field associated with a subsurface body than the vertical component of the field. This is because the total field is the vector sum of the earth's field and that of the buried body. It is customary to assume that the magnetization of the subsurface source is in the same direction as the earth's field. But if the magnetization is remanent rather than induced, such an assumption could lead to an erroneous interpretation of the magnetic data. Techniques for interpreting total fields will be considered in Chap. 16.

14-5   INSTRUMENTS USED FOR MAGNETIC MEASUREMENTS

Schmidt-Type Magnetic Field Balance

In the early years of magnetic prospecting for petroleum, when all work was on land, the magnetic field balance was the standard instrument employed for magnetic field measurements. Along with the dip needle, it was extensively used in mining exploration although it has now been essentially replaced by vertically oriented flux-gate magnetometers and by portable proton instruments, both discussed later in this section. We describe its design here because of the large amount of data from it still in active files and because of the simplicity for pedagogic purposes of the principle upon which its operation is based.

Originally developed by Schmidt, the magnetic balance consists of a magnet pivoted near, but not at, its center of mass, so that the magnetic field of the earth

---

3. Vertical dike of finite depth, striking north
$$V = 2kZ_0(\phi_1 - \phi_2 - \phi_3 + \phi_4)$$

4. Vertical dike infinitely deep, striking east
$$V = 2kZ_0\left[\frac{H_0}{Z_0}\ln(r_3/r_1) + (\phi_1 - \phi_3)\right]$$

5. Vertical dike of finite depth, striking east
$$V = 2kZ_0\left[\frac{H_0}{Z_0}\ln\frac{r_2 r_3}{r_1 r_4} + (\phi_1 - \phi_2 - \phi_3 + \phi_4)\right]$$

6. Infinite inclined dike, striking north
$$V = 2kZ_0\sin\delta\left[\cos\delta\ln\frac{r_3}{r_1} + \sin\delta(\phi_1 - \phi_3)\right]$$

7. Infinite inclined dike, striking east
$$V = 2k\sin\delta\left[(H_0\sin\delta + Z_0\cos\delta)\ln\frac{r_3}{r_1} - (H_0\cos\delta - Z_0\sin\delta)(\phi_1 - \phi_3)\right]$$

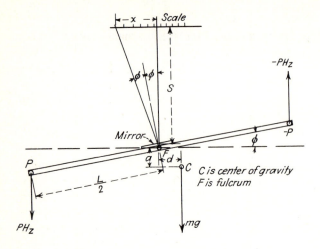

FIGURE 14-24
Principle of Schmidt-type vertical magnetometer. The deflection ϕ is expressed by the formula $\tan \phi = (MH_z - mgd)/mga$ where M is the magnetic moment.

creates a torque around the pivot that is opposed by the torque of the gravitational pull upon the center. The angle at which equilibrium is reached depends on the strength of the field. To attain high sensitivity a great deal of precision is required in the design and construction of the mechanical and optical systems. Magnetic field balances do not measure absolute fields, but they respond to changes in vertical or horizontal components as small as 1 gamma under favorable conditions. Separate models are available for measuring vertical and horizontal fields.

The operation of the vertical balance is illustrated schematically in Fig. 14-24. Assume an approximately horizontal magnet oriented perpendicular to the magnetic meridian so that the horizontal component of the earth's field exerts no effect. The magnet is balanced on a knife-edge displaced from the center of gravity C a horizontal distance d and a vertical distance a. The vertical component of the earth's magnetic field, acting on the poles as shown, tends to cause counterclockwise rotation, while the gravitational field causes a clockwise rotation. The position of equilibrium is indicated on a graduated scale by a light beam reflected from a mirror attached to the magnet. If the vertical field changes, as at a different measuring location, the position of equilibrium shifts and the difference in scale readings gives a measure of the difference in the vertical fields.

A sensitivity of 10 gammas per scale division is obtainable on this type of instrument. Readings are generally to the nearest tenth of a division.

The Flux-Gate Magnetometer

Initially used for detecting submarines from aircraft during World War II, the flux-gate magnetometer was the first type of instrument to be used for magnetic measurements from a fixed-wing aircraft. It has also been employed, but to a lesser extent, for magnetic surveys on the ground. This instrument makes use of a ferromagnetic element of such high permeability that the earth's field can induce a magnetization that is a substantial proportion of its saturation value (see Fig. 14-3). If the earth's field is superimposed upon a cyclic field induced by a sufficiently large alternating current in a coil around the magnet, the resultant field will saturate the core. The place in the energizing cycle at which saturation is reached is observed, and this gives a measure of the earth's ambient field.

Principle of operation   Figure 14-25 illustrates how the field is measured by elements of this type. Two parallel cores, each with a magnetization curve of the type shown at the upper left, are aligned with their axes in the direction of the earth's field. Primary coils in series magnetize the two cores with the same flux density but in opposite directions, since they are wound oppositely around the respective cores. The earth's field at a given instant reinforces the field set up by one of the coils and opposes the field of the other. Each coil has, in addition, a secondary winding, the two secondaries being connected to a voltmeter that reads the *difference* of the two outputs.

Consider the magnetization of a single core in the absence of an ambient field

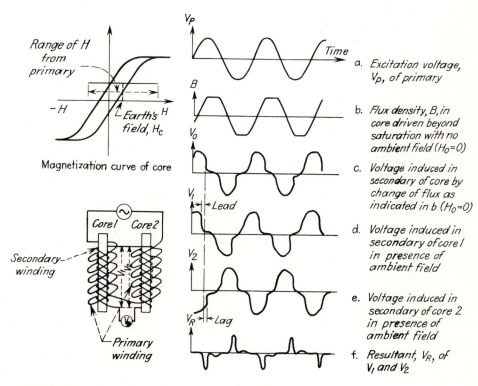

a. Excitation voltage, V_p, of primary

b. Flux density, B, in core driven beyond saturation with no ambient field ($H_0 = 0$)

c. Voltage induced in secondary of core by change of flux as indicated in b ($H_0 = 0$)

d. Voltage induced in secondary of core 1 in presence of ambient field

e. Voltage induced in secondary of core 2 in presence of ambient field

f. Resultant, V_R, of V_1 and V_2

FIGURE 14-25
Principle of flux-gate magnetometer.

(as when the core's axis is normal to the earth's field). The sinusoidal exciting field H (see curve a of Fig. 14-25) drives the core past saturation at the top and bottom of each cycle (as shown by the truncated peaks in curve b). The secondary voltage is proportional to the rate of change of magnetic flux and thus will dip toward zero during the portion of the cycle when the core is saturated (as in curve c).

Now if an ambient field is introduced which aids the magnetization from the exciting current, the saturation point, as indicated by the dip in the secondary voltage, is reached earlier in the cycle than it would be in the presence of an ambient field opposing the exciting field (curves d and e). If the voltage outputs of both coils are connected in opposition, the resultant output (curve f) consists of pairs of pips as shown. The height of the pips is, within reasonable limits, proportional to the ambient magnetic field.

Airborne instruments of the flux-gate type Two types of flux-gate magnetometers are in use for airborne surveys. Both were employed for detection of submarines during World War II. One is the Gulf Airborne Magnetometer,[9,10] which operates on peak voltages; the other, which responds to second-harmonic signals, is exemplified by the AN/ASQ-3A[11] designed by the Naval Ordnance Laboratory and the Bell Telephone Laboratories. The McPhar 555AB instrument, made in Canada, is also of this type. Development of the Gulf instrument had already been started by Victor Vacquier as a tool for magnetic prospecting when the entry of the United States into World War II led him to adapt the device to submarine detection. In it the output signal is amplified, and the height of the pulse thus obtained determines the dc voltage, which is recorded by a moving-tape potentiometer. The system is illustrated in Fig. 14-26. The record from this instru-

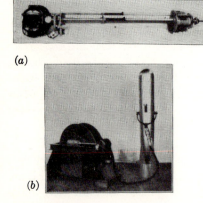

(a)

(b)

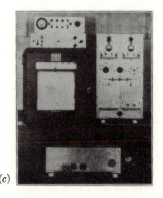

(c)

FIGURE 14-26
Gulf airborne magnetometer: (a) complete magnetometer head unit with housing removed; servomotor for orientation at right end, flux-gate unit at left; (b) housing and reeling mechanism for detector element; (c) power supply (*bottom*), high-speed pen recorder (*left*), and associated control equipment. (*Gulf Research & Development Co.*)

ment shows the total magnetic field versus time. To increase sensitivity, most of the ambient field is balanced out by compensating coils. The axis of the sensing element is kept in orientation with the total field by a magnetic vane mounted on a disk which rotates at 3600 r/min in a plane perpendicular to the axis. If the field is not aligned with the axis, a ripple is generated by the vane which causes servomotors to drive gimbals oriented in mutually perpendicular directions that align the axis with the field.

The AN/ASQ-3A system makes use of a core which is magnetized to saturation many times a second, back emf being set up which has even harmonics proportional to the strength of the external field. These harmonics of the voltage are amplified and rectified, and the output is recorded on a moving-tape milliammeter. The orientation system works from the outputs of three identical magnetometer coils with axes mutually perpendicular. The output actuates servomotors which turn two of the coils into null position, leaving the third to record the total field.

The sensitivity of flux-gate magnetometers for airborne use is about 1 gamma, adequate for most applications to oil exploration except where gradiometers are used or where especially high precision is required.

Flux-gate instruments for use on the ground   Flux-gate magnetometers for ground surveys are manufactured in Canada and Finland. Their characteristics have been tabulated by Hood.[12] They are designed for easy portability and rapid operation, being hand-held with a strap around the operator's neck. All of them measure vertical fields. Precision of the readings from different instruments ranges from 0.5 gamma (Sharpe MF-1-100) to 12.5 gammas (Jolander 46-65), most types being close to 5 gammas. Each has a series of ranges extending generally from 0 to 1000 gammas at the low end and from 0 to 100,000 at the high end.

Nuclear Resonance (Proton) Magnetometers

The proton magnetometer, based on the phenomenon of nuclear magnetic resonance, was introduced in 1954 by Packard and Varian.[13] The first paper discussing its applicability to exploration was published by Waters and Phillips[14] in 1956. An early review of its performance appeared in 1958.[15] The most widespread use of this instrument has been in airborne surveys, but systems have also been developed for operation on land and for towing behind a ship.

Principle of nuclear magnetic resonance   Most chemical elements have a magnetic moment. The nuclei of these elements may be looked upon as minute magnets in the form of spheres spinning about their magnetic axes. According to the laws of quantum mechanics, such spheres will tend to align themselves either parallel or perpendicular to any external magnetic field. Thus, the randomly oriented nuclei of this type will separate into two spinning groups, one parallel and one antiparallel, when the external field is applied. The nuclei in the latter group will have a higher energy level than those in the former. For this reason, a larger number of the nuclei will point in the antiparallel direction, and there will be a resultant magnetic force in that direction.

The simplest nucleus having this property is the proton, or hydrogen nucleus. Because oxygen has no magnetic moment, a water sample can be regarded as an assemblage of protons so far as its nuclear resonance characteristics are concerned.

Consider a bottle of water in the earth's field alone. Suddenly an external field of much greater magnitude is applied in a direction approximately perpendicular to the earth's field. The orientation of the proton's moment will shift until it points in the direction corresponding to the resultant of the two fields. If the external field is 100 or more times as great as that of the earth, the resultant field generated in the bottle will be in approximately the same direction as the applied field, but it will not reach its full value immediately upon application of the external field. It will approach it exponentially with a time constant of approximately 3 s.

When the external magnetic field is removed, the magnetic moment will return to its original direction, which is that of the earth's field, by precessing around that field at an angular velocity $\omega = \gamma_p H$, where γ_p, the gyromagnetic ratio of the proton, is a constant and H is the strength of the earth's field. The mechanism is similar to the precession of a gyroscope. The frequency of precession is about 2000 Hz.

The precessional oscillation will induce an electric potential in a coil wound around the water sample. To determine the earth's total field, it is only necessary to measure the frequency of this induced voltage. The most convenient way to measure this frequency is to count a predetermined number of cycles of the precessional voltage and to time these cycles precisely with an appropriate electronic system.

Operation of nuclear magnetometers   Proton magnetometers are available in three models, one for airborne surveys, one for towing behind a ship, and the third for use on the ground. All measure the total magnetic field of the earth rather than its components. The precision of the airborne or ship-towed unit is substantially greater than that of the ground magnetometer.

AIRBORNE PROTON MAGNETOMETER   In a typical airborne instrument such as the Varian, a polarizing field of about 100 Oe is applied to the towed sensing head, which is an electrostatically shielded bottle of water. Frequencies of the precessing nuclei are measured once each second over an interval of about $\frac{1}{2}$ s. Figure 14-27 is a simplified block diagram showing the operation of the frequency-measuring system. After the polarizing field is cut off, the precessional signal induced in the coil causes an electronic gate to open. This gate closes after a specified number of sine waves (about 500 for a $\frac{1}{2}$-s interval) have passed through the circuit. While the gate is open, it passes a signal from a standard-frequency (100-kHz) oscillator whose sine waves are in turn counted by the *fast counter*. A binary digital recorder registers the number of cycles of the high-frequency signal that have passed through the gate during the time corresponding to the specified number of cycles from the pro-

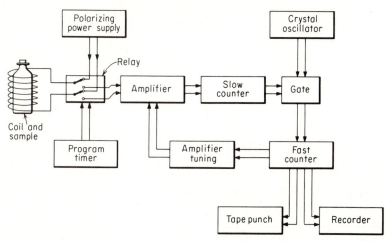

FIGURE 14-27
Block diagram of Varian nuclear-resonance magnetometer. (*Varian Associates.*)

cessional oscillation. This number is converted into a voltage and presented as a pen-written line on a moving paper chart. It may also be stored digitally on tape. Since the count of the oscillator signal is accurate to the nearest cycle, and since 50,000 cycles will pass through the counter in $\frac{1}{2}$ s, the accuracy of measurement is 1 part in 50,000, or about 1 gamma in most parts of the world. A precision of 0.5 γ or better is claimed for some models.

The entire instrument weighs less than 120 lb, so that it can be installed in any light single-engine airplane. The instrument is free from drift, and the sensing element does not require the complicated mechanism needed for orienting towed elements of the type used in a flux-gate system.

GROUND MAGNETOMETER   The nuclear magnetometer used on the ground operates on a similar principle but is designed to be portable. A number of modifications of the airborne instrument have been introduced with this object in view. Five commercial systems for ground observation are described by Hood,[12] most of them having similar properties.

A typical instrument has the following characteristics. Total weight, including batteries, is less than 16 lb. All electronic parts are transistorized, and the accuracy ranges from 1 to 10 gammas, even the higher value being adequate for most mining applications. Readings, which are taken at 6-s intervals, are made visually on meters and are recorded on data sheets by the operator.

Six plug-in tuning units are provided, each covering a different range of 12,000 gammas total field. The unit to be used is selected for the value of field appropriate to the part of the world where the work is being done. A vibrating-reed frequency meter, calibrated in gammas, makes it possible to read the frequency directly. A total of 53 reeds, spaced in 20-gamma steps, gives a full-scale range of more than 1000 gammas. There is also a 12-step range selector, accounting for total range of 12,000 gammas for each plug-in unit. An earphone enables the operator to monitor the signal audibly and to tell from the change in pitch when the earth's magnetic field is varying along his traverse.

SHIP-TOWED NUCLEAR-PRECESSION MAGNETOMETERS   The nuclear-precession magnetometer has been adapted for use at sea by installation in a waterproof container designed to be towed behind a ship by a marine cable. The cable is several hundred feet long to avoid interference from fields originating on the ship. Sampling rate ranges from one reading in 3 s to one in 60 s, with 6 s the most usual interval. Geophysical reconnaissance of previously unexplored offshore shelf areas has increased since the early 1960s and it has become customary in surveys carried out for such a purpose to obtain gravity and magnetic data simultaneously with seismic reflection, and occasionally sonobuoy refraction information. The principle of operation of the marine proton magnetometer is the same as that employed in airborne surveys. Figure 14-28 illustrates a widely used commercial sensor designed for towing from a ship.

FIGURE 14-28
Varian proton magnetometer designed for towing from a ship being lowered into the water. (*Varian Associates of Canada, Ltd.*)

Optical-Pumping Magnetometers (Rubidium and Cesium Vapor)

During the early 1960s a type of magnetometer was introduced which was capable of measuring the earth's total field with a substantially greater precision than was possible with previous instruments. The principle on which it operates makes use of a development in radio-frequency spectroscopy called *optical pumping*. A description of this process designed for the technically oriented layman was written by Bloom.[16] For accounts at a higher technical level, the reader is referred to Bloom[17] and Duffus.[18]

Energy states in atoms   The energy of an atom is governed by the orbits in which its valence electrons revolve about its nucleus and also by the direction of the spin axes of the electrons themselves. The orbits which an electron can occupy as well as its axial orientation are limited to a small number, each corresponding to a specific level of energy. An electron can drop from a higher energy level to a lower one with the emission of a packet, or quantum, of electromagnetic radiation, which has an energy equivalent to that lost by the electron. Also, irradiation with energy from an external source can raise an electron from an orbit with lower energy to one having higher energy, in which case a quantum of the external radiation having the energy thus gained is absorbed by the atom. In either case the energy change ΔE is related to the frequency of the radiation γ by Planck's relation $\Delta E = h\gamma$, where h is Planck's constant. This means that light emitted by an atom when an electron drops from an orbit having a given energy to another orbit with an energy ΔE lower will have a frequency equal to $\Delta E/h$. By the same token, an electromagnetic wave packet with a frequency of $\Delta E/h$ irradiating the atom will raise the energy of an atom at the lower level an amount ΔE by shifting the electron from a lower-energy to a higher-energy orbital state. For such a transition to take place the radiation must have exactly the frequency corresponding to the difference in orbital energies. No other will bring about such a reaction.

Mechanism of optical pumping   Figure 14-29a shows the mechanism involved in the transitions that result in optical pumping. In the absence of a magnetic field, the valence electron of an alkali-metal atom (such as rubidium or cesium) has two states, B, the normal or ground level, and A, the excited level. In the presence of a magnetic field like that of the earth each of the states splits up into pairs of levels, A_1 and A_2 and B_1 and B_2, respectively, the energies of which are much closer to one another than those of A and B, as the diagram shows. A_1 and B_1 are the energies when the magnetic moment due to electron spin is parallel to the field. A_2 and B_2 are the energies when it is antiparallel. The respective separations in energy between A_1 and A_2 and B_1 and B_2 are proportional to the total magnetic field. If one can measure the energy absorbed or emitted in a transition between level B_1 and B_2, one can determine the magnitude of the field. This energy is proportional to the frequency of the radiation associated with the transmission. While the energy represented by the transition from A to B is in the optical or visible range, the

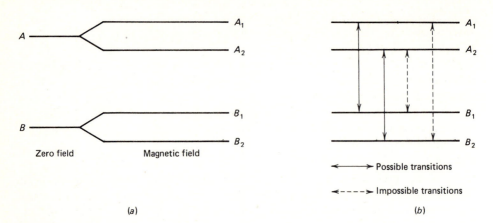

(a)   (b)

FIGURE 14-29
Change of energy levels of electrons in alkali metals brought about by introduction of magnetic field: (a) levels with (right) and without (left) external
field; (b) allowed and forbidden transitions when atoms are illuminated with
light having a frequency of $(A - B)/h$.

smaller amount of energy corresponding to the transition from A_1 to A_2 or from
B_1 to B_2 falls in the much lower-frequency radio-wave spectrum.

Under normal illumination there is a constant shifting of electrons from one orbit
to another, and the number of atoms at each energy level is essentially equal, so that
any transitions balance. In order to observe radiation associated with the transitions related to the magnetization it is necessary to put an abnormal number of
electrons into levels A_2 or B_2. This is the purpose of the *pumping*, which is done by
irradiation with visible light.

The reason the magnetic field splits each orbital energy level into two sublevels
is that the spin axis of the electron going around the orbit is parallel to the axis
of the orbit in the higher-energy state and antiparallel to it in the lower. If we
illuminate the electrons with light polarized so that it can raise the energy levels
only of electrons with either the parallel or antiparallel spin axis but not both, we
can bring about an abnormal distribution of energy states.

Figure 14-29b illustrates how this occurs. Suppose the atoms are illuminated with
the light having a frequency $(A - B)/h$ which has a polarization that affects only
the electrons with parallel spin. The only transition that is then possible is from
B_1 to A_1. Atoms raised to A_1 in this way will generally remain in that state a
short time and then emit energy, dropping back to the B_1 level. Atoms in the
B_2 state cannot be raised to a higher level by the irradiating light, and hence they
remain in B_2, but those at the B_1 level can be raised. Eventually, all the atoms at
the B_1 level will be shifted to A_1 and the B_1 level will become empty.

Each quantum of light that raises an atom from a lower to a higher energy level
is removed from the light beams by absorption. Such removal, which involves
many atoms, causes a readily measurable drop in the intensity of the light. But when

all atoms have been removed from the energy level B_1 at which such absorption takes place, the vapor containing the atoms becomes entirely transparent to the light.

If at this point the atoms can be returned to the B_1 state, some of the light quanta will be captured by them and their energy will be raised to the A_1 level. This will have the effect of decreasing the intensity of the light beam when it passes through the container holding the vapor. Atoms in the B_2 state can be lifted to the B_1 state if they are irradiated with radio waves having a frequency that corresponds to the energy difference between B_1 and B_2. To determine this difference and hence the magnetic field, it is only necessary to pass radio waves of continually varying frequency through the medium and to note the frequency at which the intensity of the light beam passing through the vapor suddenly drops, as indicated by a photocell registering the change. The intensity decreases because transitions from B_1 to A_1 have now resumed. This is the principle of the optical-pumping magnetometer.

Detection systems The most commonly used detection system is the one employed in the Varian cesium-vapor magnetometer. This is a single-cell self-oscillator, as illustrated in Fig. 14-30; the active constituent is cesium vapor. To understand the operation of this system one must picture the radio waves as causing a wobble of the spinning electrons in the cesium atoms; this has the effect of an alternating shutter, which produces a flicker in the transmitted light with a frequency the same as that of the radio waves themselves. This flicker is detected on a photocell, amplified, and fed back into radio-frequency coils around the container holding the vapor.

In the Varian system illustrated on Fig. 14-30, either rubidium or cesium vapor could be used, although the rubidium lamps and cells are now obtainable only for replacement. Currently manufactured units operate only with cesium vapor.

A sensitivity of 0.005 gamma is claimed for the Varian cesium-vapor instrument. The frequency range of the sensing element is 70 to 280 kHz. This corresponds to a

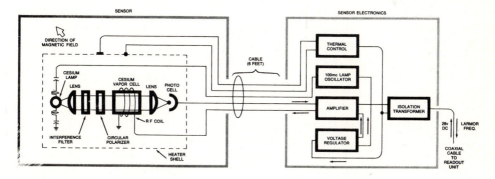

FIGURE 14-30
Block diagram of Varian cesium-vapor magnetometer. (*Varian Associates of Canada, Ltd.*)

range in magnetic field of 20,000 to 80,000 gammas, a range that can be covered in 10 steps of 6000 gammas, each requiring a different crystal.

The rubidium-vapor magnetometer, or the cesium-vapor unit which is now supplanting it, is generally used where high precision is needed. If the magnetic relief is very small, or if the object of the survey is to observe magnetic effects from sedimentary rocks which are superimposed upon larger anomalies originating in the basement, the extra precision of the optical-pumping instrument can be most valuable. Such precision is critically needed in magnetic gradiometers, described in the following paragraphs.

The Magnetic Gradiometer

The advantages of measuring gradients in the earth's field between nearby detecting elements for magnetic exploration are considerable, and efforts to develop gradiometers for this purpose were reported as early as 1954, when Wickerham[19] described some unsuccessful experiments with flux-gate instruments. Gradients are not affected by diurnal variations, and the complex operating procedures needed to remove their effects in conventional surveys are not required. Moreover, the differences in field between two nearby points emphasize anomalies from shallow sources having steep gradients at the expense of low-gradient variations of the type one generally wishes to eliminate as regional background.

The gradiometers most widely used in exploration employ pairs of optical-pumping (rubidium- or cesium-vapor) sensing elements. The difference between the readings of the two magnetometers divided by their separation (about 100 ft)

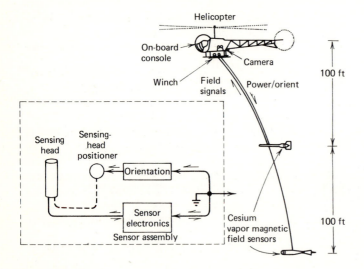

FIGURE 14-31
Arrangement of sensors in magnetic gradiometer. Inset shows control elements for each sensing unit. (*After Slack et al.*[20])

yields the gradient of the total field in the direction of the line separating the sensors. The detectors are ordinarily suspended from a helicopter or fixed-wing aircraft so that one is about 100 ft above the other. An instrumental system for measuring such gradients, the *geomagnetic gradiometer*, is described by Slack et al.[20]

Figure 14-31 illustrates the configuration of the system when a helicopter transports it. The sensors, installed in streamlined birds, are cesium-vapor magnetometers. A winch allows the birds to be lowered or raised when the helicopter is in flight. In the cockpit of the helicopter there are controls for orienting the sensing elements within the birds, both for calibration and for recording the outputs. The individual magnetometer readings are recorded on a strip record calibrated to give a reading of 20 to 100 gammas full scale. The difference channel has a sensitivity an order of magnitude greater, 2 to 10 gammas full scale.

Special gradiometers employing rubidium- and cesium-vapor sensors have been developed for subsea exploration. In experiments reported by Breiner[21] each magnetometer was placed in a fiber-glass dinghy, the two boats being attached to each other catamaran style with a separation of 12 ft. The assemblage was towed 100 ft behind a launch containing the recording equipment. The gradients transverse to the profile were observed in this way. This system has been used off the Alaska coast in experiments to test its suitability for locating minerals in placer deposits.

REFERENCES

1. Strangway, David W.: "History of the Earth's Magnetic Field," McGraw-Hill, New York, 1970.
2. Runcorn, S. K.: The Magnetism of the Earth's Body, pp. 498–533, in J. Bartels (ed.), "Handbuch der Physik," vol. 47, 1956.
3. Mooney, Harold M.: Magnetic Susceptibility Measurements in Minnesota, I: Technique of Measurement, *Geophysics*, vol. 17, pp. 531–543, 1952.
4. Barret, W. M.: A Semi-Portable A. C. Susceptibility Meter, pp. 17–24, in "Early Geophysical Papers," Society of Exploration Geophysicists, Tulsa, Okla., 1947.
5. Slichter, L. B.: Certain Aspects of Magnetic Surveying, *Trans. Am. Inst. Min. Met. Eng.*, vol. 81, Geophysical Prospecting, 1929, pp. 238–260, 1929.
6. Stearn, N. H.: A Background for the Application of Geomagnetics to Exploration, *Trans. Am. Inst. Min. Met. Eng.*, vol. 81, Geophysical Prospecting, 1929, pp. 315–344, 1929.
7. Lindsley, D. H., G. E. Andreasen, and J. R. Balsley: Magnetic Properties of Rocks and Minerals, pp. 543–552, in Sydney P. Clark, Jr. (ed.), "Handbook of Physical Constants," rev. ed., *Geol. Soc. Am. Mem.* 97, 1966.
8. Cook, Kenneth L.: Quantitative Interpretation of Vertical Magnetic Anomalies over Veins, *Geophysics*, vol. 15, pp. 667–686, 1950.
9. Muffly, Gary: The Airborne Magnetometer, *Geophysics*, vol. 11, pp. 321–334, 1946.
10. Wyckoff, R. D.: The Gulf Airborne Magnetometer, *Geophysics*, vol. 13, pp. 182–208, 1948.
11. Jensen, Homer: Geophysical Surveying with the Magnetic Airborne Detector AN/ASQ-3A, *U.S. Nav. Ordnance Lab. Rep.* 937, 1945; L. H. Rumbaugh and L. R.

Alldredge: Airborne Equipment for Geomagnetic Measurements, *Trans. Am. Geophys. Union*, vol. 30, pp. 836–849, 1949.

12. Hood, Peter: Magnetic Surveying Instrumentation: A Review of Recent Advances, pp. 3–31, in "Mining and Groundwater Geophysics,/1967" *Geol. Sur. Can., Econ. Geol. Rep.* 26, Ottawa, 1970.

13. Packard, M., and R. Varian: *Phys. Rev.*, vol. 93, p. 941, 1954.

14. Waters, G. S., and G. Phillips: A New Method of Measuring the Earth's Magnetic Field, *Geophys. Prospect.*, vol. 4, pp. 1–9, 1956.

15. Anonymous: The Nuclear Magnetometer, *Petrol. Times*, vol. 62, pp. 46–51, 1958.

16. Bloom, A. L.: Optical Pumping, *Sci. Am.*, vol. 203, pp. 72–80, 1960.

17. Bloom, A. L.: Principles of Operation of the Rubidium Vapor Magnetometer, *Appl. Opt.*, vol. 1, pp. 61–68, 1962.

18. Duffus, H. J.: Techniques for Measuring High-Frequency Components of the Geomagnetic Field, in S. K. Runcorn (ed.), "Methods and Techniques in Geophysics," vol. 2, Interscience, New York, 1966.

19. Wickerham, W. E.: The Gulf Airborne Magnetic Gradiometer, *Geophysics*, vol. 19, pp. 116–123, 1954.

20. Slack, Howard A., Vance M. Lynch, and Lee Langan: The Geomagnetic Gradiometer, *Geophysics*, vol. 32, pp. 877–892, 1967.

21. Breiner, Sheldon: Ocean Magnetic Measurements, *IEEE Ocean Electron. Symp., Honolulu, Aug. 29, 1966.*

MAGNETIC SURVEYING TECHNIQUES

Magnetic prospecting may be carried out on land, from aircraft, and from ships. The field techniques are of course different for the three types of surveys, although the airborne and marine operations generally employ very similar detecting and recording equipment. Land operations require instrumentation that would not be suitable for the mobile type of surveys undertaken in airborne or marine work.

On land, magnetic observations are usually made at fixed positions comparable to the discrete stations used for gravity surveys. In aerial and marine surveys, magnetic fields are recorded continuously from sensors in motion. For a long time the most widely used tool for land surveys was the Schmidt-type magnetic balance designed for measuring the vertical component of the earth's field or the horizontal component at low magnetic latitudes. In recent years flux-gate nuclear-precession (proton) magnetometers have been put to increasing use for measurements on land.

The field operations in magnetic prospecting for oil are generally set up somewhat differently from those in mineral exploration. As the targets in oil exploration are deeper than in mineral prospecting, the station spacing in land work or the line spacing in airborne or marine surveys will generally be greater in searching for oil than in mining exploration. Ground magnetometers are rarely used in the former type of survey, as virtually all magnetic prospecting for oil is carried out from aircraft or ships.

In this chapter, we shall consider the acquisition and reduction of magnetic data from land, airborne, and marine surveys. Chapter 16 will be concerned mainly with interpretation and geological applications of the data.

15-1 MAGNETIC SURVEYS ON LAND

Positioning and occupation of stations   In land surveys for petroleum, stations are usually spaced about 1 mi apart; in sectionized parts of the United States and Canada, they are placed at section corners. For mining exploration, the spacing of stations is much closer, separations being as little as 25 ft. Regardless of the objective, stations should be set up at safe distances from all iron objects that might interfere with the normal field. There should be no railroad tracks within 125 yd, no automobiles within 30 yd, and no wire fencing (particularly in the north-south direction) within 35 yd. Power lines, bridges, culverts, and houses should be avoided. Moreover, the operator should carry a minimum of magnetic material about his person.

If a single instrument is used on the survey, a base station is chosen at the beginning of the day's work and a schedule is arranged that will permit a return to the base station every 2 or 3 h during the course of the day's readings. The purpose of the reoccupations is to keep track of diurnal variations in the earth's field, the principle being much like that employed in gravity prospecting. Another method of eliminating diurnal variation is to use two instruments, a base-station magnetometer and a field magnetometer. This way it is only necessary to return the field magnetometer to the base for checking at the end of the day, as the base magnetometer is adapted for continuous recording of the variation at the base station. Any difference between relative readings of the two instruments at the beginning and end of the day is distributed among the stations that were occupied in between. Many geophysicists find this technique unsatisfactory because a lack of linearity is sometimes observed in instrumental drift characteristics as well as in the diurnal variation itself.

Field operations with portable nuclear-precession magnetometers designed for land work were described on pages 511–512.

Reduction of land data   Before magnetic readings taken in surveys on the ground can be mapped, several corrections must be applied, different ones being required for different instruments. A diurnal correction must always be made and in some types of survey a "normal" correction for irregularities in the earth's field is needed.

Diurnal correction   The diurnal variation of the earth's magnetic field, discussed in the last chapter, may have an amplitude as great as 100 gammas, and it must therefore be taken into account in reducing magnetic data taken in the field. There are several techniques by which diurnal effects can be removed. In rough terrain, where frequent reoccupation is not practical, it is possible to determine the approximate background field at any time from published variometer curves taken at the nearest permanent magnetic observatories. Since the curves are often tens of gammas different at places only a few hundred miles apart, this approach would be unreliable for precision work anywhere but in the vicinity of a magnetic observatory although it might be quite adequate for reconnaissance surveys. A procedure more commonly used for single-instrument operations is to return to the base station every 2 h or so and to construct a variation curve for each day's work by plotting the readings at this station against time. Irregularities as great as 10 gammas might be missed during intervals between reoccupations of the base, so that this technique is not reliable if a precision of a few gammas is necessary.

Where two instruments are employed no farther than 50 mi apart and where the one at the base station records continuously, the diurnal correction should be accurate to a few gammas.

Use of a diurnal-variation curve for correcting field observations requires, of course, that the time of each reading at the field station be noted. A reference time is chosen on each curve, and all values are corrected to this time.

When variations in the earth's field are large and highly irregular, as in magnetic storms, it is not feasible to correct for them at all and field work must be discontinued until conditions return to normal. A recording magnetometer set up in a central location such as a regional office is often used to monitor such disturbances.

Normal corrections   In the last chapter it was noted that there are continual changes in the magnitude and direction of the earth's main field as one goes from one place to another. These changes correspond in a way to the variations of the earth's gravity with latitude. The correspondence is not exact because the changes in magnetic field over comparable distances constitute a much larger proportion of the earth's total field than with corresponding changes in gravity. They are also less easily predictable.

Such variations are in many ways similar to the large-scale regional anomalies observed in gravity work, but they can seldom be associated with known geology. Correction for such variations is made by methods quite similar to those used for removing regional trends in gravity interpretation. In the United States the regional magnetic fields can be picked from maps and tables published by the National Oceanographic and Atmospheric Administration*, based on measurements of the earth's magnetic field at stations having an average grid spacing of 10 mi (see Howe and Knapp[1]). This spacing is much too great for locating anomalies from small subsurface structures, but it is small enough to allow the mapping of regional magnetic trends.

The standard maps based on these surveys are contoured with an interval of 1000 gammas. It may be desirable to work with contour lines having a much

---

* The U.S. Coast and Geodetic Survey, which formerly published such maps, is now incorporated into NOAA.

smaller spacing. In such cases, normal contours with intervals as small as 10 gammas can be drawn across the area covered in a magnetic survey by interpolation of the published contours or by determination of their gradients.

The principles and techniques involved in making normal corrections are so similar to those that enter into the removal of regional magnetic fields that it is sometimes hard to differentiate between the two operations. The principal difference is in the origin of the changes in field being removed. If there is a geological basis for them, they should be looked upon as regional effects which could in themselves have significance, and techniques of anomaly separation should be applied. If their origin cannot be associated with crustal sources and appears to be deeper in the earth, procedures such as those just discussed for making normal corrections should be employed.

Normal corrections can often be made by a computer using data from the observations being reduced provided that the area of the survey is large enough. After the readings and their position coordinates are put into computer storage, a program is introduced for fitting the data to a low-order polynomial surface. Residuals are obtained by subtracting values calculated for this surface at each station from the observed magnetic field at that station.

Sample reduction   The reduction procedure can be illustrated by a calculation sheet (Table 15-1) of the kind used in an actual survey with a Schmidt-type magnetometer. Although the instrument is now obsolete, the reduction procedures for it illustrate the principles employed in the corresponding calculations for modern field magnetometers. The instrument has a sensitivity k of 16.2 gammas per scale division. The diurnal-variation curve obtained by a second instrument at

Table 15-1   SAMPLE REDUCTION CALCULATION

	Station					
	Base	1	2	3	4	Base
Time	9:00	10:00	11:30	1:15	2:30	3:00
A, east reading	5.1	6.9	7.6	9.0	9.5	5.8
west reading	5.4	6.7	7.2	8.7	9.7	5.8
Average A	5.2	6.8	7.4	8.9	9.6	5.8
kA	84	110	120	144	156	94
Diurnal variation	−15	−22	−40	−25	−12	−6
Regional correction	−316	−328	−340	−352	−364	−316
Base correction	545	545	545	545	545	545
Z, gammas	298	305	285	312	323	316
Instrument closure error	0	−3	−8	−11	−14	−16
Adjusted Z	298	302	277	311	309	310

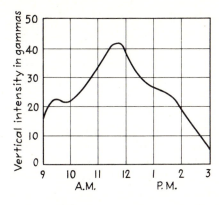

FIGURE 15-1
Diurnal-variation curve for sample
computation of Table 15-1.

a base station is shown in Fig. 15-1. The regional corrections can be interpolated from government maps of vertical magnetic intensity. The base correction is simply the vertical magnetic field strength established at the base station by ties with a preexisting net. Any difference between values for successive occupations must be considered as a closure error and distributed among all observations within the loop.

15-2   SURVEYS WITH AIRBORNE MAGNETOMETERS

The vast majority of the magnetic surveys that have been made over the world has been carried out with airborne instruments. The advantages in economy and data quality of measuring magnetic fields from the air were recognized long before appropriate equipment became available, and experiments with airborne observations were carried out as early as 1921, when a captive balloon was used for magnetic measurements over a known ore body in Sweden. Even before magnetic submarine detectors using flux-gate elements were developed, efforts were being made to increase the sensitivity of existing flux-gate magnetometers so that they could be adapted for prospecting from aircraft. One of the pioneers in this research was Victor Vacquier, who perfected a flux-gate sensor in 1941 that was put to successful use for submarine detection during World War II. The first exploration work with this instrument was done in 1945 over the Naval Petroleum Reserve No. 4 in northern Alaska, and during the next 10 years only flux-gate magnetometers were used for airborne magnetic surveys.

The nuclear-magnetic-resonance (proton) magnetometer first went into service in airborne prospecting during the middle 1950s, and the optical-pumping type (rubidium and cesium vapor) in the early 1960s. Both types were first put to use in marine exploration during the later 1960s.

Virtually all airborne detecting systems measure total fields rather than individual components of the earth's field. Thus interpretation of airborne magnetic data is more complex than it is for data obtained on land with instruments which record only the vertical (or horizontal) component of the field.

Position location For mapping aeromagnetic data it is necessary to correlate all total-field readings with the position of the plane at the instant they are taken. A number of techniques may be used for this purpose. One involves radio devices such as Shoran; another makes use of aerial photography. The widely used doppler system is based on the doppler frequency effect that is observed in radio signals sent out from a moving plane and returned from the ground.

Shoran is a line-of-sight electronic system for measuring the distances from each of two ground stations to the plane. Its principle of operation was discussed on page 139 in connection with seismic operations at sea. A microwave transmitter-receiver unit operating at two frequencies (one for each ground station) is installed on the plane and constitutes the master station of the system. The slave units are located at the ground stations. Continuous-wave radio systems involving the use of ground-based stations are Lorac, Raydist, and Decca. They are primarily employed in marine surveys, and the principle behind their operation is discussed on pages 139 to 142. The doppler equipment makes use of the doppler frequency shift of waves emanating from a moving source. In this system, pulsed or continuous waves are sent diagonally downward fore and aft (see Fig. 15-2), and the frequencies of the returned beams are compared with them to obtain the true ground speed. The heading is obtained from a special kind of magnetic compass and is maintained by a directional gyro, which is also used as an integrating device. Over land the precision of positioning is theoretically better than 1 part in 1000, but the system may not be reliable at flying elevations less than 350 ft.

Where positions are determined by aerial photography, continuous-strip cameras are generally used. In one type of camera an electronic intervalometer is used for marking time intervals on all records. This permits fiducial marks at any desired interval ranging from 1 to 120 s. The marks are also impressed on the moving paper strips containing the record of magnetic intensity and the altimeter record. Over water or featureless terrain such as Arctic wastes or rain-forest areas, aerial photography is of course not suitable, and electronic techniques are mandatory.

Where mapped features are widely recognizable from the aircraft, it is possible to plot positions on the map by direct observation from the plane. Fiducials marked on the magnetometer record correspond to the mapping locations. Recognizable features are seldom observed frequently enough to permit the use of such a simple positioning approach except when the magnetometer is transported by helicopter or slow-flying aircraft at low ground clearances.

Location of magnetometer head with respect to the plane Several arrangements have been used for the placing of the magnetometer head during flight. The principal objective in each case is to remove magnetic effects from iron in the plane's engines and elsewhere in the aircraft.

The most usual procedure is to tow the magnetic element (referred to as the bird) at the end of a cable (Fig. 15-3). The cable is about 80 to 100 ft long, and at this distance the magnetic effect of the plane should be negligible. The outer case is streamlined for maximum stability and minimum air resistance and is equipped

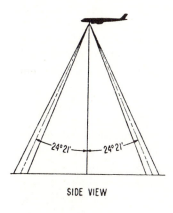

SIDE VIEW

FRONT VIEW

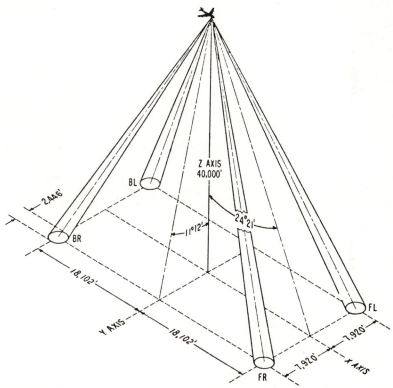

FIGURE 15-2

Principle of the doppler system. Four radar beams are sent out, two ahead and two behind the plane. The relation between the doppler frequency shifts for the waves scattered back to the plane from the four areas where the beams intercept the ground makes it possible to determine the speed and direction of the plane's travel. The position is obtained by integrating the speed. (*The Bendix Corporation, Avionics Division.*)

FIGURE 15-3
An airplane towing a magnetometer during inflight recording. (*Aero Service Corporation.*)

with fins to prevent rotation. The bird is usually lowered through a trap door in the floor of the cabin after the takeoff.

Sometimes the magnetic detecting element is mounted on the wing tip or at the tip of the airplane's tail. When the magnetometer is attached to the plane, compensation must be introduced in the aircraft itself. This arrangement is required when another receiver is trailed at the same time for electromagnetic or other airborne measurements. Another arrangement is to put the detector inboard. This can be done if compensation is provided for the magnetic effects of the iron in the aircraft itself.

Instrument operation Operation of the magnetometer in flight is relatively simple. With all pen-recording instruments in current use, the magnetometer observer need only adjust the sensitivity, since there is an automatic mechanism for

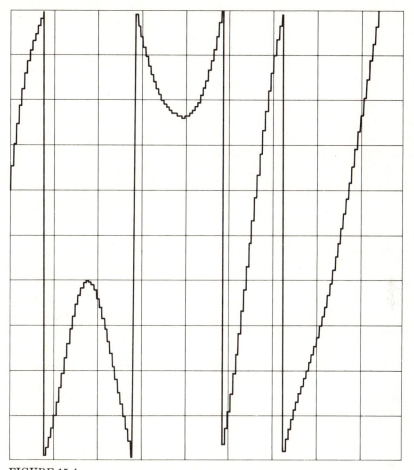

FIGURE 15-4
Tracing of typical chart record of total magnetic field made in flight. Full-scale
deflection is 100 gammas, each division representing 10 gammas. The horizontal
scale is four divisions per nautical mile. When the pen is about to go off scale,
it is automatically shifted to the opposite edge of the tape. (*Aero Service
Corporation.*)

shifting the recording pen to keep it from going off scale. Figure 15-4 illustrates a
typical tape record with a magnetic signature that has been shifted by this
mechanism.

The Varian proton magnetometer records the total field with a pen on moving
paper and also displays it visually in gammas on a set of emitting diodes, one for
each digit. The data may be simultaneously recorded in digital form on punch tape
or magnetic tape, the data being printed out in numerical form after the flight is
completed. Such digital systems require no scale adjustment or scale shifting, nor is
there danger of going off scale. Also there is no practical limitation to the precision

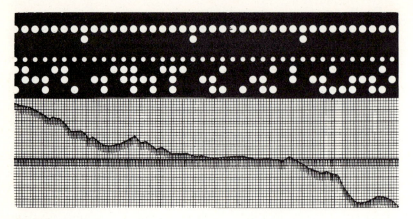

FIGURE 15-5
Sample tapes from Varian nuclear-resonance magnetometer. Lower strip shows
conventional analog presentation. Upper strip is equivalent digital record in
punch-mark form. (*Sparton Aerial Surveys, Ltd.*)

obtainable where the magnetic relief is large except that it is hard to detect sharp
or narrow anomalies in this way. Figure 15-5 shows a typical punch tape recorded
during a proton-magnetometer survey.

Selection of flight patterns and flying elevations   The flight trajectory and
elevation are governed by the information desired. The most usual pattern consists of a rectangular grid, the lines in one direction having a considerably closer
spacing than those in the other. At all but very low magnetic latitudes the more
closely spaced lines will be laid out perpendicular to the magnetic strike of the
region or feature being mapped. This strike will generally follow the structural trend
of the crystalline or igneous rocks in the area. Deviations as much as 30° from the
normal to the strike direction should not be detrimental. Near the magnetic equator,
the earth's magnetizing field is approximately horizontal, and the poles of the
induced magnetization will be distributed along surfaces trending in the east-west
direction. If the geologic trends are north-south, the magnetic anomalies will not
reflect the structure very well, as elongated surfaces in vertical planes oriented in
this direction will not accumulate an appreciable density of magnetic poles. Such
effects of magnetic latitude on total-field anomalies will be discussed further in
Chap. 16.

The spacing of the lines is governed by the shapes, dimensions, and depths of the
targets. A narrowly localized source, e.g., a vertical pencil-shaped plug or a small
ore body having the shape of a sphere, will be indicated by an anomaly with circular
contours, and the half width of the anomaly will be about half as great as the depth
of the sphere's center. A line spacing equal to the depth should result in at least one
line showing an amplitude for the anomaly that is no less than half its peak value.
For depth estimation, however, this much attenuation of maximum amplitude

could lead to appreciable errors. Elongated anomalies are usually observable on
more than one line if the traverses are across strike. Agocs[2] has published a statistical
study of the spacing necessary to detect an anomaly of a given size. To point out an
example, the Mattagami ore body in Quebec did not show up on the map made from
an aeromagnetic survey with flight lines $\frac{1}{2}$-mi apart because it happened to fall
halfway between the lines. A subsequent survey with $\frac{1}{4}$-mi spacing gave a very
much better indication. Figure 15-6 illustrates the magnetic anomalies observed
over the feature on the maps obtained with the respective spacings of flight lines.

The height of the flight path may vary from 200 ft to many thousands of feet
above the ground surface. In mining surveys the pilot endeavors to fly at a constant
height above the ground. As it is not feasible to hold a specified clearance above the
terrain within even as much as 50 ft, an effort is made to keep within a predetermined *mean* height above the ground surface. The elevation is chosen on the basis
of line spacing. If the spacing is $\frac{1}{4}$ mi, a clearance of 500 ft will generally be specified.

In petroleum exploration, the sources of magnetic anomalies, which are almost
always in the basement rocks, are much deeper than in mining surveys, and there is
ordinarily little advantage in maintaining a constant elevation above ground level.

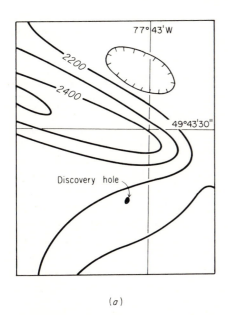

(a)

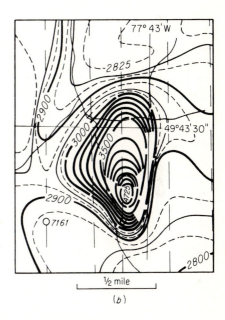

½ mile

(b)

FIGURE 15-6
Comparison of magnetic pictures obtained over Mattagami sulfide area in
Quebec with $\frac{1}{2}$- and $\frac{1}{4}$-mi flight spacings: (a) Survey with $\frac{1}{2}$-mi spacing gave
nose-shaped high of 200-gammas reversal trending from west; (b) $\frac{1}{4}$-mi spacing
shows high with almost 2000-gammas closure that was missed in other survey.
Flight lines are shown only on map (b). The anomaly in (b) was responsible for
the subsequent discovery of one of the largest sulfide deposits in Canada.
(*Geological Survey of Canada and Hunting Airborne Geophysics, Ltd.*)

It is customary to fly, so far as is feasible, at a uniform elevation above sea level but within a range of 750 to 1500 ft above the earth's surface. Such elevations make it possible to avoid magnetic effects from localized near-surface sources such as structures containing iron.

The actual configuration of the flight pattern is determined by the nature of the anomalies being sought, by the locations of ground stations for reference in positioning, and particularly by the magnitude of the diurnal variations causing misclosures around loops.

The most common type of pattern in current use is illustrated in Fig. 15-7. This consists of mutually perpendicular sets of parallel legs separated from one another by a distance which usually ranges from 5 to 10 mi. All the north-south legs are flown consecutively, after which the east-west legs are traversed (or vice versa).

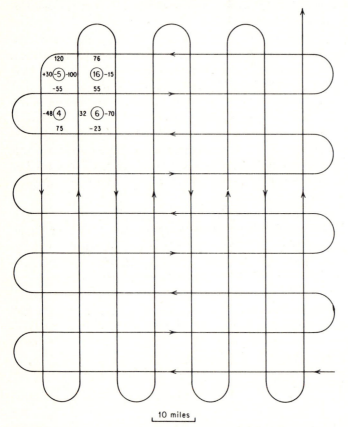

|← 10 miles →|

FIGURE 15-7

Standard flight pattern to facilitate elimination of diurnal variation and other errors. Numbers around rectangles in upper left illustrate method of adjustment for errors. Numbers along sides are differences in gammas of total field between adjacent corners. Numbers in center show misclosures around each loop. Values at corners are adjusted by least squares to minimize misclosures.

Adjustment for diurnal variation at crossings is made by least squares to minimize the misclosures around the individual loops. This is illustrated in the figure for the four northwesternmost loops of the diagram. Differences in observed magnetic field are determined from the tape records for each side of each of the rectangles. If there were no drift or diurnal change, the differences would add up to zero around the loop. The actual misclosures written down inside the rectangles are stored in a computer and adjustments for them as determined by least squares are distributed smoothly over the entire flight trajectory. One computational technique for doing this has been described by Gibson.[3]

It is evident that this procedure does not correct errors caused by diurnal changes taking place while a traverse between tie points is being flown. Reford and Sumner[4] show how the maximum error from such changes can be predicted in terms of tie-line spacing and aircraft speed. These parameters can be changed to keep the adjusted data within the desired accuracy.

For most operations, a continuously recording ground magnetometer located at a base station near the area of the survey monitors ambient geomagnetic variations. This could be used for diurnal corrections, but it is not generally employed for this purpose, as the diurnal variation can change over distances that are short compared with most flight trajectories. One of the functions of the magnetometer is to monitor magnetic storms. When such a storm begins, recording from the air is discontinued until conditions return to normal. Where high precision is required, the ground monitor is used to ensure that the rate of diurnal variation is linear during the flight so that the differences around the loops can be distributed without excessive error. If the departure from linearity exceeds specifications, the survey may have to be repeated.

Surveys from helicopters   In the early days of aeromagnetic work, Lundberg[5] conducted surveys over the Canadian shield from a helicopter using a specially designed earth inductor as magnetometer. More recently, flux-gate and nuclear-resonance instruments have been used in helicopters for mining surveys. Helicopters are more suitable than fixed-wing aircraft for magnetic surveys over very rugged terrain because they can maintain more uniform elevation above the ground surface. Variable ground clearance from airplanes over such terrain may result in strong herringbone patterns on the magnetic maps. Also, it is easier to follow a flight line of limited length with precision from a helicopter. This could be an important factor in detailed mining surveys.

15-3   MAGNETIC SURVEYS AT SEA

In making magnetic measurements from ships it is almost always necessary to place the sensor far enough away from the ship to prevent interference from magnetic materials on board. A common arrangement for accomplishing this objective is to tow the magnetometer, installed in a waterproof casing, behind the ship, usually from 500 to 1000 ft behind the fantail. The recording equipment in the instrument room is connected to the sensing element by a cable attached to the tow line.

Although any total-field sensor might be employed for such operators, virtually all marine exploration surveys make use of nuclear-precession (proton) magnetometers adapted for towing from a ship. The Varian marine magnetometer (Fig. 14-28) is the system that is most widely used for this purpose.

The sensor contains a special noise-canceling toroidal coil; a low-noise towing cable is provided that is given enough positive buoyancy to maintain the sensor near the water surface at slow speeds. Any risk that the head of the device might be snagged by obstructions at the water bottom is minimized by this arrangement.

Magnetic measurements are almost always taken simultaneously with a marine seismic survey or gravity survey or with a combination of the two. The magnetometer is generally towed behind the ship near the seismic energy source (which except for the sparker has little magnetic effect). The ship's trajectory is seldom laid out in a pattern that will minimize error due to diurnal variation as such a course would generally not be very suitable for the seismic operations. The ship's speed is almost always set to yield optimum quality of the seismic data even though a much higher speed would be permissible if only the magnetic data were being obtained.

REFERENCES

1. Howe, H. H., and D. G. Knapp: U.S. Magnetic Tables and Magnetic Charts for 1935, *U.S. Coast Geod. Surv. Ser.* 602, 1938.
2. Agocs, W. B.: Line Spacing Effect and Determination of Optimum Spacing Illustrated by Marmora, Ontario Magnetic Anomaly, *Geophysics*, vol. 20, pp. 871–885, 1955.
3. Gibson, M. O.: Network Adjustment by Least Squares: Alternative Formulation and Solution by Iteration, *Geophysics*, vol. 6, pp. 168–179, 1941.
4. Reford, M. S., and J. S. Sumner: Aeromagnetics (review article), *Geophysics*, vol. 29, pp. 482–516, 1964.
5. Lundberg, H.: Results Obtained by the Helicopter-borne Magnetometer, *Trans. Can. Inst. Min. Met.*, vol. 50, pp. 392–400, 1947.

INTERPRETATION OF MAGNETIC DATA

The techniques employed for interpreting magnetic data have many similarities to those on which gravity interpretation is based. This is to be expected in view of the fact that the laws of potential theory are fundamental to both methods. The similarities are greatest when one compares gravity data (for which the acceleration is always vertical) with magnetic data from land magnetometers of the type that record the *vertical* component of the earth's magnetic field.

Airborne and ship-towed magnetometers, as well as land instruments of the nuclear-magnetic-resonance type, record total fields, which are intrinsically more difficult to interpret than their horizontal or vertical components. The challenge of developing effective total-field interpretation methods and the large amount of airborne and shipborne data available for such analysis have stimulated the development of many ingenious techniques for this purpose. The literature on the subject has been particularly voluminous since the middle 1960s.

In this chapter, we shall consider single-component (vertical or horizontal) and total-field interpretation separately. From a practical standpoint, the differences are often small, particularly at high magnetic latitudes (where total fields are not very far from vertical) and at low magnetic latitudes (where they can be approximated as horizontal). Over almost all the United States and Canada, for example, the inclination of the earth's field is greater than 60°. Thus for buried sources having induced magnetism, total-field anomalies over most of North America should be at least qualitatively similar to vertical anomalies from the same sources.

16-1 QUALITATIVE INTERPRETATION OF MAGNETIC DATA

A large amount of the effort put into interpreting data obtained in magnetic exploration goes no farther than a qualitative evaluation of magnetic maps or profiles. The presence or absence of a fault or an intrusive body may be of much more importance than its shape or its depth of burial, neither of which can be uniquely determined from magnetic data anyway. In many magnetic surveys, the objectives are to ascertain the presence of sedimentary basins surrounded by areas where the basement is shallow and to map their approximate boundaries. Such information can often be obtained by visual inspection of a magnetic map. A basin is characterized by smooth contours and low magnetic relief, while the surrounding platform area shows steep gradients and high relief in the magnetic contours. Often a well-defined boundary between zones with appreciably different degrees of relief can indicate the presence of a major basement fault. Figure 16-1 illustrates two boundaries of this kind observed in southwestern Oklahoma at the edge of the Anadarko Basin.

A diapiric feature observed in a seismic reflection section may be interpreted as either a salt dome or an igneous intrusive. The best way to identify its composition may be to run a magnetic survey across the feature. This could indicate the lithologic nature of the diapir (whether it is salt or igneous material), information that probably could not be obtained from the seismic data at all.

Significance of Magnetic Contours

Magnetic contours can look so deceptively similar to contours of subsurface structure that it is easy for those not familiar with magnetic maps to interpret the magnetic features indicated by the contours as if they were structural. In areas where the hard rocks are shallow and uniformly magnetized such an assumption might give useful qualitative information on the structure of the rocks, particularly when linear trends are observed. Magnetic maps over sedimentary basins, where the magnetized rocks are deep, can almost never be relied on for reliable information about the structure of the basement surface.

The magnetic relief observed over sedimentary basin areas is almost always controlled by the *lithology* of the basement rather than by its topography. Changes in lithology which give rise to lateral contrasts in susceptibility show up in the magnetic contours more conspicuously than topographic features on the basement surface. Changes in the magnetization of basement rocks a mile or more deep may result in anomalies up to several thousand gammas in magnetic readings at the surface. At the same depth, structural relief on the basement surface as great as 500 ft would seldom produce anomalies larger than 50 gammas. Figure 16-2 illustrates why anomalies due to polarization can be so much larger than those resulting

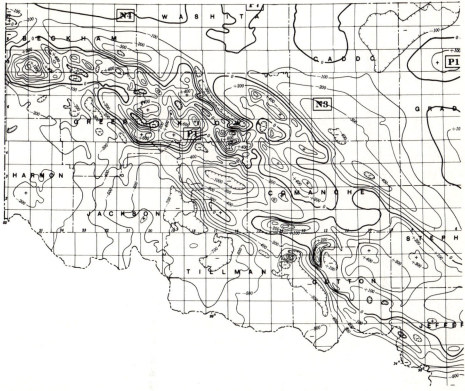

FIGURE 16-1
Magnetic map of southwestern Oklahoma showing sudden changes in mag-
netic relief. One is at the southern edge of Anadarko Basin to the north of the
Wichita Mountain belt, in which the magnetic gradients are high. The other
transition is between the mountain belt and the basin to the south of it. (*From
V. L. Jones, Vertical-Intensity Magnetic Map of Oklahoma, Proc. Geophys. Soc.
Tulsa, vol. 8, p. 43, 1961–1964.*)

from topography. The vertical effect of a field of intruded magnetic material having
a susceptibility contrast of 0.005 cgs units with respect to the rocks alongside it is
$2\pi \times 0.005$ cgs units, or about 3000 gammas. An erosional remnant 1000 ft high and
5000 ft in diameter on an otherwise flat andesite surface covered by 5000 ft of
nonmagnetic sediments produces an anomaly of only about 120 gammas. This
example indicates how relatively small deep-seated susceptibility changes not associated with structure on the basement surface can cause an anomaly much greater
than a topographic feature on this surface having a large amount of relief.

Use of Magnetic Data in Mapping Surface Geology

One of the most useful geological applications of magnetic surveys is to map
structural trends by following lineations in magnetic contours. In some cases the
lineations reflect the strike lines of elongated intrusive features or the surfaces of

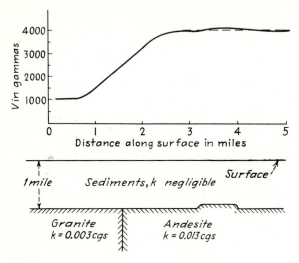

FIGURE 16-2
Comparison of magnetic effect of lateral susceptibility change in basement
with effect of structural feature on basement surface.

large faults reflected in the basement topography or lithology. Such features are often concealed under sedimentary cover and show up only on magnetic maps. The U.S. Geological Survey has been carrying out detailed aeromagnetic surveys over many parts of the United States. These have been of particular value for mapping geology in areas where igneous and metamorphic rocks occur near the surface but are hidden from view because of vegetation or other cover. Even where the surface geology is well exposed, geological maps can be made with a minimum of expense for field work if only scattered data points from surface outcrops are connected by use of aeromagnetically observed trends. The magnetic data facilitate interpolation of the field measurements, making a high density of ground observations unnecessary. This approach can be particularly fruitful in remote areas or those where surface access is difficult, provided that at least some of the geologic lineations have magnetic expression.

Figure 16-3 shows an aeromagnetic map of the Southern Cross area of Western Australia, which has well-defined magnetic anomalies indicating the trend, width, and dip of banded iron formations which act as marker beds to trace the folding axes of concealed metavolcanics and metasediments. The map is from a paper by Rayner.[1]

In mining exploration it is common to plot contacts, dips, strikes, faults, and similar features observed at the surface on aerial photographs along with identification of all outcropping rocks, usually by dabs of color at the positions where they are mapped. Magnetic contours superimposed on the same photographs make it easier to fill in the gaps between outcrops.

Often the general shape of a magnetic anomaly will help in determining the basic

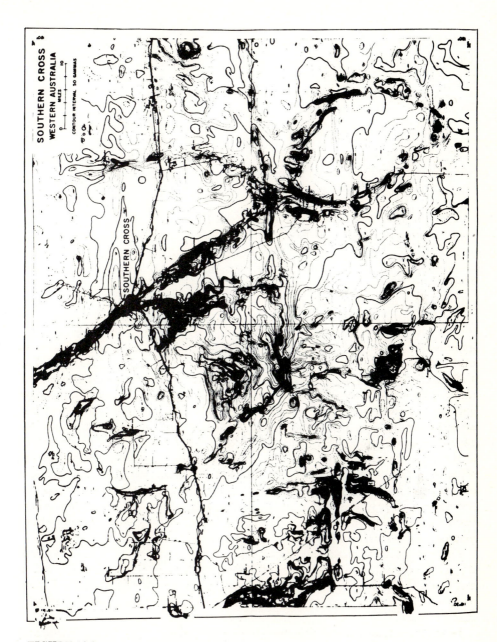

FIGURE 16-3
Magnetic trends in Southern Cross area of Western Australia associated with presence of banded iron formations, which show structure of concealed metavolcanics and metasediments. (*From J. M. Rayner*,[1] *"Mining and Groundwater Geophysics," The Queen's Printer, Ottawa, 1967.*)

geologic nature of its buried source. If the contours show circular symmetry where the magnetization is vertical, the body may be a plug. If closed contours are elongated, the source may be a dike and the direction of elongation should indicate its strike. If there is an elongated zone of steep gradient without well-defined closures, it is quite possible that this pattern results from subsurface faulting which has displaced magnetized rocks.

The delineation from surface magnetic measurements of contacts between different rock types can be helpful in planning other, more detailed geophysical surveys, e.g., induced-polarization or electromagnetic. The work can be restricted to those areas where the rock type is indicated as favorable.

Effect of Flight Elevation on Observed Fields

When an aeromagnetic survey is run at a number of different altitudes above the ground, the data obtained at the lower elevations show more sharply defined and better-resolved magnetic anomalies than those made at the greater heights. Since there is virtually no difference between the magnetic effect of most sediments and that of the air, an aeromagnetic profile made at a height h above the ground showing an anomaly from a magnetized body buried at a depth z should be identical to a profile that would be obtained with a total-field ground magnetometer over the same feature if it were buried under a layer of sediments having a depth $z + h$. As the flight elevation increases, the anomaly from any buried magnetized feature is attenuated in amplitude and spread out over a wider area. Figure 16-4 shows three profiles flown over the Benson Mines magnetite deposit in St. Lawrence County, N.Y., at heights ranging from 1000 to 10,000 ft above the ground. The very sharp magnetic peak recorded near the center of the profile from the 1000-ft altitude becomes almost unobservable at 10,000 ft.

The resolution of individual anomalies from separate buried sources depends on the height of the flight line above the level of the sources. Figure 16-5 shows how flying at a low elevation will indicate two anomalies while at a greater elevation

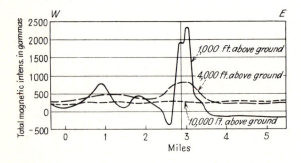

FIGURE 16-4
Effect of flying at different elevations over Benson Mines. (*U.S. Geological Survey.*)

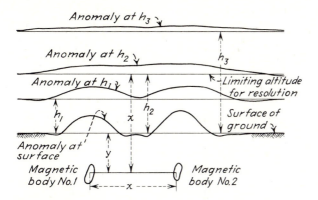

FIGURE 16-5
Resolution of anomalies from two buried magnetic bodies a distance x apart.
Magnetic profiles shown at surface and at three flight elevations (h_1, h_2, and
h_3). Anomalies will not generally be resolved if flight-line altitude is greater
than distance $x - y$.

the two magnetic features merge into one because of the spreading of the magnetic
effect from each source with increasing distance in accordance with the inverse-
square law.

Dependence of Total Field Anomalies upon Magnetic Latitude and Orientation of the Source

The magnetic anomaly observed from a buried magnetized body will depend upon
the magnetic latitude as well as upon the orientation (with respect to the magnetic
meridian) of any axis of elongation associated with the body. In the previous
chapter it was pointed out that the flight lines are generally oriented in a direction
normal to the geologic strike where that is known. If the flight paths make an
oblique angle with the strike or axis of elongation, the interpretation of anomalies
thus recorded can be very complicated.

**North-south flight line perpendicular to axis of elongated body with
square cross section**   Where the flight line is normal to the axis of elongation of
a magnetized body, the anomaly obtained from it near the magnetic equator will
appear quite different from that at middle latitudes or at the magnetic poles.
Consider the two-dimensional magnetized body with its height equal to its width
illustrated in Fig. 16-6. Its magnetization is entirely of the induced type. Its
east-west axis, normal to the north-south flight line, is perpendicular to the plane
of the cross section and is infinitely long. The total-field anomalies that would be observed over the body are shown for the magnetic north pole, the magnetic equator,
and for a magnetic latitude of 45°N.

If the magnetization is vertical, as at the north pole (Fig. 16-6a), the flux lines
associated with the induced magnetization are symmetrical about a vertical axis

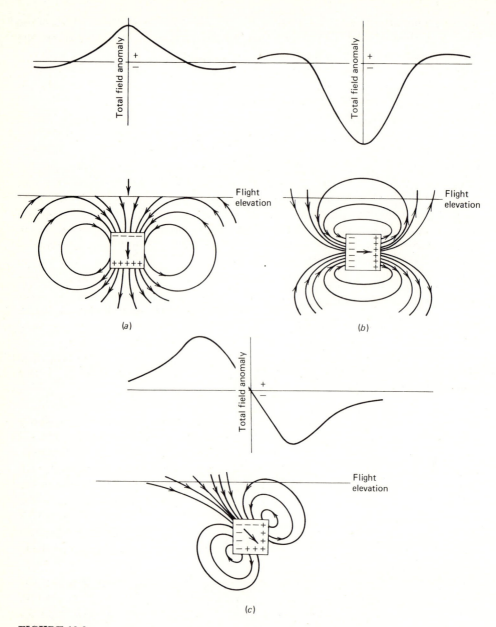

FIGURE 16-6
Total field anomalies observed when flight line is perpendicular to axis of
buried body with square cross section elongated perpendicular to page: (*a*) at
north magnetic pole; (*b*) at magnetic equator; (*c*) at magnetic latitude of 45°N.
All magnetization is induced. Anomaly is positive when the field of the buried
body reinforces the earth's field and is negative when the field opposes the
earth's field.

through the center of the body. Where they are directed downward at the level of the flight line, they reinforce the earth's field and give a positive anomaly. Where the field is tangent to the horizontal line of flight (as it is on both sides of the body), the anomaly is zero. Where the field has an upward component when it crosses this line (as is the case still farther in each direction from the center of the body), it opposes the earth's field and the anomaly becomes negative.

When the earth's field is horizontal, as it is at the magnetic equator, the lines of force from the magnetized body will oppose the earth's field over the center of the body, giving rise to a negative anomaly. If the axis of the body trends in the magnetic east-west direction and the flight line is in the north-south direction, the total-field anomaly will have the character indicated in Fig. 16-6b. At the two positions along the flight line where the lines of force cross it at right angles, the anomaly becomes effectively zero, and at greater distances on either side the field from the magnetic source will reinforce that of the earth. Thus, as the figure shows, the anomaly at the magnetic equator has about the same form as it does at the north pole, but it will be inverted in sign.

If the same two-dimensional body, oriented in the east-west direction, is at a magnetic latitude of 45° (Fig. 16-6c) the poles will be distributed symmetrically with respect to the direction of the field, the two faces that are closest to the north pole containing positive poles and the two farthest from it containing negative poles. The lines of force are symmetrical about a diagonal through the cross section and are perpendicular to the earth's field at a point on the surface above the center of the body. As a field having this direction neither increases or decreases the earth's field, the anomaly is zero at this point.

Effect of orientation of axis of body on anomaly obtained when flying perpendicular to axis   The direction of the axis of a buried magnetized body will also affect the total-field anomaly it produces along a flight line across this axis. Let us consider a buried slab having the form of a rectangular parallelepiped, with its long direction horizontal and with its horizontal dimension perpendicular to the axis substantially greater than its height. The flight line is always normal to the long axis (perpendicular to the paper) regardless of its orientation. Figure 16-7 shows the geometry for the three cases where the axis is respectively north-south (Fig. 16-7a), east-west (Fig. 16-7b), and N45°E(Fig. 16-7c), the flight lines being perpendicular to the respective directions.

For the slab with its axis oriented north-south, the anomaly is positive over the body itself and negative on either side. The amplitude of the anomaly becomes greater as the magnetic latitude becomes higher. When the axis of the slab is in the east-west direction, the anomaly depends on the magnetic inclination. If the magnetic latitude is 45°, the anomaly is unsymmetrical, being positive over the updip end of the inducing field and negative over the downdip end. If there is no inclination, the field is symmetrical and negative directly over the body. When the axis of the slab has an orientation of 45° with the north-south direction, the anomaly is rather similar to that observed across the body when oriented east-west at a

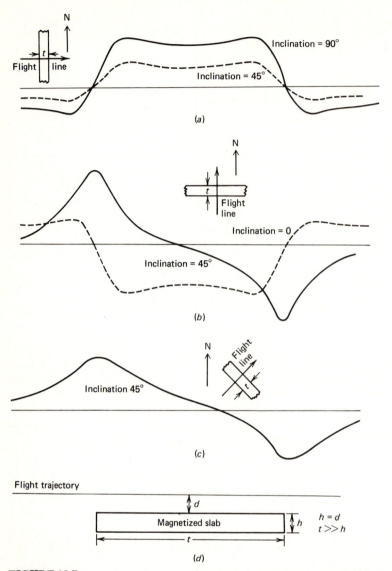

FIGURE 16-7
Total-field anomalies observed over a two-dimensional magnetized slab with long axis horizontal and horizontal dimension perpendicular to axis much greater than thickness (vertical dimension). Flight lines always perpendicular to long axis. (*a*) Orientation of axis north–south; (*b*) orientation of axis east–west; (*c*) orientation of axis N45°E. (*d*) Cross section showing position of slab in vertical plane-of-flight line.

magnetic inclination of 45°, although the amplitude is lower and the point where the field changes direction is not quite the same.

16-2 QUANTITATIVE INTERPRETATION OF VERTICAL-FIELD DATA

The quantitative interpretation of vertical-field data, as we have noted, is generally more manageable than in the case of total-field data, but the applicability of such an interpretation can be quite limited in areas where the actual fields deviate substantially from the vertical.

For most work at high magnetic latitudes (say greater than 60°) the errors involved in assuming vertical fields are usually less than those that might result from incorrect assumptions regarding the true direction of magnetization, particularly when the relative distribution of induced and remanent components is not known.

Representation of buried magnetic bodies as simple magnets A vertical or steeply inclined vein of magnetic ore which is so long compared with the depth of its top that the effect of its lower pole may be neglected can sometimes be represented as an isolated magnetic pole. Such a source should give a vertical-field profile similar to that shown in Fig. 14-15. The variation in the vertical field with horizontal distance would then be the same as that obtained for the gravitational effect of a sphere. We can thus find the depth of the isolated pole from the half width (as we do in gravity calculations), which is known to be 0.768 times the depth z. For linear ore bodies that are shorter one must treat the body as a buried dipole or elementary magnet. If the length and inclination of the dipole are great enough to allow isolation of the anomalies from the positive and negative poles, depths can be estimated from each by use of the isolated-pole formula. It is rare, however, for the two individual anomalies to be so readily separable that this approach can be effective. Empirical formulas for dipoles like those presented by Heiland[2] are generally more useful for interpretation.

Extended sources When the anomaly is of such a nature that a more complex and extended source is indicated, it is often useful to compare the profile or the contours with theoretical profiles or contours from simple geometrical forms, as is done in gravity interpretation. Formulas for the vertical magnetic anomalies from several such models were presented in Chap. 14 (pages 495 to 500). These include spheres, cylinders, and slabs with vertical walls.

From the sphere or horizontal cylinder, relations can be readily ascertained between the depth z_c to the center of the source and the half width of the anomaly curve. For a sphere, z_c is twice the half width, while for a horizontal cylinder, it is 2.05 times the half width.

The two-dimensional, i.e., elongated, vertical or inclined slab is generally a useful geometrical model for an igneous dike, a feature of particular interest in mining exploration as a potential source of ore. Cook[3] has applied formulas derived by Haalck and others to compute anomalies in vertical fields from a large variety of model dikes: vertical, inclined, infinitely deep or of finite depth, striking north, striking east, etc. Many of these are shown in Fig. 14-23.

Where the shape of the buried magnetic body cannot be represented adequately by generalized geometric forms, graphical devices similar to the graticules described for gravity calculations have been used.

Peters' methods   Peters[4] has devised two rule-of-thumb techniques for depth estimation, the *slope method* and the *error-curve method*; they are simple and applicable to vertical anomalies that may not be necessarily associated with simple geometrical forms. The first method has been widely used to determine the depth to the top of the basement surface from the anomaly observed over a portion of the basement having a different susceptibility from the rest. The second method requires a contour map and a program for continuing the mapped data upward and downward. The depth to the top of the body can be estimated from the results.

Figure 16-8 illustrates the mechanics of applying the slope method. The anomalous source is a vertical slab of thickness t extending indefinitely downward from the top of the "magnetic surface." It is assumed that the slab has a susceptibility k and that neither it nor the surrounding material has any remanent magnetization. It is infinitely long in the direction perpendicular to the profile (in this case, perpendicular to the page), thus being two-dimensional. The top of the slab is at a depth h below the magnetic surface. The magnetization kH of the slab is I', while the polarization of the surrounding material in which the slab is embedded is I_0. Thus the effective magnetization of the slab is $I = I' - I_0$. It is evident that this model would apply in the case where a dike of width t with vertical sides has intruded into crystalline rocks having a different magnetic susceptibility.

Peters' slope method makes it possible to compute the depth h to the top of the intrusive from the magnetic anomaly it produces. Taking the profile of vertical

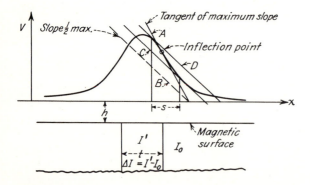

FIGURE 16-8
Peters' slope method for calculating depth to magnetic surface from anomaly
curve for vertical slab having anomalous magnetization.

intensity observed over the slab, one locates the inflection point corresponding to maximum slope and measures the slope of the tangent (line A) to the curve at that point. Then one constructs a line (designated as B) with half this slope and draws the two tangents C and D (both parallel to B) at the upper and lower inflection points of the anomaly curve as shown. The horizontal separation s between the two points of tangency can be related to the depth h by the formula

$$s = 1.6h$$

if h and t are of comparable magnitude. The same rule can be applied to estimate the depth to the top of a thin vertically magnetized dike projecting vertically upward into the sedimentary section, provided the depth of its top is shallow compared with the depth of the basement. Here the magnetization contrast could be taken as the magnetization of the dike itself.

The second interpretation technique introduced by Peters, the error-curve method, is based on the fact that a magnetic anomaly can be continued downward mathematically in accordance with the procedures discussed in Chap. 12 for gravity anomalies. The magnetic field is transformed from the surface where it is measured to various progressively greater depths from which it is continued upward to the surface. The field doubly transformed in this way should be theoretically identical to the original field as measured. Actually, errors in the measurements and limitations in the sampling prevent the computed anomaly from being identical to the observed anomaly in practice. But the average or rms difference between the two maps should be small for all depths of continuation down to the depth of the top of the magnetized body constituting the source.

Downward continuation of an anomaly observed at the surface to a depth greater than that of its source will result in an unstable oscillatory field. When such a downward-continued anomaly is continued upward to the surface, it will be quite different from the original observed anomaly. The greater the depth of continuation below the top of the source, the larger the rms difference between the two fields.

Figure 16-9 illustrates how this principle can be employed to determine the depth to the top of a magnetized body. First one locates the point of maximum curvature observed in the vertical intensity. The double continuation itself is carried out by drawing a series of circles about this point having radii of $b_n h$, where the b_n's are

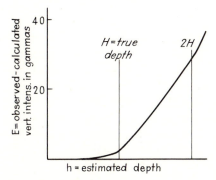

FIGURE 16-9
Peters' error-curve method for establishing depth of burial of magnetic anomaly. (*After Peters.*[4])

specified constants ranging from 0.1 to 3.75 and h is the assumed depth of the anomaly. Plotting the deviation between computed and observed intensity (the error E) for a range of h values, one locates the true depth of burial at the value of h where E shows a sharp inflection.

16-3 INTERPRETATION OF TOTAL-FIELD DATA

Most of the data obtained in magnetic surveys for prospecting purposes are of the total-field type. The mileage covered in aeromagnetic and ship-towed magnetometer surveys, in which total fields have been recorded, is greater by far than that mapped in land surveys with vertical-field or horizontal-field instruments.

The quantitative interpretation of total-field data can be so complex, however, that rigorous analysis is carried out on a routine basis only when the subsurface sources can be represented by simple, usually approximate, models.

Because of its relative simplicity as well as its practical applications, the interpretation of magnetic fields from inclined slabs with infinite lateral extent and parallel sides has been studied extensively. Magnetic effects from vertical prisms of rectangular cross section have also received much attention. In the most general case, the remanent magnetization of the source is not in the same direction as the induced magnetization produced by the present earth's field, nor will the intensities associated with the two types be the same. The true magnetization is the vector sum of the induced and remanent components.

Where a dike or prism is interpreted to be the source of an anomaly on a magnetic map or profile, the information of greatest practical interest is the depth to its top. In exploring sedimentary basins, the assumption is generally made that intrusives within the basement are truncated by erosion at its surface. Where this is the case the depth to the top of the dike or prism is equal to the thickness of the sedimentary section. The intrusive may not extend upward as far as the basement surface; therefore any sedimentary thickness so computed is generally considered to be a minimum.

Vertical Prisms

One of the earliest published works on aeromagnetic interpretation is by Vacquier et al.,[5] who established criteria for determining the depths to the tops of vertical prisms having the form of rectangular parallelepipeds from the total fields mapped over them. The deeper the top of the prism, the gentler the gradients of the field as they fall off in different directions from the prism over and beyond its edges. The gradients are measured at a number of characteristic positions on the magnetic contour map, and the depths are determined from the gradients by factors dependent on the elongation of the anomaly, the inclination of the earth's field, and the orientation of the principal axis of the mapped anomaly with respect to the horizontal component of the field. The method as originally published applies only to induced magnetization or to remanent magnetization with polarization in the same direction as the earth's field.

The gradients are expressed as characteristic distances in the manner illustrated in Fig. 16-10. Maps of observed intensity or of the curvature (second derivative) of the intensity can be used to obtain such distances, but each of these index distances must be multiplied by appropriate constants to obtain depths.

With the maps of intensity, the length of the region along a line perpendicular to the contours where they are steepest can be considered as an inverse measure of the gradient. As the gradient itself is a kind of inverse measure of depth to the source, it is reasonable to expect that such a length should be approximately proportional to this depth. Some proportionality constants established for typical index distances are indicated in Table 16-1. The dimensions indicated on the first two columns are in units of depth to the top of the prism.

The corresponding distances on the curvature map extend from the position of maximum curvature (the maximum flexure), which is roughly at the point of tangency for the upper half-maximum-slope lines of Peters' slope method (see Fig. 16-8), to the zero-curvature position, the point of tangency of Peters' maximum slope. The proportionality constants between index distances so determined and the depths to the top of the prism, some of which are given in Table 16-1, are of the same order of magnitude as the numerical constant .60 used to obtain depth from a comparable distance along the magnetic profile in the Peters' method.

Vacquier et al.[5] give a large number of model anomalies for rectangular parallelepipeds having different kinds of geometry and with sides having different orientations with the magnetic north. Anomalies from such sources are presented for the entire range of magnetic latitudes. Both intensity and curvature contours are mapped for all models.

**Table 16-1 DEPTH INDICES FOR SQUARE-SHAPED MODELS
AT INCLINATION 60°**

Horizontal dimensions	Vertical thickness	A	B	C	D	E	F	G
1 × 1	∞	1.2		0.6		0.8	0.5	0.5
2 × 2	∞	1.3	1.4	0.7	1.9	0.9		0.9
4 × 4	∞	1.0	0.8	0.8	1.5	0.9	0.8	0.9
8 × 8	∞	0.9	0.7	0.9	1.4	1.3	1.2	1.3
8 × 8	7	0.8	0.6	0.9	1.6	1.3	1.3	1.4
8 × 8	3	0.8	0.4	0.9	1.3	1.0	1.2	1.1
8 × 8	1	0.7	0.4	0.7	1.2	0.9	1.2	1.2

Depth to top is unit for dimensions.   SOURCE: Vacquier et al., *Geol. Soc. Am. Mem.*, 47, 1951.

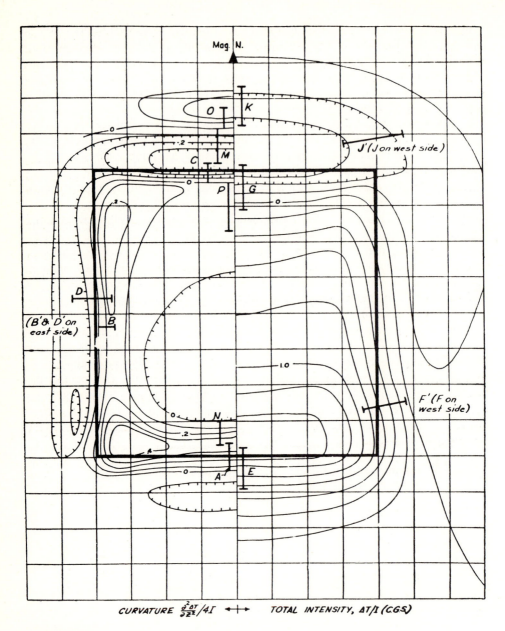

CURVATURE $\frac{\partial^2 \Delta T}{\partial z^2}/4I$ ←|→ TOTAL INTENSITY, $\Delta T/I$ (CGS)

FIGURE 16-10
Definition of indices for determining depths to source of total-field anomalies
from characteristic distances on maps of total magnetic intensity or of curvature of magnetic field using method developed by Vacquier et al.[5]

In actual interpretation of magnetic maps, the ratio of horizontal dimensions and the orientation of the sides with respect to magnetic north are obtained from the contours of the observed intensities or of the computed curvatures. The magnetic latitude is assumed known. The interpreter would relate the characteristics of the observed maps to the appropriate models and thus determine the correct index factors for his depth calculations.

Andreasen and Zietz[6] have adapted the Vacquier et al. method to the case where the direction of magnetization is not parallel to the earth's field, i.e., where the prismoidal body has remanent as well as induced magnetization. Intensity maps (but not curvature maps) are computed for various prism configurations and directions of magnetization. As the resultant direction of magnetization will normally be unknown, it is necessary to deduce this vector from the magnetic map itself, and a procedure is presented for such a determination in cases where the inclination of the earth's field is around 75° and the dip of the magnetization is low. The angle between the magnetic north and the line connecting the maximum and minimum values observed over the prism is an approximate measure of the true azimuth of the magnetization. The dip of the magnetization can sometimes be determined from the ratio between the maximum and minimum anomalies associated with the prism using a table given in the paper.

Inclined Dikes

The great number of variables which determine the field of an inclined dike is indicated by a formula of McGrath and Hood,[7] which covers the case where the magnetization of the dike is not necessarily in the same direction as the earth's total field. Their equation for the total-field anomaly $\Delta T(x)$ observed on a profile in the x direction over an inclined dike having a strike in the y direction at a distance x from a point on the flight line directly over the center of the top surface of the dike is

$$\Delta T(x) = 2Jbc \sin \theta \left[\sin \alpha \left(\tan^{-1} \frac{x+d}{h} - \tan^{-1} \frac{x-d}{h} \right) \right.$$

$$\left. - \frac{\cos \alpha}{2} \log \frac{(x+d)^2 + h^2}{(x-d)^2 + h^2} \right] \qquad (16\text{-}1)$$

where J = intensity of magnetization of dike
$b = (\sin i)/(\sin \psi)$
$c = (\sin I)/(\sin \lambda)$
i = inclination of magnetization i
I = inclination of earth's field
$\alpha = \lambda + \psi - \theta$

The other quantities including the angles ψ, θ, and λ, are defined in Fig. 16-11.

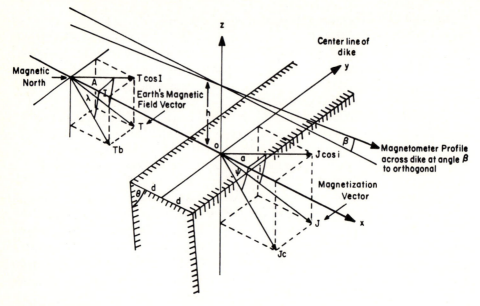

FIGURE 16-11
Geometry of inclined dike in magnetic field having a direction different from
that of the dike's magnetization. Angles defined here are used in Eq. (16-1).
The equation is for a flight line in the x direction. (*From McGrath and Hood.*[7])

The principal objective in the interpretation of magnetic data over a dipping dike
is to determine the position, depth, and width of the dike and the angle of its dip.
The directions of the total magnetic field and of the magnetization are needed to
determine the dip angle. The general approach is to construct model curves based on
Eq. (16-1) assuming different combinations of variables and to compare model
profiles with those observed. McGrath and Hood[7] present a computer-based method
of matching the curves. The techniques of Hutchison and Gay, discussed later in
this section, involve the same approach.

Werner method This method makes use of a semiempirical calculation which is
quite easy to apply. Werner's original publication is not readily accessible, but the
technique has been summarized by Hartman et al.[8]

Consider a dike with infinite strike length and depth extent. The total anomalous
field $\Delta T(x)$ at a distance x along a measuring line perpendicular to the dike can be
expressed in the empirical form

$$\Delta T(x) = \frac{A(x - x_0) + Bz}{(x - x_0)^2 + z^2} \tag{16-2}$$

where x_0 is the distance to a point immediately over the center of the dike's top and
z is the depth to the top. A and B are functions (to be determined) of the field

strength, dip, and strike as well as the inclination and declination of the dike's magnetization.

Equation (16-2) has four unknowns, A, B, x_0, and z. If it is put into the form

$$a_0 + a_1 + b_1 F + b_1 x F = x^2 F \qquad (16\text{-}3)$$

where

$$a_0 = -Ax_0 + Bz, \; b_0 = -x_0{}^2 - z^2, \; a_1 = A, \; b_1 = 2x_0, \; \text{and} \; F = \Delta T(x)$$

it is easy to show that

$$x_0 = \frac{b_1}{2} \qquad \text{and} \qquad z = \tfrac{1}{2}\sqrt{-4b_0 - b_1{}^2}$$

It is only necessary to substitute the $\Delta T(x)$'s at four different values of x to obtain a_0, a_1, b_0, and b_1. From them one solves for x_0 and z.

Methods involving separation of anomaly into symmetrical and asymmetrical components   Several techniques for determining the geometrical and magnetic characteristics of an inclined dike from total-field data involve separation of the field into its symmetric and antisymmetric parts. A somewhat different basic approach, first introduced by Baranov,[9] is *reduction to the pole*, which through an appropriate convolution operator makes it possible to determine, from the observed total field, the position and depth of the pole that has a magnetic effect equivalent to that of an extended source with inclined magnetization.

NAUDY METHOD   A method published by Naudy[10] combines the concepts of separation of the field into odd and even functions and of reduction to the pole. Any function of x, such as $y(x)$, can be separated into a symmetrical component $S_1(x)$ and an asymmetrical component $S_2(x)$ about an arbitrarily chosen center so that

$$S_1(x) + S_2(x) = y(x) \qquad (16\text{-}4)$$

$S_1(x)$ does not change its value when x is replaced by $-x$ while $S_2(x)$ reverses sign if the sign of x is changed. It is not feasible to obtain a meaningful interpretation of the $S_2(x)$ part of the curve unless it is converted to a symmetrical function by reduction to the pole. This is done by convolving the measured profile with an appropriate set of coefficients determined by Baranov's procedures.

Naudy's method involves the determination of $S_1(x)$ curves both from the observed profile and from the transformed profile obtained by convolution of the measured curve. Then a theoretical anomaly curve over each of a series of symmetrical model dikes having assigned characteristics is correlated with both curves. The model dikes cover a large range of depths and widths, and the model that yields the greatest correlation coefficient is used for mapping the depth under the center of each anomaly on the profile under investigation.

If there should be many dikes along the same line at different depths, Naudy's procedure makes it possible to map them individually by correlation with a series of models covering the expected range of depths. This technique has the important advantage that it can be used with profiles having anomalies from dikes at different depths which cannot be reliably separated from one another by inspection.

HUTCHISON METHOD   Where dikes give anomalies that stand out on magnetic maps without troublesome interference from other dikes or comparable magnetic sources, a method proposed by Hutchison[11] may be most convenient to use. Although designed primarily for vertical-field data, the method is adaptable to total-field analysis if the magnetic profile is perpendicular to the strike of the dike. Like the Naudy technique, the method involves separation of the anomaly into its symmetrical and asymmetrical parts.

The symmetrical function is plotted on logarithmic paper after subtraction of the magnetic base level. The asymmetric function is plotted logarithmically in the same way. The resultant curves are matched to a set of master curves, which are reproduced in Hutchison's paper. The curves of each set that show the best match correspond to the actual geometry of the dike giving the anomaly. The positions of the two top edges of the dike are determined by this approach. Thus the dike's depth and the component in a known direction of its breadth can be obtained but not its angle of dip.

Actually, three families of curves are plotted in the paper. One is for a dike, one for a scarp, and one for a thin bed.

GAY'S MASTER CURVES   Gay[12] has published a set of master curves for interpreting anomalies (vertical, horizontal, and total-field) over inclined dikes, both thin and thick. Superposition of observed normalized profiles on the master curves makes it possible to find the model that should correspond most closely in its characteristics to the dike producing the anomaly. The curves are plotted for different dips, strikes, inclinations of the earth's field, and (for thick dikes) ratios of width to depth of the top.

Two-dimensional Bodies with Polygonal Cross Section

The interpretation methods we have considered until now for two-dimensional magnetized bodies have been applicable only to vertical or inclined slabs having the general form of dikes with bases very much deeper than their top surfaces. A method introduced in 1972 by Nabighian[13] makes it possible to deduce the depth and form of any two-dimensional body whose constant cross section is perpendicular to its axis. Such a cross section can be approximated by polygons, as shown in Fig. 16-12, for an elongated sill bounded by curved surfaces. If enough lines such as AB, BC, etc. are drawn, the cross section of the sill can be closely approximated by a polygon such as $ABCDEFGHI$. The various faces of the resulting prismoidal body can be considered to be infinitely long strips with traces AB, BC, CD, etc., on a plane perpendicular to the elongation axis of the body. For a uniform magnetization within its interior the magnetic effect of the sill will be the same as that of a hollow prism of the same shape with a proper distribution of poles on the surfaces.

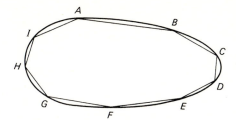

FIGURE 16-12
Representation used by Nabighian for
determining magnetic anomaly from
two-dimensional body with curved outline. The shape of the body is approximated as a polygon constructed by
drawing chords across closely spaced
points on boundary.

The horizontal derivative of the field observed over the body is equivalent to that
which would be obtained by adding the effects of the thin magnetized sheets that
constitute the bounding polygon. The derivative can be transformed into an
analytical form which is the summation of individual symmetrical bell-shaped
anomalies from sources at each corner of the polygon. The individual anomalies can
often be separated by eye; in more complicated models trial-and-error can be an
effective means of separation. From the characteristics of the curves after separation it is possible to determine the depth of each corner, the dips of the faces, and
the susceptibility of the magnetized body. The process is illustrated in Fig. 16-13,
where individual anomalies originating from each of the corners of the trapezoid are
so weighted that their sums will best match the observed derivative curve.

An alternative procedure for obtaining the configuration of the body by reduction
to the pole is also introduced in Nabighian's paper. This procedure is often more
convenient than the convolution method of Baranov for the same transformation.

16-4   GENERAL CONSIDERATIONS IN MAGNETIC INTERPRETATION

Regionals and residuals   The removal of regional effects to obtain residual
anomalies is of course as important in magnetic interpretation as it is for gravity
data. The mathematical methods for obtaining residual gravity fields discussed in
Chap. 13 are all applicable (directly or with modifications) to magnetic interpretation and may be used either with vertical or with total-field data. The techniques
actually employed, however, may be somewhat different. In mining exploration
with the ground magnetometer, the fields from magnetic minerals in relatively
shallow ore bodies are probably so large that the background field will not affect the
desired results significantly. In the analysis of anomalies from such sources it is
seldom worthwhile to remove the regional field except when regional variations are
large, in which case they can often be removed visually.

Where large areas are covered, it is often feasible to remove regional effects by
incorporating them into the normal correction. The simplest approach is to fit the
magnetic data over as large an area as possible around the zone of interest to a
polynomial of appropriate order and to subtract the polynomial surface from the observed surface. The principal problem in removing a magnetic regional, as in the

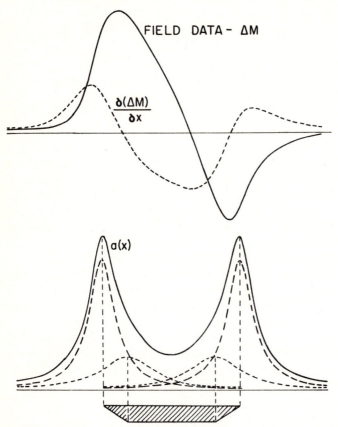

FIGURE 16-13
Interpretation by Nabighian of total-field data (shown in the upper part of the diagram) using a trapezoidal model for which amplitudes of anomalies from individual corners are summed after being weighted in a way that will best simulate symmetrical portion of observed anomaly. (*From Nabighian.*[13])

corresponding gravity case, is one of interpretation. It may not be possible to determine whether an apparent residual feature has structural significance or is the result of a lithologic change deep within the basement.

Ambiguity of magnetic data   The statements made in Chap. 13 on the ambiguity of gravity data hold for magnetic data as well. The laws of potential theory that tell us why gravity data can never be accounted for by a single, unique interpretation apply also to the magnetic case. Here again independent geological information is necessary if the interpreter is to choose the most reasonable of an indefinite number of subsurface pictures that might fit the magnetic observations equally well. Just as it is often impossible to differentiate between structure and density contrast in determining the source of a gravity anomaly, it may be infeasible for the same reason to say whether an observed magnetic anomaly results from structural relief or from a lateral change in susceptibility.

Even when concentrations of magnetic material are accurately located by magnetic methods, it is not generally possible to evaluate their economic possibilities from magnetic data alone. There is often such wide divergence in the susceptibilities of various minerals that very small quantities of one may have more effect than very large quantities of another. Because this fact was not recognized, the early magnetic prospecting for iron deposits in Wisconsin, Michigan, and Minnesota was spectacularly unsuccessful. A large-scale program of magnetic exploration begun about 1875 showed up innumerable magnetic anomalies which were suggestive of commercial iron deposits. When these were investigated further, it was found that in most cases no economic concentrations of iron were present. Other areas exhibiting no magnetic anomalies at all often turned out to be highly productive. This puzzle remained unsolved for 50 years until Mason, Slichter, and Hay began to investigate the fundamental magnetic properties of various minerals associated with iron deposits. Discovering that magnetite has a much greater susceptibility than any other iron-bearing mineral, they were the first to realize that magnetic data reflect percentages of magnetite, but not of most other ferrous constituents. A minute amount of magnetite disseminated in nonferrous minerals could give a greater magnetic anomaly than a large, commercially productive deposit of hematite, which is a much less magnetic ferrous mineral. This discovery immediately led to new and more successful interpretation techniques in magnetic prospecting for iron. Weaver[14] has cited this example to show how empiricism not supported by proper analytical reasoning can lead to trouble in exploration.

16-5   INTERPRETATION OF DATA FROM TYPICAL MAGNETIC SURVEYS

In this section we shall consider the results of some typical magnetic surveys, ground, airborne, and ship-towed, with particular reference, where the information is available, to the relationship between the magnetic data and the geology of the area involved. Examples will be presented both from mineral and petroleum exploration.

Magnetic exploration for minerals is generally undertaken with one of three objectives in mind: (1) to look for magnetic minerals, such as iron ores rich in magnetite, directly; (2) to use associated magnetic minerals as tracers in the search for nonmagnetic minerals; or (3) to determine the depth, size, or structure of mineralized zones for which there is no surface evidence because of alluvial or vegetative cover. Faults and similar structural features that have a bearing on the depths of ore deposits can often be shown up in this way.

There are a few rare cases where the occurrence of oil in an individual pool can be related to a magnetic anomaly, such as one from a buried basement ridge causing entrapment in overlying sediments draping over the magnetic source. Except in such cases, magnetic surveys are useful in oil exploration only for determining basement

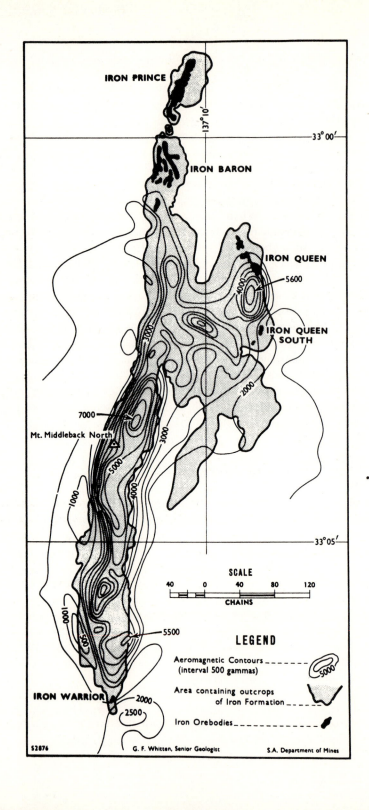

IRON PRINCE

137° 10′

33° 00′

IRON BARON

IRON QUEEN

5600

IRON QUEEN
SOUTH

2000

7000

Mt. Middleback North

3000

1000

5000

5000

33° 05′

5500

IRON WARRIOR

2000

2500

SCALE

40   0   40   80   120

CHAINS

LEGEND

Aeromagnetic Contours _____
(interval 500 gammas)

5000

Area containing outcrops
of Iron Formation _____

Iron Orebodies _____

S2876

G. F. Whitten, Senior Geologist

S.A. Department of Mines

depths so as to cast light on the existence and geometry of sedimentary basins, particularly in little-explored areas.

Magnetic exploration for iron ore   Approximately 90 percent of the world's iron production comes from sedimentary sources which are primarily oolitic or siliceous in composition. The remainder is from igneous rocks, the ores generally being in close proximity to intrusive masses from which the iron was derived or from weathering of nonferrous constituents of the igneous material. Ores associated with igneous rocks are likely to have a high magnetite-hematite ratio, and they can be detected by magnetic measurements directly. Hematitic ore bodies are non-magnetic, but they can often be related genetically or structurally to surrounding formations containing magnetite. Magnetic surveys in such cases would then be used for indirect location of the ore.

The applicability as well as the limitations of magnetic surveys for locating economic deposits of hematitic ores are illustrated in Fig. 16-14, which shows the total-field magnetic contours over the Northern Middleback Range of South Australia, the major source of iron in the continent. The shaded area bounded by heavy lines is the area within which iron-bearing formations outcrop. The black patches represent ore bodies where the iron occurs in economic quantities. The magnetic contours indicate the general area where iron is found, but the magnitudes give no clue to the locations of the ore bodies themselves. Two important deposits, the Iron Prince and Iron Baron, are somewhat outside the zone of magnetic anomalies. The ore bodies do yield diagnostic gravity anomalies, however, and the standard exploration procedure here has been to block out areas on the magnetic map where iron is indicated and then to look for economic concentrations with the gravimeter.

Where the iron is taconitic, the state of oxidation has a considerable effect on its magnetic character. Figure 16-15 shows the magnetic field over a portion of the western Mesabi Range. Here the oxidized portions of Biwabik iron formation are much less magnetic than the taconitic rocks that are not oxidized. This is because the oxidation breaks down magnetite. As the less-oxidized ore is easier to process, areas with magnetic highs should be more prospective as sources of commercial ore.

Figure 16-16 shows the aeromagnetic contours observed over Pea Ridge, Mo., where an ore body which was expected to produce 2 million tons a year was discovered at a depth of about 1250 ft. The drilling that led to the discovery was undertaken to investigate the source of the 3200-gamma anomaly indicated on the map. The ore body is about 3000 ft across with its center slightly north of the magnetic peak. The iron bodies are hydrothermal fracture fillings and replacements in Precambrian volcanics.

FIGURE 16-14
Total-field magnetic contours over Northern Middleback Range of South Australia. Note correlation of anomalies with shaded areas, where iron formations outcrop. (*From Webb.*[15])

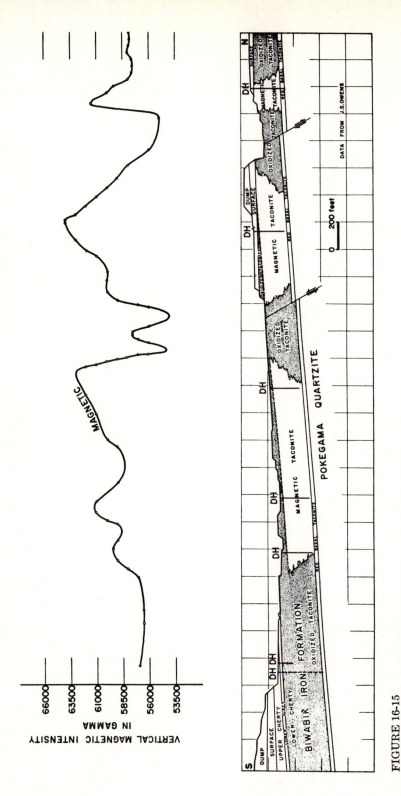

FIGURE 16-15
Vertical magnetic intensity over Biwabik iron formation in Mesabi Range. Note magnetic lows over oxidized zones. DH indicates drill hole. (*From Leney.*[16])

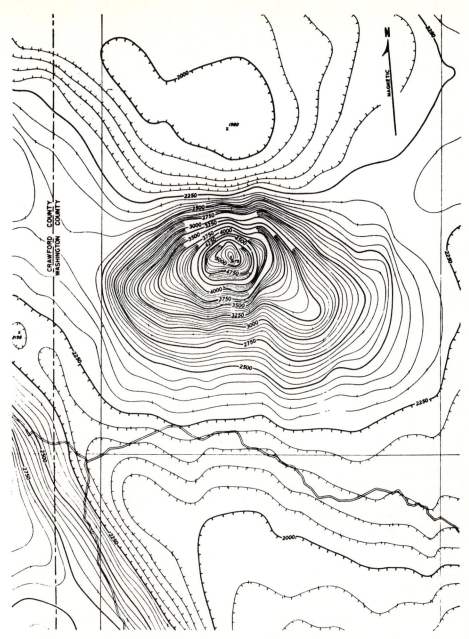

TOTAL INTENSITY AEROMAGNETIC MAP OF PART OF SULLIVAN QUADRANGLE, MISSOURI

RELATIVE TO ARBITRARY DATUM

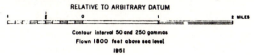

Contour Interval 50 and 250 gammas
Flown 1800 feet above sea level
1951

FIGURE 16-16

Aeromagnetic contours over Pea Ridge, Mo., ore body. The ore, 1250 ft deep
and 3000 ft in horizontal dimensions, is located in the area where the magnetic intensity is highest. (*From Leney.*[16])

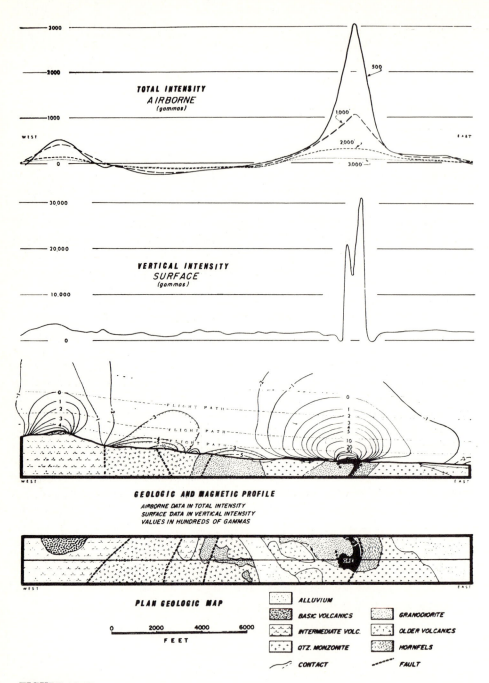

FIGURE 16-17
Airborne profiles of total intensity at various flying heights compared with
vertical intensity measured on surface over Dayton ore body in Nevada for
which geological map and section are shown. (*From Riddell.*[17])

Figure 16-17 shows the surface vertical magnetic profile and the aeromagnetic total-field profile over the Dayton ore body in Nevada, which was formed by replacement of metamorphosed limy sediments. A granodiorite intrusion adjacent to the body apparently folded and fractured the sediments, allowing the replacement to take place. The effect of flight elevation on the aeromagnetic anomaly is pronounced.

Magnetic exploration for other minerals Although not magnetic themselves, base metals (e.g., nickel and copper) and gold are often found in association with magnetic minerals such as magnetite and pyrrhotite. Some metallic ores are found in the margins of batholiths and other intrusives which can be detailed under drift by magnetic surveys.

Figure 16-18 shows the vertical magnetic anomaly observed by Galbraith[18] over a nickel and copper sulfide ore body beneath a norite contact in the Sudbury district of Canada. The ore was discovered by a borehole, located on the basis of the magnetic indications, which penetrated 100 ft of drift before reaching the desired target.

Magnetic surveys have frequently been employed in prospecting for diamond-bearing kimberlite pipes in various areas. This type of exploration has been described by Gerryts[19] in a review of geophysical methods of exploring for diamonds in various parts of the world. Kimberlite, an ultrabasic rock which is a common source of diamonds, contains magnetite and ilmenite as primary constituents. Magnetic surveys for such bodies have been reported in the United States and the U.S.S.R. as well as in South, East, and West Africa. Figure 16-19 illustrates the aeromagnetic anomaly observed over some pipes in the Koffiefontein area of South Africa.

Joesting[20] has used the magnetometer to locate gold in Alaska, taking advantage of the close association of gold and magnetite in gravels. Figure 16-20 shows the good correlation observed between the magnetic field and the gold values over a muck-covered placer at Portage Creek. The magnitude of the magnetic field does not appear to show any relation to the thickness of gravel or the depth to the basement.

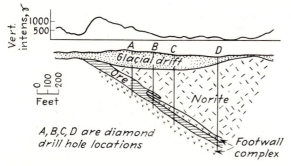

FIGURE 16-18
Magnetic field over drift-covered sulfide-ore body at Sudbury, Ontario. (*After Galbraith.*[18])

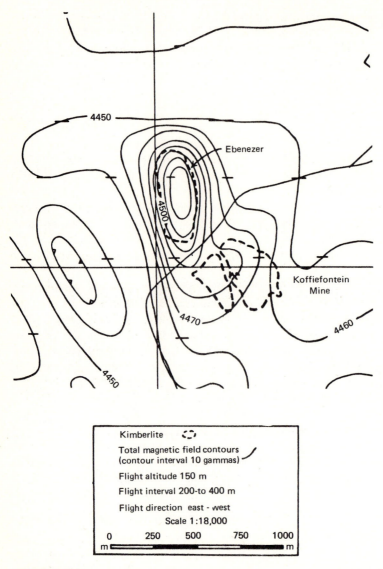

FIGURE 16-19
Aeromagnetic anomaly over kimberlite pipes in Koffiefontein area, South Africa.
(*From Gerryts.*[19])

Magnetic exploration for oil The most widespread application of magnetic prospecting to oil exploration since World War II has been to determine the present extent and thickness of sedimentary basins in previously unexplored areas. This is done by mapping the depths to the tops of intrusives in the basement from aeromagnetic or ship-towed magnetic data by methods discussed earlier in this chapter.

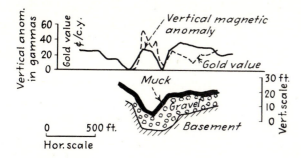

FIGURE 16-20
Magnetic field, gold values, and geologic structure along profile at Portage
Creek, Alaska. (*After Joesting.*[20])

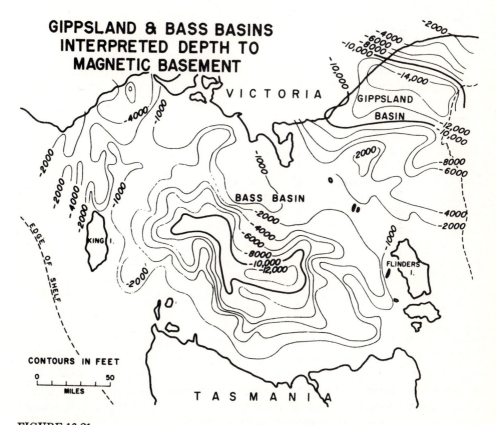

FIGURE 16-21
Depths to magnetic basement in Gippsland and Bass basins south of Victoria,
Australia as interpreted from aeromagnetic data. (*From Wallis.*[22])

Steenland[21] reports that the discrepancy between basement depths estimated by this means and those indicated by drilling in northern Alberta and northeastern British Columbia has been on the average no greater than 5 percent.

Figure 16-21 shows contours of basement depth over the Bass Strait area between the Australian mainland and Tasmania based on analysis of magnetic anomalies. The map, made before any offshore seismic work or drilling had been carried out, indicates two underwater basins which were considered to be prospective for oil. One was an offshore extension of the then nonproductive Ottway Basin, previously located on land by drilling in southeastern Victoria. The existence of the other basin, in the center of Bass Strait, had not been known before the magnetic survey. The information on basin geometry thus obtained made it possible to program subsequent seismic lines over areas where the sedimentary basins appeared to be best developed. On the basis of this shooting, large quantities of gas have been discovered in the offshore portion of the Ottway Basin in fields that are well placed for supplying the energy needs in population centers such as Sydney and Melbourne.

Even in known basin areas, magnetic surveys can show topographic trends in the basement that might have an important bearing on overlying sedimentary structures. Figure 16-22 shows a map of total fields from an aeromagnetic survey over the Wichita Mountain area of southwestern Oklahoma. The northwest-southeast lineation of magnetic contours follows the trend of the Wichita Mountains, but no igneous rock is exposed along this axis. An analysis by Vacquier et al.[5] leads to the conclusion that the zone of high intensity is not the axis of the buried Wichitas but a linear band of gabbro about 3 mi wide buried along the southwest slope of the mountains. A depth of burial of 2300 ft is estimated from the magnetic gradients. A purely qualitative analysis would put the highest ridge of the buried Wichitas along the axis of the magnetic high several miles from that determined by Vacquier's computations. Such a displacement could have a significant bearing on conclusions that might be reached by a petroleum geologist looking for sedimentary structures in the area that might be controlled by buried basement topography.

A number of oil fields have been discovered entirely or primarily through magnetometer surveys. In most of them the production is from porous serpentine plugs which have a high magnetic susceptibility compared with surrounding sediments. These fields are all quite small. Sometimes a sedimentary oil trap is formed by draping over a structural high on the basement surface. The Cumberland Field, in Oklahoma (Fig. 16-23) is located along the axis of a magnetic high with a 100-gamma reversal. The cross section in the figure was inferred by Cram[23] on the basis of subsurface and magnetic information. The Jackson Field in Mississippi is located over a magnetic high with the closing magnetic contour of lowest intensity 20 mi in diameter. The greatest magnitude of the anomaly is more than 1200 gammas. The origin of the anomaly is believed to be an igneous plug, very possibly cryptovolcanic, but it has never been penetrated by the drill. The largest field ever discovered directly by the magnetometer is the Hobbs Field of Lea County, N.M.

TOTAL INTENSITY AEROMAGNETIC MAP RELATIVE TO ARBITRARY DATUM
MANGUM, OKLAHOMA

Flown 1935' above sea level

FIGURE 16-22
Aeromagnetic contours over area of Wichita Mountains near Mangum, Okla.
(*From Vacquier et al.[5]*)

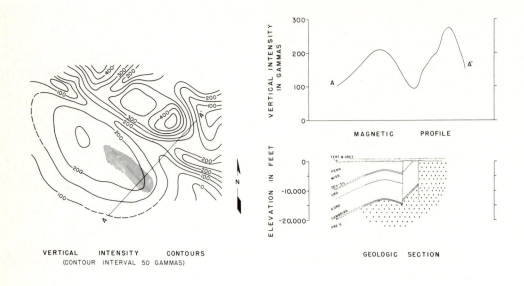

VERTICAL INTENSITY CONTOURS
(CONTOUR INTERVAL 50 GAMMAS)

GEOLOGIC SECTION

CUMBERLAND FIELD

BRYAN & MARSHALL COUNTIES, OKLAHOMA

HORIZONTAL SCALE IN MILES

0   3

FIGURE 16-23
Magnetic intensities observed over Cumberland Field, Okla., along with geologic section. (*After J. W. Peters, Mines Mag., 1949.*)

The discovery well was drilled in 1927 on the basis of magnetic and torsion-balance surveys.

Weak negative anomalies are to be expected over salt domes because of the contrast between the slightly negative susceptibility (about -0.5×10^{-6} cgs units) of the salt or cap rock and the small positive susceptibility (10 to 50×10^{-6} cgs units) of the surrounding sediments. So far as is known, no oil has been associated with salt domes exhibiting such a magnetic effect, although several experimental surveys have revealed anomalies of the predicted size.

REFERENCES

1. Rayner, J. M.: The Role of Geophysics in the Development of Mineral Resources, pp. 259–266, in "Mining and Groundwater Geophysics/1967," *Geol. Surv. Can. Econ. Geol. Rep.* 26, Ottawa, 1970.
2. Heiland, C. A.: "Geophysical Exploration," Prentice-Hall, New York, 1940.
3. Cook, Kenneth L.: Quantitative Interpretation of Magnetic Anomalies over Veins, *Geophysics*, vol. 15, pp. 667–686, 1950.
4. Peters, L. J.: The Direct Approach to Magnetic Interpretation and Its Practical Applications, *Geophysics*, vol. 14, pp. 290–320, 1949.

5. Vacquier, V., Nelson C. Steenland, Roland G. Henderson, and Isidore Zietz: Interpretation of Aeromagnetic Maps, *Geol. Soc. Am. Mem. 47*, 1951.

6. Andreasen, G. E., and I. Zietz: Limiting Parameters in the Magnetic Interpretation of a Geologic Structure, *Geophysics*, vol. 27, pp. 807–814, 1962.

7. McGrath, P. H., and Peter J. Hood: The Dipping Dike Case: A Computer Curve-matching Method of Magnetic Interpretation, *Geophysics*, vol. 35, pp. 831–848, 1970.

8. Hartman, Ronald R., Dennis J. Teskey, and Jeffrey L. Friedberg: A System for Rapid Digital Aeromagnetic Interpretation, *Geophysics*, vol. 36, pp. 891–918, 1971.

9. Baranov, V.: A New Method for Interpretation of Aeromagnetic Maps: Pseudogravimetric Anomalies, *Geophysics*, vol. 22, pp. 359–383, 1957.

10. Naudy, Henri: Automatic Determination of Depth on Aeromagnetic Profiles, *Geophysics*, vol. 36, pp. 717–722, 1971.

11. Hutchison, R. D.: Magnetic Analysis by Logarithmic Curves, *Geophysics*, vol. 23, pp. 744–769, 1958.

12. Gay, S. Parker, Jr.: Standard Curves for Interpretation of Magnetic Anomalies over Long Tabular Bodies, "Mining Geophysics," vol. 2, pp. 512–548, Society of Exploration Geophysicists, Tulsa, Okla., 1967.

13. Nabighian, Misac N.: The Analytic Signal of Two-dimensional Magnetic Bodies with Polygonal Cross-Section: Its Properties and Use for Automated Anomaly Interpretation, *Geophysics*, vol. 37, pp. 507–517, 1972.

14. Weaver, P.: The Relative Place of Empirical and Analytic Methods of Geophysical Interpretation, *Geophysics*, vol. 7, pp. 281–292, 1942.

15. Webb, J. E.: The Search for Iron Ore, Eyre-Peninsula, South Australia, "Mining Geophysics," vol. 1, pp. 379–390, Society of Exploration Geophysicists, Tulsa, Okla., 1966.

16. Leney, George W.: Field Studies in Iron Ore Geophysics, "Mining Geophysics," vol. 1, pp. 391–417, Society of Exploration Geophysicists, Tulsa, Okla., 1966.

17. Riddell, Paul A.: Magnetic Observations at the Dayton Iron Deposit, Lyon County, Nevada, "Mining Geophysics," vol. 1, pp. 418–428, Society of Exploration Geophysicists, Tulsa, Okla., 1966.

18. Galbraith, F. M.: The Magnetometer as a Geological Instrument at Sudbury, *Trans. Am. Inst. Min. Met. Eng.*, vol. 164, Geophysics, 1945, pp. 98–106, 1945.

19. Gerryts, E.: Diamond Prospecting by Geophysical Methods: A Review of Current Practice, pp. 439–446, in "Mining and Groundwater Geophysics/1967," *Geol. Surv. Can., Econ. Geol. Rep. 26*, Ottawa, 1970.

20. Joesting, H. R.: Magnetometer and Direct-Current Resistivity Studies in Alaska, *Trans. Am. Inst. Min. Met. Eng.*, vol. 164, Geophysics, 1945, pp. 66–87, 1945.

21. Steenland, N. C.: An Evaluation of the Peace River Aeromagnetic Interpretation, *Geophysics*, vol. 28, pp. 745–755, 1963.

22. Wallis, W. E.: Offshore Petroleum Exploration, Gippsland and Bass Basins—Southeast Asia,* *Proc., 7th World Petrol. Congr.*, vol. 2, pp. 783–791, Elsevier, London, 1967.

23. Cram, Ira H.: Cumberland Oil Field, Bryan and Marshall Counties, Okla., pp. 341–358, in "Structure of Typical American Oil Fields," American Association of Petroleum Geologists, 1948.

* This title undoubtedly contains an error. "Asia" should be "Australia."

17
ELECTRICAL PROSPECTING METHODS

Electrical prospecting is far more diversified than the other geophysical methods we have studied. Some electrical techniques, e.g., the self-potential, telluric-current and magnetotelluric, depend on naturally occurring influence fields and in this respect resemble gravity and magnetic prospecting. Other methods require electric currents or fields which are introduced into the earth artificially and in this respect are similar to the seismic techniques. In the latter category belong the potential-drop methods, such as resistivity, as well as electromagnetic and induced-polarization prospecting. Because of the diversity of electrical prospecting techniques and their limited use compared with the other methods we have studied, they will not be treated with the same detail as seismic, gravity, and magnetic methods. Further information on electrical prospecting can be found in Keller and Frischknecht[1] and Parasnis.[2]

Electrical methods are much more frequently used in searching for metals and minerals than they are in exploring for petroleum. This is because most of them have proved effective only for shallow exploration, seldom giving information on subsurface features deeper than 1000 or 1500 ft. Some of the methods, such as spontaneous polarization, are valid only for locating ores in the neighborhood of the water table. Others, such as the dipole technique for measuring resistivity and the magnetotelluric method, have sufficient penetration for mapping the basement surfaces underlying sedimentary basins. Both are widely used for oil exploration in the U.S.S.R.

Electrical prospecting methods are being employed to an increasing extent in engineering geology, where resistivity measurements are used for finding the depth to bedrock and also in geothermal exploration.

17-1   ELECTRICAL PROPERTIES OF ROCKS

Electrical prospecting makes use of three fundamental properties of rocks: (1) the resistivity, or inverse conductivity, governs the amount of current that passes through the rock when a specified potential difference is applied; (2) the electro-chemical activity with respect to electrolytes in the ground is the basis for the self-potential and induced-polarization methods; (3) the dielectric constant gives information on the capacity of a rock material to store electric charge and governs in part the response of rock formations to high-frequency alternating currents introduced into the earth by conductive or inductive means.

Resistivity   The electrical resistivity of any material is defined as the resistance of a cylinder with a cross section of unit area and with unit length. If the resistance of a conducting cylinder having a length l and cross-sectional area S is R, the resistivity ρ is expressed by the formula

$$\rho = \frac{RS}{l} \qquad (17\text{-}1)$$

The generally accepted unit of resistivity is the ohm-meter, although the ohm-centimeter is sometimes used. The current I is related to the impressed voltage V and the resistance R by Ohm's law:

$$I = \frac{V}{R} \cdot \qquad (17\text{-}2)$$

The conductivity σ of a material is defined as $1/\rho$, the reciprocal of its resistivity.

In most rock materials the porosity and the chemical content of the water filling the pore spaces is more important in governing resistivity than the conductivity of the mineral grains of which the rock itself is composed. The salinity of the water in the pores is probably the most critical factor determining the resistivity. When porous rocks, particularly those with large concentrations of magnetite or graphite, lie above the water table at shallow depths, or when they occur at such great depths that all pore spaces are closed by ambient pressure, the conduction through them takes place within the mineral grains themselves. Under these conditions, the resistivity of the rock will depend on the resistivity of the grains. When the pores are saturated with fluids, it will be governed by the fluid resistivity as well.

The range of resistivities among rocks and rock materials is enormous, extending

Table 17-1   RESISTIVITIES (IN OHM-METERS) FOR WATER-BEARING ROCKS OF VARIOUS TYPES

Geologic age	Marine sand, shale, graywacke	Terrestrial sands, claystone, arkose	Volcanic rocks (basalt, rhyolite, tuffs)	Granite, gabbro, etc.	Limestone, dolomite, anhydrite, salt
Quaternary, Tertiary	1–10	15–50	10–200	500–2000	50–5000
Mesozoic	5–20	25–100	20–500	500–2000	100–10,000
Carboniferous	10–40	50–300	50–1000	1000–5000	200–100,000
Pre-Carboniferous Paleozoic	40–200	100–500	100–2000	1000–5000	10,000–100,000
Precambrian	100–2000	300–5000	200–5000	5000–20,000	10,000–100,000

SOURCE: G. R. Keller, in "Handbook of Physical Constants," rev. ed., *Geol. Soc. Am. Mem.* 97, 1966.

from 10^{-5} to 10^{15} Ω-m. Rocks and minerals with resistivities from 10^{-5} to 10^{-1} Ω-m are considered good conductors; those from 1 to 10^7 Ω-m, intermediate conductors, and those from 10^8 to 10^{15} Ω-m poor conductors. Table 17-1 lists the ranges within which resistivities have been observed for several types of water-bearing rocks.

Figure 17-1 shows the effect of geologic age upon the resistivity of sedimentary rocks. The horizontal lines show the range of resistivities measured around radio stations for sedimentary rocks with apparently similar lithologic characteristics having ages which cover the entire range of geologic time. Normally one would expect a fairly uniform increase of resistivity with geological age because of the greater compaction associated with increasing thickness of overburden. But the anomalously high resistivities of the Tertiary rocks reflect the fact that the deposition at this time was mainly in fresh-water rather than in saltwater, as was the case during the Mesozoic.

There is no consistent difference between the *range* of resistivities of igneous and sedimentary rocks, although metamorphics appear to have a higher resistivity, statistically, than either of the other types. Certain rock materials, including some that are sought in mining exploration, tend to have anomalously low resistivities (high conductivities) with respect to surrounding rocks. This makes it possible to locate them by measuring resistivity with instruments on the surface.

Electrochemical activity   The electrochemical activity of rocks depends on their chemical composition and on the composition and concentration of the electrolytes dissolved in the groundwater with which they are in contact. Such activity governs the magnitude and sign of the voltage developed when the rock material is in equilibrium with an electrolyte.

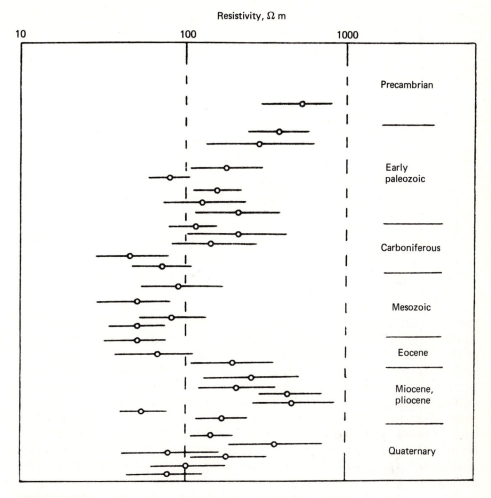

FIGURE 17-1
Average resistivity for sedimentary rocks of various geologic ages. Each bar
indicates the range within which 95 percent of the values fall for the group
indicated. (*From Keller and Frischknecht.*[1])

Dielectric constant   The dielectric constant, which corresponds to permeability
in magnetic materials, is a measure of the polarizability of a material in an electric
field. The polarization, or electrical moment per unit volume, P is proportional to the
impressed electrical field E, and the proportionality constant is χ, the electrical
susceptibility. The total electrical flux per unit area (corresponding to magnetic
flux density) is $E + 4\pi P$ or $(1 + 4\pi\chi)E$. The quantity $1 + 4\pi\chi$ is designated as ϵ,
the dielectric constant, which is analogous to magnetic permeability.

   This property determines the effective capacitance of a rock material and
consequently its static response to any applied electric field, either direct or alternat-

ing. The dielectric constant of a vacuum is unity. For most hard rocks it ranges from about 6 to 16 electrostatic units (esu). For wet soils and clays it is somewhat greater than this, extending up to 40 or 50 esu. The dielectric constant is generally independent of frequency below 100 Hz, but at higher frequencies a dependence is observed. It is also quite sensitive to temperature, the value of ϵ increasing as the rocks become hotter.

17-2 THE SELF-POTENTIAL METHOD

The self-potential method involves measurement at the surface of electric potentials developed in the earth by electrochemical action between minerals and the solutions with which they are in contact. No external electric fields are applied with this technique. When different portions of an ore body are in contact with solutions of different composition, chemical reactions take place which result in different solution pressures along the respective areas of contact. The difference in solution pressure gives rise to a potential difference which causes current flow in the ground.

Source of potential The potential anomaly observed over a sulfide or graphite body is invariably negative. Earlier theories attributed the anomaly in sulfides to oxidation of the portion of the body penetrating the aerated zone above the water table. The process was presumed to generate sulfuric acid. Neutralization of this acid would form salts with an equilibrium potential that would cause current to flow through the pyrite to solutions below it with a different potential. The theory of Sato and Mooney[3] is now more widely accepted. The mechanism they have proposed is not based on oxidation of the sulfide. The ore body, being a good conductor, carries current from oxidizing electrolytes above the water table to reducing ones below it without being oxidized itself. The flow pattern is shown in Fig. 17-2, and it is detected by the negative potential the current introduces as it passes along the surface.

While the strongest potentials of this kind are excited in ores such as pyrites, a number of other minerals, e.g., pyrrhotite and magnetite, generate diagnostic self-potential patterns. Kruger and Lacy[4] describe an alunite body which was responsible for a 1700-mV potential anomaly at Cerro de Pasco in Peru. The polarization was attributed by them to free acid released during alunitization of the country rock, a mechanism that does not conform to more recent thinking. Not all near-surface sulfide bodies exhibit anomalous potentials, since there are many surface conditions that inhibit oxidation. Beneath the water table or permafrost, oxidation is virtually absent.

Spurious sources of potential often obscure effects of subsurface electrochemical action. Elevation changes (on account of normal atmospheric potential gradient) cause voltages in the earth which may not be easy to predict. Telluric currents (natural earth currents of global extent flowing through the earth's crust) may result in potential differences which are sometimes difficult to separate from

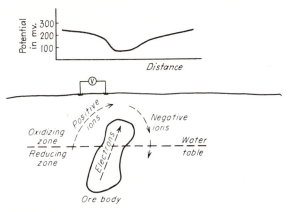

FIGURE 17-2
Natural potential profile and pattern of current flow
where sulfide body penetrates oxidizing zone above
water table.

electrochemical potentials. Streaming potentials can also cause anomalies, the
largest of which are greater than those associated with oxidation of an ore body.

Field procedure and interpretation   Self-potential anomalies, often hundreds
of millivolts in magnitude, can be detected by nonpolarizing porous electrodes
connected to the respective terminals of a millivoltmeter. The potentials can be
measured along profiles with pairs of such electrodes maintained at uniform
separation. With this arrangement, gradients are usually mapped rather than actual
potential differences. Equipotential lines are sometimes determined by maintaining
one electrode in a fixed position and finding the line along the surface for which no
potential difference is observed between it and a movable probe.

The theory of interpretation is quite similar to that for magnetic prospecting
since dipole potential fields are involved in both cases. Theoretical studies by
Petrovski,[5] Stern,[6] de Witte,[7] and Yüngül[8] have led to quantitative techniques for
interpreting potential anomalies. These are useful only when the polarized body can
be represented by a sphere or some other simple geometrical form.

Self-potential surveys have often led to the discovery of sulfide-ore bodies at
shallow depths of burial. Among these are several at Noranda and one near
Sherbrooke, Quebec. The possibilities of the technique have been investigated in
other areas, such as the tristate zinc and lead district (see Jakosky et al.[9]), where
no correlation was found between natural potentials and lead and zinc
mineralization.

Yüngül[8] has described a self-potential survey in the Sariyer area of Turkey which
led to the discovery of a sulfide mass containing zones with copper concentrations as
high as 14 percent. A cross section of this ore body, as determined from boreholes,
shafts, and tunnels, is shown in Fig. 17-3 along with the potential profile observed

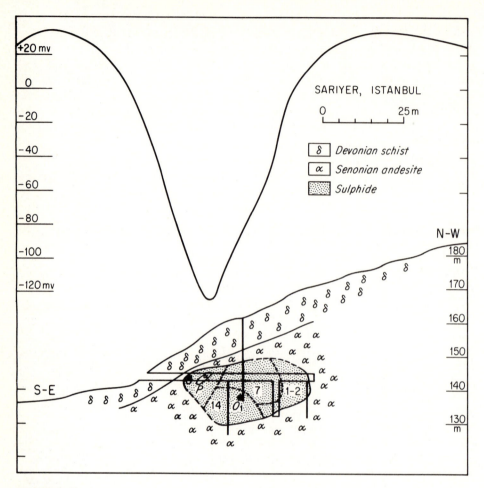

FIGURE 17-3
Self-potential profile and geologic section, Sariyer, Turkey. (*After Yüngül.*[8])

on the overlying surface. The steep topography required the use of terrain corrections which resulted in a displacement of the predicted center of the ore body northwest-ward from the polarization minimum that was measured on the ground. A drill hole at the position of minimum potential would have encountered only pyrite, as was learned subsequently when the adit shown in the figure was excavated.

17-3 DC (AND LOW-FREQUENCY CURRENT) RESISTIVITY METHODS

As resistivity is a fundamental electrical property of rock materials closely related to their lithology, the determination of the subsurface distribution of resistivity from measurements on the surface can yield useful information on the structure or composition of buried formations. The most common method for carrying out such measurements involves the passage into the earth of direct (or very low-frequency) current. Four-electrode arrays are generally used at the surface, one pair for introducing current into the earth, the other pair for measurement of the potential associated with the current. The potential-drop-ratio method is a variation on this procedure also used for determining resistivity. Other techniques, such as the magnetotelluric and dipole methods, measure resistivity in quite different ways.

A comprehensive study of resistivity interpretation which covers both the physical theory behind it and geological applications has been made by Van Nostrand and Cook.[10] A more compact review of a tutorial nature was published by Ginzburg.[11]

Current Flow and Potentials between Electrodes at Surface; The Equipotential-Line Method

If two electrodes are inserted in the ground, and if an external voltage is applied across them, there will be a flow of current through the earth from one electrode to the other. The lines of flow are always perpendicular to lines along which the potential is constant, the latter being referred to as equipotential lines. This relationship is illustrated by the cross section shown in Fig. 17-4a. The potential difference (or voltage) impressed across electrodes A and B is distributed along the space in between them as indicated by the dotted lines. In a homogeneous conductor, the potential with respect to A along a vertical plane cutting the surface at C, which is midway between A and B, will be half as great as its value at B. If one could measure the potential underground, one would observe that the potential is the same as at any surface point such as D wherever the ratio of distances from the point to A and to B respectively is the same as the ratio at the surface point. In the case of D, this ratio is one-third. The full line extending downward from D and bending back under A is the trace of this *equipotential surface* on the vertical plane containing A and B. Figure 17-4b shows where a family of such surfaces intersects the horizontal plane containing A and B. The equipotential lines must always be perpendicular to the lines of current flow, since no component of the current at any point can flow along the equipotential line at that point.

The configuration of equipotential lines in the earth, although not observable directly, can be deduced by measurements with potential electrodes at the surface. Lateral inhomogeneities in the conductivity of subsurface materials cause distortions in the current flow which give rise to corresponding irregularities in the equipotential lines. Figure 17-5 illustrates how bodies of anomalously high or low resistivity deflect the flow of current in their vicinity. Changes in the potential field associated with these deflections can be observed by potential electrodes at the surface. This is the basis for the equipotential-line method, a qualitative technique which is sometimes employed to locate ore bodies having anomalous resistivity.

Variations of resistivity in the vertical direction also affect the pattern of equipotential lines, causing distortions which are likely to show up as anomalies. It

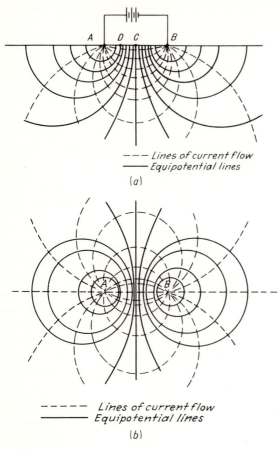

FIGURE 17-4
Equipotential lines and flow lines: (a) below the earth's surface in vertical
plane of electrodes; (b) on plane of earth's surface; electrodes at A and B.

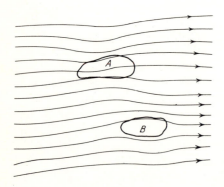

FIGURE 17-5
Distortion of current flow lines by
bodies having (A) anomalously high or
(B) anomalously low conductivity.

is not feasible, however, to map such variations simply by determining the equi-potential-line configuration at the surface. A more quantitative approach must be used, one that is based on the ratio between the current introduced into the earth and the potential differences it causes between selected points along the surface. The resistivity method, which involves this type of measurement, is designed to allow three-dimensional mapping of the resistivity within the earth. If the variation of resistivity is entirely vertical (as when the pattern consists of a number of discrete layers, each of different resistivity, separated by horizontal interfaces), the technique is relatively simple even though there will be limitations in the precision and reliability with which the actual resistivity pattern can be determined. If discrete laterally bounded bodies of anomalous resistivity are to be detected and their depths and shapes determined, the method becomes more difficult to apply.

Apparent Resistivity

All resistivity techniques in general use require the measurement of apparent resistivity. To illustrate this concept, let us consider a semi-infinite solid with uniform resistivity ρ. Assume that a current I is introduced in this material through electrodes at positions A and B on its surface (Fig. 17-6). Assume also that the potential gradient associated with this current is measured across two other electrodes at positions C and D on the same surface. The potential at electrode C will be

$$V_C = \frac{I\rho}{2\pi}\left(\frac{1}{r_1} - \frac{1}{r_2}\right) \qquad (17\text{-}3)$$

where r_1 is the distance from potential electrode C to current electrode A and r_2 is the distance from this electrode to current electrode B. Similarly, the potential at electrode D is

$$V_D = \frac{I\rho}{2\pi}\left(\frac{1}{R_1} - \frac{1}{R_2}\right) \qquad (17\text{-}4)$$

where R_1 is the distance from D to A and R_2 the distance from D to B.

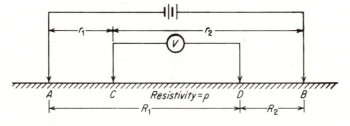

FIGURE 17-6
Arrangement of current electrodes (A and B) and potential electrodes (C and D).

The potential difference V that would be measured by a potentiometer across electrodes C and D is simply $V_C - V_D$. Subtracting Eq. (17-4) from Eq. (17-3) and solving for ρ, we obtain a value which we designate as

$$\rho_a = \frac{2\pi V}{I} \frac{1}{1/r_1 - 1/r_2 - 1/R_1 + 1/R_2} \tag{17-5}$$

The result is independent of the positions of the electrodes and is not affected when the current and potential electrodes are interchanged.

The value ρ_a obtained from Eq. (17-5) is designated as the *apparent resistivity*, which is equivalent to the true resistivity only when the latter is uniform throughout the subsurface. Otherwise it must be looked upon as the most convenient way to represent the actual distribution of laterally homogeneous resistivity in the subsurface on the basis of surface measurements. If the electrodes are laid out along a line and their separations are increased in a systematic manner, the change in the apparent resistivity [as defined in Eq. (17-5)] with electrode spacing makes it possible to determine the variation of resistivity with depth within limits of precision that depend on the subsurface layering configuration.

To illustrate how this concept is applied, let us assume that the subsurface consists of two layers separated by a horizontal boundary, the upper layer having a resistivity ρ_1 and the lower one having a resistivity ρ_2, which is less than ρ_1. The current between electrodes A and B will no longer flow along approximately circular arcs as it did in a homogeneous earth as shown in Fig. 17-4. The lines of flow are moved downward as shown in Fig. 17-7 because the lower resistivity below the interface results in an easier path for the current within the deeper zone. For the same reason, the total current is greater than it would be if the upper material extended downward to infinity. Moreover, the deeper the interface the smaller the increase in current flow, while the greater the electrode separation in proportion to the depth of the interface the greater the effect of the low-resistivity substratum on the current that flows between the electrodes.

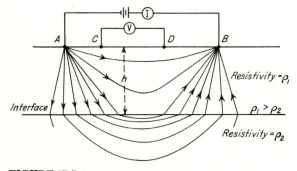

FIGURE 17-7
Lines of current flow between electrodes A and B in two-layered earth with higher conductivity in deeper layer. Compare with flow lines for homogeneous earth (Fig. 17-4).

The distortion of the lines of current flow from the simple pattern shown on Fig. 17-4 for a homogeneous earth to the more complicated one of Fig. 17-7 for a two-layer subsurface results in a corresponding distortion of the equipotential lines which are mapped where they intersect the surface. As the apparent resistivity depends only on the potential difference between the voltage electrodes divided by the current that flows into the earth, any irregularities in the positions where the equipotentials intercept the surface indicate departures from homogeneity in the subsurface resistivity.

It is easy to see that the apparent resistivity would be virtually the same as the resistivity ρ_1 of the upper medium when the separation of current electrodes is very small compared with the thickness h of the upper layer. This is because very little current would penetrate to the substratum below the interface. At spacings which are very large compared with h, the apparent resistivity approaches ρ_2 because the portion of the current confined to the surface layer becomes negligible. Figure 17-8 shows a schematic curve of apparent resistivity versus the ratio of electrode spacing to depth for the two-layer case.

Electrode Arrangements and Field Procedures

In actual practice a number of different surface configurations are used for the current and potential electrodes. In most arrangements, both sets of electrodes are

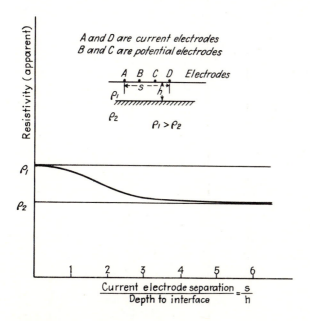

FIGURE 17-8
Apparent resistivity as function of electrode separation for two-layer case illustrated in Fig. 17-7 (schematic).

laid out along a line. The current electrodes are generally placed on the outside of the potential electrodes, although the opposite arrangement should be equivalent in principle. The most widely used configurations will be described in the paragraphs that follow.

Wenner arrangement One of the most common electrode arrangements for resistivity measurement in English-speaking countries is the Wenner configuration illustrated in Fig. 17-9a. Here each potential electrode is separated from the adjacent current electrode by a distance a which is one-third the separation of the current electrodes. For this geometry, Eq. (17-5) becomes

$$\rho_a = 2\pi a \frac{V}{I} \qquad (17\text{-}6)$$

Schlumberger arrangement In the Schlumberger configuration, the operator expands the electrode spacing by increasing the distance between current electrodes

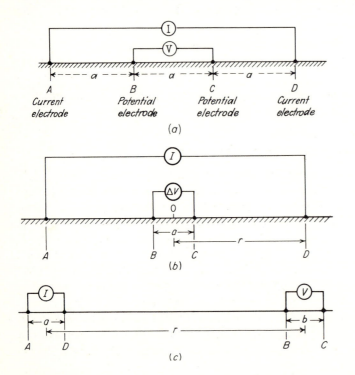

FIGURE 17-9
Electrode configurations in common use: (a) Wenner arrangement; a is spacing as used in resistivity formulas; (b) Schlumberger arrangement; a is constant, and r is increased during the measurement; (c) dipole arrangement; separation r between centers of respective electrode pairs is changed during the measurement.

or that between potential electrodes, but only one at a time, during the course of a measurement. The potential electrodes are assumed to be an infinitesimal distance apart, and observed values of potential are adjusted by extrapolation to fit this assumption. The Schlumberger electrode arrangement is illustrated in Fig. 17-9b.

The apparent resistivity at the center of a Schlumberger array is

$$\rho_a = \frac{\pi r^2}{a} \frac{V}{I} \tag{17-7}$$

where r is defined in the diagram. For accurate results a should be less than one-fifth of r.

Dipole method   The dipole method, more recently introduced than the Wenner or Schlumberger method, is widely used in the U.S.S.R. for deep penetration. It is illustrated in Fig. 17-9c. The current electrodes are usually well separated from the potential electrodes. If the separations a and b are equal and the distance between the centers of the respective pairs is r, the apparent resistivity determined by this arrangement is

$$\rho_a = \pi \left(\frac{r^2}{a} - r \right) \frac{V}{I} \tag{17-8}$$

This holds when the dipole array is collinear. Where it is not, a more complicated expression is used to obtain the apparent resistivity.

Cancellation of Spurious Potentials at Electrodes

In addition to the potential difference associated with the current introduced into the earth by the current electrodes, the voltage reading may include spurious electrochemical potentials between the electrodes and electrolytes in the earth. Nonpolarizing electrodes (such as copper–copper sulfate porous pots) are often used to avoid such effects. A long established technique for eliminating contact potentials of this kind is that of Gish and Rooney.[12] It makes use of a commutator which reverses the directions of the current flow as well as the potential polarities about 30 times per second. Any polarizing potentials at the electrodes, being reversed in sign with each half revolution of the commutator, will cancel out. More recently, systems have been put to use in which dc pulses of 1 to 8 s duration are introduced into the earth with alternately reversed polarity. Spurious electrode potentials can also be canceled by using low-frequency alternating current, 5 Hz being a typical frequency.

Fixed versus Variable Electrode Separations

Two general approaches are used for making resistivity measurements in the field. With the first, the electrode spacings are fixed, and the array of electrodes is moved laterally along a profile with constant separation, the apparent resistivities being plotted at the midpoints. This is called *continuous profiling*. Any ore body having anomalous conductivity which is shallower than the depth of maximum effective penetration should show up as an anomaly on the resulting map. In the second type of measurement the center of the electrode spread remains fixed, but the separation of electrodes is progressively increased. The depth to which information is obtained is approximately the same as the electrode separation in the Wenner arrangement. This technique is called *vertical electric sounding*.

Interpretation

The quantitative interpretation of resistivity data has been the subject of mathematical studies for several decades. Most of the theory developed for this purpose is intended for application to simple models. Among these are multiple layers (up to four at the most) separated from one another by plane (preferably horizontal) interfaces, faults, or hemispherical filled sinks. In recent years workable techniques have been developed for predicting resistivity effects from bodies having an arbitrary shape that cannot be described analytically.

Many geological features of economic interest can be represented quite adequately by a simple layering picture. Where there is an unconsolidated overburden over bedrock, it may be possible to estimate the depth to solid rock rather closely from resistivity measurements using the formulas for a two-layer configuration. Most resistivity surveys carried out for engineering purposes are designed to yield this type of information.

Some ore bodies with anomalous conductivity can be represented as spheres or spheroids, and their dimensions and depth of burial can be calculated from resistivity data using mathematical expressions worked out for these geometrical forms. The same approach can be applied to the study of filled sinks.

Layered media with horizontal interfaces The simplest configuration for which theoretical solutions are available is that involving a small number of discrete layers, each having a uniform but different resistivity and separated from adjacent layers by plane horizontal interfaces.

Hummel[13] has worked out the theory for the two- and three-layer cases using the method of images. This technique is based on the premise that there are current sources at all positions of the mirror images of the electrodes with respect to the various interfaces across which there is a discontinuity in resistivity. Multiple reflections result in an infinite number of such images for each interface. Since each reflection involves a loss of intensity (as with partly silvered mirrors) and successive reflections correspond to increasingly distant sources, it is only necessary to consider the effect of the first few multiples to obtain a usable value for the potential. For a surface layer of thickness h and resistivity ρ_1 overlying an infinitely thick substratum of resistivity ρ_2, the apparent resistivity ρ_a is

$$\rho_a = \rho_1\left\{1 + 4\left[\frac{k}{\sqrt{1 + (2h/a)^2}} - \frac{k}{\sqrt{4 + (2h/a)^2}}\right.\right.$$

$$+ \frac{k^2}{\sqrt{1 + (4h/a)^2}} - \frac{k^2}{\sqrt{4 + (4h/a)^2}}$$

$$\left.\left.+ \frac{k^3}{\sqrt{1 + (6h/a)^2}} - \frac{k^3}{\sqrt{4 + (6h/a)^2}} + \cdots\right]\right\} \qquad (17\text{-}9)$$

where a is the electrode separation for the Wenner arrangement and k, the resistivity *reflectivity*, is $(\rho_2 - \rho_1)/(\rho_2 + \rho_1)$. Equation (17-9) expresses ρ_a as an infinite series with the nth term having the form

$$4\rho_1\left[\frac{k^n}{\sqrt{1 + (2nh/a)^2}} - \frac{k^n}{\sqrt{4 + (2nh/a)^2}}\right] \qquad (17\text{-}10)$$

Since k is always less than unity, the series converges; i.e., the terms approach zero as n increases, and only a limited number of terms (each corresponding to a successive *multiple reflection*) is needed in carrying out the summation.

Hummel's work was based on that of Stefanescu et al.,[14] who had previously developed a more general integral expression suitable for the computation of master curves.

With the Wenner arrangement of electrodes, the apparent resistivity ρ_a as measured in the field is simply $2\pi a(V/I)$. When ρ_a is determined over a range of a values, the quantities h and k can be found by matching the curves of ρ_a versus a, as observed, with theoretical curves based on various assumed layering conditions.

Master curves   The most common device for interpreting resistivity data for a small number of horizontal layers is an assemblage of master curves. Each such curve is a plot of apparent resistivity versus electrode separation for the arrangement of electrodes employed in the field and for a specified layering configuration, various thicknesses and resistivity ratios being assumed for the individual layers.

TWO-LAYER CASE   In the two-layer case involving a single layer of specified thickness h overlying an infinitely thick homogeneous substratum, a family of curves is plotted for different values of h and k. The apparent resistivity (actually $2\pi a\,V/I$ for the Wenner configuration of electrodes) is plotted versus a at the same scale as the master curves, and the curve of observed data is matched with the theoretical master curves. The correct values of h and k are established from the characteristics of the master curve giving the best match.

Originally the master curves were plotted on a linear scale as demonstrated by Tagg.[15] It is now customary to plot such curves on a logarithic scale like that shown in Fig. 17-10 for the Schlumberger configuration. The abscissa is the logarithm of a/h, the ratio of the potential electrode separation to layer thickness. The ordinate is the logarithm of the ratio of the apparent resistivity to its limiting value ρ_1. If the assumption of two horizontally bounded layers is correct, the only unknown is h, the layer thickness. The advantage of using log-log plots for the observed data and the master curves is that the experimental curve giving the best fit to the field measurements will be parallel to the applicable master curve. When h is not known, as is generally the case, it is only necessary to assume a value for the depth in plotting the experimental points. The a/h value of the master curve which most closely parallels the plot of observed data enables one to determine h.

Let us illustrate the use of two-layer curves with an example. Assume that the apparent resistivity obtained with a Wenner configuration varies with electrode

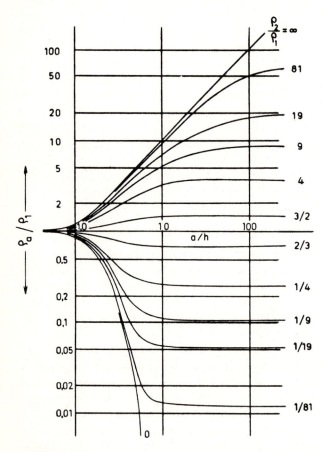

FIGURE 17-10
Typical master curves for Schlumberger electrode arrangement when layer of resistivity ρ_1 and thickness h overlies substratum of resistivity ρ_2. (*From Parasnis.*[2])

**Table 17-2   APPARENT RESISTIVITY VERSUS
ELECTRODE SPACING**

a, ft	ρ_a, $\Omega \cdot$m	ρ_0/ρ_a
300	89.6	0.75
400	107.4	0.623
500	123.2	0.545
600	138.6	0.483
700	152.2	0.440
800	164.8	0.407

spacing as shown in Table 17-2. The ρ_a at $a = 0$, which is designated ρ_0, has been determined by extrapolation to be 67.2 $\Omega \cdot$m.

We shall plot these data on a set of curves for two layers such as those in Fig. 17-11, designed for a Wenner arrangement, where the resistivity of the uppermost layer is less than the apparent resistivity. For convenience, we first assume h to be 300 ft and plot a/h versus ρ_a/ρ_0 on the coordinate scale of the master curve. We find that we can get the best match by shifting the curve thus obtained a lateral distance to the right which represents a factor of 1.075 on the logarithmic scale for a/h. The assumed value of h must be divided by this factor to obtain the correct thickness, which turns out to be 280 ft. The ρ_1 corresponding to a k value of 0.7 is 381 $\Omega \cdot$m. We have thus determined the thickness of the surface layer and the resistivity of the substratum by fitting the data to the master curves.

CURVES FOR THREE OR MORE LAYERS   The Compagnie Générale de Géophysique[16] has published a set of 480 master curves corresponding to this number of three-layer resistivity configurations. The relative thicknesses of the different layers and their relative resistivities are varied systematically over the ranges of thickness and resistivity for each layer that are expected to be of practical interest. The curves have been computed for the Schlumberger electrode arrangement on the assumption that only the spacing of the current electrodes has been increased.

Figure 17-12 illustrates how the three-layer curves are used to interpret field observations. The data points for ρ_a versus $AB/2$ (where AB is the current electrode separation) are plotted on semitransparent log-log paper. The graph paper is then superimposed on the sheet containing the set of curves chosen for comparison, and its position is shifted horizontally and vertically to obtain the best possible fit. In the example shown in Fig. 17-12, the data points appear to fit a three-layer curve for the case where the second layer has a resistivity 1/39 times that of the first and a thickness ½ times as great and where the infinitely thick third layer has infinite resistivity.

More elaborate sets of curves covering four-layer configurations as well as

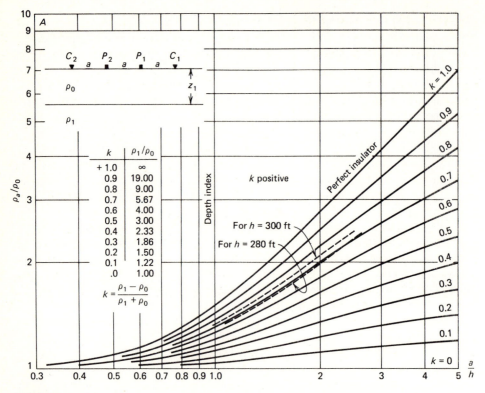

FIGURE 17-11

Two-layer master curves of normalized apparent resistivity versus electrode spacing assuming Wenner electrode configuration for case where apparent resistivity ρ_a is greater than resistivity ρ_0 of surface layer. k is the fraction $(\rho_1 - \rho_0)/(\rho_1 + \rho_0)$, where ρ_1 is resistivity of infinite substratum. Points plotted on the chart are for set of apparent resistivity values observed at different spacings. Upper dashed line shows plot assuming layer thickness of 300 ft. Lower dashed line shows position of data points after shifting to get best fit to master curves. This indicates actual thickness to be 280 ft. (*Adapted from Roman by Van Nostrand and Cook.*[10])

three-layer have been published by Mooney and Wetzel[17] and by Orellana and Mooney.[18] More recently the European Association of Exploration Geophysicists[19] has made curves available that allow greater flexibility in the choice of resistivity patterns.

PARTIAL CURVE MATCHING   For certain types of layering, it is most expeditious to interpret three-layer configurations by fitting the data to two theoretical curves, each for two layers and each covering a different portion of the observed apparent resistivity-versus-electrode-spacing plot. Figure 17-13 shows how apparent-resistivity data observed over a subsurface consisting of two horizontal slabs overlying an infinite substratum can be analyzed by use of this method. The curve for the experimental data is slid horizontally and vertically until the best fit is obtained for its left side to a theoretical curve in a family of curves computed for various two-layer configurations. This comparison yields the thickness of the uppermost layer as well as the k value corresponding to the resistivity contrast between this layer and the one directly below it. The thickness of the middle layer and the resistivity of its substratum can be determined by fitting the right side of the curve to an entirely different family of theoretical curves, which corresponds to the relation between the resistivities of the two lower materials. The accuracy of this approach is best when, as is the case in Fig. 17-13, the middle layer has a resistivity substantially lower than those of the other two media and when its thickness is 3 or 4 times that of the top layer. It is least valid for a middle layer approximately as thick as the upper one. If the middle layer has a higher resistivity than the upper layer, the electrode separations necessary to find its depth might have to be so large that this approach would no longer be practical or economic.

Appropriate master curves can be constructed during the course of an interpretation for any sequence of layers when there is ready access to a digital computer. Argelo[21] has prepared a computer program for this purpose which improves on the results obtainable from partial curve matching as described in the preceding paragraph. The Watson and Johnson approach shown in Fig. 17-13 gives an ap-

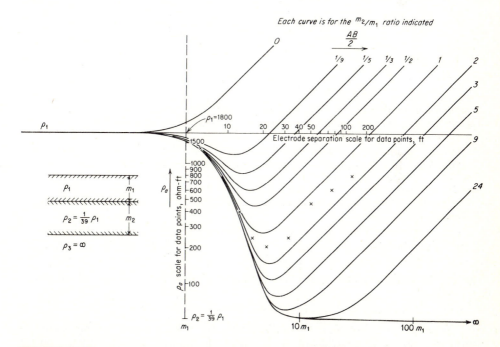

FIGURE 17-12
Typical set of logarithmic three-layer curves computed by Compagnie Générale de Géophysique. High-resistivity (ρ_1) upper layer overlies very low resistivity ($\frac{1}{39}\rho_1$) middle layer, with bottom layer having infinite resistivity. Hypothetical data points indicate that $m_2 = 1\frac{1}{2} m_1$. (*Compagnie Générale de Géophysique.*[16])

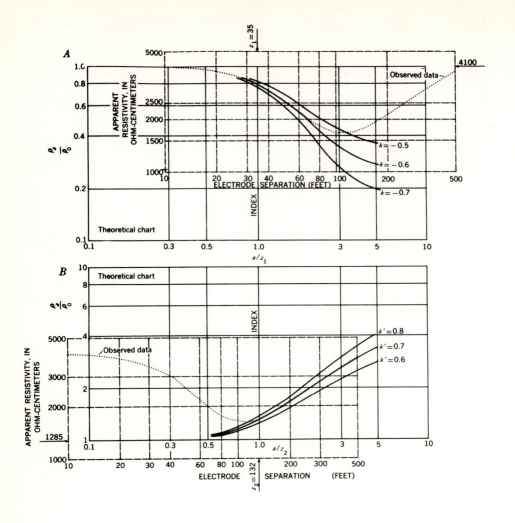

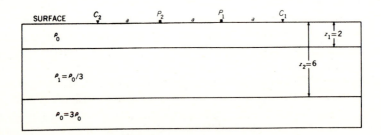

FIGURE 17-13

Use of a pair of two-layer curves to interpret resistivity data over a three-layer configuration. The two-layer model for the uppermost layer and its substratum (the middle layer) is used for shorter electrode separations, and the other model, involving the top two layers (averaged as one) and the third layer, is used for the greater electrode separations. (*After Watson and Johnson.*[20])

proximate interpretation, and a master curve can be drawn from the layering configuration thus deduced. Upon comparison of the new master curve and the curve based on field observations, the model can be revised and new curves computed iteratively until the best possible fit is observed. In recent years computer-based inversion of this kind has largely replaced curve matching for routine interpretation.

The theoretical curves we have discussed are all for plane *horizontal* interfaces between layers. The theory for an *inclined* interface between two layers is much more complicated. A number of papers have appeared in the literature on the interpretation of apparent resistivity data for dipping beds. Perhaps the most useful curves for this purpose can be found in a paper by Unz,[22] who made use of image theory. A direct solution of the problem has been published by Maeda.[23]

The limiting case of the inclined interface is the *vertical* discontinuity in resistivity, one that may be approached in nature over outcrops of dikes or near-vertical faults. Logn[24] has worked out theoretical curves for single vertical planes of discontinuity in resistivity and also for two vertical planes not far apart which bound a material having anomalous resistivity such as a vertical dike.

The Inverse Problem in Resistivity Interpretation

The resistivity-interpretation methods thus far considered are indirect, like those most widely used in gravity and magnetic interpretation. A series of models is assumed, and the model for which the predicted apparent-resistivity-versus-electrode-spacing curve shows closest agreement with the curve based on field measurements is considered as the correct solution. As with gravity and magnetics, direct solutions have been devised for certain cases. The solutions for the electrical parameters, unlike those involving the other types of potential fields, are unique, at least in theory. Here the resistivity distribution is determined as a function of depth by mathematical operations on measured potentials or apparent resistivities. The methods thus far developed apply only to cases with horizontal bedding.

Slichter[25] published a solution to the basic potential equation for which resistivity varies continuously with depth. This case, which can theoretically be used for direct interpretation, has unfortunately little practical application in prospecting. Van Nostrand and Cook,[10] following Pekeris,[26] have developed a method for solving the problem where the variation of resistivity with depth is stepwise. This, of course, corresponds to a multilayered earth. Koefoed[27] has published papers on a method, also based on Pekeris' theory, for rapid computation of the layering configuration from an apparent-resistivity curve. Marsden[60] has shown how to use computers to obtain electric soundings directly from apparent-resistivity-versus-electrode-spacing data obtained in the field.

Effects of Discrete Rock Masses Having Anomalous Resistivity

The mathematical analysis of the potential-field distortions caused by discrete buried bodies such as ore masses is considerably more complicated than it is for layered media. It is generally necessary to represent the buried mass as a generalized geometrical form such as a sphere or spheroid. Van Nostrand[28] has worked out a solution for the apparent resistivity, as measured by a Wenner electrode configuration, over a buried conducting sphere. From his analysis he concludes that a sphere with a top at a greater depth than its radius cannot be detected by conventional resistivity measurements.

The same theoretical approach was extended by Cook and Van Nostrand[29] to the mapping of filled near-surface sinks, a problem of considerable practical importance. The sinks were represented by hemispheres and hemispheroids with their bounding planes at the surface of the earth. Theoretical curves were worked out both for horizontal profiling (with electrode spacing constant but the spread shifted horizontally along a traverse) and for vertical electric sounding.

In exploratory drilling for highly conductive ore bodies, such as sulfides, it is sometimes desirable to estimate the overall extent of such a deposit after it has first been penetrated by the drill. Resistivity measurements where one or more electrodes make contact with the ore body in the borehole are often used for this purpose. Such a procedure is most practical if the mass can be approximated analytically by a geometrical form such as a spheroid. Seigel[30] worked out apparent-resistivity curves for a buried oblate spheroid having a long horizontal axis which is penetrated by a borehole through its vertical axis of revolution, the electrode array straddling the center of the mass. Clark[31] has published similar curves for a conducting spindlelike ore body having the shape of a prolate spheroid, both for the case where the long axis is vertical and for that where it dips at 45°.

Discussion

In applying techniques that involve fitting observational data points to theoretical curves, one should always realize that the same data might equally well fit a number of such curves representing widely different subsurface configurations. Unique solutions can theoretically be obtained for simple subsurface configurations, particularly for cases that involve horizontal layers. Even here independent geologic controls are desirable to narrow the range of different interpretations which could be equally well justified on the basis of the electrical data alone.

Examples of Resistivity Surveys

The simplest kind of resistivity survey is the *horizontal* type using a constant electrode spacing in which subsurface features are recognized qualitatively from anomalies in the apparent resistivities along a profile. The resistivity value is plotted versus distance along the profile at the position of the center of the electrode spread. Ginzburg[11] has reported on a survey of this type in Colombia to test the value of resistivity methods for locating a limestone aquifer interrupted by faulting. Figure 17-14 illustrates the type of information obtainable in this way. The interpretation he shows was subsequently confirmed by drilling.

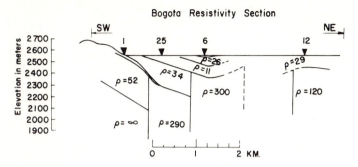

FIGURE 17-14
Resistivity cross section obtained near Bogota, Colombia, to test value of resistivity for locating faulted aquifers. (*After Ginzburg.*[11])

Cook and Van Nostrand[29] have compared the horizontal resistivity profiles
observed over a shale sink in Cherokee County, Kan., with theoretical curves for a
sunken hemispherical boss having a plane boundary flush with the free surface.
Figure 17-15 indicates the close resemblance that was shown between their theoretical and observed curves.

A problem which is often attacked by resistivity measurements is to locate the
depth at which saltwater invades fresh water in aquifers. Figure 17-16 illustrates the
results obtained in a resistivity survey carried out for this purpose in Israel. The
apparent resistivity in the calcareous sandstone aquifer was observed to decrease
rapidly with increasing electrode spacing, and the particularly rapid drop when the
spacing became greater than 50 m corresponds to a discontinuity in resistivity
from 60 $\Omega \cdot$m above to 10 $\Omega \cdot$m below. This boundary was later shown by drilling to
be the freshwater-saltwater contact. Other soundings, also confirmed by drill holes,
showed the boundary to be dipping 10 m/km. The survey thus made it possible to
map the useful portion of the aquifer quite precisely.

17-4 TELLURIC, MAGNETOTELLURIC, AND AFMAG METHODS

In resistivity prospecting one measures the potential drop on the surface caused by
an externally applied current. The penetration obtained with the method is
approximately proportional to the separation of the current electrodes. In 1939
Schlumberger[32] reported on some experiments in France, where artificially applied
currents were dispensed with entirely and natural earth currents used instead. These
currents, often referred to as *telluric currents*, flow everywhere along the surface of
the earth in large sheets. For some time, prospecting with telluric currents was the
electrical method most widely used in oil exploration. Although experimental surveys
have been carried out in the United States (see Dahlberg[33] and Boissonnas and
Leonardon[34] for examples), the most extensive application of telluric-current
prospecting was in Europe and North Africa. The technique has been used only to a

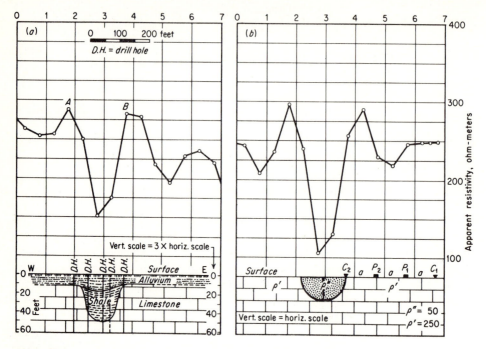

FIGURE 17-15
Comparison of observed and theoretical horizontal resistivity profiles over filled sink for Wenner configuration: (a) observed field curve with geologic cross section; (b) theoretical field with assumed data points only at electrode positions. Diameter of sink is $3a/2$. Assumed value of $\rho''/\rho_1' = \frac{1}{5}$. (*After Cook and Van Nostrand.*[29])

very limited extent for oil exploration in the United States because it is primarily intended for reconnaissance of unexplored areas and at the time it was introduced there were few areas in the United States for which it was considered suitable.

There are a number of techniques based on the use of telluric currents, and their application to oil exploration has been discussed by Vozoff et al.[35] The earliest involves the use of telluric currents alone. The more recently introduced magneto-telluric method, in which natural electric currents are related to their associated magnetic oscillations, has now come into more widespread use. The AFMAG method measures natural oscillations of the earth's magnetic field at audio frequencies.

The Telluric-Current Method

It is generally believed that telluric currents are induced just below the earth's surface by ionospheric currents which correlate with the diurnal changes in the earth's magnetic field. The currents cannot of course be measured directly, but the horizontal potential gradients they produce at the surface are easily measurable.

Earth currents vary geographically, diurnally, and seasonally. As with magnetic diurnal variation, the pattern of earth currents appears fixed with respect to the sun,

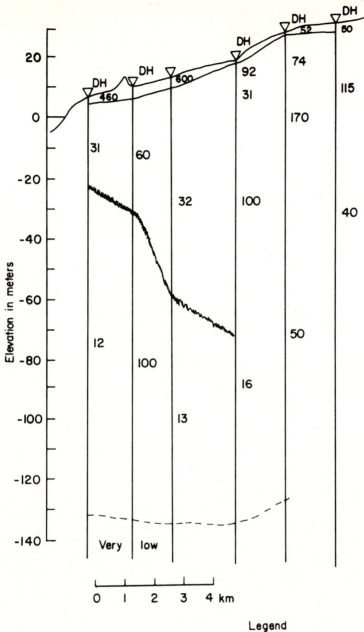

FIGURE 17-16

Cross section showing resistivity change at freshwater-saltwater interface. Numbers indicate resistivities. Survey was at Ashkelon, Israel. (*After A. Ginzburg.*[11])

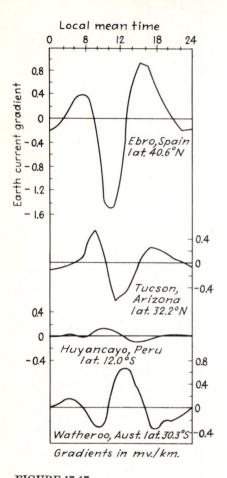

FIGURE 17-17
Diurnal variation in earth-current potential gradients in north-south direction at four latitudes. (*By permission after W. J. Rooney, "Terrestrial Magnetism and Electricity," McGraw-Hill Book Company, 1939.*)

shifting along the earth's surface as the earth rotates. The diurnal curve for any location can be deduced from a world chart; sample curves for four stations at different latitudes are reproduced in Fig. 17-17.

Application to prospecting Telluric currents flow along the earth's surface in large sheets coinciding with conducting layers that cover the outer portion of the earth's crust. The distribution of current density within the sheets will depend on the resistivity of the formations carrying the currents. Thus if a poorly conducting salt dome penetrates more highly conducting formations, the lines of current flow will tend to bypass the salt and cause distortions in the potential gradients at the surface which are associated with the current. These distortions, when measured by appropriately located electrode pairs and properly interpreted, should make it possible to locate the dome.

Earth currents are often oscillatory with periods of a few seconds to a few minutes. The oscillations are observed to be homogeneous in waveform (although not in amplitude or direction) over wide areas. Figure 17-18 shows a typical record of such oscillation as observed at points several miles apart but recorded on the same tape. Since individual oscillations can be correlated in this way, it is possible to map variations in earth currents from place to place by comparing amplitudes of simultaneous oscillations at two locations. The amplitude of the oscillation is measured rather than the absolute potential difference; thus any polarization effects at the electrodes should not influence the results. The voltage differences are measured in two perpendicular directions simultaneously at two locations: a base station and a mobile station. It is customary to compare amplitudes at a number of different frequencies, sometimes up to a dozen. The frequencies can be separated by inspection, by analog filters in the recording equipment, or by subsequent processing of the field data.

Typical telluric-current surveys Boissonnas and Leonardon[34] have reported upon a telluric-current survey over the Haynesville salt dome in Texas. They maintained a base station in continual operation while moving a second electrode setup to various observation points. The variations in magnitude and direction between base and field station were computed and used as the mapping parameter. Separations of base and field stations ranged from 1 to 75 mi. The ratio of the average potential gradient at the field station to that at the base was plotted at each field location, and the values for all the stations were contoured. The resulting contours conformed quite closely to the outlines of the Haynesville dome, indicating that the telluric currents tend to be deflected by the high-resistivity salt into the more readily conducting formations, 1200 ft thick, above the dome.

Mainguy and Grepin[36] have described a number of telluric-current surveys which they carried out as part of a program to explore for oil in southeastern France. The

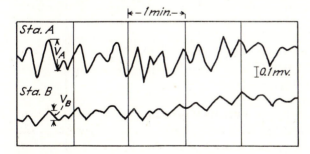

FIGURE 17-18
Earth-potential records made simultaneously at two stations 10 mi apart.

horizons of greatest interest consisted of formations with a high resistivity such as dense limestone overlain by conducting strata. In the Ales-Maruejols Basin, seismic surveys gave a geologic picture quite different from that indicated by the telluric currents. Subsequent drilling, which resulted in the discovery of oil, supported the telluric-current picture.

The Magnetotelluric Method

Most telluric currents are induced in the earth by changes with time in the earth's magnetic field. Those of interest in exploration are of high frequency and might be looked upon as noise so far as conventional magnetic measurements are concerned. The electric currents and the magnetic fluctuations which cause them are, as shown in Fig. 17-19, very similar in form, although there is a phase lag between the two. In 1953 Cagniard[37] showed how the ratio of the alternating magnetic field to the alternating electric field, plotted as a function of frequency, can yield information on the variation of resistivity with depth within the earth. The same conclusion was reached independently by geophysicists in the U.S.S.R. Vozoff[38] has published a review article on this method which summarizes the state of the art as of 1971.

Magnetic oscillations of low frequency (less than 1 Hz) are associated with currents flowing in the ionosphere, an electrified layer surrounding the earth beyond its atmosphere. Higher-frequency signals are usually associated with meteorological phenomena such as thunderstorms and tornadoes. Some of the noise comes from man-made sources such as electrified trains.

Magnetotelluric operations involve the simultaneous measurements of the oscillating electric and magnetic fields at the same location with recording over a frequency range dependent on the depth of the target. The lower the frequency the greater the depth of penetration. From measurements of the relative amplitudes of the respective signals at each component frequency, one can calculate the apparent resistivity to the greatest depth in the earth to which energy at that frequency penetrates. The respective fields for which the ratio is determined are perpendicular to one another and are usually measured in the north-south, east-west, and vertical directions. To obtain a complete description of the resistivity variations in the vertical direction and in mutually perpendicular horizontal directions as well, one measures two horizontal components of both fields and the vertical component of the magnetic field. Five parameters must thus be recorded.

Instrumentation and field techniques The instruments used for magnetotelluric measurements of resistivity are simple in principle, consisting of separate sensors and recorders for the alternating electric and magnetic fields. The higher the frequencies the weaker the fields to be measured, the upper frequency limit generally used being in the neighborhood of 10 Hz. The newly introduced audiomagnetotelluric (AMT) technique uses higher frequencies, sometimes extending into the kilohertz range.

The oscillatory electric fields are recorded by nonpolarizing porous-pot electrodes made up of cadmium immersed in cadmium chloride solution. The magnetic signals

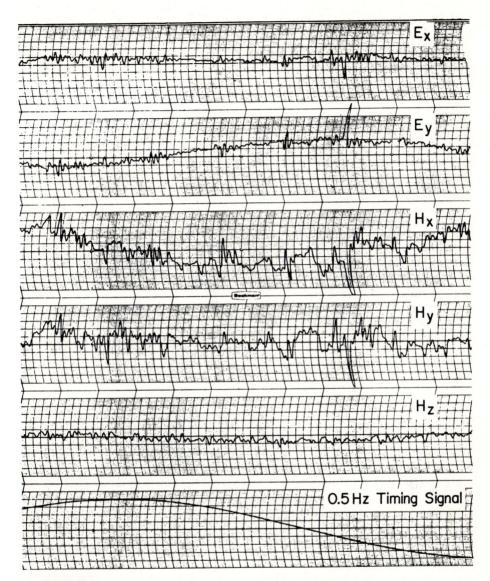

FIGURE 17-19
Typical magnetotelluric signals showing, from top to bottom, electric field
variation in the x and y directions and magnetic field variations in the x, y, and
z directions. The time interval from peak to trough of the sine wave at the
bottom is 1 s. (*After Vozoff.*[38])

are sensed by induction coils containing up to 30,000 turns of copper wire wound
around a core of molybdenum and permalloy 6 ft long and ½ in in diameter.
A special chopper-type preamplifier is used in recording the magnetic oscillations.
The outputs for both fields are recorded digitally with a resolution of about 60 dB.

In a typical field arrangement, two pairs of electrodes 2000 ft apart are laid out along mutually perpendicular lines. Three magnetometers are oriented at right angles to one another and are buried in the ground to minimize any movement of the sensors that might result in spurious signals. The magnetic sensors are located near the crossing point of the lines between the respective pairs of electrodes.

Analysis of data In processing the digitally recorded signals, usually by computer, it is necessary to express amplitudes in the frequency domain. This requires Fourier analysis of the recorded signals. The directions of the fields can also be determined by comparison of the observed components. The various ratios may be printed out in tabular form or presented graphically as functions of frequency if a proper plotter is available.

Interpretation The object of magnetotelluric interpretation is to obtain the distribution of resistivity within the earth in one, two, or three dimensions, the number depending on the complexity of the subsurface structure. Horizontal layering would require information in the downward direction only. A laterally uniform subsurface configuration, like a vertical fault perpendicular to the traverse, would require a two-dimensional presentation, while a more complex interpretation technique would be necessary if the resistivities varied in all three directions.

Conversion of frequency f to depth of penetration δ for a resistivity ρ is based on the *skin-depth* relation

$$\delta = \frac{1}{2} \sqrt{\frac{\rho}{f}} \qquad (17\text{-}11)$$

The apparent resistivity ρ_a is obtained from the electric field E_x in the x direction and the magnetic field H_y in the y direction from the formula

$$\rho_a = \frac{1}{5f} \left(\frac{|E_x|}{|H_y|} \right)^2 \qquad (17\text{-}12)$$

where E_x is expressed in millivolts and H_y in gammas.

The variation of ρ_a with frequency can be expressed in terms of resistivity versus depth by fitting the ρ_a-versus-f curves from the magnetotelluric measurements to theoretical curves calculated for various layering configurations. The procedure is quite similar to that involved in using master curves for interpreting resistivity data obtained with artificially applied currents.

Applications Vozoff[38] describes a magnetotelluric survey in the Anadarko Basin of Oklahoma for which the structure, deduced from drilling and surface data, is shown in Fig. 17-20. A traverse of 13 stations with an average separation of 6 mi was run parallel to the section along a line about 25 mi to its east. The shallow basement at the southern end of the area consists of Wichita Mountain granites. A fault just north of the mountains bounds a thick sedimentary section which has its deepest portion a short distance beyond it, after which it thins to the north.

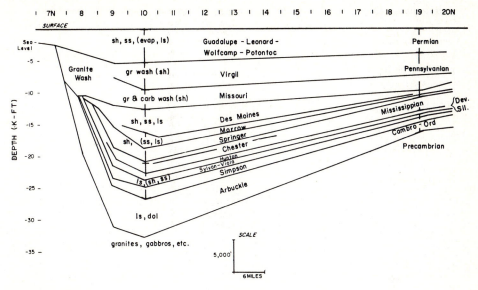

FIGURE 17-20

Geological cross section of Anadarko Basin area where magnetotelluric traverse
was run. The sections is in the N–S direction. (*From Vozoff*.[38])

The magnetotelluric results (Fig. 17-21) confirm the major features of the
geological section in Fig. 17-20 except that the major fault at the southern end of the
line would have to be moved to the south. The decrease in resistivity with depth at
this end suggests overthrusting of the less-conducting Wichita Mountain formations
over the better-conducting sediments below.

The magnetotelluric method is usually most valuable for reconnaissance studies
of large areas which have not been extensively explored by other means. New basins
can be delineated rapidly and economically by this technique. This is why the
method has been widely used in the U.S.S.R. to determine the depth and extent of
previously unexplored sedimentary basins. For large-scale regional studies, the
method is capable of giving information from greater depths than any other geophysical technique.

Where a high-resistivity surface cover, such as the thick volcanic material
blanketing the Columbia Plateau, makes conventional resistivity techniques
unsuitable for deep exploration, magnetotelluric methods will sometimes yield data
on the resistivity of formations beneath the cover.

Prospecting with Natural Alternating Magnetic Fields (AFMAG)

Closely related to the telluric and to the magnetotelluric methods is a technique
based on the measurement of natural oscillating magnetic fields at audio and
subaudio frequencies. Known as AFMAG (for audio-frequency magnetic fields),
this method was introduced in 1958 by Ward.[39]

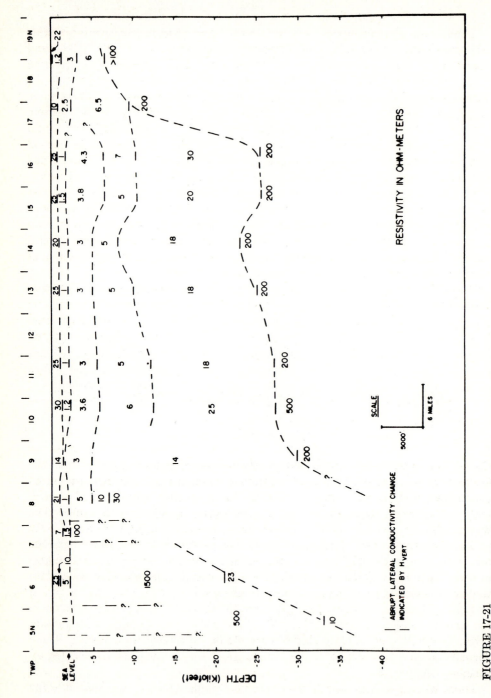

FIGURE 17-21
Resistivities interpreted from magnetotelluric data obtained along line 25 mi west of geologic section shown in Fig. 17-20.

The natural electromagnetic field of the earth has components covering a wide frequency band. Those in the audio-frequency range (a few tens to a few thousand hertz) result from electric disturbances in the atmosphere. At large distances from their sources, the fields propagate in such a way that the electric components are almost vertical and the magnetic components are almost horizontal. Where there is a conducting zone with a vertical boundary (as in the case of a dike penetrating to or almost to the surface), the plane of polarization of the normally horizontal magnetic field will be tilted in such a way that it will have an appreciable vertical component. Such a tilt of the magnetic vector should be diagnostic of lateral discontinuities in resistivity of the type often observed at the edges of an ore body.

In field operations, detecting coils are used to measure the inclination (tilt) and in most cases the azimuth of the alternating magnetic field. Azimuths are determined to the nearest 10° or so, and the tilt is measured along the azimuth with an accuracy of 1 or 2°. Two frequencies are employed, 150 and 510 Hz being typical values. The ratio of the responses at the respective frequencies gives a measure of the conductivity of the anomalous feature being detected.

The AFMAG method has a greater horizontal and vertical range of detection than methods using artificial energy sources, and it is therefore particularly suitable for detecting large deep-seated bodies. Faults and buried vertical dikes have been located in this way. As of the middle 1970s the method has been largely supplanted by the AMT technique.

17-5   ELECTROMAGNETIC PROSPECTING OF THE INDUCTIVE TYPE

Principles

One of the most widely used electrical techniques in mineral exploration is the electromagnetic method. It is based on the induction of electric currents in buried conductors, such as certain ore bodies, by the magnetic components of electromagnetic waves generated at the earth's surface or in aircraft above its surface. The waves originate from alternating currents at frequencies ranging from a few hertz to a few megahertz which are passed through loops of wire on the ground or in the air. With some methods, current is introduced into the earth by grounded lengths of wire. The oscillating magnetic fields thus generated are propagated as waves into the earth, attenuating at a rate that depends on the frequency and the electrical properties of the rock material through which they travel. The higher the frequency, the greater the rate of attenuation. When the waves pass through a conducting body, they induce alternating electric currents in the conductive materials. These currents become the source of new electromagnetic waves which can be detected by suitable pickup coils.

There is such a large selection of specific techniques which are available for ground and airborne electromagnetic surveys that it will not be possible to discuss them in any detail here. For further information on such techniques the reader is referred to Keller and Frischknecht,[1] Parasnis,[2] Grant and West,[40] Bosschart,[41] and Ward.[42]

An early application of inductive methods is the location of buried metal, e.g., military mines, by sappers with search coils which respond to waves induced in metallic materials by high-frequency currents. Current sources are generally circular loops in either the horizontal or the vertical plane or elongated horizontal wires. The most commonly used frequencies are in the neighborhood of 500 Hz.

During the 1930s, Peters and Bardeen[43] worked out the theory for determining the depth of penetration of electromagnetic waves as a function of frequency. They showed that there is an optimum frequency f that will give the greatest strength of returned signal from a conductor at a depth h where the resistivity in the overlying material is ρ. It can be determined from the relation

$$h \sqrt{\frac{f}{\rho}} = 10 \qquad (17\text{-}13)$$

where h is in meters, f in hertz, and ρ in ohm-centimeters.

For a depth of 100 m and a resistivity of 10^2 $\Omega \cdot$m, this equation shows that the best frequency is 100 Hz, while penetration to 300 m is best accomplished at 10 Hz. If the resistivity is abnormally high, say 10^4 $\Omega \cdot$m, a conducting layer 1000 m deep would give the best energy return with a 100-Hz signal, while a layer of 100 $\Omega \cdot$m resistivity would have to be 100 m deep to have its maximum response at this frequency. Presently available techniques of measuring electromagnetic waves at very low frequencies allow penetrations as deep as many kilometers, even through sedimentary sections.

Depth Sounding and Horizontal Profiling

Most inductive methods were originally designed to locate the horizontal positions of buried conductors, but techniques have been developed for estimating their depth as well. The two kinds of methods are analogous in many ways to the horizontal and vertical sounding used in dc (or low-frequency) resistivity prospecting.

Depth sounding In applying the electromagnetic method to depth sounding, one can either vary the spacing between the transmitting loop and receiving coil or vary the frequency while maintaining fixed positions of the transmitter and receiver. Varying the frequency is generally preferred, as any change in the positions of the surface units may introduce undesired effects. If the boundaries to be located occur over a range of depths, it may be necessary to work with a number of spacings on the surface, as the best resolution will be obtained at a separation of coils which depends on the depth of the target. Either loops or long wires may be used as sources. A long grounded wire can usually be considered as an infinite linear source; very large rectangular loops can be treated in the same way if the measuring coil is much nearer to one side of the loop than it is to any of the others.

The theory by which vertical sounding data are interpreted is not difficult to apply where simple models can be fitted to the subsurface structure. The depth to the top of a thick, perfectly conducting layer underlying an infinitely resistant overburden can be ascertained by measuring the ratio of the amplitudes of the horizontal to the vertical magnetic fields observed on the surface. The depth h of a thin horizontal conducting sheet can be determined by comparing this ratio at various values of x, the horizontal source-receiver distance, using a logarithmic curve computed for different ratios of h to x.

SUNDBERG METHOD   The Sundberg method, described by Zuschlag,[44] gives data which are interpreted by use of the equations and curves computed for one or more thin conducting sheets. The current flows through an insulated copper cable, connected to a source of alternating current, which runs along the surface either as a long grounded wire or as a large insulated loop which forms a rectangle. A number of transverse receiving profiles are laid out perpendicular to the grounded wire or to one side of the loop. The magnetic component of the induced electromagnetic field is measured at discrete points along the profile by special search coils consisting of several hundred turns of wire. At least two frequencies are employed, and depth soundings can be made by comparing results at the different frequencies. The quantity actually mapped is the ratio of field intensity at each measuring point to the current in the source loop.

Figure 17-22 illustrates an application of the Sundberg method. Here the loop constituting the source is a rectangle 4000 by 1500 ft in dimensions. The position of the body is indicated, as shown, by the zone over which an observable magnetic response is recorded. If the ore body could be approximated as a sheet, its depth could be determined by measuring the respective components of the field in phase and out of phase with the source signal.

The Sundberg method was used to explore for oil at shallow depths before seismic methods were developed for effective mapping of shallow structures. The Bruner fault in Texas was mapped by this technique. Here a discontinuity was traced in

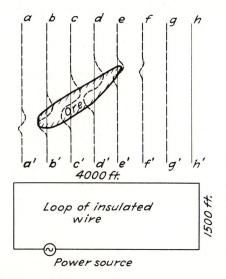

4000 ft.

Loop of insulated wire

1500 ft

Power source

FIGURE 17-22
Detection of conducting ore body under shallow overburden by electromagnetic induction. Current flows through ground loop, and search coils are moved along traverse lines such as aa', bb', etc. Dashed lines show relative response of vertical search coil as function of position along traverses.

a thin conducting formation about 500 ft deep; this was later shown to coincide with the boundary of the oil-producing zone associated with the fault. The method is now employed only for mining exploration.

TURAM METHOD   The Turam method is similar to the Sundberg method but differs from it in that the ratio of field intensities at two points is observed rather than the intensity at a single point. As with the Sundberg method, two frequencies within the 100- to 800-Hz range are employed. A cable several kilometers long is laid out as a source parallel to the expected strike and is grounded at both ends. Measuring coils are read in pairs along lines perpendicular to the cable, the spacing between them generally being about 25 m. In-phase and out-of-phase ratios are observed as well as amplitude ratios and phase-angle differences.

Horizontal profiling   Most electromagnetic surveys are carried out with the object of locating the horizontal positions of buried conducting bodies rather than of determining their depths. Both fixed and moving sources are used in surveys conducted for this purpose. The fixed-source methods are generally designed in such a way that the transmitter is kept at a single location and the receiving loop is moved over the area being explored. The moving-source method involves shifting both source and receiver, usually in a way that maintains a fixed separation between the two. The principal complication that arises in the moving-source method is variable coupling between the source and the conductor. This could cause variations in the response which, not being associated with the subsurface conductor, might be hard to interpret.

The moving-source method requires smaller and more portable generators and source loops than the fixed-source technique, which often employs large insulated loops or long grounded wires.

Some methods are designed to measure only the direction of the induced magnetic field; they can be highly sensitive and do not require complicated instruments. They avoid the necessity for measuring a reference signal associated with the source itself, a difficult requirement in determining relative amplitudes of induced fields. The ratio of the fields observed at two locations or the ratio of two components observed at the same place will give information on subsurface resistivity although the results are more difficult to interpret. The direction of the field is most generally measured by the dip-angle method (also called the vertical-loop method). The source loop is in the vertical plane, and the receiving loop is linked to an inclinometer which is used to measure the angle with the horizontal of the plane in which it lies when the voltage induced in it is a maximum. The azimuth for minimum response of the receiving loop when it is vertical is first measured in the manner illustrated in Fig. 17-23. This technique can be used either with a fixed-source or a moving-source configuration. Dip-angle measurements when properly interpreted can yield information on horizontal location, strike, dip, depth, length, and conductivity of a buried conducting sheet.

The *Slingram* method uses portable loops similar to those employed for measuring dip angles with a moving source but the ratio of in-phase and out-of-phase compo-

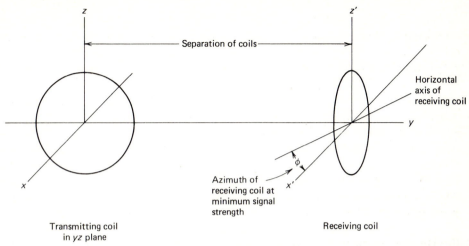

Stage 1: Determining azimuth for dip-angle measurement.

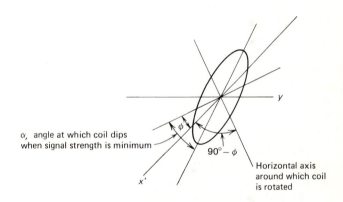

Stage 2: Measuring dip-angle for minimum response.

FIGURE 17-23
Configuration of transmitting and receiving coils for the vertical-loop or dip-angle method of electromagnetic surveying. Measurement is made in two stages as shown.

nents of the induced voltage is observed with a ratiometer. Both loops are generally horizontal. One or two fixed frequencies are employed, and the loop spacing is kept constant. Traverses are usually laid out normal to regional trends. The information obtainable from Slingram measurements is quite similar to that indicated in the previous paragraph for the dip-angle method when the conductor is a single dipping sheet or a series of parallel, closely spaced sheets equivalent in electrical properties to a wide dike.

The greatest source of difficulty in the interpretation of horizontal profiling data is the effect of conductive material in the rock laterally adjacent to prospective ore bodies and in the overburden above them. Variations in the conductivity of the overburden can give rise to anomalies which resemble those from ore bodies and which might conceal effects from deep conductors of economic interest.

Airborne Electromagnetic Prospecting

Since the early 1950s an increasing proportion of electromagnetic surveys has been carried out with airborne instruments. The primary objective of most airborne inductive surveys is to locate sulfide ores. The advantage of airborne surveys in speed of coverage and consequent economy is obvious, but technical problems not encountered in ground surveys make it more difficult to obtain useful electromagnetic data from the air than from the earth's surface. Among these problems are (1) the limited separation between source and receiver compared with that possible on the ground, (2) the exceptional sensitivity required in the measuring system to pick up the weak signals returned to the surface when the source of the primary electromagnetic energy must, for reasons of safety, be from 200 to 500 ft above the ground surface; and (3) the higher noise level introduced by motion of the receiving loop in the earth's magnetic field.

A review article by Pemberton[45] describes methods for obtaining the necessary sensitivity in airborne surveys. The simplest arrangement is to attach both the transmitting and receiving loops to opposite wing tips. Another is to receive with a loop towed in a bird connected with the transmitting aircraft by a long cable. The receiving loop may also be in a second plane flying close behind the one containing the transmitter. Still another system is designed to send out transient pulses intermittently, reception being measured during the intervals when the primary field is cut off.

A variety of systems is employed for airborne electromagnetic operations. The Törnqvist[46] method makes use of two transmitting loops attached to the aircraft, one in the horizontal plane and the other in the vertical. Two receiving loops, towed in a bird from the same or a second plane, are oriented in the same respective directions. Figure 17-24 illustrates this arrangement as adopted by International Nickel Company. The two transmitters send out alternating magnetic fields of the same frequency and amplitude but 90° different in phase. A rotating magnetic field is generated in this way. The signal from one receiving element is shifted in phase 90° with respect to that from the other, and any difference between them is registered by a pen recorder. If there were no conductor below the plane, the two signals would be identical, so that the difference after shifting would be zero and the recorder trace would show only noise. A conductor such as an ore body would cause the signals to be different from each other and the difference would be observed as a signal on the record. The difference amplitude and the phase shift are recorded by separate pens, and any conducting ore body below the aircraft should show up on both traces.

Using a single plane, it is generally necessary to tow the receiver 500 ft or so behind the aircraft. To keep the bird containing the receiver from hitting the ground, the plane must fly at a height of at least 400 ft. The two-plane arrangement makes it possible to trail the bird by a much shorter cable, say 50 to 60 ft, so that

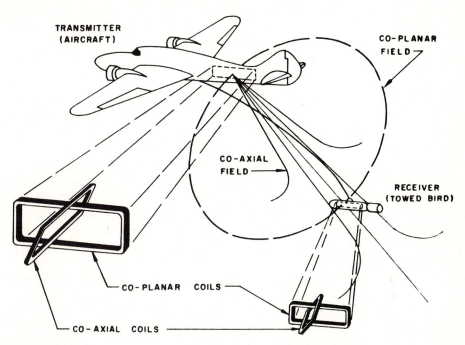

FIGURE 17-24
Arrangement of coils in airborne electromagnetic recording as carried out by
International Nickel Co. (*After Dowsett.*[47])

a flying height as low as 300 ft is possible. At this height, with an 800-ft separation
between planes, conducting bodies as deep as 1500 ft below the ground surface can
be detected on the signal traces.

A number of other airborne electromagnetic techniques are in common use. The
Slingram system, for example, has been adapted for aircraft, providing a highly
sensitive measurement technique. Fixed loops are attached either to opposite
wingtips (for vertical orientation in the same plane) or to the nose and a boom
projecting from the tail (for a vertical coaxial arrangement). The method makes
use of a single frequency in the range from 300 to 4000 Hz. The input system
records on six channels simultaneously, each representing a different range of
frequencies. Because of noise the data are generally not as good as with ground
measurements, and interpretation is likely to be more qualitative. This limitation is
generally acceptable, as the principal objective of such a survey is to separate
anomalies which have possible economic potential from those which do not, thus
localizing interesting areas for further investigation.

Typical Results of Electromagnetic Surveys

After the data obtained in electromagnetic prospecting are reduced, they are
generally presented in the form of cross sections showing the amplitudes of the

signals recorded by the receiving element versus the horizontal position. The latter is expressed in terms of distance in ground surveys and in terms of time in airborne surveys. Generally the components of the signal that are in phase and out of phase with the primary signal are plotted separately. Comparison of the two signatures facilitates interpretation.

Figure 17-25 illustrates the data obtained over the Brunswick sulfide body, consisting of massive pyrite and sphalerite. A highly magnetic iron formation is adjacent to the sulfide ore body. Respective in-phase and out-of-phase profiles are shown both for 400- and 1000-Hz signals. Curves are presented both for coplanar and coaxial coils. These responses were obtained by mixing received and primary signals in such proportion as to null the outputs. A number of important lead-copper-zinc bodies such as this were discovered in northern New Brunswick by surface electromagnetic surveys in 1952 and 1953. Both sulfide and graphite bodies give rise to electromagnetic anomalies in this mineral province.

The coplanar coil configuration shows a positive response over the iron and a negative one over the sulfide body. The coaxial coils show in-phase effects with signs the reverse of these, but the out-of-phase field gives a positive response over the sulfides and none over the iron. This combination of effects makes it possible to identify the sulfide deposits from the signatures made with the different recording configurations.

Figure 17-26a shows a series of electromagnetic profiles crossing a sulfide body in the Muskeg prospect of northwestern Quebec. The survey was carried out from the air using horizontal coplanar coils. Ore bodies in this area are covered by clay and glacial material averaging 50 ft in thickness. Although the outline of the conductor is easily traceable across the profiles, it is not possible to tell from the electromagnetic data alone that the conducting material changes from sulfides to graphite as one goes north from line 4W to line 8W. This information can be obtained from magnetic readings. Figure 17-26b relates the electromagnetic data to the geological results obtained from diagonal drill holes at the center of profile O and on profile 8W. The sulfides here are in the form of andesites, which are embedded in tuffs and rhyolite.

A third example comes from the Pipe nickel mine in north central Manitoba, which has been discussed by Dowsett.[47] Here the ore body is long, thin, and inclined at a large angle with the horizontal. It consists primarily of pyrrhotite, which is both conductive and magnetic. Alongside the pyrrhotite is an iron formation which is only weakly magnetic compared with the sulfide body. Figure 17-27 illustrates the relationship between the electromagnetic profiles recorded on the ground using a vertical loop and the magnetic profile obtained over the same traverse. Both are plotted over a geologic section obtained from five drill holes in close proximity. The magnetic anomaly appears better defined than the electromagnetic response.

16-6 INDUCED POLARIZATION

The induced-polarization method, first used in the late 1940s has been put to extensive use in the search for disseminated sulfide ores. During the 1960s this

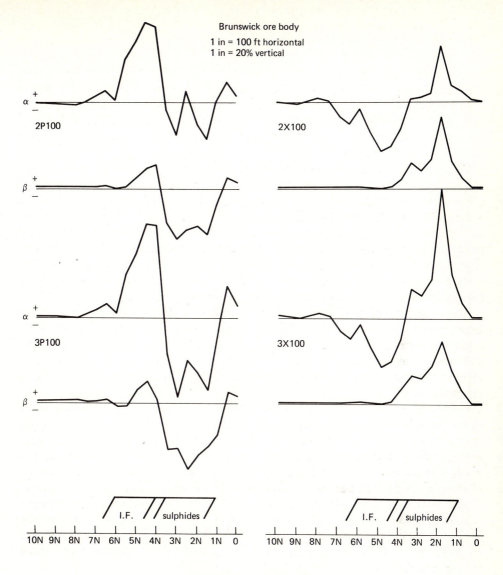

Brunswick ore body
1 in = 100 ft horizontal
1 in = 20% vertical

2P100

2X100

3P100

3X100

I.F. sulphides

I.F. sulphides

10N 9N 8N 7N 6N 5N 4N 3N 2N 1N 0

10N 9N 8N 7N 6N 5N 4N 3N 2N 1N 0

α   in-phase field in % of primary field
β   out-of-phase field in % of primary field

1 150 Hz   50-50 ft spacing
2 400 Hz   100-100 ft spacing
3 1000 Hz   200-200 ft spacing
4 3600 Hz
X coaxial coils
P coplanar coils

FIGURE 17-25
Electromagnetic profiles over New Brunswick sulfide body and adjacent iron formation (I.F.). Top set and third set of curves from top show in-phase responses to primary field. The other two show out-of-phase responses. The 2 × 100 label indicates, from legend below, that the frequency was 400 Hz, that the coils were coaxial, and that the spacing was 100 ft. Other symbols can be related to legend in a similar way. (*After Brant et al.*[48])

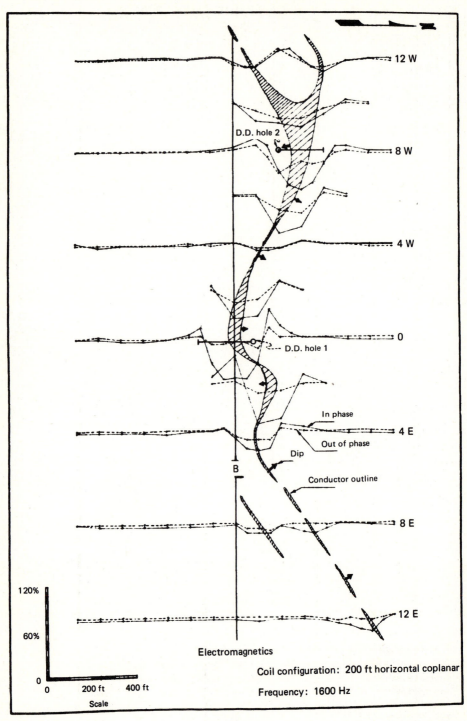

(a)

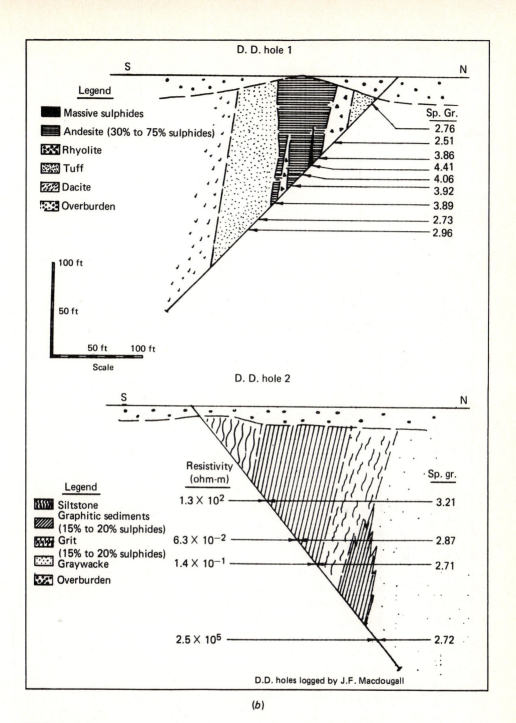

FIGURE 17-26

Results of electromagnetic survey over sulfide body in Muskeg area, north-western Quebec: (*a*) electromagnetic profiles along lines crossing ore body. Crosshatched areas show interpreted configurations of conducting bodies; (*b*) geology indicated by two slanting drill holes at locations shown in (*a*). (*After White.*[49])

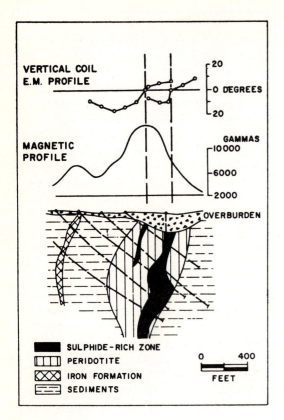

FIGURE 17-27
Electromagnetic and magnetic profiles
obtained over the Pipe nickel mine,
Manitoba. (*From Dowsett.*[47])

method became the most widely employed of all ground-based techniques in mining
geophysics. For more detailed information on induced polarization, review papers
by Madden and Cantwell[50] and by Seigel[51] as well as the discussions in Parasnis[2]
and Keller and Frischknecht[1] are suggested.

Principles When a current is passed through earth materials not containing
metallic minerals, the amount of current is related to the driving potential only by
the ohmic resistance of the formations involved. When the formations contain
metallic minerals, the currents give rise to an exchange of ions at the surface of
contact between the minerals and the electrolytes dissolved in the fluid filling the
intergranular pore spaces. This electrochemical exchange creates a voltage which
opposes the current flow through the material, and an added voltage is necessary to
overcome the barrier thus created. The extra voltage necessary to drive current
through the barrier is sometimes referred to as the *overvoltage*. When the externally

applied current is turned off, the electrochemical voltages at the metallic grain surfaces are dissipated, but not instantaneously. The decaying voltages can be measured for a time after the current is switched off. The voltage is observed to vary with time, as shown in Fig. 17-28. The ratio between the amplitude of the overvoltage just after the current stops to that just before gives a measure of the concentration of metallic minerals in the material through which the current flows.

If an alternating current is put into the earth, the overvoltage observed at the metallic surface will decrease with increasing frequency because the buildup of the opposing voltage to its full value requires a longer time than the period between changes in direction of the applied current. As the frequency of the alternating current increases, the peak overvoltage it induces attains a maximum value that represents a decreasing proportion of the ac amplitude. The ratio of the alternating induced polarization potential at two different frequencies, as well as that of the ac to that of dc, is related to the concentration of metallic minerals along the path of the current.

Field procedures   The techniques used in the field to measure induced polarization are similar in many ways to those employed for resistivity measurements. Current is introduced into the earth with two electrodes, and a potential is measured across the other two electrodes after the current is shut off. Generally the geometry of the electrode configuration is kept uniform while the position is changed laterally along a profile. This procedure is better adapted for reconnaissance of new areas than the vertical-sounding field arrangement, which, as with resistivity and electromagnetic measurements, involves changing the electrode spacing while maintaining a fixed center position.

The current introduced into the earth may be in the form of pulses, generally shaped as square waves, or of low-frequency (1 Hz or less) alternating currents. With the second arrangement data are usually compared at a variety of frequencies.

Two pulsing techniques are employed; one involves a single abrupt interruption of a direct current flowing through the earth, with subsequent measurement of its decay characteristics. The current flows for 1 to 5 min before it is cut off, the duration of the pulse being measured precisely. The transient voltage is read at closely spaced intervals after termination of the current when it is not recorded continuously. Either way it is customary to measure the area beneath the voltage-time curve (up to the time that the voltage is no longer detectable) to determine the

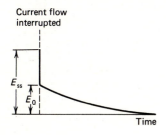

Current flow interrupted

E_{ss}

E_0

Time

FIGURE 17-28
Decay of electric field strength in rock sample when external voltage is shut off. (*Keller and Frischknecht.*[1])

overvoltage created while the current was flowing. Sometimes special integrating elements (such as *RC* circuits or fluxmeters) are employed to give a direct reading of the area.

A somewhat different approach makes use of a series of identical pulses repeated at short, uniform intervals. The form of the signal may not be quite the same as that for the single-pulse arrangement unless the interval between the repeated pulses is large. This is because the decay signal from the cutoff of one pulse may be super-imposed on the corresponding signal from the subsequent pulse or pulses, the amount of superposition depending on the time constant. In any case the voltage observed during the period between pulses is averaged, and this average is used in the interpretation.

The *variable-frequency method* is designed to measure changes in apparent resistivity when the frequency is changed. Polarization opposing externally im-pressed potential drops would have the same effect as a resistance in series with the actual resistance of the formations except that the effect would vary with frequency. In rocks without minerals that give rise to induced polarization, there will be a very small decrease in resistivity with increasing frequency, usually less than 1 percent. But where there is induced polarization, the decrease of resistivity will be much larger, sometimes becoming as great as 10 to 20 percent for a tenfold frequency increase. The percentage P of decrease may be expressed in the form

$$P = \frac{\rho_2 - \rho_1}{\sqrt{\rho_2\rho_1}} \times 100\%$$

where ρ_2 is the resistivity measured at a particular frequency and ρ_1 the resistivity at another frequency 10 times as high. According to Marshall and Madden,[52] rocks with concentrated sulfides would give rise to a P greater than 10 and porphyry copper ores (2 to 10 percent sulfides) to a P between 5 or 10. Rocks with a trace of sulfide mineralization would have a P value from 2 to 5, sandstones and siltstones from 1 to 3, basalt from 1 to 2, and granite from 0.1 to 0.5.

Applicability of the method   In addition to its primary use in exploration for disseminated sulfides, induced polarization has been proposed as a tool for ground-water exploration. Vacquier et al.[53] have carried out polarization experiments with sand-clay mixtures saturated with different electrolytes. Clay particles can cause polarization by blocking the flow of electrolyte-bearing fluid through pore spaces. Ionic exchanges in the clay molecules can cause the clay-sand mixture to act as a distributed electronegative membrane. Proper interpretation of induced-polariza-tion data can show the depth at which these effects take place, thus establishing the position of the water table. Frische and von Buttlar[54] give formulas for computing the depth of the water table from such data.

Rogers[55] has endeavored to evaluate the induced-polarization method as a means of locating disseminated sulfide ores. He concludes that within certain limits of size, depth, and sulfide concentration, the method can establish the presence or absence of sulfides showing a metallic luster in 80 percent of the cases where it is used alone

and in a higher percentage of cases when other criteria are available. The lateral
extent of the sulfide ore body can be determined to an order of magnitude while the
percent of sulfides by volume, said to be the least reliable and generally least
important estimate of all, can "sometimes be squeezed out of the data." Rogers
acknowledges that the induced-polarization method finds many sulfide deposits
that are uneconomic. All that can be done about this, he says, is to test all induced-
polarization indications that appear geologically favorable and play the statistics
in the hope that if enough prospects are tested, a good one will eventually be found.
"This is essentially done with all exploration techniques," he concludes.

Examples Figure 17-29 illustrates the type of data obtained by induced polarization over a sulfide body in Arizona. The method of presenting the induced-polarization results below the horizontal line of stations at the center is quite common for
surveys in which the separation between the pair of current electrodes and the pair
of potential electrodes is systematically increased. The polarization ("the metallic
conduction factor") based on the difference in resistivity at two frequencies is
plotted at a depth of half the separation of the pairs below a point midway between
them. As the separation increases, the "depth" at which the value is plotted
increases. It is not intended for such a plotting of depth to be looked upon as
anything but a rough approximation.

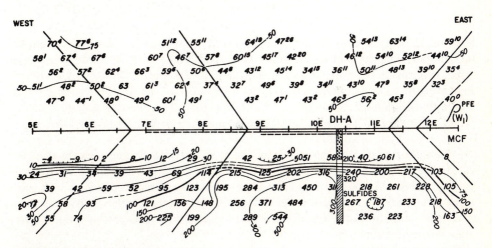

FIGURE 17-29
Presentation of induced-polarization data as illustrated by plot along traverse
over sulfide body in Arizona. Above the horizontal line, conventionally determined apparent resistivities are plotted. Below the line a polarization factor
based on the difference between apparent resistivities at two frequencies is
shown. The exponents for the resistivity values show polarization effects expressed in a different way from those indicated below the line. It is evident
that the induced-polarization data are more diagnostic of the presence of
sulfides than are the resistivity data. (*After Rogers.*[55])

Above the line of stations at the center are contours of the apparent resistivity, determined by conventional techniques, plotted in the same way (with the polarization presented in terms of percentage difference of resistivity at the two frequencies shown as an exponent of the resistivity value). Increasing penetration here is indicated as an increasing distance *above* the horizontal midline. The resistivity contours appear much less diagnostic of anomalous bodies below the surface.

Drilling has confirmed the indication in the polarization data that there is a zone of sulfide mineralization between stations 12E and 9E. The log of the drill hole shown near $10\frac{1}{2}$E shows 210 ft of gravel below the surface, a zone of oxidation in the bedrock from 210 to 320 ft, and a zone with 3 to 5 percent of sulfides by volume from 320 to 600 ft.

Results in a similar form are presented in Fig. 17-30 from the Krain copper deposit in British Columbia. The induced-polarization data gave better definition of the copper ore than magnetometer or self-potential data did, according to a report on the investigation by Hansen and Barr.[56] The definition of the metallic conduction factor, which is contoured on the same basis as in Fig. 17-29, is indicated in the figure. In this case also the resistivity contours above the baseline do not appear to be as diagnostic of the copper concentration as do those of induced polarization.

17-7 USE OF ELECTRICAL PROSPECTING FOR OIL EXPLORATION

Many of the electrical methods discussed in this chapter have been experimented with at one time or another in exploring for oil. Some of them have been used rather extensively for this purpose in the U.S.S.R. But in the United States and in most other areas of the western world, the application of electrical techniques in oil exploration is minimal.

Dix has said that if electrical-resistivity measurements had been tried in the Gulf Coast during the middle 1920s instead of the torsion balance or the refraction fan-shooting surveys that turned out to be so effective in finding piercement-type salt domes, the entire history of American exploration geophysics might have been different from what it actually was. The resistivity contrast between salt and sediments should have made it quite easy to detect the shallow domes by this technique, and the development of seismic and gravity prospecting might have been delayed for some years.

The applicability of electrical prospecting to oil exploration is limited because there are few diagnostic contrasts in the electrical properties of sedimentary rocks to provide structural or lithologic information that is useful in the search for hydrocarbons. Anhydrites and massive limestones have anomalously high resistivities, but the dimensions (particularly the thicknesses) of bodies composed of these materials are generally so small compared with the depths of interest in petroleum prospecting that intrinsic limits of resolution make the prospect of mapping them by electrical means look unpromising.

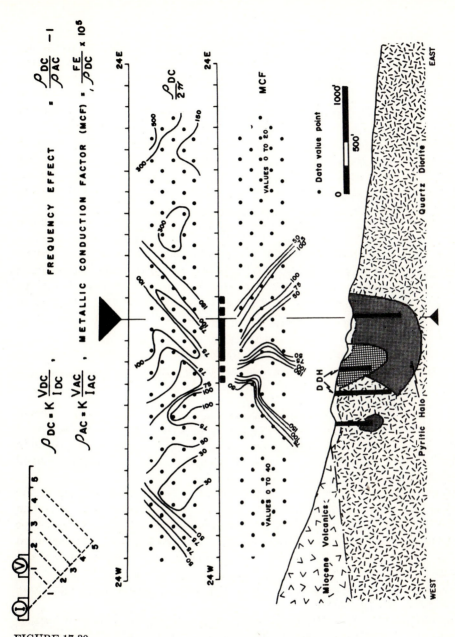

FIGURE 17-30

Induced-polarization profile over Krain sulfide deposit, British Columbia. Contours have the same significance as in Fig. 17-29. The metallic conduction factor (MCF), contoured below the line, is defined at the top of the figure. Ore body is indicated by coarse cross-hatching. (*After Hansen and Barr.*[56])

Determination of basement depth In general the sedimentary section has a range of resistivities that is substantially lower than the range for basement rocks so that sedimentary basins can be modeled as a layer of low resistivity (the sedimentary column) overlying a semi-infinite substratum having a much higher resistivity (the basement). Thus any electrical prospecting technique that can respond to resistivity changes at the base of the sedimentary section should yield information on the depth to the basement. In many little-explored areas of the world, the first stage in the exploration process is to determine whether or not a sedimentary basin exists and if it does to determine its approximate geometry.

In the U.S.S.R., where it is believed that not all sedimentary basins have yet been discovered and defined, two electrical techniques are employed on an extensive scale to determine the existence, approximate depth, and boundaries of new basin areas. One is the dipole method, the other the magnetotelluric.

The dipole method involves the introduction through electrodes of a large current into the earth in the form of a square wave. With this technique, a switch between the generator and current electrodes is closed for about 1 min and then opened. At distances as great as 20 km away, a detector picks up a voltage associated with the current pulse. The signal thus received will no longer look like a square wave but will be distorted in shape and spread out in time. The nature of the distortion will depend upon the distribution of resistivity in the subsurface down to a depth which increases with the separation of source and receiver (assuming the strength of the current is enough to allow the necessary penetration). The waveforms can be computed for assumed variations of resistivity with depth, and a catalog of waveshapes can be built up for a wide range of layering configurations. The resistivity structure in the subsurface can be determined by matching the observed waveform with the one from the catalog that gives the closest fit. If the model corresponds to a single layer over a substratum of higher resistivity, the thickness of the layer is assumed (in the absence of geological indications to the contrary) to be the depth of the sedimentary section along the profile between the source and receiver.

Magnetotelluric techniques are particularly suitable for this type of exploration because no external energy source is needed and frequencies low enough to provide the desired penetration are generally present in natural alternating fields.

Other applications The magnetotelluric method has also been used to map structures in the sedimentary section that cannot be readily observed by conventional petroleum-exploration techniques. In the Columbia River Plateau of the Pacific Northwest, lavas at or near the surface make it virtually impossible to obtain information on the configuration of underlying sediments by seismic, gravity, or magnetic methods. But magnetotelluric surveys carried out in the area have reportedly provided such information. The precise detection of high-frequency components of the magnetotelluric fields that are necessary for obtaining such shallow penetration requires equipment that is almost as expensive as the digital recording systems used in seismic work.

Keller[57] suggests that resistivity surveys carried out on a regional scale should make it possible to map paleoenvironments and give clues to the geological history of a petroliferous area that could otherwise be obtained only by stratigraphic drilling. Because of the limited resolution of resistivity measurements at the depths involved, only gross changes in resistivity affecting a thousand feet or more of the sedimentary section are likely to be significant. Information of this kind can be most useful in the search for stratigraphic oil. Such environmental effects, for example, as a transition over a large thickness of section from a predominantly sand facies to a predominantly shale facies could be detected in this way. Keller has shown from electric log data that there is a major depositional hinge line in the Denver-Julesberg Basin of eastern Colorado of a type that could have been mapped by surface electrical techniques if the wells had not been available.

Actually, the most widespread use of electrical exploration techniques in oil exploration is in electric logging. Well logging is an area that transcends the boundaries ordinarily established between geology, geophysics, and petroleum engineering. Although it falls within the conceptional framework of geophysics, most development of well-logging techniques as well as interpretation of the logs is carried out in the United States by people who look upon themselves as petroleum engineers or geologists rather than as geophysicists. In the U.S.S.R., on the other hand, well logging is classified as a branch of applied geophysics. In fact, the Russian term for well logging is translatable into English as *industrial geophysics*. Limitations of space prevent us from covering electric logging in this book. Many useful textbooks on this subject, e.g., Wyllie[58] or Pirson,[59] are available to those who wish to apply the principles that have been discussed in this chapter to the interpretation of electrical measurements in boreholes.

REFERENCES

1. Keller, George V., and Frank C. Frischknecht: "Electrical Methods in Geophysical Prospecting," Pergamon, London, 1966.
2. Parasnis, D. S.: "Principles of Applied Geophysics," Chapman & Hall, London, 1972.
3. Sato, M., and H. M. Mooney: The Electrochemical Mechanism of Sulfide Self-Potentials, *Geophysics*, vol. 25, pp. 226–249, 1960.
4. Kruger, F. C., and W. C. Lacy: Geological Explanation of Geophysical Anomalies near Cerro de Pasco, Peru, *Econ. Geol.*, vol. 44, pp. 485–491, 1949.
5. Petrovski, A.: The Problem of a Hidden Polarized Sphere, *Phil. Mag.*, vol. 5, pp. 334–353, 914–933, 1928.
6. Stern, Walter: Relation between Spontaneous Polarization Curves and Depth, Size and Dip of Ore Bodies, *Trans. Am. Inst. Min. Met. Eng.*, vol. 164, Geophysics, 1945, pp. 180–196, 1945.
7. de Witte, L.: A New Method of Interpretation of Self-Potential Field Data, *Geophysics*, vol. 13, pp. 600–608, 1948.
8. Yüngül, Sulhi: Interpretation of Spontaneous Polarization Anomalies Caused by Spherical Ore Bodies, *Geophysics*, vol. 15, pp. 237–246, 1950.

9. Jakosky, J. J., R. M. Dreyer, and C. H. Wilson: Geophysical Investigations in the Tri-State Zinc and Lead Mining District, *Univ. Kans. Eng. Exp. Stn., Bull.* 24, 1942.

10. Van Nostrand, Robert G., and Kenneth L. Cook: Interpretation of Resistivity Data, *U.S. Geol. Surv. Prof. Pap.* 499, 1966.

11. Ginzburg, Avihu: Resistivity Surveying, *Geophys. Surv.*, vol. 1, pp. 325–355, 1974.

12. Gish, O. H., and W. J. Rooney: Measurement of Resistivity of Large Masses of Undisturbed Earth, *Terr. Mag. Atm. Electr.*, vol. 30, pp. 161–188, 1925.

13. Hummel, J. N.: A Theoretical Study of Apparent Resistivity in Surface Potential Methods, *Trans. Am. Inst. Min. Met. Eng.*, vol. 97, Geophysical Prospecting, 1932, pp. 302–422, 1932.

14. Stefanescu, S., C. Schlumberger, and M. Schlumberger: Sur la distribution électrique potentielle autour d'une prise de terre ponctuelle dans un terrain à couches horizontales homogènes et isotopes (The distribution of electrical potential about a point electrode in an earth of horizontal, homogeneous, and isotropic beds), *J. Phys. Radium*, vol. 1, pp. 130–141, 1930.

15. Tagg, G. F.: Interpretation of Resistivity Measurements, *Trans. Am. Inst. Min. Met. Eng.*, vol. 110, Geophysical Prospecting, 1934, pp. 135–147, 1934.

16. Compagnie Générale de Géophysique: Abaques de sondage électrique, *Geophys. Prospect.*, vol. 3, suppl. 3, pp. 1–7 plus charts, 1955.

17. Mooney, H. M., and W. W. Wetzel, "The Potentials about a Point Electrode and Apparent Resistivity Curves for a Two-, Three-, and Four-Layer Earth," University of Minnesota Press, Minneapolis, 1956.

18. Orellana, E., and H. M. Mooney: "Master Tables for Vertical Electrical Soundings over Layered Structures," Interciencia, Madrid, 1966.

19. Compagnie Générale de Géophysique: Master Curves for Electrical Sounding, European Association of Exploration Geophysicists, The Hague, 1963.

20. Watson, R. J., and J. F. Johnson: On the Extension of Two-Layer Methods of Interpretation of Earth Resistivity Data to Three or More Layers, *Geophysics*, vol. 3, pp. 7–21, 1942.

21. Argelo, S. M.: Two Computer Programs for the Calculation of Standard Graphs for Resistivity Prospecting, *Geophys. Prospect.*, vol. 15, pp. 71–91, 1967.

22. Unz, M.: Apparent Resistivity Curves for Dipping Beds, *Geophysics*, vol. 18, pp. 116–137, 1953.

23. Maeda, Katsuro: Apparent Resistivity for Dipping Beds, *Geophysics*, vol. 20, pp. 123–139, 1955.

24. Logn, O.: Mapping Nearly Vertical Discontinuities by Earth Resistivities, *Geophysics*, vol. 19, pp. 739–760, 1954.

25. Slichter, L. B.: The Interpretation of the Resistivity Prospecting Method for Horizontal Structures, *Physics*, vol. 4, pp. 307–322, 407, 1933.

26. Pekeris, C. L.: Direct Method of Interpretation in Resistivity Prospecting, *Geophysics*, vol. 5, pp. 31–42, 1940.

27. Koefoed, O.: "The Application of the Kernel Function in Interpreting Geoelectrical Resistivity Measurements," Borntraeger, Berlin, 1968.

28. Van Nostrand, Robert G.: Limitations on Resistivity Methods as Inferred from the Buried Sphere Problem, *Geophysics*, vol. 18, pp. 423–433, 1953.

29. Cook, Kenneth L., and Robert G. Van Nostrand: Interpretation of Resistivity Data over Filled Sinks, *Geophysics*, vol. 19, pp. 761–790, 1954.

30. Seigel, H. O.: Ore Body Size Determination in Electrical Prospecting, *Geophysics*, vol. 17, pp. 907–914, 1952.
31. Clark, A. R.: The Determination of the Long Dimension of Conducting Ore Bodies, *Geophysics*, vol. 21, pp. 470–478, 1956.
32. Schlumberger, Marcel: The Application of Telluric Currents, *Trans. Am. Geophys. Union*, vol. 20, p. 271, 1939.
33. Dahlberg, R. S., Jr.: An Investigation of Natural Earth Currents, *Geophysics*, vol. 10, pp. 494–506, 1945.
34. Boissonnas, Eric, and E. G. Leonardon: Geophysical Exploration by Telluric Currents with Special Reference to a Survey of the Haynesville Salt Dome, Wood County, Texas, *Geophysics*, vol. 13, pp. 387–403, 1948.
35. Vozoff, K., R. M. Ellis, and M. D. Burke: Telluric Currents and Their Use in Petroleum Exploration, *Bull. Am. Assoc. Petrol. Geol.*, vol. 48, pp. 1890–1901, 1964.
36. Mainguy, M., and A. Grepin: Some Practical Examples of Interpretation of Telluric Methods in Languedoc (Southeastern France), *Geophys. Prospect.*, vol. 1, pp. 233–240, 1953.
37. Cagniard, L.: Basic Theory of the Magneto-Telluric Method of Geophysical Prospecting, *Geophysics*, vol. 18, pp. 605–635, 1953.
38. Vozoff, Keeva: The Magnetotelluric Method in the Exploration of Sedimentary Basins, *Geophysics*, vol. 37, pp. 98–141, 1972.
39. Ward, S. H.: AFMAG: Airborne and Ground, *Geophysics*, vol. 24, pp. 761–789, 1959.
40. Grant, F. S., and G. F. West: "Interpretation Theory in Applied Geophysics," McGraw-Hill, New York, 1965.
41. Bosschart, Robert A.: Ground Electromagnetic Methods, pp. 67–80, in "Mining and Groundwater Geophysics/1967," *Geol. Surv. Can. Econ. Geol. Rep.* 26, Ottawa, 1970.
42. Ward, Stanley H.: Airborne Electromagnetic Methods, pp. 81–108, in "Mining and Groundwater Geophysics/1967," *Geol. Surv. Can. Econ. Geol. Rep.* 26, Ottawa, 1970.
43. Peters, L. J., and John Bardeen: Some Aspects of Electrical Prospecting Applied in Locating Oil Structures, *Physics*, vol. 2, pp. 103–122, 1932; reprinted in "Early Geophysical papers," pp. 145–164, Society of Exploration Geophysicists, Tulsa, Okla., 1947.
44. Zuschlag, Theodor: Mapping Oil Structures by the Sundberg Method, *Trans. Am. Inst. Min. Met. Eng.*, vol. 97, Geophysical Prospecting, 1932, pp. 144–159, 1932.
45. Pemberton, Roger H.: Airborne Electromagnetics in Review, *Geophysics*, vol. 27, pp. 691–713, 1962.
46. Törnqvist, G.: Some Practical Results of Airborne Electromagnetic Prospecting in Sweden, *Geophys. Prospect.*, vol. 6, pp. 112–126, 1958.
47. Dowsett, John S.: Geophysical Exploration Methods for Nickel, pp. 310–321, in "Mining and Groundwater Geophysics/1967," *Geol. Surv. Can., Econ. Geol. Rep.* 26, Ottawa, 1970.
48. Brant, A. A., W. M. Dolan, and C. L. Elliot: Coplanar and Coaxial EM Tests in Bathurst Area, New Brunswick, Canada, 1956, "Mining Geophysics," vol. 1, pp. 130–141, Society of Exploration Geophysicists, Tulsa, Okla., 1966.
49. White, P. S.: Airborne Electromagnetic Survey and Ground Follow-up in Northwestern Quebec, "Mining Geophysics," vol. 1, pp. 252–264, Society of Exploration Geophysicists, Tulsa, Okla., 1966.
50. Madden, T. R., and T. Cantwell: Induced Polarization: A Review, "Mining Geo-

physics," vol. 2, pp. 373–400, Society of Exploration Geophysicists, Tulsa, Okla., 1967.

51. Seigel, H. O.: Mathematical Formulation and Type Curves for Induced-Polarization, *Geophysics*, vol. 24, pp. 547–565, 1959.

52. Marshall, D. J., and T. R. Madden: Induced-Polarization: A Study of Its Causes, *Geophysics*, vol. 24, pp. 790–816, 1959.

53. Vacquier, V., C. R. Holmes, P. P. Kintzinger, and Michel Lavergne: Prospecting for Ground-Water by Induced Electrical Polarization, *Geophysics*, vol. 22, pp. 660–687, 1957.

54. Frische, R. H., and H. von Buttlar: A Theoretical Study of Induced Electrical Polarization, *Geophysics*, vol. 22, pp. 688–706, 1957.

55. Rogers, George R.: An Evaluation of the Induced-Polarization Method in the Search for Disseminated Sulfides, "Mining Geophysics," vol. 1, pp. 350–358, Society of Exploration Geophysicists, Tulsa, Okla., 1966.

56. Hansen, Don A., and David A. Barr: Exploration Case-History of a Disseminated Copper Deposit, "Mining Geophysics," vol. 1, pp. 306–312, Society of Exploration Geophysicists, Tulsa, Okla., 1966.

57. Keller, George V.: Electrical Prospecting for Oil, *Q. Colo. Sch. Min.*, vol. 63, no. 2, 1968.

58. Wyllie, M. R. J.: "The Fundamentals of Electric Log Interpretation," Academic, New York, 1949.

59. Pirson, S. J.: "Handbook of Well Log Analysis," Prentice-Hall, Englewood Cliffs, N.J., 1963.

60. Marsden, D.: *Eur. Assoc. Explor. Geophys.* 34th Meet., Paris, 1972.

Names of the authors cited in the text are not included in the index. They are listed in the bibliographies at the end of each chapter.